Boston Studies in the Philosophy and History of Science

Volume 335

Editors
Alisa Bokulich, Boston University
Jürgen Renn, Max Planck Institute for the History of Science
Michela Massimi, University of Edinburgh

Managing Editor
Lindy Divarci, Max Planck Institute for the History of Science

Editorial Board
Theodore Arabatzis, University of Athens
Heather E. Douglas, University of Waterloo
Jean Gayon, Université Paris 1
Thomas F. Glick, Boston University
Hubert Goenner, University of Goettingen
John Heilbron, University of California, Berkeley
Diana Kormos-Buchwald, California Institute of Technology
Christoph Lehner, Max Planck Institute for the History of Science
Peter McLaughlin, Universität Heidelberg
Agustí Nieto-Galan, Universitat Autònoma de Barcelona
Nuccio Ordine, Universitá della Calabria
Sylvan S. Schweber, Harvard University
Ana Simões, Universidade de Lisboa
John J. Stachel, Boston University
Baichun Zhang, Chinese Academy of Science

The series *Boston Studies in the Philosophy and History of Science* was conceived in the broadest framework of interdisciplinary and international concerns. Natural scientists, mathematicians, social scientists and philosophers have contributed to the series, as have historians and sociologists of science, linguists, psychologists, physicians, and literary critics.

The series has been able to include works by authors from many other countries around the world.

The editors believe that the history and philosophy of science should itself be scientific, self-consciously critical, humane as well as rational, sceptical and undogmatic while also receptive to discussion of first principles. One of the aims of Boston Studies, therefore, is to develop collaboration among scientists, historians and philosophers.

Boston Studies in the Philosophy and History of Science looks into and reflects on interactions between epistemological and historical dimensions in an effort to understand the scientific enterprise from every viewpoint.

More information about this series at http://www.springer.com/series/5710

Jochen Büttner

Swinging and Rolling

Unveiling Galileo's unorthodox path
from a challenging problem to a new science

Jochen Büttner
Max Planck Institute for the History
of Science
Berlin, Germany

ISSN 0068-0346   ISSN 2214-7942   (electronic)
Boston Studies in the Philosophy and History of Science
ISBN 978-94-024-1592-6   ISBN 978-94-024-1594-0   (eBook)
https://doi.org/10.1007/978-94-024-1594-0

Library of Congress Control Number: 2018966146

© Springer Nature B.V. 2019
This work is subject to copyright. All rights are reserved by the Publisher, whether the whole or part of the material is concerned, specifically the rights of translation, reprinting, reuse of illustrations, recitation, broadcasting, reproduction on microfilms or in any other physical way, and transmission or information storage and retrieval, electronic adaptation, computer software, or by similar or dissimilar methodology now known or hereafter developed.
The use of general descriptive names, registered names, trademarks, service marks, etc. in this publication does not imply, even in the absence of a specific statement, that such names are exempt from the relevant protective laws and regulations and therefore free for general use.
The publisher, the authors, and the editors are safe to assume that the advice and information in this book are believed to be true and accurate at the date of publication. Neither the publisher nor the authors or the editors give a warranty, express or implied, with respect to the material contained herein or for any errors or omissions that may have been made. The publisher remains neutral with regard to jurisdictional claims in published maps and institutional affiliations.

This Springer imprint is published by the registered company Springer Nature B.V.
The registered company address is: Van Godewijckstraat 30, 3311 GX Dordrecht, The Netherlands

. . . post diuturnas mentis agitationes . . .
Galileo Galilei – Discorsi (1638)

Foreword

The Historical Epistemology of Mechanics

The Historical Epistemology of Mechanics studies the long-term development of mechanical knowledge. Mechanical knowledge concerns material bodies in time and space, their motions, and the forces that cause or resist such motions. Mechanical knowledge enables us to predict how bodies change their position with time as long as we know their current state and the forces acting upon them. Mechanical knowledge of this kind played a special role in the process of transformation from natural philosophy to modern science. Natural philosophy from its very inception in the works of Aristotle constructed conceptual systems to represent pictures of the world as a whole. But, in contrast to such global intentions, the origins of mechanical knowledge have to be sought in the much more down-to-earth practical activities of achieving the specific tasks of everyday life.

Over a long historical period, the development of mechanical knowledge and its transmission from one generation to the next remained an inherent dimension of such activities, unrelated to any cognitive endeavors aimed at constructing a mechanical worldview. It was only after the first attempts in classical antiquity to include mechanical knowledge in the conceptual systems of natural philosophy that its assimilation to them and the corresponding accommodation of such systems to mechanical concepts led to conflicts between mechanical knowledge and knowledge about nature as a whole. It was only after the growing body of mechanical knowledge became a vital resource of early modern societies that mechanical knowledge within its own conceptual systematization started to compete with natural philosophy by constructing its own worldviews. This finally resulted in early modern times in what has been called the "mechanization of the world picture."

The main goal of the series under the heading *The Historical Epistemology of Mechanics*, conceived in analogy to the four-volume set on *The Genesis of General Relativity*, is to explain the development and diffusion of mechanical knowledge in terms of historical-epistemological concepts. The studies presented within the series are based on a research project centered at the Max Planck Institute for the History of Science in Berlin. While the emphasis of the research has been on the period of the Scientific Revolution, the analysis also takes into account the long-term development of mechanical knowledge without which neither its emergence nor the consequences of this period can be adequately understood. Just as the reconstruction of the relativity revolution in *The Genesis of General Relativity* takes Einstein's work as the point of reference for a thorough contextualization of his achievements, the reconstruction of the transformation of mechanical knowledge during the Scientific Revolution similarly refers to Galileo's work as a point of departure for outlining a historical epistemology of mechanics.

The development of an adequate theoretical framework provides a common basis for the investigations constituting *The Historical Epistemology of Mechanics*. The longevity of mechanics makes it particularly clear that large domains of human knowledge accumulated by experience are not simply lost when theories are revised, even if this knowledge does not explicitly appear in such theories. Thus, formal logic is of little use for a description of the multilayered architecture of scientific knowledge that allows both the continuous and the discontinuous aspects of the transmission of mechanical knowledge to be accounted for. In order to explain structural transformations of systems of knowledge, it is furthermore necessary to take into account the collective character and the historical specificity of the knowledge being transmitted and transformed as well as to employ sophisticated models for reconstructing processes of knowledge development. Concepts such as "mental model," "shared knowledge," "challenging object," and "knowledge reorganization" have turned out in our work to be pivotal for such explanations.

We conceive of mental models as knowledge representation structures based on default logic, which allow inferences to be drawn from prior experiences about complex objects and processes even when only incomplete information on them is available. Mental models relevant to the history of mechanics belong either to generally shared knowledge or to the shared knowledge of specific groups. Accordingly, they can be related either to intuitive, to practical, or to theoretical knowledge. They are, in any case, characterized by a remarkable longevity—even across historical breaks—as becomes clear when considering examples such as the mental models of an atom, of a balance, of the center of gravity, or of positional weight. Their persistence in shaping the shared knowledge documented by the historical sources becomes particularly apparent in the consistency of the terminology used, a consistency that offers one important element for an empirical control in the reconstruction of mental models and their historical development. The concept of mental model is particularly suited to study the role of practical knowledge for the transformation of mechanics in the early modern period. Conceiving a body in terms of the intuitive mental model of gravitation, for instance, implies that a heavy body falls down in "natural motion" if its motion is not inhibited or deflected by a force, that it makes an impact when it falls, that the force of this impact is the larger the longer it falls, but also—at a later stage of development—that the body has a weight that can be measured by a balance. Whenever the question of a quantitative measure of impact arises—as it does in the early modern period—this model of intuitive and practical knowledge could be and was further extended in reaction to new experiences with the challenging objects of contemporary technology.

More generally, we conceive of challenging objects as historically specific material objects, processes, or practices entering the range of application of a system of knowledge without the system being capable of providing a canonical explanation for them. Examples run from mechanical devices challenging Aristotelian dynamics, via artillery challenging early modern theories of motion, to blackbody radiation challenging classical radiation theory. In reaction to such challenges, knowledge systems are typically further elaborated, occasionally to the extent that they give rise to internal tensions and even inconsistencies. Such explorations of their limits may then become starting points for their reorganization where often previously marginal insights take on a central role in an emerging new system of knowledge. Such processes of reorganization may be exemplified by the emergence of theoretical mechanics from Aristotelian natural philosophy in ancient Greece, the transformation of preclassical into classical mechanics in early modern times, or the emergence of quantum theory from classical physics at the turn of the last century.

The investigations constituting *The Historical Epistemology of Mechanics* build on this theoretical framework, centering on the role of shared knowledge, of challenging objects, and of knowledge re-organization. The first study, Matthias Schemmel's *The English Galileo: Thomas Harriot's Work on Motion as an Example of Preclassical Mechanics*, investigates the shared knowledge of preclassical mechanics by relating the work of Thomas Harriot on motion, documented by a wealth of manuscripts, to that of Galileo and other contemporaries. Harriot and Galileo indeed exploited the same shared knowledge resources in order to approach the same challenging objects. The study of Harriot's parallel work thus allows the exploration of the structure of the shared knowledge of early modern mechanics to perceive possible alternative histories and to distinguish between individual peculiarities and shared structures of early modern mechanical reasoning. The second study of the series, Matteo Valleriani's *Galileo Engineer*, looks more closely at the role of Galileo as a practical mathematician and engineer-scientist. It focuses on his intellectual development in the frame of the interaction between natural philosophy and the challenging objects provided by technological developments. It analyzes Galileo's contribution to the practical science of machines as well as his role as a teacher involved in the contemporary art of war. The results of this analysis highlight Galileo's profile as a military engineer. The book develops a model according to which new scientific knowledge was generated on the basis of the interaction between theoretical knowledge—basically Aristotelian—and the practical knowledge Galileo shared with his contemporaries. Galileo's work is reinterpreted in its entirety against the background of a historiographical investigation concerning the early modern figure of the engineer-scientist.

This third contribution to this series, Jochen Büttner's study *Swinging and Rolling: Unveiling Galileo's unorthodox path from a challenging problem to a new science of motion*, looks more closely at the reorganization of mechanical knowledge that took place in the course of Galileo's research process. This process pursued within a network of exchanges with his contemporaries and documented by a vast collection of research notes is reconstructed in minute detail. It is shown that a problem arising from the encounter of two challenging objects, the pendulum and the inclined plane, motivated and shaped Galileo's thinking and became a mainspring for the formation of Galileo's comprehensive theory of naturally accelerated motion. Tensions between the existing frameworks and the new insights engendered by his investigation of the relation between the swinging of pendulums and the rolling of balls along inclined caused Galileo to reorganize his knowledge, thus effectively creating his new science of motion, a milestone of the transition from preclassical to classical mechanics.

The fourth upcoming and conclusive volume will articulate more extensively the theoretical foundations of a historical epistemology of mechanics, still with a focus on preclassical mechanics, its prehistory, and its contexts. The framework presented in this volume makes it possible to reconstruct the long-term development of mechanical knowledge from its anthropological origins via the formation of a mechanical worldview to the transformations of classical mechanics in modern physics. It also investigates the societal conditions fostering the integration of this knowledge in the emergence and expansion of preclassical mechanics and the far-reaching consequences it had on the development of physical knowledge.

Berlin, Germany   Jürgen Renn

Preface

In 1994, the Max Planck Institute for the History of Science was founded in Berlin. In the following year, while I was still studying physics and philosophy at the Freie Universität, I was given the opportunity to start working there as a student assistant in the department of Jürgen Renn. At that time, in a pioneering attempt, an online edition of Galileo's so-called *Notes on Motion* was being prepared in a joint effort of the department together with the Istituto e Museo di Storia della Scienza and with the Biblioteca Nazionale Centrale, both situated in Florence. I was assigned the task of overseeing and completing what, at the time, appeared to be merely some last remaining nonissues before the edition could go online. As can easily be imagined, this turned out to be a misjudgment, and before long, I found myself deeply and profoundly immersed in studying the content of the manuscript. This, paired with the intellectual stimulus coming from the historical epistemology as it was and still is practiced in Department I of the Max Planck Institute for the History of Science, triggered my academic transformation into a historian of science. This transformation, in a sense, was completed in 2009 when I defended my Ph.D. thesis *Galileo's Challenges: The Origin and Early Conceptual Development of Galileo's Theory of Naturally Accelerated Motion on Inclined Planes* at the Humboldt University in Berlin. It is this thesis, in thoroughly revised form, which more than a decade after it was initially conceived I present to a wider audience for the first time in the form of this book.

Due to the long time that has passed since I first directed my attention to studying Galileo's working notes, the list of persons that I owe gratitude and thanks has grown beyond comprehension while, at the same time, memory also begins to fade. Hence, first of all, I would like to thank all those friends and colleagues who greatly deserve thanks but who I unavoidably and also inexcusably will forget to mention here. Jürgen Renn acted as advisor for the dissertation which resulted in this book, and I thank him for this. Much more importantly, however, as my teacher he had great impact on my intellectual formation and as mentor supported and encouraged me in many ways for which I am extremely grateful. Gerd Graßhoff provided council on the first steps of my investigations into Galileo's science of motion, and I much appreciate the direction this gave to my later studies. For my work on Galileo's

Notes on Motion, I have been allowed to use data compiled and edited by a working group at the Max Planck Institute for the History of Science. I would like to thank the members of this group, in particular Simone Rieger and the late Peter Damerow for courteously making this material available to me. Lindy Divarci, publication manager at the institute, has been of great help in getting this book published. Her always friendly support is much appreciated. I am much obliged to Sarah Kühne for taking on the major task of language editing the manuscript.

I would like to thank all my colleagues at the institute for supporting me in so many different ways and plainly for being excellent colleagues. Special thanks goes to Matteo Valleriani, long-time colleague and friend, whose incredible aptitude for keeping up a good spirit against all adversity has more than once helped me to not hang my head and to take things somewhat more lightly in times of difficulty. I would like to thank the many colleagues with whom I have had the opportunity to share my thoughts on Galileo. Their feedback, whether at conferences, in direct discussions or in written exchanges has provided important stimuli that have contributed in shaping and occasionally first provoked the thoughts that I present here. Special thanks to Christian Barth, Philipp Strauß, and Martin Klein on account of keeping me entertained while revising this manuscript.

I fancy that no non-historian of science ever had to listen to so many details about Galileo, as Jochen Schneider, former research coordinator at the Max Planck Institute for the History of Science. I thank him for his patience and his acute responses, not at all restricted to questions regarding Galileo and not the least for teaching and still playing Go with me. I would like to thank Norbert Palz, for being a great friend and for granting me refuge in his office space when I needed a change of scene while writing this manuscript; Peggy Haines for being a role model in responsible scholarship and for her faith in my capabilities, unsupported but unswerving; and my mother for her patience with her son and for her continued support of me and my family. I want to commemorate my father who sympathetically and always reassuringly accompanied my transformation from the physicist I was trained to be into a historian of science but who sadly didn't live to see it completed. Last but not least, I want to thank my wife Sonja, my daughter Toni, and, my son Mattes. If it weren't for them, this book would certainly have been completed much earlier. Yet of all the reasons that may have delayed publication, this is the one, which I can without hesitation state, has made my life and certainly also this book better.

Berlin, Germany
March 2017

Jochen Büttner

Contents

1	**Introduction**	1
	References	20
Part I	**Novel Insights About Accelerated Motion: The Challenge of Swinging and Rolling**	
2	**Before Natural Acceleration**	25
	2.1 Velocity, Motion as a Changing Quality, and Galileo's Early Account of Acceleration	28
	2.2 Motion Along Inclined Planes and Dynamics	42
	References	57
3	**A Glimpse at a Challenging Research Agenda: Galileo to Guidobaldo del Monte in 1602**	61
	References	72
4	**Sparking the Investigation of Naturally Accelerated Motion: The Pendulum Plane Experiment**	75
	4.1 Experimenting with Swinging and Rolling	77
	4.2 The Experiment in Context	82
	4.3 The Single Chord Hypothesis as Starting Point	86
	4.4 Summarizing Conclusion	93
	References	94
5	**Squaring the Pendulum's Arc: Motion Along Broken Chords**	97
	5.1 Toward a Proof of the Law of the Broken Chord	99
	5.2 Expanding on the Law of the Broken Chord	116
	5.3 Tackling the Problem of Motion on an Arc: The Broken Chord Approach	120
	5.4 Summarizing Conclusion	140
	References	143

6 Swinging and Rolling Revisited: Motion Along Broken Chords and the Pendulum Plane Experiment ... 145
- 6.1 Reevaluating the Experimental Data Under a New Conceptualization ... 147
- 6.2 The Pendulum Plane Experiment from a Modern Perspective ... 152
- 6.3 Swinging and Rolling: From Promising Start to Dead End ... 159
- References ... 167

7 Accumulating Insights: The Problem of Motion Along Broken Chords Driving Conceptual Development ... 169
- 7.1 A New Technique for the Diagrammatical Representation of Motion ... 170
- 7.2 The Length Time Proportionality as a Spin-Off ... 182
- 7.3 Bent Planes Traversed in Least Time ... 188
- 7.4 Galileo's Considerations on Motion Along Bent Planes as the Source of New Propositions ... 196
- 7.5 Summarizing Conclusion ... 204
- References ... 207

8 Toward a Foundation: The Ex Mechanicis Proof of the Law of Chords ... 209
- 8.1 The Formation of the Ex Mechanicis Proof of the Law of Chords ... 211
- 8.2 The Role of the Ex Mechanicis Proof for the New Science of Motion ... 227
- 8.3 The Origin of the Law of Chords as a Heuristic Assumption ... 233
- 8.4 Summarizing Conclusion ... 237
- References ... 239

9 Whatever Happened to Swinging and Rolling: Faint Echoes and a Late Insight ... 241
- 9.1 The Brachistochrone Argument ... 243
- 9.2 A Reply to Baliani Drafted ca. 1638: An Unexpected Twist of Swinging and Rolling ... 252
- 9.3 Summarizing Conclusion ... 265
- References ... 267

Part II Novel Insights and Old Concepts: From Exploration to Formalization

10 Toward a New Science: Gathering Results and the Rise and Demise of a Dynamical Foundation ... 271
- 10.1 The Star Group Folios: A First Step Toward a New Science ... 273
- 10.2 The Rise and Demise of a Dynamical Foundation ... 307
- 10.3 Summarizing Conclusion ... 314
- References ... 316

11	**Toward a New Science: Axiomatization and a New Foundation**	317
	11.1 1602: Immersed in Foundational Problems	319
	11.2 The 1602 Foundational Situation in Context	337
	11.3 Toward a New Foundation in 1604	349
	11.4 Summarizing Conclusion	354
	References	357
12	**Conclusion: The Emergence and Early Evolution of Galileo's New Science of Motion**	359
13	**Appendix: Folio Pages**	369
	13.1 Folio 90	370
	13.2 Folio 115	372
	13.3 Folio 121	374
	13.4 Folio 129	376
	13.5 Folio 130	379
	13.6 Folio 148	386
	13.7 Double Folio 149, 150	392
	13.8 Folio 151	396
	13.9 Folio 153	400
	13.10 Folio 154	402
	13.11 Folio 155	405
	13.12 Double Folio 156, 157	408
	13.13 Folio 166	413
	13.14 Folio 167	420
	13.15 Folio 176	423
	13.16 Folio 183	427
	13.17 Folio 184	429
	13.18 Folio 185	435
	13.19 Folio 186	439
	13.20 Folio 189	442
	13.21 Folio 192	454
	References	457
14	**Appendix: Documents**	459
	14.1 Letter from Paolo Sarpi to Galileo, 2 September 1602	459
	14.2 Letter from Galileo to Guidobaldo del Monte, 29 November 1602	461
	14.3 Letter to Paolo Sarpi, 16 October 1604	465
	14.4 Draft of a Reply to Baliani Composed ca. 1640	467
	References	470
Index Locorum		471

Chapter 1
Introduction

If a small bronze ball tied to a string and suspended from a pivot is displaced from its equilibrium position and then let go, the period of swing is independent of the initial displacement. This is commonly referred to as the isochronism of the pendulum. If, instead, the ball rolls down along an inclined plane connecting a point on the arc of pendulum swing to its equilibrium position, the time of motion is independent of the inclination of the plane. This statement will henceforth be referred to as the *isochronism* or *law of chords*.

Moreover, if the pendulum's string is lengthened or shortened, the period of the pendulum increases or decreases in proportion to the root of the length of the string. This is commonly referred to as the *law of the pendulum*. If, on the other hand, one lets the ball roll down over a longer or a shorter distance on the inclined plane, the time of motion is found to increase or decrease with the root of the distance fallen. This is commonly referred to as the *law of fall*.

On the phenomenological level, swinging and rolling thus seem to manifest a remarkable similarity. They certainly did for Galileo. Is this similarity a mere coincidence from which nothing can be learned? Galileo certainly didn't think so. For him, the challenging similarity he perceived between swinging and rolling bestowed the key to deeper insights.[1] Despite immense effort, Galileo was never able to meet the challenge and recover the similarity he perceived between the two types of motion theoretically. Yet it was his inquiry into the relation between swinging and rolling which gave rise to and essentially shaped his new science of motion. This is what this book sets out to demonstrate.

The new science of motion was published in 1638 as part of the *Discorsi e dimostrazioni matematiche intorno a due nuove scienze attenenti alla mecanica & i*

---

[1] In his considerations regarding the kinematics of the descent of heavy bodies, Galileo did not distinguish whether a body was rolling, sliding or falling freely because he did not realize that this difference mattered. For him, naturally accelerated motion was simply "quello che generalmente è esercitato da i mobili gravi descendenti (EN VIII, 196)."

movimenti locali, or in short, the *Discorsi*.² In the book, Galileo indeed put forward a comprehensive treatment of local motion in which the naturally accelerated motion of freely moving heavy bodies, in particular, was treated in an unprecedented manner.³ As it turns out, the foundations for this new science had been laid more than 35 years earlier. How do we know this?

By fortuitous circumstances a manuscript has come down to us that documents Galileo's work on the problems of naturally accelerated motion from its very beginnings, around the turn of the seventeenth century, right up to the publication of the *Discorsi*. The manuscript comprises an unordered and undated, almost chaotic collection of calculations, drawings and texts, both in Latin and Italian. The content ranges from single numbers to complex calculations, from tests of the pen to intricate geometrical constructions, and from incomplete sentences to full drafts of propositions as they were later published almost verbatim in the *Discorsi*.⁴ Fittingly, the manuscript has come to be referred to as Galileo's *Notes on Motion*. It is primarily through an in-depth analysis of this remarkable source that the origin and the early genesis of Galileo's new science will be reconstructed here.

Despite their often idiosyncratic and piecemeal character, the *Notes on Motion* are the essential primary source for reconstructing the intellectual pathways which led Galileo to the second of his new sciences.⁵ Due not the least to its fairly

---

²Galilei (1638), referred to in short here as the *Discorsi*. Favaro's edition of the *Discorsi* is contained in volume VIII of the *Edizione Nazionale* (EN VIII, 11-362). The Edizione Nazionale is freely accessible online through the *Galileo//thek@* at: http://galileoteca.museogalileo.it/indice.html. Accessed 14 Jun 2016.

³The *Discorsi* are written as a dialog between Salviati, Sagredo, and Simplicio, who met on a number of successive days to treat different topics regarding the studies of an unnamed academician they all know (i.e., Galileo himself). The discussion takes place over four days. On the first day, a number of diverse topics were discussed, ranging from the speed of light to the question of whether there are more numbers than prime numbers. Biener (2004) argues that the unifying motive is Galileo's attempt to promote the possibility of treating matter mathematically. The discussion of the second day revolves around the first new science concerning the strength of materials. The new theory of motion is introduced on the third and the fourth days. Latin text passages are interleaved with the dialog, allegedly taken from a treatise by the academician, entitled *De motu locali* and read by Salviati. By availing himself of the literary scheme of, as it were, a book in a book, Galileo thus manages to present his new science as a deductive treatise while the dialog in which it is embedded allows for elucidating excursions. *De motu locali* comprises three books, the first two of which are discussed on the third day, while the Third Book is read on the fourth day. The First Book deals with the basic rules of kinematics of uniform motion, the Second Book with free falling motion along inclined planes, and the Third Book is concerned with projectile motion.

⁴Galileo's *Notes on Motion* are part of a larger manuscript bundle preserved in the Biblioteca Nazionale Centrale di Firenze as codex 72 of the Galilean collection (Ms. Gal. 72). The manuscripts can be accessed online at: http://teca.bncf.firenze.sbn.it/ImageViewer/servlet/ImageViewer?idr=BNCF0003760961#page/1/mode/2up. Accessed 23 Feb 2017. In addition to the *Notes on Motion*, comprised on folios numbered 33 to 194, the manuscript bundle contains miscellaneous material, most notably two copies of Galileo's *Le Meccaniche*.

⁵The importance of the *Notes on Motion* has not always been as univocally accepted as it is today. McMullin (1967, 10–11), for instance, stated on the occasion of Galileo's 300th birthday that we "shall never be able to reconstruct the sequence of steps that lead him [Galileo] to the

convoluted history, the manuscript has not always been as readily available as it is today, and this is indeed reflected in the historiography of the document.[6] A first, albeit partial, rendition of the manuscript was included by Antonio Favaro in Volume VIII of his masterful 20-volume *Edizione Nazionale*, which appeared in regular succession between the years 1890 and 1909. Favaro's rendition of the content of the manuscript is endowed with serious shortcomings, and the *Notes on Motion* essentially remained in wait of their rediscovery by the history of science.[7] In the 1970s, prompted by the concurrence of Galileo's 300th birthday and the publication of a second edition of the *Edizione Nazionale* in the decade before, Galileo scholarship proliferated. A new edition of the *Notes on Motion* was published in 1979 by Drake as a special volume of the *Annali di Storia della Scienza*.[8] This edition, defective in its own way, has today been superseded by an electronic representation of Galileo's *Notes on Motion*, which in a pioneering achievement in 1999 was made freely available on the Internet and has remained accessible online ever since.[9]

The *Discorsi* marks the endpoint of the development whose origin is being traced here and thus naturally represents a second crucial source for the interpretation to

---

Discorsi; ...Historians of science have tended to read their own philosophy of science into the methodology that guided those crucial and creative years; the temptation to do so is great, since the evidence available in scattered remarks in letters or in the *Dialogo* and *Discorsi* is scant and very ambiguous." Conspicuously, the *Notes on Motion* are not even mentioned in his list of sources.

[6] A summarizing account of history of the manuscript is provided at: http://www.mpiwg-berlin.mpg.de/Galileo_Prototype/MAIN/HISTORY.HTM. Accessed 12 Jun 2016.

[7] Material from the *Notes on Motion* has found its way into the *Edizione Nazionale* in various forms. These range from redrawn black and white facsimiles of a small number of pages, to footnote annotations to the published text of the *Discorsi*. A number of entries together with associated diagrams have been reproduced, sometimes erroneously, in transcription. The edition contains, in particular, only transcriptions of texts of a certain length which are sufficiently self-contained. Calculations have been omitted almost completely; frequently, so have the diagrams contained in the manuscripts. Favaro's arrangement of the material is not based on its position on the folios of the manuscript but follows a questionable logic. Wisan (1974, 126) commented that "Favaro's treatment of the manuscript seems almost calculated to discourage further investigation of the fragments." For a reappraisal of the *Edizione Nazionale*, see Camerota and Castagnetti (2001).

[8] See Drake (1979). Wisan (1982) provides an astute critique of this work. Regarding the origin of the proliferation of Galileo scholarship in the 1970s, Segre (1998, 405) remarked "[i]n his article ['Galileo's Experimental Confirmation of Horizontal Inertia: Unpublished Manuscripts,' Isis, 64 (1973), pp. 291–305.], Drake attempted to reconstruct Galileo's inclined-plane experiment on the basis of unpublished manuscripts. Drake's work initiated a new rich trend of Galilean studies, in attempts to interpret Galileo's working notes."

[9] The electronic representation of Galileo's *Notes on Motion* was instituted through a joint effort of the Biblioteca Nazionale Centrale di Firenze, the Max Planck Institute for the History of Science and the Istituto e Museo di Storia della Scienza. It can be accessed at http://www.mpiwg-berlin.mpg.de/Galileo_Prototype/INDEX.HTM. Accessed 12 Jun 2016. In the electronic representation, the individual contributions of the collaborators have not been designated explicitly. Results comprised in the electronic representation, which are based on my own intellectual work, such as the analysis of the deductive structure of the propositions contained in the Second and the Third Book of the Third Day of the *Discorsi*, are used unreservedly throughout this book.

be established. Beyond that, my interpretation is based on pertinent passages from Galileo's earlier writings, some of which were published during his lifetime, some of which weren't, and on his extensive correspondence. All these sources are contained in the *Edizione Nazionale* and are today conveniently accessible online together with additional material through the *Galileo//thek@*.[10]

The literature touching on aspects related to Galileo's *Notes on Motion* that has accumulated up to the present day is vast and complex and cannot be presented, let alone reviewed here. Yet just as much as it is based on an in-depth analysis of the primary sources, the interpretation that will be presented owes to earlier accounts, even if in the end, it appears to throw doubts on many of them. Thus, a brief survey is given of those works in which the *Notes on Motion* are treated comprehensively and which had a paramount influence on my research.

In her Ph.D. thesis, W. Wisan provided a groundbreaking analysis of the manuscript. Published under the title *The New Science of Motion: A Study of Galileo's De motu locali*, it certainly ranks first and, in most respects, still stands uncontested among the extant studies of the *Notes on Motion*.[11] Wisan organized her work starting from an analysis of the treatment of naturally accelerated motion that Galileo gave in the *Discorsi* and from there sought to identify and interpret entries in the manuscript documenting considerations leading to the establishment of propositions as they were later published. Approaching the content of the *Notes on Motion* mainly from the perspective of the final outcome, she occasionally misjudges the importance Galileo attached to arguments in the publication as being an indication that these arguments played an equally important role in the conceptual development of the new science and disregards some strands of thought, which can be discerned in the manuscript but did not find their way into Galileo's published work.

Wisan relied rather extensively on arguments drawn from the evolution of Galileo's terminology without, however, providing a systematic analysis thereof. Her assumptions in this respect, rather, appear to be mostly based on the intuitive understanding that almost inevitably develops when working intensively with material such as that in the *Notes on Motion*. However, for virtually all examples of expressions that she advances supposedly used by Galileo exclusively at either an early or a late stage, an often significant number of exceptions can be identified. Thus, those of Wisan's conclusions which are based on clues from Galileo's use of language alone often appear somewhat controversial. Yet, in general, her work is thoroughly reliable and remains an indispensable source for any fresh attempt to analyze the content of the manuscript.[12]

---

[10] Consult http://galileoteca.museogalileo.it/indice.html. Accessed 14 Jan 2017.

[11] A revised version of Wisan's Ph.D. thesis was published in the *Archive for History of Exact Sciences*. See Wisan (1974).

[12] Wisan's accomplishments are rendered even more remarkable when it is acknowledged that she did not have today's facile access to the manuscript and mainly had to work with Favaro's edition.

Similarly essential, but at the same time more problematic, is Drake's oeuvre centering on Galileo's life and work. In his second career as a historian of science, Drake published more than 130 books and articles on Galileo, which continue to have a tremendous impact. Of principal importance, with respect to the *Notes on Motion*, is his edition of the manuscript, which brought the attention of historians of science back to the *Notes on Motion*, as well as *Galileo at Work: His Scientific Biography*, in which Drake consolidated the results of his innumerable studies concerning details of Galileo's work into an overarching account. The underlying individual studies themselves are frequently rather hermetic and every so often characterized by a "Byzantine complexity."[13]

The approaches Drake took and the methods he employed are occasionally controversial and many of the detailed interpretations he has advanced have been proven untenable. Yet his knowledgeability concerning Galileo's life and work should not be questioned. His work, not least by its sheer volume, contains innumerable gems of information indispensable for all manner of critical reexamination of Galileo's work.

With regard to the *Notes on Motion*, inferences drawn from the types of paper Galileo used were an important source for Drake's interpretations. Although his approach is agreeable in principle, it has meanwhile been shown that in doing so, Drake committed some serious methodological blunders and hence his arguments based on inferences of this kind cannot be relied on.[14]

In his seminal book *Momento*, Galluzzi traces the history of Galileo's science. At the center of his investigation is Galileo's attempt to unify static and dynamic principles in mechanics, which Galluzzi analyzes by focusing on the concept of "moment," tracing the development this complex notion took in Galileo's works. As such the book is not devoted primarily to an analysis of the *Notes on Motion* as a whole. Yet the concept of "moment" figures prominently in a number of considerations documented in the manuscript, most of which were written down by Galileo during the period of work under examination here. In consequence, a great number of entries in the manuscript, which Galluzzi has analyzed in considerable detail, will likewise be discussed in this book. Yet, due to his different focus, Galluzzi did and could understandably not contextualize these entries with the other material contained in the manuscript in a way comparable to what is ventured here. Thus from the perspective developed in this book, his discussion of the content of the manuscript occasionally remains somewhat eclectic. A thorough contextualization necessitates a reevaluation of some aspects, yet overall, Galluzzi's findings are absolutely sound and in line with the interpretation presented here.

---

[13] The expression "Byzantine complexity" to characterize Drake's detailed studies was coined by Hill (1994).

[14] Using the example of the folios which Drake dated to 1602 based on watermark evidence, in his Ph.D. thesis, Hooper discusses the shortcomings and even blunders of Drake's approach to conclude: "This is an unsettling picture of Galileo's working style and the image of Galileo's intellectual development that emerges under this scheme is also tenuous and obscure. (Hooper 1992, 311)."

In his Ph.D. thesis, *Galileo and the Problems of Motion*, W. Hooper aimed at a more or less comprehensive account of Galileo's work on problems of motion from its earliest beginnings in the 1580s to the final publication of the *Discorsi*. In view of the broad scope of his project and the multiplicity of topics touched upon, his discussions of distinct stages of the development of Galileo's considerations regarding bodies in motion are comparatively short. With respect to the period under investigation here in particular, in his attempt at a reconstruction Hooper relied on a rather limited number of entries in the manuscript which, in the extant literature, had already been discussed as pertaining to an early stage of Galileo's considerations regarding naturally accelerated motion. Where his readings deviate from the interpretations established earlier, they are often rather speculative.[15] In his critique of existing interpretations Hooper is sharp-witted and his exposition of methodological flaws in earlier accounts has provided important stimulus for my own work.

The last book to be mentioned here, *Exploring the Limits of Preclassical Mechanics*, is a collective work.[16] Chapter 3, *Proofs and Paradoxes: Free Fall and Projectile Motion in Galileo's Physics*, which is the one most pertinent to my own account, was authored mainly by Jürgen Renn, who I will indeed address as the author when referring to arguments made in this chapter. Renn's attempt to reconstruct Galileo's pathways in a conceptual space spanned by the existing bodies of knowledge, which eventually led to the emergence of a new theory, has served as a role model for my own work. Whereas Renn focused mainly on the conceptual development of Galileo's theory of projectile motion, the focus in this work is on the development of Galileo's science of naturally accelerated motion on inclined planes. Renn has provided some intriguing detailed studies of individual pages of the manuscript. His account, limited to a single chapter of a book, makes some pre-assumptions in selecting and interpreting the material, which are, however, predominantly supported by my interpretation.

Besides the works mentioned above in which the development of Galileo's consideration regarding motion are treated extensively and in which, as a rule, larger portions of the manuscript are discussed, a vast number of contributions, dealing essentially with single aspects of Galileo's considerations and/or single entries or small groups of entries in the manuscript, are extant.[17]

---

[15] In his discussion of the *Notes on Motion*, Hooper concentrates mainly "on the meaning and significance of five documents ... folios 147, 164, 179, and 152. (Hooper 1992, 312)" According to his interpretation, the transition from the theory of motion of the *De Motu Antiquiora* to naturally accelerated motion hinged crucially on an experiment he surmised Galileo conducted but for which he does not produce a persuasive argument.

[16] See Damerow et al. (2004).

[17] No clear-cut distinction can be drawn between contributions which have treated the *Notes on Motion* more comprehensively and those dealing essentially with single aspects of the manuscript. Naylor, for instance, dealt in a series of articles (Naylor 1974, 1975, 1976, 1977, 1980), with individual aspects analyzing only a limited number of entries in the manuscript. Taken together,

These studies are typically concerned with rather specific questions, which for the most part originated not from the study of the manuscript itself, but from an inquiry into other sources, such as Galileo's correspondence or his published works. For the most part, the interpretations established offer only a restricted contextualization of the material discussed. A typical example are the numerous studies that have sought to identify and reconstruct Galileo's proof of the law of fall from an erroneous principle, a proof he mentioned finding but did not quote in a letter of 1604, as well as his recognition of the error contained therein. As Palmieri fittingly remarked, such "interpretations of single folios of manuscript 72, though interesting, raise mostly unresolvable questions since they are based on reconstructions which are so under-determined that a verdict cannot be reached."[18] In their sum, however, these studies represent an indispensable knowledge base for any study of the manuscripts.

With such considerable effort invested into studying the *Notes on Motion*, in the past a number of pertinent questions have been answered. At the same time, "recent work with Galileo's manuscript notes has reopened the problems of the development of his work on motion in a number of most interesting ways."[19] Studying the genesis of Galileo's new science of motion based on the *Notes on Motion* is thus in some sense a historiographical legacy in its own right.

The study of a single manuscript, written by a single individual, is undeniably an exercise in micro-history yet one that is distinguished from a mere case study in that it aims to contribute to a broader historical understanding.[20] The paramount influence of Galileo's *Discorsi* on the development of scientific thought in the seventeenth century and the history of science in general, although arguably still insufficiently understood, are beyond controversy.[21] Yet it may seem that in order to explore the impact of Galileo's science of motion, an analysis of the *Discorsi* would suffice, as the concrete way in which Galileo achieved the results published therein had virtually no influence on their reception.

Against this it can be maintained that analyzing the origin and conceptual genesis of Galileo's new science allows better recognition and a more precise location and understanding of the intellectual borders Galileo crossed, so that we might appreciate where he ventured untrodden paths and where he stayed on the safe ground of traditional knowledge. This not only contributes to an understanding of Galileo's achievements but at the same time to our understanding of the traditional

---

however, these articles present a comprehensive treatment of the question of the origin and development of Galileo's theory of projectile motion.

[18] See Palmieri (2005, footnote 7).

[19] See Hooper (1992, 35).

[20] The relation between micro- and macro-histories has been the subject of meta-historiographical debates. See, for instance, Schlumbohm (2000) or Levi (1992).

[21] For the reception of Galilean science in early modern Europe the reader is referred to the contributions collected in Palmerino and Thijssen (2004). For an illuminating account of the historiography of the recognition of Galileo as one of the pioneers of the modern scientific method, see Segre (1998, 396–398).

systems of knowledge with which the new science, according to common reception, competed. Revealing how Galileo arrived at his results will thus eventually also help to elucidate how Galileo's successors, in dealing with these results, brought about the radical changes that are commonly attributed to Galileo himself.[22]

The question of micro- versus macro-histories raises another issue every historical study has to confront, namely, that of how to confine ones subject without doing an injustice to its larger historical contexts. Evidently Galileo's considerations can only be interpreted based on an understanding of the wider contexts in which his life and work were situated and, furthermore, taking into account the various influences he was exposed to. Such contexts have been unraveled in countless studies and cover a wide spectrum, for instance, ranging from social norms in a system of patronage to the teachings of the professors of the Jesuit Collegio Romano, where Galileo appropriated knowledge indispensable for his own work.[23] Moreover, fundamental bodies of knowledge—such as precalculus Euclidean geometry and in particular the existing conceptual frameworks by means of which phenomena of motion were described and analyzed at the time, such as Aristotelian physics or the middle science of mechanics in its different manifestations—defined, to a large extent, the conceptual space in which Galileo pursued a particular intellectual trajectory. The genesis of his new science of motion is indeed conceivable only against the backdrop of these bodies of knowledge.[24]

At the same time, as the interpretation given here will show, in his considerations regarding naturally accelerated motion, Galileo ventured into untrodden territories early on. In consequence, his successive ponderings, while still based on more fundamental knowledge structures, were fairly self-sustained and in fact hardly fed at all on new external knowledge resources as they progressed. Wisan, in her analysis

---

[22] Cf. Renn (1993).

[23] With regard to Galileo's life and work, Biagioli (1993, 2) identified "the shift from the artisan's workshop to the princely court as a crucial site for the new sciences." It is, in contrast, demonstrated here that Galileo's new science essentially emerged in the first half of the first decade of the seventeenth century and thus in closer relation to what Biagioli refers to as the "artisan's workshop" than to Galileo's activities as Mathematician and Philosopher to the Grand Duke of Tuscany. The influence which the works of the professors at the Jesuit Collegio Romano exerted on Galileo's early writings is analyzed in detail in Wallace (1984, 1991) and Galilei and Wallace (1992). Büttner et al. (2001) identify different bodies of shared knowledge which influenced Galileo at various points and without which neither his own theoretical investigations nor their reception can be adequately understood. Duhem can rightfully be accredited a special role in unearthing Galileo's intellectual prerequisites. Segre (1998, 402) has described Duhem's role as follows: "Pierre Duhem...studied the development of science as constant modification to theories, known as the 'conventionalist' approach to the history of science. Duhem published a monumental study of Galileo's predecessors, emphasizing the former's debt to their work."

[24] Schemmel (2008) has compared the work of Galileo and Harriot with special focus on the question whether and to what extent commonalities in their work can be explained by the fact that both made recourse to the same shared knowledge resources. According to his analysis these resources span a conceptual space creating "possible alternative pathways through this knowledge" without, of course, the individual trajectories being determined.

of the manuscript, has considered "numerous ancient, medieval, and Renaissance sources upon which Galileo seems to have relied, but reached the conclusion that Galileo's work on motion was 'novel in its conception and execution'."[25] It is not least this novel character of large parts of Galileo's new science which lends justification to an approach that, while acknowledging Galileo's intellectual repertoire, focuses almost exclusively on the analysis of a single document.

Galileo's *Notes on Motion* are working notes in the truest sense. The distribution of individual entries over the folios of the manuscript is often extremely chaotic. Sometimes different entries on one and the same page are oriented in four different directions. The character of the entries varies over a wide spectrum, from scribbles to test the pen to complete drafts of propositions that were reproduced almost verbatim in the *Discorsi*. Virtually all of the entries in the manuscript are undated.

Entries, be they calculations, diagrams, or separate paragraphs of text, are regarded here as the basic unit of analysis.[26] More often than not individual entries on a page combine to form larger units of meaning. A number of successive text paragraphs, for instance, often represents a proposition, i.e., a statement concerning naturally accelerated motion that is equipped with a proof and to which, as a rule, a proof diagram is associated.

Due to the nature of the manuscript, unambiguously referring to an entry is not always easy. In the electronic representation of Galileo's *Notes on Motion*, as the only complete edition of the manuscript extant, individual entries are identified and equipped with unique identifiers. This, in principle, allows the reference of entries by online citation as is done here in the footnotes.[27]

Yet the electronic representation of Galileo's *Notes on Motion* has been an early pioneering attempt to exploit new media and, in particular, the Internet, for the

---

[25] See Palmieri (2003, 230).

[26] The identification of individual calculations, diagrams or textual entries is not unambiguous and to a certain extent involves interpretation.

[27] Due to the vast number of references to the electronic edition, identification of the referenced tokens by their full *URL* would have been inappropriate, not least for aesthetic reasons. Instead, the following conventions are employed which allow unambiguous location of the reference items: individual folio pages are referenced by the folio number followed by the postfix "recto" or "verso". In rare cases where the same page number is assigned to consecutive folios by the original page numbering, or where a sheet has been pasted on another page, small letters following the folio number have been used for disambiguation. The easiest way to access the referenced page in the online edition is to consult the list of folios in the *Notes on Motion* at http://www.mpiwg-berlin.mpg.de/Galileo_Prototype/MAIN/LIST.HTM (accessed 14 Jun 2016) and navigate to the respective page from there. Individual entries on the pages are referenced by specifying the identification assigned to them in the electronic representation of Galileo's *Notes on Motion*. For the reader of a paper copy of this document, the easiest way to access the referenced information is to consult the so-called working-level representation of the folio page on which the entry is contained. The image of the folio displayed on a working-level is an active map which allows single entries to be accessed by clicking on the part of the page that contains the entry. For example, in order to access text paragraph T2B on folio page 35 recto, access http://www.mpiwg-berlin.mpg.de/Galileo_Prototype/HTML/F035_R/M035_R.HTM and click on the part of the image map that links to http://www.mpiwg-berlin.mpg.de/Galileo_Prototype/HTML/F035_R/T2B.HTM.

representation and dissemination of a manuscript source. It has been online for almost 20 years and, as it is hosted by stable institutions, I expect it to remain accessible for much longer. But how long this will be and by what it will be succeeded cannot currently be judged. Thus, when referring to an individual entry on a folio page in the main text, I have attempted to provide the information necessary to, at least in principle, allow the identification of the entry on the folio page even without access to the electronic representation of Galileo's *Notes on Motion*.

Besides the single entry, the single folio page is the second most important unit of analysis. Folios can, based on material criteria of which more below, be related independently of their content. However, a folio page can certainly comprise entries related to different contexts, unrelated to each other, possibly even written at widely differing times.[28] Yet prior studies have shown, and this is confirmed here, that the content of individual pages more often than not pertains to the elaboration of just a single problem and that, even if Galileo treats more than one problem on a page, it can still as a rule be assumed that the entries on a page were written more or less contemporaneously, and the question concerning a possible relation among them is thus almost always appropriate.[29] Thus every entry discussed is always placed within the context of the page on which it is contained.[30]

Over time Galileo retained earlier notes, occasionally adding new notes, sometimes even time and again to the same page. In revisions and reorderings of the manuscript over the course of more than 30 years, he sorted out material which at the time appeared to be obsolete. As concerns number and order of the folios of the manuscript as it has come down, it is unclear to what extent the manuscript as it stands today represents the state in which Galileo bequeathed it to his disciples.[31]

---

[28] As prior studies have revealed, in some very rare cases, Galileo did indeed reuse folios pages or rework their content at a much later date than their original composition. When Drake (1979) rearranged the entries of the manuscript into what he conceived as their most likely order of composition, he, in contrast, ended up frequently attributing entries contained on one and the same page to widely separate periods of time. With regard to this, Hooper (1992, 312) has remarked that: "a more cautious approach would attach greater importance to the integrity of the documents, and would cut them only with good reason."

[29] When Galileo elaborates separate problems on the same page, this is often reflected in the layout. For instance, Galileo often turned the page 180 degrees before he started working on a new problem.

[30] In the Appendix in Chap. 13 the folio pages relevant for the interpretation are analyzed in full. In the text, results are summarized in such a way that the reader, as a rule, does not have to refer to the detailed analyses in the appendix in order to follow the arguments but can do so in order to check the underpinning presumptions.

[31] The twisted history of the manuscript is epitomized in a particular episode: Nelli, one of the first collectors of Galilean autographs, had bought some mortadella from a butcher and found it wrapped in a paper bearing notes in Galileo's hand. He immediately sent an attendant to the butcher to buy the remaining manuscripts, which had been used as wrapping paper. How these manuscripts, formerly in possession of the brothers Panzanini, got into the hands of the butcher is unclear. Cf. Favaro (1885, 55). Even though it is debatable whether the *Notes on Motion* were among these manuscripts, stains on their pages have come to be referred to jestingly as "mortadella spots."

1 Introduction

This mingled, provisional, and triaged character of the notes presents some serious hermeneutical problems for the undertaking of a rational reconstruction of the conceptual development of Galileo's new science.

To begin with, virtually all of the criteria at our disposal which allow for inferences concerning the relative ordering of the entries are weak, i.e., to be effective they require additional assumptions whose validity, in turn, needs to be vindicated on the basis of an assessment of the very material to which these assumptions are applied. In consequence, the conclusions which can be drawn are likewise weak and moreover often equivocal. Caution needs to be exerted not to fall victim to circular reasoning and to implicitly presuppose what must properly be the result of interpretation.

This can be illustrated by the question of how and what conclusions can be drawn from the fact that different types of paper can be discerned in the *Notes on Motion*, mainly because many of the folios bear watermarks. In some cases these watermarks allow the conclusion that two or more folios stem from the same paper mill. This, in turn, suggests that they were acquired by Galileo at the same time or at least over a limited period of time.

However, this certainly does not imply that Galileo wrote on such folios contemporaneously. He may well have used one folio to take notes immediately after purchase and kept the other in a drawer to pull it out only months, years, or even decades later. We can certainly assume that Galileo used up batches of paper he bought in a rather brief period of time, yet this is an assumption about his praxis, whose validity can and must be vindicated based on an analysis of the manuscript. But even if it can be demonstrated for a significantly large group of folios that the entries on them were drafted in a limited period of time, this merely lends plausibility to the conjecture but does not demonstrate beyond doubt that the content of an additional folio sharing the same watermark was likewise drafted in that same period.

Inferences concerning the rational and inner logic relating Galileo's individual entries ultimately remain the most important criterion for ordering and interpreting his notes.[32] The conclusions that can be gleaned this way are likewise usually weak. A first note may express what, from our perspective, appears to be a necessary and natural consequence of a thought expressed in a second note. Yet in no strong sense does this imply that the entries in question were written down in this order, nor in fact that Galileo necessarily realized one as being a consequence of the other. As reconstructing the conceptual development of Galileo's science calls for recovering the inner logic relating his considerations, it is important to be careful not to tacitly presuppose what one is trying to infer.

Whether used as wrapping paper or not this episode draws our attention to the fact that we cannot expect to find the manuscript in the state in which Galileo had left it.

[32] By inner logic relating entries, I not only refer to relations expressed by logical inferences but to any relationships whatsoever between the content of different entries, for instance, copying relations.

Inferences regarding the relation of Galileo's considerations cover a wide spectrum. They range from conclusions based on secure low-level assumptions, such as when it concluded that one calculation is related to and in fact succeeded another because the result of the latter is used as an input value in the former; or when from the layout of the entries on a page their order of composition can be inferred; to speculative high-level interpretations, such as when it is conjectured that an insight expressed in one entry gave rise to a consideration expressed in another.[33] Throughout this work I have attempted to base my overarching interpretations of the latter type on the former type of conclusions.

In view of what was stated above, the attempt to interpret the *Notes on Motion* can be compared to assembling a "Shmuzzle," a puzzle whose single parts are Escher tiles. Almost every single tile of such a puzzle fits together with any given other tile, and a limited number of tiles of the puzzle can be combined coherently in an almost limitless number of ways. The more parts that are added, however, the more difficult it becomes to find an appropriate combination. Finally, it is extremely difficult to find a solution that allows the last tile to be added seamlessly. This analogy, to the extent that it applies, entails a number of consequences.

Firstly, as many entries of the manuscripts should be considered for an interpretation as possible, more evidence sets narrower boundaries for imagination. Secondly, all available criteria should be exploited. Thirdly, the often equivocal local interpretations of groups of notes need to be probed and mutually related in order to arrive at a coherent overall picture.[34]

Unfortunately, it was impossible to resort to all criteria potentially available for an interpretation, at least not in the desired manner. The reason for this is the lack of corresponding systematic prior studies of the manuscript under particular criteria. Among these, materiality, in particular that of the types of paper and ink used, ranks first. A restricted pilot study of ink types in the manuscripts yielded promising results but, alas, was not followed up.[35]

---

[33] A typical example of an inference concerning the temporal order of entries from layout are those cases in which one entry is written around another one and must thus have been composed later. Diagram D01A on folio 147 recto, for instance, must have been put on the page before the text paragraphs T1A and T1B as they clearly have been written around the diagram. For an example of how the reoccurrence of a unique number on some of the pages of the manuscript allows for a grouping, resulting in a consistent interpretation, the reader is referred to Büttner (2001).

[34] Many extant studies of the manuscript have taken into consideration only a very limited number of entries and presented hermetically self-contained accounts that are speculative, resting heavily on the selection of the material considered and the criteria applied in the analysis. The selection often appears to have been made to support a preconceived interpretation. As a result, the reader may accept the interpretations advanced *bona fide*, whereas it is only the recognition of interpretational alternatives against which a given account emerges as being justified.

[35] In the 1990s a PIXE analysis of the ink types in Galileo's *Notes on Motion* was conducted as a collaborative effort by the Biblioteca Nazionale Centrale, Florence; the Istituto Nazionale di Fisica Nucleare, Sezione di Firenze, Florence; the Istituto e Museo di Storia della Scienza, Florence; and the Max Planck Institute for the History of Science, Berlin. Promising preliminary results were published in Pirolo et al. (1996). The bulk of the data that was collected lies, however, fallow and

1 Introduction

Drake has studied the paper in the manuscript, in particular attempting to exploit watermark information for the ordering of entries in the manuscript. Comparing these watermarks to those in Galileo's correspondence and other documents, he ventured a dating of the material in the manuscript. Hooper has demonstrated Drake's approach to contain serious problems.[36] In consequence, here watermark evidence is used sparingly and mainly as a heuristic.[37] I express my hope that a systematic study of the paper types in the manuscript exploiting all modern means of paper analysis will be carried out in the future.

A particular type of information pertinent to paper has, however, been brought to bear on the analysis rather systematically. It regards the so-called double folios. Whereas the larger sheets from the paper press were usually cut twice, thus producing four quarto folios, double folios resulted when the folded sheet was cut only once.[38] Galileo usually seems to have folded these in the middle. Many of the double folios were at some later point cut but some remain uncut to the present day. In the former case, it is occasionally possible to infer that two folios, now separated, once formed a double folio, namely, when content, once united, is now distributed over two folios running over their edges. Folios recognized as once having formed half of a double folio are routinely discussed together with the respective other half.

Further criteria that should systematically be studied, to refine, control, and hopefully support the results presented here, concern particularities of Galileo's way of doing and systematic changes potentially observable therein. Examples of this are, for instance, submitting his changing styles of handwriting for graphological analysis or an analysis of spelling errors as has been carried out for another of his

---

awaits its final evaluation. This is all the more regrettable as these results may eventually allow for a comparison of the inks used in the *Notes on Motion* and in Galileo's housekeeping book, the so-called *Ricordi Autografi*, in which the entries are dated.

[36] By considering the concrete praxis of paper making in the seventeenth century, Hooper (1992, 287–291) has identified a number of problems that must be solved before watermark information can be exploited for ordering and dating the *Notes on Motion*. He has thus argued, for instance, that the watermarks of all papers produced as part of the same batch and thus likewise probably sold together as a batch should vary slightly owing to the process of manufacturing. Finding exactly identical watermarks on two folios thus should not be taken as an indication that these folios were acquired at the same time as is usually done. Furthermore, a sheet, as produced in the paper mill and later cut into four quarto folios, would be expected to contain two watermarks. See Hunter (1978), in particular p. 268. Given an arbitrary sample of cut quarto folios, such as the *Notes on Motion*, about half of the pages should bear watermarks. Yet the actual number of folios with watermarks in the manuscript is significantly higher.

[37] In Chap. 10 a group of folios will be analyzed where the criteria for grouping (and naming) the folios is their common watermark. However, this shared characteristic on the material level is shown to correspond to unique features that the folios of this group share with regard to content.

[38] Today the manuscript is bound, and it is difficult and in some cases impossible to judge whether two individual folios are part of an uncut double folio. I owe my information, not only as to which folios of the manuscript form double folios but also that on other aspects regarding the materiality of the folio pages, to a group of scholars from the Max Planck Institute for the History of Science, who had the opportunity to examine the manuscript when it was still unbound and who have been so kind as to let me use the results of their analysis.

manuscripts.[39] As diachronic changes in hand or orthography should be virtually independent of what is being written, such analysis would allow comparison to other documents written by Galileo and thus open the prospect of verifying and refining dating.

I hope that the results presented in this book will contribute to a reawakening of interest in the manuscript and thus assist in stimulating studies such as those alluded to above, which currently remain a desideratum. I myself eagerly await the integration of the results of such studies into a more coherent picture, a picture that will necessarily expand on but also, I optimistically expect, confirm the findings presented here.

This book does not trace the development of Galileo's new science of motion from its inception to the final publication, nor does it treat the new science in its entirety. My analysis remains restricted to one branch of Galileo's new science, the descent of heavy bodies along inclined planes, including the case of vertical free fall. I start with the reconstruction of a challenging research program Galileo was pursuing in the second half of 1602 and end with a reconstruction of Galileo's first attempt to transform the results achieved into a new science, which as will be argued, happened in 1604. This time span, as short as it may be, marks the formative phase of the new science of motion. Corresponding to the limited period covered, only a portion of the almost 170 folios of the manuscript are included in the interpretation. This selection is, however, based on a more comprehensive analysis of the manuscript, further results of which I hope to be able to publish in the future.

The fictitious treatise *De motu locali*, which the three dialog partners of the *Discorsi* discussed, is divided into three separate books or chapters, each treating a different type of motion—uniform motion, naturally accelerated motion and finally the motion of projectiles. In the opening passage of the Third Day of the *Discorsi*, the distinction is made explicit:

> Tripartito dividimus hanc tractationem: in prima parte consideramus ea quae spectant ad motum aequabilem, seu uniformem; in secunda de motu naturaliter accelerato scribimus; in tertia, de motu violento, seu de proiectis (EN VIII, p, 190.)

---

[39] With respect to handwriting, Drake claimed to be able to relate individual entries to specific rheumatic attacks Galileo suffered and that are documented in his correspondence. Cf., for instance, Drake (1978, 76). Even if this is clearly an exercise in over interpretation, it is not ruled out that a trained graphologist could infer valuable information from Galileo's changing styles of writing. With regard to the entries considered in this book, my own attempts to identify significant differences in Galileo's handwriting failed consistently, which could be seen as confirming that these entries, in fact document a rather short period of work. With regard to spelling errors, in a pioneering study, Hooper tried to reconstruct the order of the various parts of the text collection assembled under the heading *De Motu Antiquiora*, based on arguments drawn from a statistical analysis of such errors in different parts of the text. His conclusions have, however, since been convincingly refuted, for instance, in Camerota (1992), calling into question the validity of the method Hooper employed.

I follow Galileo's own terminology and will refer to the considerations delineated by the scope of this second book as the new science of naturally accelerated motion or of natural descent.

The manuscript contains notes pertinent to all three chapters. However, just as in the *Discorsi* also in the manuscript, considerations regarding natural descent remain rather tidily separated from Galileo's considerations concerning projectile motion. Only a very limited number of pages exist on which considerations relating to motion on inclined planes mix with those relating to projectile motion, facilitating a, for the time being, separate treatment of Galileo's considerations related to the two types of motion.[40]

The folios of the manuscript can, moreover, rather unambiguously be distinguished and grouped into those which Galileo bought in his Paduan period and those that he purchased during his time in Florence.[41] Of the folios attributable to the Paduan period, the great majority have been considered in this interpretation.

Ultimately, the restrictions on content and period are not desirable and I hope that my account will yield a brick to lay in a future more comprehensive reconstruction and interpretation of the development of Galileo's science of motion, one which does due justice to the close conceptual relation between its two major branches and which would take into account the entire period up to the publication of the *Discorsi* and possibly beyond.

No tangible overall account of how Galileo's new science of naturally accelerated motion emerged exists. Yet a standard understanding has asserted itself, which rather than being addressed as an interpretation, is more appropriately addressed as a chronology. It draws, in particular, on two letters that Galileo sent in the first decade of the seventeenth century, in which he made rather detailed reference to his current work regarding motion on inclined planes.

---

[40] Whereas Galileo's considerations concerning motion along inclined planes do not presuppose his considerations concerning projectile motion, the opposite is not the case. Considerations concerning projectile motion are directly related to motion along inclined planes—for instance, in Galileo's so-called theorem of equivalence (cf. Renn 1990), or else in his attempt to construct the trajectory of projectile motion from a superposition of a decelerated motion upward, an inclined plane corresponding to motion along the line of shot and a naturally accelerated motion downward (see Damerow et al. 2004, 216–220). Assumptions regarding motions along inclined planes and those regarding projectile motion finally come together in the theoretical considerations that accompany Galileo's projection experiments. In these experiments, a ball first rolls along an inclined plane and is then projected, either horizontally after deflection or at an oblique angle downward directly from the plane. The folios comprising the records of the experimental are 81 recto, 114 verso, and 116 verso. Hahn (2002) provides a survey of the experiments including references to the relevant literature.

[41] This distinction is made possible because of watermarks, and even in the absence of a watermark, paper from the Paduan period can usually be distinguished from that of the Florentine period as the quality of the former is usually inferior. After his move to Florence, Galileo reviewed the material that had gathered up and had two of his disciples, Mario Guiducci and Niccolò Arrighetti, copy selected entries to separate sheets of paper. Arguably the originals from which the copies were made are all attributable to the Paduan period.

The first letter was sent to Guidobaldo del Monte on 29 November 1602. Galileo communicated to his friend and mentor that he was seeking a demonstration of the alleged isochronism of the pendulum and that he was trying to obtain such a demonstration based on assumptions about the motion of heavy bodies along inclined planes. The second letter was sent to Paolo Sarpi on 16 October 1604. Here Galileo essentially announced that he had found an indubitable principle on the basis of which he was able to prove the law of fall and other things.

According to the standard understanding, the letter of 1604 testifies to Galileo's discovery of the law of fall and thus more generally marks the onset of his considerations regrading naturally accelerated motion that eventually resulted in the formation of a new science. It has been concluded in consequence that the work Galileo alluded to in the letter of 1602 must have been carried out under the earlier assumption that motion along inclined planes is, in principle, uniform. No evidence to further support this conclusion was, however, disclosed in Galileo's oeuvre.[42]

Damerow et al. have addressed what has been presented above as the standard understanding and have stressed that it is not adequately based on an analysis of the available sources. As the authors phrase it, it "emerges as a peculiarity of recent Galileo historiography that the dates at which Galileo supposedly made his major discoveries have remained largely unchallenged, in spite of the relatively weak direct evidence available for this dating."[43]

In this book the standard understanding is challenged and ultimately overthrown. The two letters of 1602 and 1604 do indeed mark milestones of the early development of the new science of naturally accelerated motion. Yet as will be demonstrated, it is the earlier and not the later letter which testifies to the beginning of Galileo's systematic considerations regarding naturally accelerated motion. What the letter of 1604 testifies to is an alleged breakthrough in his attempt to transform the results achieved by these considerations into the nucleus of a new science, mainly by relating them to the extant theoretical frameworks employed to describe motion and its causation.

What is summarized in just two sentences above is unfolded and argued for in great detail in the chapters that follow. The book is divided into two parts. The first part, *Novel insights about accelerated motion: the challenge of swinging and rolling*, succeeding this introduction, is opened by Chap. 2. This chapter, *Before Natural Acceleration*, discusses aspects of Galileo's thinking about motion and mechanics prior to his conceptual shift toward the assumption that motion of fall is naturally

---

[42] Drake (1978, 67), for instance, claimed, regarding a group of entries in the manuscript that "[t]hese probably all belong to 1602, when it appears that Galileo having revised and expanded his *Mechanics*, decided to write a new treatise on motion. Several propositions for this were neatly written out before he fully realized the importance of acceleration, and the sheets bearing these were mutilated in the course of the subsequent revisions of the projected treatise." His claim cannot be upheld as it will be shown that what Drake claimed to be subsequent revisions were in fact part of Galileo's original considerations conducted from the start, under the conceptualization of motion along inclined planes as naturally accelerated.

[43] See Damerow et al. (2001, 305).

accelerated, i.e., essentially prior to 1602. It serves a propaedeutic function for, and the selection of the material discussed is motivated entirely by, the interpretation presented in the succeeding chapters. In Chap. 3, *A Glimpse at a Challenging Research Agenda: Galileo to Guidobaldo del Monte in 1602*, a detailed rereading of the famous letter sent by Galileo to Guidobaldo del Monte in 1602 is advanced. In this letter Galileo outlined the challenging research agenda on swinging and rolling his work on which the succeeding chapters identify in the manuscript.

The fourth chapter, *Sparking the Investigation of Naturally Accelerated Motion: The Pendulum Plane Experiment*, reconstructs and interprets an experiment in which Galileo related the motion of a pendulum to the rolling of a ball downward along an inclined plane. Galileo tested the straightforward hypothesis that it would take a pendulum the same time to swing from rest at a deflected position to the lowest point of its circular path as it would take a heavy object to fall or roll along an inclined plane spanning this same arc as a chord. Hypothesis and experiment, it is argued, stood at the very beginning of Galileo's investigations into the relation between swinging and rolling. Despite the fact that the experimental record has survived intact, the experiment has virtually been ignored in the literature.[44] Galileo's hypothesis was not confirmed, and in consequence he advanced his research program by modifying the underlying conceptualization of the relation between swinging and rolling.

Central to his new approach was the idea that swinging and rolling along the same arc were kinematically equivalent and that the latter motion could be represented by motion along paths made up of a series of conjugate chords of increasing number inscribed into the arc. Galileo's work under the new approach is analyzed in Chap. 5, *Squaring the Pendulum's Arc: Motion Along Broken Chords*, subsumed under the title *broken chord approach* in allusion to its central idea. As a first substantial result, Galileo was able to show that motion along a path made up out of two conjugated chords was completed in less time than motion along a single chord spanning the same arc of a circle and comprising either its apex or nadir. He mentioned this proposition in the letter to Guidobaldo written toward the end of 1602, where he claimed to have already found a proof. Work on the broken chord approach is dated to 1602 by an argument that is independent of the information conveyed in the letter. Despite the fact that Galileo was able to solve a number of intricate problems that his new approach had bestowed him with, he was ultimately not able to come up with an argument that would have allowed an inference regarding the limiting case of motion along the arc.

As discussed in Chap. 6, *Swinging and Rolling Revisited: Motion Along Broken Chords and the Pendulum Plane Experiment*, Galileo's elaboration of the broken chord approach allowed him to resort to his earlier experimental data and to reinterpret it in light of the results achieved under his new conceptualization. From the time measured for rolling along an inclined plane, Galileo calculated the time it would take a body to fall along different polygonal paths inscribed into the arc

---

[44]For the experimental turn in Galileo scholarship, see, for instance, Segre (1980).

of the swing of the pendulum, whose period he had measured in the experiment. According to his new conceptualization, the time of motion along a polygonal path and the time to swing along the arc should converge as the path approached the arc. Yet his calculations showed that a rather large, and from an empirical point of view, unsatisfactory deviation remained. As Galileo was not able to disentangle the factors which, from a modern perspective, contributed to this deviation and as in addition his considerations in the context of the broken chord approach had yielded indications of a problem with the underlying conceptualization, the research on swinging and rolling had run into an impasse and indeed came to a halt.

While they did not lead to the desired goal, Galileo's grapplings with the broken chord approach had given rise to a considerable number of new problems but had also generated new insights concerning naturally accelerated motion along inclined planes. Chapter 7, *Accumulating Insights: The Problem of Motion Along Broken Chords Driving Conceptual Development*, discusses the most important results, spin-offs as it were, that had emerged and accumulated this way. It is demonstrated that Galileo at that time formulated a number of propositions equipped with proof that he would go on to publish more than 35 years later, some almost verbatim, in the *Discorsi*. This not only marks the pheno-kinematics of naturally accelerated motion thus beginning to unfold as the subject matter of his new science.[45] It is, moreover, shown that Galileo, as a spin-off of the broken chord approach, devised a new technique for the diagrammatical representation of motion by which distances traversed and times elapsed during motion are represented in one and the same diagram. This technique was soon to become an indispensable prerequisite for Galileo's arguments concerning naturally accelerated motion.

In Chap. 8, *Toward a Foundation: The Ex Mechanicis Proof of the Law of Chords*, the sequence of considerations that led Galileo to establish the so-called ex mechanicis proof of the law of chords is reconstructed and discussed. The ex mechanicis proof which Galileo mentioned in his letter to Guidobaldo in 1602 is traditionally understood as having emerged under his older conceptualization, according to which motion on inclined planes is characteristically uniform. It is demonstrated, in contrast, that the proof emerged as another spin-off of Galileo's work on swinging and rolling. Its establishment thus succeeded the conceptual shift toward the assumption of natural acceleration. Moreover, the argument of the proof is, in contrast to what is commonly assumed, unrelated to how Galileo gained insight into the law of chords, which he had known and relied on long before the proof was constructed. As the name indicates, the ex mechanicis proof was based on a dynamical argument and Galileo's construction of the proof, it is argued,

---

[45] In classical mechanics kinematics includes the concepts of velocity and acceleration. Galileo's propositions, in contrast, remain restricted to stating relations between spaces traversed and times elapsed in naturally accelerated motion. Distances and times are directly accessible to sense perception as opposed to velocity and acceleration which are derived concepts only. Hence I have chosen the term pheno-kinematics to refer to the aspect of kinematics covered by Galileo's approach. Cf. Chap. 2.

bears witness to the beginning of a new stage of work characterized by the attempt to transform the new results into a demonstrative science. This stage of work is discussed in the second part of the book.

The first part closes with an outlook on the further fate of Galileo's research regarding the relation of swinging and rolling in Chap. 9, *Whatever Happened to Swinging and Rolling: Faint Echoes and a Late Insight*. On the one hand, Galileo's argument regarding the brachistochrone, that is, the curve along which a body moves from a given point to another in shortest time, is discussed, versions of which are contained in the *Dialogue*, the *Discorsi*, and also in Galileo's private communication.[46] Rather than being the result of a systematic search and a mainspring for the emergence of the new science as has been claimed in the literature, the argument is shown to have emerged from Galileo's investigation into the relation of swinging and rolling. The argument is flawed, and Galileo was, according to all evidence, aware of its flawed character. The chapter, moreover, looks into an episode in the aftermath of the publication of the *Discorsi*. In his *De motu naturali gravium solidorum*, published only a few months before the *Discorsi*, Giovanni Battista Baliani had derived the properties of naturally accelerated motion on inclined planes from assumptions about pendulum motion and thus apparently achieved what Galileo hadn't been able to accomplish with his program on swinging and rolling. In a letter drafted in response to this publication, harking back to his own earlier insights, Galileo criticized Baliani's approach. As a result of this criticism, Galileo was able to devise a proof of the law of the pendulum from assumptions about naturally accelerated motion and thus to, at least partially, solve the problem he had been carrying around for more than 35 years. As marginal notes in Galileo's own copy of the *Discorsi* show, he intended to include this new result in a revised edition.

The second part of this book, "Novel insights and old concepts – A new science in the making," comprises only two chapters. In Chap. 10, *Toward a New Science: Gathering Results and the Rise and Demise of a Dynamical Foundation*, a group of folios, designated as the star group folios due to their shared watermark, a small star, is analyzed. On these folios, Galileo collected and systematized the results of his research on swinging and rolling. His motivation for doing so was not so much to structure the material that had accumulated for the purposes of his own further investigations, rather his motivation, it is argued, was to prepare a potential future publication, which required an appropriate expounding of the material that had accumulated. In doing so he started to analyze his new insights about the pheno-kinematics of naturally accelerated motion in terms of concepts anchored in the traditional frameworks employed in arguing about motion and its causation. This opened the prospect of a dynamical argument, which with the ex mechanicis proof, was soon found. Yet further analysis led to the recognition of a serious problem in the conceptual underpinnings of the dynamical proof.

---

[46]Galileo's *Dialogo ... sopra i due massimi sistemi del mondo ...* of 1632 will be referred to here plainly as the *Dialogue*.

Chapter 11, *Toward a New Science: Axiomatization and a New Foundation*, includes but one additional folio, folio 147, into the analysis, the content of which is shown to have been drafted contemporaneously to that of the star group folios. It thus transpires that Galileo was concerned with an axiomatization of the new results engendered by the research on swinging and rolling, i.e., he was rather systematically exploring which of his newly found propositions regarding the pheno-kinematics of naturally accelerated motion could serve as a minimal, yet strong enough, set of statements to act as a fundament for the deductive structure of a new science. It thus becomes manifest that Galileo's analysis of his new findings, making recourse to concepts of the theoretical frameworks traditionally employed in the description of motion and its causation, aimed at constructing proofs from fundamental principles for the propositions upon which he aspired to root the deductive tree. Yet Galileo's new insights about the pheno-kinematics of naturally accelerated motion, which by that time had assumed a rather incontestable status, turned out not to be seamlessly integrable with the existing frameworks, which they indeed conflicted with. A way out offered itself when Galileo resorted to the doctrine of the configurations of motion for the foundation of his new science, which he emphatically communicated to Paolo Sarpi in 1604. Galileo's 1604 attempt at a foundation, preconfigured by the insights gained but also by the problems encountered around 1602, is reconstructed.

Famously, with the principle of acceleration formulated in the letter to Paolo Sarpi, Galileo was on the wrong track. It took him time to realize this and even more time to finally harvest the fruit of the new science as it was published in 1638, from the seed that had been planted with his research on the challenging relation between swinging and rolling so much earlier. Tracing the development of the new science from 1604 onward, however, remains reserved for a future publication.

The law of fall and the law of chords. Isochronism of the pendulum and the law of the pendulum. An astonishing, extraordinary similarity. Galileo was never able to theoretically recover the similarity he perceived between pendulum motion and naturally accelerated motion on an inclined plane and thus to solve his challenging problem of swinging and rolling. From a modern perspective, it is amply clear why not. Pendulum motion is not isochronous; the similarity Galileo conceived was a chimera. According to all evidence, however, Galileo never forfeited his conviction for the similarity and remained convinced that his research program could, in principle, be crowned with the success not granted to him. Almost coincidentally, as a spin-off he was granted with a much larger success. The attempt to save his firm conviction, almost in passing, bestowed him with a new science, which he has been praised for by the history of science. This, in a nutshell, is the thesis of this book.

References

Biagioli, M. (1993). *Galileo, Courtier: The practice of science in the culture of absolutism*. Chicago: The University Press.
Biener, Z. (2004). Galileo's first new science: The science of matter. *Perspectives on Science, 12*(3), 262–287.

References

Büttner, J. (2001). Galileo's cosmogony. In J. Montesinos & C. Solís (Eds.), *Largo campo di filosofare: Eurosymposium Galileo 2001* (pp. 391–401). La Orotava: Fundación Canaria Orotava de Historia de la Ciencia.

Büttner, J., Damerow, P., & Renn, J. (2001). Traces of an invisible giant: Shared knowledge in Galileo's unpublished treatises. In J. Montesinos & C. Solís (Eds.), *Largo campo di filosofare: Eurosymposium Galileo 2001* (pp. 183–201). La Orotava: Fundación Canaria Orotava de Historia de la Ciencia.

Camerota, M. (1992). *Gli Scritti De Motu Antiquiora di Galileo Galilei: Il Ms. Gal. 71. Un'analisi storico-critica*. Cagliari: Cooperativa Universitaria.

Camerota, M., & Castagnetti, G. (2001). Antonio Favaro and the Edizione Nazionale of Galileo's works. In J. Renn (Ed.), *Galileo in context* (pp. 357–361). Cambridge: Cambridge University Press.

Damerow, P., Renn, J., & Rieger, S. (2001). Hunting the white elephant: When and how did Galileo discover the law of fall? In J. Renn (Ed.), *Galileo in context* (pp. 29–150). Cambridge: Cambridge University Press.

Damerow, P., Freudenthal, G., McLaughlin, P., & Renn, J. (2004). *Exploring the limits of preclassical mechanics*. New York: Springer.

Drake, S. (1978). *Galileo at work: His scientific biography*. Chicago: University of Chicago Press.

Drake, S. (1979). Galileo's notes on motion arranged in probable order of composition and presented in reduced facsimile. In *Annali dell'Istituto e Museo di Storia della Scienza Suppl. Fasc. 2, Monografia n. 3*. Florence: Istituto e Museo di Storia della Scienza.

Favaro, A. (1885). Documenti inediti per la storia dei manoscritti galileiani della Biblioteca Nazionale di Firenze. *Bullettino di bibliografia e di storia delle scienze matematiche e fisiche, XVIII*, 1–112, 151–230.

Galilei, G. (1638). *Discorsi e dimostrazioni matematiche: Intorno à due nuoue scienze attenenti alla mecanica i movimenti locali*. Leyden: Appresso gli Elsevirii.

Galilei, G., & Wallace, W.A. (1992). *Galileo's logical treatises: A translation, with notes and commentary, of his appropriated Latin questions on Aristotle's Posterior analytics* (Boston Studies in the Philosophy of Science, Vol. 138). Dordrecht: Kluwer.

Hahn, A.J. (2002). The pendulum swings again: A mathematical reassessment of Galileo's experiments with inclined planes. *Archive for History of Exact Sciences, 56*, 339–361.

Hill, D.K. (1994). Pendulums and planes: What Galileo didn't publish. *Nuncius Ann Storia Sci, 2(9)*, 499–515.

Hooper, W.E. (1992). *Galileo and the problems of motion*. Dissertation, Indiana University.

Hunter, D. (1978). *Papermaking: The history and technique of an ancient craft* (1st ed.). New York: Dover.

Levi, G. (1992). On microhistory. In P. Burke (Ed.), *New perspectives on historical writing* (pp. 93–113). University Park: Pennsylvania State University Press.

McMullin, E. (1967). Introduction. In E. McMullin (Ed.), *Galileo man of science* (pp. 3–51). New York/London: Basic Books, Inc.

Naylor, R.H. (1974). The evolution of an experiment: Guidobaldo Del Monte and Galileo's "Discorsi" demonstration of the parabolic trajectory. *Physics, 16*(4), 323–346.

Naylor, R.H. (1975). An aspect of Galileo's study of the parabolic trajectory. *ISIS, 66*, 394–396.

Naylor, R.H. (1976). Galileo: The search for the parabolic trajectory. *Annals of Science, 33*, 153–172.

Naylor, R.H. (1977). Galileo's theory of motion: Processes of conceptual change in the period 1604–1610. *Annals of Science, 34*, 365–392.

Naylor, R.H. (1980). Galileo's theory of projectile motion. *ISIS, 71*, 550–570.

Palmerino, C.R., & Thijssen, J.M.M.H. (Eds.) (2004). *The reception of the Galilean science of motion in seventeenth-century Europe* (Boston Studies in the Philosophy of Science, Vol. 239). Dordrecht: Kluwer.

Palmieri, P. (2003). Mental models in Galileo's early mathematization of nature. *Studies in History and Philosophy of Science, 34*, 229–264.

Palmieri, P. (2005). Galileo's construction of idealized fall in the void. *Historia Scientiarum, 43*, 343–389.
Pirolo, P., Trucci, I., Del Carmine, P., & Lucarelli, F. (1996). *Pilot Study for a systematic PIXE analysis of the ink types in Galileo's Ms. 72 – Project report no. 1*. Preprint 54, Max Planck Institute for the History of Science.
Renn, J. (1990). Galileo's theorem of equivalence: The missing keystone of his theory of motion. In T.H. Levere & W.R. Shea (Eds.), *Nature, experiment, and the sciences: Essays on Galileo and the history of science in Honour of Stillman Drake*. Dordrecht: Kluwer.
Renn, J. (1993). Einstein as a disciple of Galileo: A comparative study of concept development in physics. *Science in Context, 6*, 311–341.
Schemmel, M. (2008). *The English Galileo Thomas Harriot's work on motion as an example of preclassical mechanics* (Boston Studies in the Philosophy of Science, Vol. 249). Dordrecht: Springer.
Schlumbohm, J. (Ed.) (2000). *Mikrogeschichte – Makrogeschichte. Komplementär oder inkommensurabel?* Göttingen: Jürgen Schlumbohm im Auftrag des Max-Planck-Instituts für Geschichte in Göttingen.
Segre, M. (1980). The role of experiment in Galileo's physics. *Archive for History of Exact Sciences, 23*(3), 227–252.
Segre, M. (1998). The never-ending Galileo story. In P. Machamer (Ed.), *The Cambridge companion to Galileo* (pp. 388–416). Cambridge: Cambridge University Press.
Wallace, W.A. (1984). *Galileo and his sources: The heritage of the Collegio Romano in Galileo's science*. Princeton: Princeton University Press.
Wallace, W.A. (1991). *Galileo, the Jesuits and the medieval Aristotle* (Collected Studies series, CS 346). Aldershot: Ashgate.
Wisan, W.L. (1974). The new science of motion: A study of Galileo's De motu locali. *Archive for History of Exact Sciences, 13*, 103–306.
Wisan, W.L. (1982). Review of: Galileo's notes on motion: Arranged in probable order of composition and presented in reduced facsimile by Stillman Drake. *ISIS, 73*(3), 471–472.

Part I
Novel Insights About Accelerated Motion: The Challenge of Swinging and Rolling

Chapter 2
Before Natural Acceleration

By way of introduction, this chapter discusses aspects of Galileo's considerations regarding motion and mechanics prior to his conceptual shift toward the assumption that motion of fall is naturally accelerated, i.e., in particular prior to 1602. The topics that will be discussed have been selected according to their relevance for the discussion in the succeeding chapters. As such, the chapter aims neither for comprehensiveness, nor does it aim to make an original contribution. It serves a propaedeutical function and accordingly Galileo experts will find little new in it. For all others, I hope the chapter will provide a fair introduction into the topic proper of the book.

No Eureka moment marked the beginning of Galileo's new science. There is no clear watershed to distinguish Galileo's considerations into those that pertain to the old and those pertaining to his new science. Indeed, it has convincingly been demonstrated that, as regards the conceptual framework upon which Galileo founded his new science, he essentially remained within the confines of preclassical mechanics.[1] It was through interdependent insights and a network of related activities that the new science Galileo was to go on to publish toward the end of his life gradually emerged.

Yet with respect to the continuous process by which the new science came into being, we can identify, at least in hindsight, a criterion according to which Galileo's earlier thinking about motion can be distinguished rather clearly from his later approach, and this is the conviction that the motion of freely falling bodies is naturally accelerated. As will be discussed below, in his earliest writings on motion, Galileo, in contrast, maintained that bodies fall uniformly with their own characteristic velocity. When Galileo started to take the notes assembled in the *Notes on Motion*, he no longer believed this to be the case. As such, the expression *before natural acceleration* is used here more or less as a vignette to refer to those of

---

[1] Cf. Damerow et al. (2004).

Galileo's considerations regarding motion preceding the ones documented in the manuscript and which will be reconstructed and interpreted in this book.

Most of the information relevant to the discussion of the current chapter comes from the *De Motu Antiquiora*, a collection of Galileo's earliest writings on the problems of local motion. Favaro assembled and included the complete corpus of these writings in the first volume of his *Edizione Nazionale* of the works of Galileo.[2] Today, most of the material in this collection is bound together and preserved as Ms. Gal. 71 in the Biblioteca Nazionale Centrale di Firenze.[3] The collection comprises six different parts, a plan of work, a treatment in dialogue form referred to as *Dialogue*, a treatise *On Motion (De Motu)* in 23 chapters, a reworking of the first two chapters of this treatise, *Memoranda* pertaining mostly to *On Motion*, and finally a shorter treatise in 10 chapters.[4]

Date and order of the individual parts have been intensely discussed. An agreement is starting to surface that the material must have been written between 1589 and 1591 and that, as concerns the order of the main parts, the *Dialogue* was written before *On Motion* followed by the shorter treatise in 10 chapters.[5] Neither absolute date nor order is, however, particularly relevant to the arguments I am going to make for which it is sufficed to agree that the material was composed during Galileo's time as a lecturer at the University of Pisa.

---

[2] Cf. Favaro's Avvertimento to *De Motu Antiquiora* in EN I, 245–246. Favaro himself referred to the collection as *De Motu*. According to Vincenzo Viviani and Giovanni Battista Nelli, Galileo himself, on a cover lost today, used the title *De Motu Antiquiora* to refer to these texts. Cf. Camerota and Helbing (2000).

[3] The content of the manuscript can be accessed online through the digital collection of the Biblioteca Nazionale Centrale at: http://www.bncf.firenze.sbn.it. Accessed 10 Feb 2017.

[4] The different parts of *De Motu Antiquiora* have been addressed by different names. Sometimes different scholars have even used the same name to refer to different parts. For disambiguation I provide the following table which gives the names I use for the parts, the page numbers in Volume I of the EN, and the shelf mark and folio numbers of the corresponding manuscript.

	EN reference	Coll. of Galilean manuscripts
On motion	I,251–340	Ms. Gal. 71, 61v–124v
First two chapters reworked	I,341–343	Ms. Gal. 71, 133r–134v
Treatise in 10 chapters	I,344–366	Ms. Gal. 71, 43r–60v
Dialogue	I,367–408	Ms. Gal. 71, 4r–35r
Memoranda	I,409–417	Ms. Gal.46, 102–129
Work plan	I,418–419	Ms. Gal. 71, 3v

[5] Fredette (1972) has compellingly made the case that *On Motion* was conceived of by Galileo as being composed of two books and that the treatise in 10 chapters represents a reworking of the first one.

As already stated, it is not my intention to analyze or discuss Galileo's considerations laid down in the *De Motu Antiquiora* as a whole. This has been done by others, and excellent literature on the subject is available.[6] Instead, in Sect. 2.1 the concept of velocity that Galileo shared with his contemporaries will briefly be introduced. The concept constituted a crucial constituent for any approach to problems of motion and as such was also fundamental for Galileo's consideration in the *De Motu Antiquiora* and remained fundamental for his later considerations regarding naturally accelerated motion. Next, the *doctrine of the configurations of motion*, developed in the Late Middle Ages, among others, to analyze motion as a changing quality, is briefly discussed. In particular, the concept of degree of velocity has its origin in this conceptual tradition. This concept does not figure in the *De Motu Antiquiora* but famously was to assume a central role in Galileo's attempts to provide a foundation for his new science of naturally accelerated motion. Finally, the account Galileo gave in the *De Motu Antiquiora* of why falling bodies initially accelerate will be inspected in some detail. This will show that, as already indicated above, when Galileo wrote *De Motu Antiquiora*, he still considered acceleration to be an accidental phenomenon of falling motion not prone to systematic treatment in a deductive, demonstrative science.

The second section, Sect. 2.2, discusses and analyzes in some detail the content of a single chapter of *De motu*. In this chapter Galileo investigated the free motion of heavy bodies along inclined planes. Just as he would later do, Galileo treated the motion along inclined planes as not distinct in principle from the free fall vertically downward, i.e., as a natural motion. By the so-called *bent lever proof*, he inferred the ratio of the effective motive forces acting on incline planes of different inclinations and assumed these forces to be in proportion to the velocities of motions resulting from these forces. As will be discussed in the succeeding chapters, at some point Galileo also applied the same dynamical argument to naturally accelerated motion along inclined planes. This resulted in a proof, found as we know from correspondence already in 1602 and included 35 years later in the *Discorsi*. Yet the investigation into the dynamics of naturally accelerated motion resulted in daunting and eventually unsolvable problems, which led Galileo to dispense with the attempt to found his new science on dynamics and eventually resulted in the foundation known to us through the *Discorsi*, which is based on concepts anchored in the doctrine of the configurations of motion.

---

[6]For more detailed treatments of the manuscript and its contents, see, e.g., Camerota (1992), Fredette (1969), Galilei (1960), Giusti (1998), Hooper (1992), or Fredette (2001). For a recent synopsis of *De Motu Antiquiora*, see Salvia (2017).

2.1 Velocity, Motion as a Changing Quality, and Galileo's Early Account of Acceleration

Kinematic velocity and the kinematic propositions In Book VI of *Physics*, Aristotle stated regarding processes of change:

> And since every magnitude is divisible into magnitudes—for we have shown that it is impossible for anything continuous to be composed of indivisible parts, and every magnitude is continuous—it necessarily follows that the quicker of two things traverses a greater magnitude in an equal time, an equal magnitude in less time, and a greater magnitude in less time, in conformity with the definition sometimes given of 'the quicker. (*Phys.* VI.2, 232a23–27, transl. R.P. Hardie and R.K. Gaye)

Applied to the case of local motion, this implies that of two motions, the one is quicker which traverses a greater distance in the same time or else which covers the same space in less time. If to be quicker is taken to mean endowed with greater velocity, these two statements can be rendered as:

$$s_1 = s_2 \text{ and } t_1 < t_2 \Rightarrow v_1 > v_2$$

$$t_1 = t_2 \text{ and } s_1 > s_2 \Rightarrow v_1 > v_2.^7$$

By the time of Galileo, the understanding of velocity expressed by Aristotle had developed into a generally shared concept. In the literature, in particular that on Galileo, this concept of velocity has variably been referred to as velocity or speed usually with an additional qualification such as holistic, total, or overall. Souffrin, who has meticulously reconstructed the history and scope of the understanding entailed by this concept of velocity, has proposed referring to it without translation simply by the Latin *velocitas* to exclude any confusion with the modern concept of velocity.[8] For reasons that will become obvious, I use the term *kinematic velocity* or, when unambiguous, simply *velocity*.[9]

---

[7] Throughout this work a great number of Galileo's (and other's) statements have been rendered into modern symbolic notation. Such an approach has been criticized by Palmieri (2003, 230): "the translation of verbal mathematical arguments into a symbolic notation inevitably carries with it the risk of missing important aspects that are inextricably linked with the original language." Even though such critique has its justification, I believe that this technique can have great benefit, but only if it is applied with an awareness of its limitations. The conventions used are more or less self-explanatory and, where they are not, will be introduced explicitly.

[8] In his own text (Souffrin 1992, 237) even puts the word between slashes and writes /velocitas/ to "représenter *toutes les formes* du latin *velocita*."

[9] From a modern perspective the difference between velocity and speed amounts to the difference between a vector quantity and a scalar quantity. Thus, for instance, the velocity of a body that has returned to its point of departure is zero as the displacement vector is zero. For Galileo and his contemporaries, it was distance traversed rather than displacement that mattered, and this seems to be the reason why many authors who have discussed the same material I discuss here have preferred speed over velocity in translations but also refer to Galileo's concept in

2.1 Velocity, Motion as a Changing Quality, and Acceleration

The preclassical concept of kinematic velocity can be characterized as follows[10]:

- Kinematic velocity entails a holistic measure of movement; it characterizes motions extended in space and time.[11] Concretely it holds that:

$$\text{First kinematic proposition: } s_1 = s_2 \Leftrightarrow v_1/v_2 \sim t_2/t_1$$

$$\text{Second kinematic proposition: } t_1 = t_2 \Leftrightarrow v_1/v_2 \sim s_1/s_2.{}^{12}$$

Kinematic velocity thus is usually not specified absolutely but, as a rule, in relation to the velocity of another motion. Alluding to *the* velocity of *a* motion made little sense, and when this was done, comparison to another motion was usually implicit.
- The comparisons of kinematic velocities by means of the two kinematic propositions as given above by no means entail that the motions compared had to be uniform.[13]

---

general. However, the early modern concept is as incommensurable with our modern concept of speed or average speed as it is with our concept of velocity or average velocity, and thus the reason to prefer the term speed over the term velocity is revealed to be void. My choice to use velocity is based simply on the fact that it is linguistically closer to Galileo's own terminology.

[10] My exposition follows closely the one given by Souffrin (1992), who provides examples of late medieval and early modern authors including Galileo, to support the reconstruction given.

[11] The holistic character needs to be emphasized mainly because in the classical framework velocity is defined as the derivative of position with respect to time and is thus an instantaneous magnitude. Average speed, which applies to distance traversed and time elapsed and which is, in this respect, somewhat closer to the preclassical concept, is understood to be a derived concept. Notwithstanding some resemblances to the modern average speed, the preclassical kinematic velocity is distinct from it in important ways. Thus what holds for the ratio of two kinematic velocities in the preclassical framework holds for the ratio of the average speed of the same motions in the classical framework. On the other hand, the definition of a magnitude as a distance traveled divided by the time elapsed is categorically ruled out in the preclassical framework.

[12] Throughout the book, primarily for ease of apprehension, I use modern symbolic notation to transcribe verbal statements. Such an approach is criticized by Palmieri (2003, 230) who states that "the translation of verbal mathematical arguments into a symbolic notation inevitably carries with it the risk of missing important aspects that are inextricably linked with the original language." This is true. Yet I believe it should not lead us to abandon such use of symbolic notation but to use it carefully and with an awareness of its limitations.

[13] Cf. Clagett (1979).

Already in antiquity, a third such kinematic proposition had been introduced by Archimedes according to which, for two motions that proceed with equal velocity, the distances traversed by these motions are in the same proportion as the times in which these distances are traversed[14]:

$$v_1 = v_2 \Rightarrow s_1/s_2 \sim t_1/t_2.$$

The reverse implication was also considered true in preclassical physics and hence[15]:

Third kinematic proposition: $v_1 = v_2 \Leftrightarrow s_1/s_2 \sim t_1/t_2.$

Archimedes himself had limited his statement to the case of uniform motion.[16] By Galileo's time this limitation had apparently been dropped, and the proportionality between spaces traversed and times elapsed thus provided an operative definition of equal velocities. Galileo indeed, as will be seen, used all three kinematic propositions in his investigations in the context of his emerging new science of naturally accelerated motion as well. Together, these three kinematic propositions define, if implicitly, the preclassical concept of kinematic velocity.

Based on the understanding of kinematic velocity as entailed by the kinematic propositions, in the First Book of the treatise *De motu locali* presented on the Third Day of the *Discorsi*, Galileo developed a deductively organized account of uniform motion, i.e., such motions in "which the distances traversed by the moving particle during any equal intervals of time, are themselves equal."[17]

The crucial role that kinematic velocity played for medieval and early modern reflections about motion is, however, not so much due to its use in investigations, which from a modern perspective can be classified as purely kinematical. Rather in a Aristotelian framework, it is the kinematic velocity which bridges between kinematics and dynamics.[18] Active causes or forces in their interplay with resistances bring about motions where velocity was understood to measure the quantity of motion and

---

[14]Proposition 1 of *On Spirals* reads: "If a point moves at a uniform rate along any line, and two lengths be taken on it, they will be proportional to the times of describing it. (Archimedes and Heath 2002, 155)".

[15]Cf. Damerow et al. (2004, 16).

[16]For the relation between the definition of velocities as equal and specification of motion as uniform, see Souffrin (1992, 238–239).

[17]Cf. Galilei et al. (1954, 154–160). The content of the first book of *De motu locali* has been extensively analyzed, for instance, by Wisan (1974), Giusti (1986), or Damerow et al. (2004).

[18]Dynamics refers to the study of motion that happens under the influence of forces. It embraces on the one hand kinematics and on the other hand kinetics, which is concerned with the effect of forces and torques on motions. For this categorization, see, for example, Burton (1890). Even though I believe this terminology to be adequate, I follow the terminology more commonly used in studies on early modern natural philosophy and distinguish dynamics and kinematics according to the distinction between the study of causes and the study of space-time effects of motion.

2.1 Velocity, Motion as a Changing Quality, and Acceleration

thus the effect.[19] If there is motion, there is a cause, and the cause must be greater, all other things being equal, if the velocity of motion is greater. Vice versa, arguing from causes to motion, a greater cause will, all other things being equal, result in motion with greater velocity. Galileo's approach to dynamics in *De Motu Antiquiora* was, as will become clear in Sect. 2.2, no different in this respect.

Pheno-kinematics Velocity figures in but one of the statements of the 38 propositions concerning naturally accelerated motion along inclined planes, which Galileo issued in the Second Book of the Third Day of the *Discorsi*. It is alluded to in the statement of proposition I theorem I, which itself is used as a premise in the proof of proposition II, i.e., the law of fall, to then never be mentioned again. All other 37 propositions make statements exclusively about the sensible qualities of spaces traversed and times elapsed. In the mathematical language of geometry and of the theory of proportions, they express lawlike relations between motions proceeding over different paths in different intervals of time, and I will refer to such statements here as statements regarding the *pheno-kinematics* of naturally accelerated motion.[20]

I use the attribution *pheno* to signal the difference to the current understanding of kinematics, which refers to a description of motion with regard to position, velocity, and acceleration (and their equivalents for rotational motion). Opposed to this, pheno-kinematical statements describe the change of a body's position in time exclusively, by resorting to the concept of spaces traversed and times elapsed. Thus, in particular, Galileo's statements and arguments, which are contingent on concepts anchored in the doctrine of the configurations of motion and which are usually attributed as kinematical, are distinctly not what I refer to as pheno-kinematical. At the same time, by using the expression pheno-kinematics, I specifically do not want to suggest, as has been done, that Galileo's insights were based on facts established by use of the senses alone. I likewise do not want to suggest, as has been done, that his approach to the new science of motion was kinematical in that it evaded regarding causes or forces.[21] It needs to be stressed that it was one of Galileo's great achievements and eventually one of the keys to his success that he found a way to submit the pheno-kinematics of naturally accelerated motion to rigorous treatment

---

[19] The literature on the medieval and early modern accounts of motion is endless. Grant (1974) and Drake and Drabkin (1969) can still be highly recommended. For a more recent work, including bibliography, see Pasnau and Trifogli (2014).

[20] For the theory of proportions as laid out in Book V of Euclid as Galileo's principal mathematical tool, see, in particular, Giusti (1992, 1993). Also Palmieri has recurrently and rightly emphasized the importance of interpreting Galileo's considerations against the background of this mathematical means. Cf., e.g., Palmieri (2005b, 345).

[21] Such a position is prevalent in many assessments of Galileo's new science. It has recently been forcefully explicated by Henry (2011).

by the tool of Euclidian geometry. Only this allowed him to establish the inferences we will encounter in the *Notes on Motion*, to formulate the proofs we read in the *Discorsi* and thus eventually to construct a demonstrative science with accelerated motion as its object.

As will be shown here, due to the specific nature of the challenging problem that spurred his efforts, Galileo indeed initially unfolded a host of pheno-kinematical statements regarding naturally accelerated motion along inclined planes. From this arose the challenge to integrate these new results within the standard dynamical framework of his day. Not surprisingly, it is in tackling this challenge that Galileo availed himself of the concept of kinematic velocity. The rather marginal role played by dynamics in the new science of motion as it later unfolded in the *Discorsi* is not due to the fact that Galileo denied forces because he saw in them an occult notion or for whatever other reason. Rather, as will be shown, his results regarding the phenokinematics of naturally accelerated motion could, for reasons all too obvious from a modern perspective, not be reconciled with the dynamical understanding of his time in which kinematic velocity played the central role as a measure of motion and thus indirectly of the active causes.

Motion as a changing quantity—the doctrine of the configuration of qualities
To inquire into acceleration, i.e., to study motion as a changing quantity, a conceptual framework offered itself that in the Middle Ages had grown out of the scholastic discussions on the intension and remission of forms, i.e., the discussions regarding problems related to change of accidental forms.[22] To cope with such problems, among others, in the fourteenth century at the Merton College in the University of Oxford, a group of scholars who have become referred to collectively as the *Oxford Calculators* coined an original doctrine encompassing a particular terminology. Later in the century in Paris, Nicole Oresme devised a system which allowed for a graphical representation of this, and it is Oresme's formative work, *Tractatus de configurationibus qualitatum et motuum*, from which I borrow the term *configurations of qualities and motions* or simply configurations of motions, to refer to the doctrine.[23]

In the early modern period, the doctrine, in particular in Oresme's interpretation, was to assume an important role in the study of motion in the works of thinkers such as Thomas Harriot or René Descartes and Isaac Beeckman.[24] Galileo, in the *Discorsi*, famously founded his new science of motion on principles formulated

---

[22]Cf. Jung (2011).

[23]See Oresme and Clagett (1968). The choice of term and my exposition lean in particular on the one contained in Damerow et al. (2004), worked out in more detail by Schemmel (2014).

[24]Cf. in particular Schemmel (2014).

with respect to degree of velocity and its change, a concept embraced and indeed defined only by this conceptual tradition.[25] According to the interpretation that will be presented in this book, after he had worked on the problem of naturally accelerated motion already for quite a while, Galileo, most likely first in 1604, started to avail himself systematically of the understanding of accelerated motion as a changing quality as it was entailed by the doctrine of the configurations of motion in an attempt to found the results he had already achieved by then in the shared and generally accepted conceptual frameworks. In entries of the *Notes on Motion* written before that time, occasional references are made to degrees of velocity revealing that Galileo's thinking about motion was indeed already informed by the doctrine. Yet in the period before 1604, Galileo resorted to the doctrine merely in regard to individual aspects of his considerations regarding motion, such as the question of what happens if an accelerated motion is diverted into a horizontal uniform motion. For the time being it did not, however, as will be seen, constitute the essential substructure of what was to turn into Galileo's new science of naturally accelerated motion.

The period for which the early genesis of Galileo's new science of motion is reconstructed in this book ends precisely at the moment when, in 1604, Galileo started to fundamentally avail himself of the doctrine in an attempt to turn his results regarding naturally accelerated motion into a new science. Hence only a cursory survey is given here, barely sufficient to convey the general understanding, against the backdrop of which Galileo's statements in this context, in particular such invoking degrees of velocity, attain their meaning. For a more in-depth exposition, the readers are referred to the vast and, in large part, excellent literature on the subject.[26]

Qualities that have a quantity, but for which no parts could be identified from which this quantity could be conceived to be composed, were considered as intensive qualities. The doctrine of the configuration of qualities allows to cope with philosophical problems arising when dealing, in particular, with the change of such intensive qualities. Change was perceived as the change of intensity of a quality,

---

[25]In his early writings Galileo explicitly refers to the "Doctores Parisienses," and he clearly "knew of their teaching from the Jesuit lecture notes (Galilei and Wallace 1992, 219)." What exactly Galileo owed to the doctrine of the configurations of motion and where he went beyond its traditional scope has been intensely debated in the history of science. On this question see, e.g., Wallace (1981) or Lewis (1980). Extreme opposite positions in this debate were taken by Duhem and Koyré. Wallace (1990, 239) pointedly represents Koyré's position: "[t]he medieval and Renaissance development that had been traced in such detail by Duhem might be of antiquarian interest, but it was not at all necessary for Koyré's understanding of Galileo and the 'new science' he had brought into being." Wallace holds that an important result of his own studies "is their vindication of Duhem, as contrasted with Koyré." As I will demonstrate, the doctrine of the configurations of motion was crucial for Galileo's attempt to provide what had been achieved independently with a conceptual underpinning.

[26]See, for instance, Sylla (1971, 1973). Jung (2011) points out that the conceptual tradition of the configurations of motions was less homogeneous than often (including owing not the least to its brevity in the account given here) portrayed.

termed a degree (*gradus, latitudo, intensio*) over an extension (*longitutdo, extensio*), which could variably be spatial or a temporal. In Oresme's diagrammatical representation, the extension of the subject, informed by the intensive quality, was represented by a straight line (usually, but not necessarily horizontal). The degrees in each point of this extension were represented by lines orthogonal to the line of extension. The proportions between the lines representing the degrees thereby represented the actual proportions of the intensity in the subject. In this graphical representation, the area of the figure defined by the line of the extension, the lines of the first and last degree and the summit line of all other degrees in between, represented the quantity of the quality.

The configuration of qualities and its graphical representation in the form of such Oresme diagrams could be applied to any kind of alteration of the intensity of a quality in a subject. In the case where the intensity of the quality does not change in the subject over its extension, the quality was said to be uniform. Since the intensity, i.e., the degree of the quality, is the same at every point of the extension, in an Oresme diagram, such a uniform quality is represented by a rectangle. If the change of intensity in the subject was proportional to the extension, the quality was said to be uniformly difform. Such a uniformly difform quality is represented by a triangle in cases where the first or the last degree is equal to zero. Other possible changes of the intensity over the extension were subsumed under the term difformly difform.

The Oxford Calculators devised a number of general rules that held for intensive qualities and their change. Most famous, arguably because of its alleged role for Galileo, is the so-called Merton rule which states that for a uniformly difform quality, the quantity of the quality is identical to that of a uniform quality over the same extension whose degrees are equal to the degree at the midpoint or, in other words, half the maximum degree of the uniformly difform quality.[27] It needs to be noted that the nature of this and comparable rules was predominantly abstract, philosophical, and logical rather than being concerned with the real physical world.

Applicable, in principle, to all kinds of change, the doctrine was chiefly applied to local motion. In the application to motion, the degree of intensity becomes the degree of velocity, which I will variably also refer to abstractly as the *intensive velocity*.[28] To refer to the intensive velocity, in principle, Galileo used two alternative expressions, "gradus velocitatis" or "momentum velocitatis," where in rare cases "celeritatis" is used instead of "velocitatis."[29] In at least one entry

---

[27]The Merton rule is often imprecisely referred to as the mean speed theorem, yet the latter more specifically refers to the application of the former to the case of local motion proceeding with a uniformly growing degree of velocity. The mean speed theorem is already found in the works of the Calculators, for instance, in Heytesbury but also in the works of Oresme. Cf. Boccaletti (2016, 32–38).

[28]For an excellent rendition of the difference between the preclassical and the classical concepts of velocity, see Souffrin (1990), in particular the schematic table he provides therein to illustrate the history of the concept.

[29]The few entries in the *Notes on Motion* which are written in Italian and in which the expression occurs always use the equivalent to the Latin "gradus velocitatis," i.e., "grado di velocità."

2.1 Velocity, Motion as a Changing Quality, and Acceleration

in the *Notes on Motion* and also in one passage in the *Discorsi*, he employs both designations concurrently and speaks of "velocitatis gradus seu momenta," showing that for him both expressions were essentially synonymous.[30] Very rarely, Galileo drops the designation of the degree altogether and speaks of a velocity in a point or points ("velocitas in punctis"), thus clearly referring to the intensive velocity.

Famously, Galileo initially believed that in naturally accelerated motion, the intensive velocity would grow in proportion to the space traversed. We know this because in a letter to Paolo Sarpi in 1604, he explicitly mentioned this assumption, which I will therefore refer to here as the *Sarpi letter principle*. Galileo later realized that the Sarpi letter principle contradicted the law of fall, and in consequence he characterized natural acceleration by a uniform difform increase of the intensive velocity over the extension of time, and this indeed would provide the definition of natural accelerated motion he would later advance in the *Discorsi*.

What about the quantity of the quality, i.e., the area in a corresponding Oresme type diagram? In applications of the doctrine of the configurations of motion to the case of motion, this was traditionally interpreted to represent the quantity of motion often, moreover, identified with the space traversed.[31] Galileo, however, never directly identified the quality of the quantity with space traversed. For him the quantity of motion, i.e., the quantity made up of the totality of the degrees, represented itself a velocity. Galileo usually referred to this by using the plural of degree of velocity in expressions, "totidem velocitatis momenta." Often, however, he dropped the designation degree and simply used the plural "velocitates" to refer to all the intensive velocities of a motion over a given distance. I will refer to the velocity, which in this sense represents the quantity of motion, as the *aggregate velocity*. As we will see, Galileo identified this aggregate velocity with the kinematic velocity as implicitly defined by the kinematic propositions.[32] Reconciling this understanding with his new insights concerning naturally accelerated motion, in particular the law of fall, was one the central problems Galileo was confronted with in his attempts to erect a foundation for his new science of naturally accelerated motion.

---

[30] According to Wallace (1984), Galileo's use of "gradus velocitatis" traces back to the Collegio Romano. Referring to Galluzzi, Wallace (1984, 268) furthermore claimed that the early use of "gradus velocitatis" was replaced by "momento velocitatis" from about 1605. In my analysis I was unable to confirm this alleged trend. On the contrary, Galileo already used "momentum velocitatis" in entries which must date from before 1604, for instance, on 163 recto.

[31] Cf. Schemmel (2014, 18).

[32] The identification of aggregate velocity with kinematic velocity would eventually force Galileo to give up on the idea that velocity was represented by the area in an Oresme diagram. This was replaced by an argument published as Proposition I of the Third Book of the Third Day of the *Discorsi*, in which equality of aggregate velocity and thus also of kinematic velocity is established arguing via pairwise correspondence of degrees of velocity instead.

Free fall and acceleration before natural acceleration For the young Galileo of *De Motu Antiquiora*, motion of fall was not naturally accelerated. It was characteristically uniform with a specific velocity depending on the material of the falling object and the medium fallen through and, also, if fall is made along an inclined plane, depending, as will be discussed below, on the inclination of the plane.[33] This distances Galileo from the position taken by the majority of his contemporaries involved in the learned discourse about motion who, following Aristotle's claim in *De caelo* that "earth moves more quickly the nearer it is to the centre ...," generally assumed that natural motion of the elements was faster toward the end, i.e., accelerated. Aristotle had, however, not made an "attempt to integrate this insight with his other views or to develop a theory of acceleration."[34] It is not least for this reason that explaining why natural motion becomes faster had become an important topic in natural philosophical debates long before the early modern period and Galileo could hardly avoid addressing this topic in *De Motu Antiquiora*.[35]

Galileo's diversion from the prevailing opinion was arguably motivated inner theoretically. His approach to falling motion in *De Motu Antiquiora* is characterized by his effort to apply Archimedes' hydrostatical theory of buoyancy to the problem of motion.[36] In this vein he identifies "the greater or lesser heaviness of the media and of the mobiles" as the cause of motion. A cause of motion identified to reside in the mobile just as much as in the medium is not easily commensurable with the Aristotelian framework, in which natural motion finds its explanation solely in the tendency of the elements to their natural place.[37] For motion caused by an external force, on the other hand, it was leaving aside for a moment the question of resistance, generally accepted that velocity as the effect is proportional to the *dynamis*, the source of motion. His approach thus almost inadvertently forced Galileo to appropriate this dynamical principle to the motion of fall and to accept

---

[33] According to Galluzzi (1979, 182), it needs not even to be particularly stressed that "la teoria galileiana del De motu, la 'dinamica pisana', come è stata definita, è una dinamica dei moti uniformi."

[34] See Gregory (2001, 3).

[35] Cf. Camerota and Helbing (2000, 346). An important contribution was made by Buridan, who had given an explanation of acceleration in terms of the concept impetus, i.e., impressed force. Alternatively, "antiperistaltic" explanations of the acceleration of natural motion were advanced. For the discussions on the cause of acceleration in the natural philosophical debates of the Late Middle Ages, see Grant (2010, 207 ff.) or Guerrini (2014).

[36] Cf. Palmieri (2005a). Benedetti had advanced a comparable account of motion through media based on Archimedean hydrostatics before. The question to what extent, if any, this influenced Galileo is still open. Grant (1966) meticulously reconstructs possible common origins of Benedetti's and Galileo's approach also pointing out the differences.

[37] Machamer (1978) specifies that "a body's movement toward its own place is movement toward its own form." See also Bodnar (2016). For Aristotle, the medium can in any case not be part of the cause (formal, efficient, or final) of the natural motion of the elements.

2.1 Velocity, Motion as a Changing Quality, and Acceleration

what he has identified as the cause of such motion as also being the cause of its "slowness and swiftness," implying that mobile bodies fall through media with uniform velocity.[38]

If motion of fall is assumed to be, in principle, uniform with a characteristic velocity, then a mobile body falling from rest will somehow first have to attain this velocity. In a chapter from *On Motion*, headed "In which is brought forth for everyone to see the cause of the acceleration of natural motion at the end, a cause far different from that which the Aristotelians assign. (EN I, 315, trans. Fredette)," Galileo submits different explanations of acceleration that had been advanced and defended to a critique.[39] Yet he also provided his own account of acceleration entailing an explanation of not only why acceleration occurs but also why it ends when the mobile body has reached its characteristic natural velocity.[40] Galileo's account, based on the concept of a self-corrupting, impressed force, has been analyzed and discussed in detail elsewhere.[41] It is summarized here in the briefest of manners primarily to indicate that, and in which way, it is incommensurable with Galileo's later approach assuming natural acceleration.

Given Galileo's central claim that velocity comes from heaviness, an increasing velocity must somehow correspond to an increasing heaviness and it is "therefore necessary that it [the mobile body falling from rest] be less heavy at the beginning of its motion than in the middle or at the end." The change in heaviness cannot be due to the medium since this is the "same at the beginning of motion as at the middle" and must therefore have an external cause, which Galileo identifies as "the force impressed by a thrower." A mobile body thrown violently straight up will indeed be "carried upward, provided the motive impressed force is greater than the resisting heaviness."[42]

> Now since this force, as has been demonstrated is continuously diminished, it will finally become so diminished that it will no longer overcome the heaviness of the mobile, and then it will not impel the mobile any further: but that impressed force will not therefore have been annihilated at the end of the violent motion, but it will only be diminished to the point that it no longer exceeds the heaviness of the mobile…But, moreover, as the impressed force diminishes in its own way, the heaviness of the mobile begins to predominate; and

---

[38] The full sentence from which the quotations in the text are taken reads: "In utroque motu ex eadem causa pendere tarditatem et celeritatem, nempe ex maiori vel minori gravitate mediorum et mobilium, mox demonstrabimus (EN I, 260)."

[39] A complete English translation of *De Motu Antiquiora* by Raymond Fredette is available online in the cultural heritage collection ECHO at http://echo.mpiwg-berlin.mpg.de/content/scientific_revolution/galileo. Accessed 15 Jan 2017. Where not noted otherwise, this and all following English quotations from *De Motu Antiquiora* are taken from his translation. For the relation of Galileo's stance regarding acceleration to the position taken by Girolamo Borro and Francesco Buonamici, his teachers at the university at Pisa, cf. Camerota and Helbing (2000).

[40] Galileo's discussion of acceleration reoccurs in similar form in the *Dialogue* of *De Motu Antiquiora*, which is now generally accepted to have been written before *On Motion*. Cf. EN I, 405–408.

[41] See, for instance, Galluzzi (1979, 182–187), Damerow et al. (2004, 147–152), or Grant (1965).

[42] See EN I, 318. Transl. Fredette.

hence the mobile starts to go down. But since at the beginning of such a descent a great deal of force impelling the body upward, which is lightness, still remains ..., it comes about that the proper heaviness of the mobile is diminished by this lightness, and, consequently, that the motion at the beginning is slower. And, moreover, since that extrinsic force is further weakened, the heaviness of the mobile is increased by having less resistance, and the mobile is moved still faster. (EN I, 319, transl. Fredette)

Galileo asserts that Hipparchus had given the same explanation without, however, missing the opportunity to declare that it had occurred to him independently.[43] Invoking a violent motion in the explanation had the advantage that resorting to the concept of a self-corrupting, impressed force did not need to be justified further, as after all the concept had originally been put forth to explain exactly such motions. On the other hand, resorting to a violent motion in an argument to explain the acceleration of natural motion would foreseeably raise objections. Thus Galileo moved on to show that essentially the same argument can also be applied to the case where "the mobile begins to be moved, from a state of rest, and not from a violent motion (EN I, 319, transl. Fredette)."

He argues that if a stone is held in the hand, the hand impresses a force just big enough to compensate for the stone's heaviness. If then, the stone is let go, the situation is essentially equivalent to the one at the turning point of motion in the previous example. Hence from thereon, by the same token of argument as before, the stone will accelerate as the impressed force diminishes.[44]

By its very nature the impressed force is not unfailing (as had indeed been argued in the chapter before) and:

[I]t will stand to reason that all the contrary force will finally be lost, and that the natural heaviness will be resumed, and that, therefore, the cause having been removed, the acceleration will cease. (EN I, 329, transl. Fredette)

According to Galileo it is not least experience that tells us so[45]:

For, in the first place, if we look at something that is not at all heavy coming from on high, a ball of wool or a feather or some such thing, we will see that it is moved more slowly at the beginning, but that nevertheless, a little later, it will observe a uniform motion. (EN I, 329, transl. Fredette)

---

[43]Drabkin in Galilei (1960, 90) traces Galileo's remark concerning Hipparchus to Simplicius *Commentary on Aristotle's De Caelo*.

[44]According to Galileo's account, the accelerated part of the downward motion is, strictly speaking, a mix of violent and natural motion and would thus, according to orthodox Aristotelian understanding, have to be classified as violent. Galileo, indeed, concedes that there is no essential difference between the decelerated forced upward motion of his first example and the accelerated downward part of motion and states "this alternative motion, while the mobile is moved from lightness to heaviness, is a single and continuous motion (EN I, 322, trans. Fredette)." In the *Memoranda*, Galileo is even more explicit and states in an entry "a violent motion precedes every case of natural motion, as we have made clear (EN I, 411, trans. Fredette)." As Wisan (1978, 9) puts it: "Galileo's acceleration is unnatural (or violent) motion."

[45]In the *Dialogue* of *De Motu Antiquiora*, Galileo even advances an argument why it is that, under certain conditions, bodies falling at a uniform rate are falsely perceived to accelerate. Cf. EN. I 407.

2.1 Velocity, Motion as a Changing Quality, and Acceleration

However, the uniform motion characteristic for free fall will not always be observed. For instance:

> [a stone] sent from merely the height of a tower, will seem to accelerate all the way to the ground; for this short distance and short time of motion are not sufficient to destroy the whole contrary force. (EN I, 239, transl. Fredette)

Generally, for Galileo in *De Motu Antiquiora*, "contrary qualities are kept longer, the heavier and the denser and the more contrary to them is the material in which they have been impressed," and thus the impressed force or lightness "recedes more easily and the more swiftly, the lighter the mobile in which it has been impressed (EN I, 335, transl. Fredette)." It is exactly for this reason that the uniformity of the velocity of falling motion can allegedly be better observed for lighter objects; their acceleration ceases quicker and thus also after a shorter distance traversed. Moreover, it also provides the reason why, of two bodies dropped from the same height at the same moment, the lighter will initially advance before the heavier body. That this should indeed be the case was intensely debated among the professors of philosophy of the University of Pisa while Galileo was enrolled there as a student, allegedly resulting in lots of things being dropped out of windows and from towers.[46] Galileo writes, for instance:

> For if one takes two different mobiles ... and then send them from the top of a tower ... the one which is lighter at the beginning of the motion will precede the heavier and will be faster. This is not the place to inquire into how these differences and, so to speak, prodigies come about (for they are accidental) ... (EN I, 273, transl. Fredette)

In *De Motu Antiquiora*, Galileo even provides concrete examples to illustrate the different rate by which an impressed force passes away in different materials. Yet these examples run into inconsistencies, at least from a modern perspective, and no systematic treatment of the accidental phenomenon of acceleration seems possible.[47]

The position, which Galileo would publicly communicate almost 50 years later in the *Discorsi*, is diametrically opposed:

> I begin by saying that a heavy body has an inherent tendency to move with a constantly and uniformly accelerated motion toward the common center of gravity, that is, toward the center of our earth, so that during equal intervals of time it receives equal increments of momentum and velocity. (EN VIII, 118, trans. Galilei et al. 1954)

---

[46] Cf. Camerota and Helbing (2000). In trials in which persons were asked to drop objects of same shape but different weight at the same time, Thomas Settle and Donald Miklich showed that the heavier object is consistently dropped with a short delay even though the actors perceive dropping the objects simultaneously. Settle (1983) relates this phenomenon, which can be explained physiologically, to the early modern debates.

[47] According to Fredette (1972, 347–348), it was the irreconcilability of his account of acceleration with the revised conceptualization of the dynamics of motion through media, as elaborated in the treatise in 10 chapters, which led Galileo to abandon the project of *De Motu Antiquiora*.

In the case of bodies falling through dense media, Galileo continued, it would be observed that acceleration ceases and motion becomes uniform. Now, however, the reason for this was entirely different:

> This quiet, yielding, fluid medium opposes motion through it which a resistance which is proportional to the rapidity with which the medium must give way to the passage of the body; which body, as I have said, is by nature continuously accelerated so that it meets with more and more resistance in the medium and hence a diminution in its rate of gain of speed until finally the speed reaches such a point and the resistance of the medium becomes so great that, balancing each other, they prevent any further acceleration and reduce the motion of the body to one which is uniform and which will thereafter maintain a constant value. (EN VIII, 118, trans. Galilei et al. 1954)

With regard to free fall, acceleration and uniform motion had changed their role as compared to Galileo's account in *De Motu Antiquiora*. Acceleration is no longer an accidental phenomenon, but by their very nature heavy bodies fall in a continuously accelerated manner. This is precisely what Galileo means when he speaks of natural acceleration. That such heavy bodies will eventually fall with uniform motion, on the other hand, is due to the external accidental hindrances ("gl'impedimenti accidentarii ed esterni"). If these external factors were removed, i.e., if falling in a vacuum, all bodies would continue to accelerate and, moreover, show exactly the same kinematical behavior as Galileo famously claims in the *Discorsi*.[48]

Bound together with the material of *De Motu Antiquiora* in Ms. Gal 71 is a text referred to as *De motu accelerato*, which presents a draft of the opening passages of the Second Book of the Third Day of the *Discorsi* in which the definition of natural acceleration is given and justified.[49] In the first paragraphs there are only little changes between this earlier draft and the published version save for one exception. The following passage of the draft was not taken over into the *Discorsi*:

> The mobile is the same, the principle of moving is the same: why should not also the rest be the same? You say: the same also the velocity. Not at all: because it is already clear, that the velocity is not the same, neither is the motion uniform: it is therefore necessary to find and to place the identity, or the uniformity as you say, or simplicity, not in the velocity, but in the velocity's increase, that is in acceleration.(EN II, 262, my translation)[50]

---

[48] Cf. Palmieri (2005b).

[49] The dating of the text is controversial. Favaro and Wohlwill have both dated it to Galileo's Paduan period, and a possible dating "to well before 1604, quite possibly to the late 1590s" has recently been proposed by Palmieri (2005b). Drake et al. (1999, 216) dates the text to 1630 or 1631. Based on comparison with material in the *Notes on Motion*, I tend to agree to a dating after Galileo's move to Florence.

[50] "Idem est mobile, idem principium movens: cur non eadem quoque reliqua? Dices: eadem quoque velocitas. Minime: iam enim re ipsa constat, velocitatem eandem non esse, nec motum esse aequabilem: oportet igitur, identitatem, seu dicas uniformitatem, ac simplicitatem, non in velocitate, sed in velocitatis additamentis, hoc est in acceleratione, reperire atque reponere (EN II, 262)."

The intellectual effort it had taken to accomplish the conceptual shift from the assumption that falling motion is characteristically uniform to the assumption that it is characteristically accelerated still clearly resonates in these sentences.

Galileo's conceptual shift is often identified with the "discovery" of the law of fall, according to which the distances traversed by a falling body are as the squares of the times elapsed during fall over these distances and it is, moreover, generally assumed that Galileo must have discovered the law in 1604 when he first mentioned it in a letter to Paolo Sarpi. Various attempts have been made to pinpoint this discovery in the *Notes on Motion*.

If we find Galileo endorsing the law of fall, this is of course a clear sign that he has begun to conceive of fall as naturally accelerated. Yet it is not at all a priori clear that it was his "discovery" of the law which forced Galileo to accept that falling motion is naturally accelerated. One can just as well argue that in order to gain insight into the law of fall and to moreover accept it as valid, one needs to have come to accept natural acceleration beforehand.[51] Due to a lack of evidence, we may ultimately never know. Indeed, the answer is not, as will be shown, to be sought in the *Notes on Motion*. Contrary to what is commonly assumed, even the earliest entries in the manuscript seem to presuppose the law of fall. The *Notes on Motion* are truly Galileo's notes on naturally accelerated motion. When Galileo took the first notes, at the latest by the end of 1602, the law of fall and with it the law of chords had apparently already acquired the status of rather well-founded heuristic assumptions.

As concerns the law of fall, Damerow et al. have argued that it was, in principle, implied when Galileo—as had been suggested by an experiment conducted most likely in 1592 together with Guidobaldo del Monte in which the two men had projected an inked ball along an inclined plane—started to assume that the projectile trajectory had a parabolic shape.[52] Currently, this appears to be the most plausible scenario of how Galileo came to accept the law of fall.[53] Whether and to what extent Galileo engaged in further exploration of the consequences of the law of fall thus implied between 1592 and 1602, or whether he possibly had even submitted the law to an empirical test before 1602, is not known.[54] The *Notes on Motion*

---

[51] Settle (1966, 148) too emphasizes that "[i]t would seem, then, that two things had to occur ...before ...Galileo could establish the foundations of a new science: a shift in primary focus from uniform motion to acceleration as the essential mode of natural motion, and the discovery of a mathematical description of that acceleration. And it would seem that we must credit systematic experimentation by Galileo with a key role in both."

[52] Cf. Damerow et al. (2001), where the authors state that "according to common historiographic criteria, Galileo must be credited with having made this discovery already as early as 1592" and that the law of fall "was merely a trivial consequence of this discovery (Damerow et al. 2001, 300)."

[53] For an account of how Galileo may have come to accept the law of chords, see Chap. 8.

[54] The experimental setup of the pendulum plane experiment which Galileo conducted in 1602, as argued in Chap. 4, is strikingly similar to the setup of the famous inclined plane experiment described by Galileo in the *Discorsi* suggesting that he indeed may have tested the law of fall before or in 1602.

may thus not tell us how Galileo arrived at these initial insights concerning the pheno-kinematics of naturally accelerated motion along inclined planes. What they do document, however, is the process through which in the investigation of a challenging problem these isolated heuristic assumptions were expanded into a set of insights regarding naturally accelerated motion along inclined planes, based on which Galileo instituted his new science of motion.

2.2 Motion Along Inclined Planes and Dynamics

Evidence that Galileo had turned his attention to the topic of the inclined plane early in his intellectual career comes from a chapter comprised in the *On Motion* treatise of *De Motu Antiquiora*. The chapter is headed "CAPUT ... in quo agitur de proportionibus motuum eiusdem mobilis super diversa plana inclinata (EN I, 293)." The answer which Galileo would go on to provide after penetrating analysis is the following: "[i]t is therefore certain that, the swiftnesses of the same mobile on different inclines are to each other inversely as the lengths of the oblique descents, provided that these hold equal right descents (EN I, 293, transl. Fredette)."[55] In the following, this statement will be referred to as the *velocity theorem*.

When, at the beginning of the last decade of the sixteenth century, Galileo wrote down the material that was later bundled together and labeled *De Motu Antiquiora*, the inclined plane was not a new topic in the history of mechanics. It had been recognized already in antiquity that by means of an inclined plane, forces can be altered and loads be lifted by a force smaller than that required to lift the load directly, making the inclined plane a prime instance of a mechanical device.[56] Heron had treated the inclined plane and included the wedge (but not the inclined plane itself) among the five simple machines from which, as he conceived, all more complicated mechanical devices were or could be composed of, and based on an understanding of which the operation of complex mechanical contrivances could thus be explained.[57] He had, moreover, provided his own theoretical account of how the force needed to draw an object up an inclined plane depends on inclination. Following the tradition, in his *Mechanicorum liber*, Galileo's friend and mentor Guidobaldo del Monte treated the wedge and the screw, the latter conceived as

---

[55]"Constat ergo, eiusdem mobilis in diversis inclinationibus celeritates esse inter se permutatim sicut obliquorum descensuum, aequales rectos descensus compraehendentium, longitudines."

[56]In the Ps. Aristotelian *Mechanical Problems*, which famously define mechanical devices as those in which "the less master the greater, and things possessing little weight move heavy weights (Aristotle and Hett 1980, 331)," the inclined plane is not investigated explicitly. Yet in Problems 17 and 19, the force altering effect of the wedge, treated by the later tradition as an instance of an inclined plane, is explored.

[57]Cf. Schiefsky (2008).

2.2 Motion Along Inclined Planes and Dynamics

being an inclined plane coiled around a cylinder, as mechanical contrivances to be understood based on an analysis of the inclined plane and Galileo would later do the same in his *Le Meccaniche*.[58]

In line with an understanding of the inclined plane as an elementary mechanical device, authors that had touched upon the issue before Galileo had commonly either considered the static case, i.e., tried to determine the force needed to keep a body steady on an inclined plane depending on the inclination or else had, in correspondence to the use dominantly made of inclined planes in technical contexts, asked for the force required to forcefully drag a body up along an inclined plane.[59]

As already indicated by the title of the relevant chapter, Galileo's approach in *De Motu Antiquiora* was different.[60] What he was inquiring into was not the forced motion of a body being dragged up (or lowered down) but the motion of heavy bodies allowed to move freely down along an inclined plane.[61] Superficially, with his analysis Galileo merely seems to add a partial aspect, that of falling, to the traditional topic of the inclined plane. Much more important, however, than adding the investigation of falling motion to the considerations regarding the inclined plane, with the approach he chose in *De Motu Antiquiora*, Galileo vice versa put the investigation of the inclined plane onto the agenda of the study of motion as one of the central topics of natural philosophical inquiry of his day. Galileo himself emphasizes:

> The question that we are about to explain has been treated thoroughly by no philosopher, as far as I know: yet, since it concerns motion, it seems that it must necessarily be examined by those who profess to hand on a treatment concerning motion that is not incomplete. (EN I, 296, transl. Fredette)[62]

As will become clear, Galileo did treat free motion along inclined planes as a motion of fall, not different in principle from the case of free fall along the vertical, i.e., natural motion according to the Aristotelian understanding and parlance.

---

[58]For Guidobaldo del Monte's *Mechanicorum liber*, see Del Monte et al. (2010).

[59]Galileo's treatment of the problem of the inclined plane in *De Motu Antiquiora* has been discussed in detail and been thoroughly contextualized in a history of mechanics by Festa and Roux (2008).

[60]In a footnote to this chapter, Drabkin remarks: "[e]ven on the specific question of the ratio of speeds Galileo can hardly claim priority (Galilei 1960, 63)." To me this remark seems to lack justification. An exception is Giovanni Marliani, who treated motion along inclined planes in his *Quaestio de proportione motuum in velocitate* of 1482. For Marliani's work, see Clagett (1941). Palmieri (2017) opposes the idea that Marliani conducted experiments involving the falling of bodies down along inclined planes.

[61]Festa and Roux (2008, 210) likewise highlight this difference when they state: "it is not a matter of using the law of the inclined plane to explain the function of the screw as it will be in the *Mecaniche*, but rather to answer two questions concerning the movement of a body along an inclined plane."

[62]Quaestio, quam nunc explicaturi sumus, a philosophis nullis, quod sciam, pertractata est: attamen, cum de motu sit, necessario examinanda videtur illis, qui de motu non mancam tractationem tradere profitentur.

Indeed, from the perspective of Galileo's approach, free vertical downward motion comes into view as the limiting case of motion along an inclined plane of ever steeper inclination.[63] Motion along inclined planes is falling motion, and Galileo would in fact state later in the chapter:

> ...a mobile, having no extrinsic resistance, will go down *naturally* [my emphasis] on a plane inclined ...below the horizon, with no extrinsic force applied; (EN I, 299, transl. Fredette)[64]

In *De Motu Antiquiora*, Galileo thus expanded the Aristotelian category of natural motion to include the free motion along inclined planes. From a modern perspective, it may be tempting to all too hastily identify as natural motions those which proceed downward under the action of gravitational forces. Such an understanding is, however, whiggish. From an Aristotelian perspective, free motion along inclined planes represents an equivocal case at best.[65] Correspondingly, the approach Galileo took concerning the dynamics of free motion down along inclined planes in *De Motu Antiquiora* was virtually unprecedented. His extension of the conditions under which natural motion takes place to include the inclined plane challenged Aristotelian dynamics, and this, in the long run, turned out to be of utmost consequence.[66] Galileo indeed retained throughout his life the idea introduced around 1590 that free motion along inclined planes is natural motion.

---

[63] From an orthodox Aristotelian perspective, natural motion, of course, includes the upward motion of light bodies. Yet Galileo began to question and eventually went on to relinquish the distinction between heavy and light in *De Motu Antiquiora*. Cf. Fredette (2001, 173).

[64] ...mobile, nullam extrinsecam habens resistentiam, in plano sub horizonte ...naturaliter descendet, nulla adhibita vi extrinseca;"

[65] For Aristotle, natural motion is not simply downward motion but the motion of bodies to their proper place, i.e., the center of the cosmos, which is incidentally the same as the center of the world. Cf. Machamer (1978). On an inclined plane a body will not descend, however, in a straight line, to its proper place.

[66] The challenge accruing to the traditional framework from the inclusion of free motion along inclined planes among natural motion starts with this very chapter of *De Motu Antiquiora*. Galileo will argue that a "mobile in plano quantulumcunque super horizontem erecto non nisi violenter ascendit: ergo restat, quod in ipso horizonte nec naturaliter nec violenter moveatur. Quod si non violenter movetur, ergo a vi omnium minima moveri poterit (EN I, 299)." Thus his identification of the motion of a heavy body down along an inclined plane as natural and that up along an inclined plane as violent leads Galileo to conclude that motion along the horizontal partakes in neither category and from this to conclude that a vanishingly small force would suffice to set a body in motion, paving the way to his principle of circular inertia. Cf. Miller (2014, chap. 5) and Damerow (2006, 15). Galileo retained the assumption (expressed by Cardano before him) that a body will be moved by a vanishingly small force on the horizontal, from *De Motu Antiquiora* onward. In the Aristotelian framework, a finite force depending on the weight of the body was required to set the body in motion along a horizontal. Laird (2001, 263), for instance, thus writes "[i]n a brilliant insight he grasped, contrary to Pappus and Guidobaldo, that ideally a body on a horizontal plane can be set in motion by any force, however small, so that the power to sustain a weight and the power to move it are effectively equal." Heron had assumed the same in the first book of his mechanics (Heron 1976, 54); however, his text was, according to present knowledge, not known in the Latin West in Galileo's day. Cf. Schiefsky (2008).

2.2 Motion Along Inclined Planes and Dynamics

Almost 50 years later it still resonated in his use of "naturaliter" in the expression "motu naturaliter accelerato" to refer to the motion along inclined planes in the *Discorsi*. Of course by then, as indicated by "accelerato" and as has been discussed above, his understanding of how falling motions proceeds had crucially changed.

In the first paragraph of the chapter, Galileo outlined the question he was attempting to answer in the following:

> For what is asked is why the same heavy mobile, in descending naturally along planes inclined to the plane of the horizon, is moved more easily and more swiftly on those that will maintain angles nearer a right angle with the horizon; and, furthermore, the ratio of such motions made on diverse inclinations is sought. (EN I, 296, transl. Fredette)[67]

Successively, Galileo clarifies the question by means of a concrete example to then, in a familiar move, transform the problem:

> ...it is manifest that what is heavy is carried downward with as much force, as would be necessary for pulling it upward by force; that is, it is carried downward with as much force as that with which it resists going up. (EN I, 297, transl. Fredette)[68]

In order to construct the dynamical argument he is striving for, Galileo needs to determine the force which causes a body to move down along an inclined plane. To do so, he introduced as compensation a model which allowed him to instead investigate the force needed to keep a body steady on an inclined plane. In turn, the force to hold a body steady on a plane was for Galileo, as we will see, virtually not distinguished from the minimal force required to pull the body up along the same plane. In the *Discorsi*, Galileo even provided a concrete example of a mechanical realization of his compensation model, a freely hanging counterweight connected by means of a rope and a pulley to the body on the inclined plane in such a manner that it could pull the latter upward in the direction of the plane (for the corresponding diagram, see Fig. 2.1):

> It is clear that the impelling force (impeto) acting on a body in descent is equal to the resistance or least force (resistenza o forza minima) sufficient to hold it at rest. In order to measure this force and resistance (forza e resistenza) I propose to use the weight of another body. Let us place upon the plane FA a body G connected to the weight H by means of a cord passing over the point F... (EN VIII, 215–216, trans. Galilei et al. 1954)[69]

---

[67] "Quaeritur enim cur idem mobile grave, naturaliter descendens per plana ad planum horizontis inclinata, in illis facilius et celerius movetur quae cum horizonte angulos recto propinquiores continebunt; et, insuper, petitur proportio talium motuum in diversis inclinationibus factorum."

[68] "...manifestum est, grave deorsum ferri tanta vi, quanta esset necessaria ad illud sursum trahendum; hoc est, fertur deorsum tanta vi, quanta resistit ne ascendat."

[69] "Qui è manifesto, tanto essere l'impeto del descendere d'un grave, quanta è la resistenza o forza minima che basta per proibirlo e fermarlo: per tal forza e resistenza, e sua misura, mi voglio servire della gravità d'un altro mobile. Intendasi ora, sopra il piano FA posare il mobile G, legato con un filo che, cavalcando sopra l'F, abbia attaccato un peso H...." This passage is part of what is known as *Viviani's Scholium*. It was dictated by Galileo to Viviani who put it into dialogue form and inserted it into the second edition of the *Discorsi* published in 1655. Cf. Halbwachs and Torunczyk (1985).

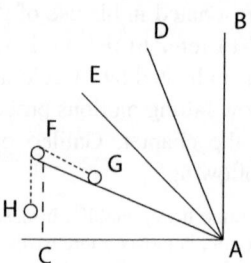

Fig. 2.1 Diagram from Galileo's *Discorsi* (EN VIII, 215, figure 51). It represents a concrete physical setup embodying the compensation model of the inclined plane. By means of a pulley *F* and a connection for the transition without loss of forces (dotted line), the natural force of the counterweight *H* outbalances the force on the inclined plane exerted by the movable body *G* on the inclined plane

With the transition to this compensation model, Galileo had moved into the realm of traditional treatments of the inclined plane as a problem in statics.[70] What was sought was a law or principle of the inclined plane determining how the force to hold a body on a plane depended on the latter's inclination.[71]

Force along the inclined plane—the bent lever proof of the law of the inclined plane With respect to Galileo's compensation model, determining the law of the inclined plane amounted to determining, depending on the inclination, the weight of a counterweight which would keep a body of given weight on the inclined plane in equilibrium. In Galileo's own words, what is sought is "how much greater will be the heaviness of this same mobile on the plane along line *bd*, than on the plane along line *be* (EN I, 297, transl. Fredette)."[72]

Based on different arguments, different solutions to this problem had been proposed before Galileo. Galileo specifically did not adopt the solution which Guidobaldo del Monte had endorsed and which had originally been put forward by Pappus. Instead, he adopted the law of the inclined plane that had been introduced

---

[70] In his resort to the compensation model, Galileo is not unique. The same compensation model is at least implicit in Jordanus' treatment of the inclined plane, and Festa and Roux (2008, 207) wonder if manuscripts of Jordanus' text may have "contained illustrations representing such an [compensation] arrangement." Stevin's proof of the law of the inclined plane of course rests fundamentally on a particular compensation model in which weights on inclined planes connected by a rope keep each other in equilibrium. Cf. Dijksterhuis (1943)

[71] For the problem of the inclined plane in the history of mechanics and contemporary to Galileo, see Gatto et al. (1996), Wisan (1974, 132–150), and in particular Festa and Roux (2008).

[72] "...quanto maior vis requiritur ad sursum impellendum mobile per lineam *bd* quam per *be* ..."

2.2 Motion Along Inclined Planes and Dynamics

by Jordanus, whose treatment of the problem had become widely known through Tartaglia's edition of *Jordani opusculum de ponderositate*.[73] Galileo settled on Jordanus' solution, yet he did not follow the latter's line of argument but provided his own proof in which he resorted in unprecedented manner to considerations regarding the positional weight of a body on a bent lever.[74] Accordingly, Galileo's proof will be referred to as the *bent lever proof*. As will be demonstrated in Chap. 8, it was precisely the specific geometrical constellation on which the bent lever proof was based that was to assume a crucial role in Galileo's early investigations regarding naturally accelerated motion along inclined planes.

Although different in detail, the majority of proofs for a law of the inclined plane that had been proposed depended in one way or another on an application of the balance-lever model, and Galileo's proof is no exception in this respect.[75] In deriving a principle of the inclined plane based on the balance-lever model, a fundamental problem poses itself. Whereas the balance-lever model allows the force to be determined at a point at a certain distance from a fulcrum depending on this distance, constructing an argument to prove a principle of the inclined plane requires determining the force on or along a plane depending on its inclination.

Pappus overcame this problem by identifying a lever within the body on the inclined plane, which he assumed to have a circular cross section. The fulcrum of the lever was supposed to be positioned above the point of contact between the body and the inclined plane, at the same height as the body's center of gravity. One lever arm extended horizontally from the fulcrum to the center of gravity, where the weight of the body was conceived to be placed as the load. The other lever arm was chosen to extend from the fulcrum to the periphery of the body (compare Fig. 2.2). The elements of the lever Pappus identified were not distinct physical entities, yet they were conceived as parts of, or embraced by, a real physical body. In other words,

---

[73] In the first edition of Guidobaldo's *Mechanicorum Liber* of 1577, Pappus' proof of the principle of the inclined plane was merely referred to. See Del Monte (1969). In Pigafetta's 1581 Italian translation of the work, the proof was added as a commentary. See Del Monte (1581). For Jordanus' proof of the law of the inclined plane, see Tartaglia (1565). All three works can be accessed through the cultural heritage online collection ECHO at http://echo.mpiwg-berlin.mpg.de/home. Accessed 15 Jan 2017.

[74] Festa and Roux (2008, 214) rightly stress that "details of his [Galileo's] demonstration prevent us from concluding, as did Caverni and Duhem, that he owes his demonstration of the law of the inclined plane to this reading [of *De ratione ponderis*]."

[75] My notion of the balance-lever model corresponds to what Festa and Roux (2008, 214) outline as follows: "In the following discussion, we take 'model of the balance' to mean the idea that all mechanical systems (and in particular the inclined plane) can be understood from the starting point of weights balanced on a balance; we note that this idea does not prejudge the manner in which this balance is itself explained." Since antiquity, the balance-lever model had indeed remained almost without alternatives for devising both qualitative and quantitative explanations of mechanical phenomena. Cf. Renn et al. (2003). The only thinker until Galileo's time to have attended to the problem of the inclined plane without making direct recourse to the balance-lever model is Stevin (1586, 41–42).

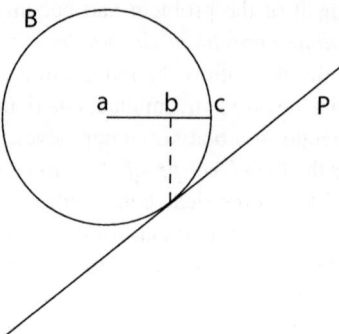

Fig. 2.2 Application of the balance-lever model in Pappus' proof of his principle of the inclined plane. For a body B, a cylinder or a sphere, on plane P, the fulcrum b is supposed to lie vertically above the point of contact between the inclined plane and body, at the same height as the body's center of gravity a. One lever arm extends from the fulcrum to the center of gravity where the weight was imagined to be placed, the other from the fulcrum to the periphery of the body at point c

Pappus' body on the inclined plane was conceived to act as if it were a lever. This can be addressed as a physical instantiation of the balance-lever model. Arguably, it is precisely by its resort to a physical instantiation that Pappus' proof and solution inflict pressing problems. His solution, for instance, depends in unwarranted manner on the shape of the body on the inclined plane. Otherwise, according to his solution, the force needed to drag a body up an inclined plane would grow above all limits as the inclined plane approached the vertical.

It may have been his realization of these problems with Pappus' solution which led Galileo to adopt the principle of the inclined plane that had been proposed by Jordanus.[76] From a modern perspective, Jordanus' principle can be identified as stating that the relation between the natural force of a body and the force of this same body along an inclined plane is given by the sine of the angle of inclination, measured with respect to the horizontal. In providing a proof for this principle, as already stated, Galileo took recourse to the model of a bent lever. The problem of the bent lever in turn had before him already been addressed by Jordanus and Benedetti, both of whom had provided a solution to the question regarding the

---

[76]The one aspect Galileo criticized explicitly with regard to Pappus' approach in his treatment of the problem of the inclined plane in *Le Meccaniche* was that the latter had tried to determine the force to move a body up along an inclined plane from the force required to move the same body on the horizontal. According to Galileo, this force is vanishingly small and, from his perspective, Pappus' approach is indeed rendered inappropriate. Jordanus' proof may well have been known to Galileo, yet the notion of positional heaviness on which it is fundamentally based is found, for example, also in the works of Cardano, Scaliger, Tartaglia, and Benedetti. Cf. Festa and Roux (2008). For the concept of positional weight or heaviness, see Damerow and Renn (2012).

2.2 Motion Along Inclined Planes and Dynamics

condition under which equilibrium is attained on a bent lever.[77] It is indeed not so much Galileo's recourse to the bent lever per se but the way in which he exploited it to solve the problem of the inclined plane which represents the decisively different and novel aspect of his proof. Instead of identifying a bent lever within the body on the inclined plane, Galileo availed himself of what can be referred to as model equivalence.[78]

Generally speaking, Galileo's argument rested on the idea that under a certain condition one compensation model—that of a body connected by means of a bent lever to, and kept in equilibrium by, another body suspended at the other end of the bent lever—can stand in for another compensation model, that of a body on an inclined plane connected by means of a rope to, and kept in equilibrium by, another body hanging freely. For an inclined plane orthogonal to the bent arm of a bent lever, so the central claim of Galileo's approach, the force along the inclined plane is the same as the positional force of the same body positioned on the end of the bent arm of the lever. Since for a bent lever the positional force or heaviness of a given object on the bent lever arm can be related to its own proper heaviness, this allows to solve the problem of the inclined plane.

With reference to Fig. 2.3, Galileo concretely claimed that on a bent lever, such as *cas* or *car*, the positional force or heaviness at the end point of the bent lever arm would be equivalent to the force along or heaviness on the corresponding inclined planes orthogonal to the bent lever arm, i.e., in the example given to the force along *gh* and *nt*, respectively.[79]

To justify his approach with regard to the body attached at the end of the bent lever arm, Galileo stated:

> But then again, when the mobile will be at point *s*, its descent at the initial point *s* will be as if on line *gh*; that is why the motion of the mobile on line *gh* will be according to the heaviness the mobile has at point *s*. (EN. I, 297, transl. Fredette)[80]

---

[77] Issued as Proposition 8 in Book I of *De ratione ponderis*, Jordanus' treatment of the bent lever replaced two propositions on the same subject in the earlier *Liber de ponderibus*. The proposition reads: "If the arms of a balance are unequal, and form an angle at the axis of support, then, if their ends are equidistant from the vertical line passing through the axis of support, equal weights suspended from them will, as so placed, be of equal heaviness (Gillispie 1981)." On the question of the attribution of various medieval mechanical manuscripts to Jordanus and for a reconstruction of their historical succession, see Clagett (1979). For Jordanus', Benedetti's, and Guidobaldo del Monte's approaches to the bent lever, see Damerow and Renn (2012).

[78] For the tendency in contemporary cognitive science to describe analogies by a structure mapping theory, see the overview articles by Gentner (1998) and more recently Gentner et al. (2001).

[79] Galileo himself refers to the contrivance in his argument as a balance (libra). In the text, I address it by the more familiar term lever and bent lever, respectively. In the abstract case of a beam not extended in space that figures in Galileo's argument, the difference between a lever supported from below and the balance supported from above vanishes.

[80] "Rursus, quando mobile erit in puncto *s*, in primo puncto *s* suus descensus erit veluti per lineam *gh*; quare mobilis per lineam *gh* motus erit secundum gravitatem quam habet mobile in puncto *s*."

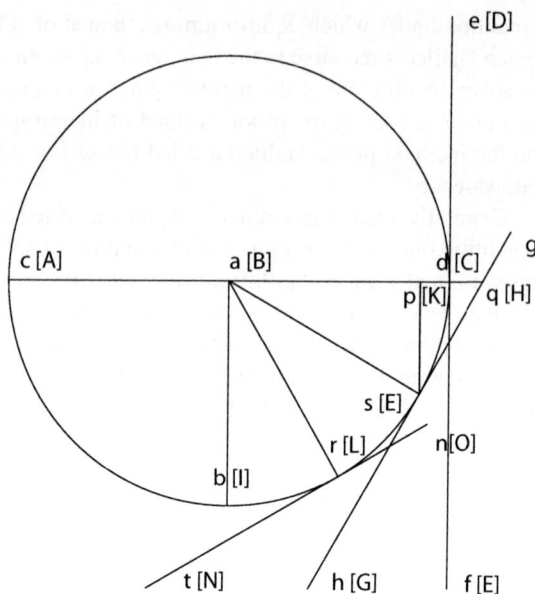

Fig. 2.3 Schema of the diagram associated with the bent lever model proof in *De Motu Antiquiora*. See EN I, p. 297. In *Le Meccaniche*, Galileo issued essentially the same diagram (cf. EN II, 181) with a different keying of the points, however. The labels for the points used in *Le Meccaniche* are given in square brackets

Thus in *De Motu Antiquiora* Galileo restricted himself to claiming that because the motion of the body on the bent lever arm will initially ("in primo puncto") be as if ("veluti") on an inclined plane tangent to the lever arm, motion along the plane will be made according to the (positional) gravity of the body at the end of the lever arm ("gravitatem...mobile in puncto"). Galileo's attempt at a clarification palpably falls short of a satisfactory explanation. In a revised version of the bent lever proof that Galileo issued in *Le Meccaniche*, he is more explicit, and thus the intuition behind his argument is revealed[81]:

> Ma il considerare questo grave discendente, e sostenuto dalli semidiametri *BF*, *BL* ora meno e ora più, e constretto a caminare per la circonferenza *CFL*, non è diverso da quello che saria imaginarsi la medesima circonferenza *CFLI* esser una superficie così piegata,

---

[81] Gatto dates *Le Meccaniche* to 1598 based on the assumption that a course Galileo gave in the academic year 1598/1599 at the University of Padua was based on the content of *Le Meccaniche*. Cf. Galilei (2002, LV). For a synopsis of *Le Meccaniche*, see the editor's introduction in Galilei (2002). *Le Meccaniche* was circulated only in manuscript copies. It was published in French translation by Mersenne (Mersenne and Galilei 1635). For the bent lever proof in *Le Meccaniche*, see EN II, 180–183. In *Le Meccaniche* the law of the inclined plane derived by the proof is applied in the treatment of the screw. Galileo had already issued the law of the inclined plane in the earlier *Delle Machine*, which is sometimes also referred to as the short version of *Le Meccaniche*. However, due to its different style, length, focus, as well as date of composition, it should be treated as an independent text. R. Gatto in Galilei (2002) dates the text to 1593 based on the assumption that it presents the lecture notes for a lecture held by Galileo at the University of Padua. Valleriani (2010), in contrast, assumes that it was used by Galileo to teach his private pupils. *Delle Machine*, of which four copies are still extant, was not included by Favaro in the EN. Today, the text has been edited and published by Gatto (Galilei 2002).

2.2 Motion Along Inclined Planes and Dynamics

e sottoposta al medesimo mobile, sì che, appoggiandovisi egli sopra, fosse constretto a descendere in essa; perché se nell'uno e nell'altro modo disegna il mobile il medesimo viaggio, niente importerá s'egli sia sospeso dal centro *B* e sostenuto dal semidiametro del cerchio, o pure se, levato tale sostegno, s'appoggi e camini su la circonferenza *CFLI*. (EN II, 182)

The argument in *Le Meccaniche* pertains to the very same constellation of bent lever and inclined plane orthogonal to the bent lever arm, which is, moreover, represented by the very same diagram as in *De Motu Antiquiora*, merely with the points labeled differently (see Fig. 2.3, where the labels for the points used in *Le Meccaniche* are given in brackets). Yet in this revised version of the argument, Galileo does not directly identify the force on the bent lever arm with the force along an inclined plane orthogonal to it. Instead, in an intermediate step, he explicitly introduces a concave surface whose curvature is defined by the path that the body on the bent arm would describe if the bent lever were turned around its fulcrum. Thus by definition, a body on the bent arm on the lever and a body placed upon the concave surface are constrained to move along the same circular path.

According to Galileo, with regard to motion along or else rest at any point on this circular path, it does not matter whether the body is supported from above by a lever arm or whether it is supported from below by a concave surface. From our modern perspective, which perceives of force as a vector quantity, Galileo's intuition can comfortably be expressed. In both cases, the gravitational force acting on the same body can be decomposed into two components. The first component is parallel to the bent lever arm and orthogonal to the curved surface, respectively. As an external force it acts on lever arm and surface, respectively, and is, in both cases, passively counteracted by tensile forces in the material. It is thus the second component of the gravitational force alone, orthogonal to the first, which needs to be considered. It produces the torque which makes the bent lever turn (or stay at rest if counteracted by a torque of same magnitude), and it moves (or keeps at rest if counterbalanced) the body on the inclined plane, respectively.

Only in a second step does Galileo progress from considering a concave surface to considering an inclined plane. Using the example of the body positioned in *F*, he states:

...ma quando il mobile è in *F*, nel primo punto di tale suo moto è come se fosse nel piano elevato secondo la contingente linea *GFH*, perciò che l' inclinazione della circonferenza nel punto *F* non differisce dall' inclinazione della contingente *FG*, altro che l' angolo insensibile del contatto. (EN II, 182)

At any given point, the circular path and an inclined plane tangent to it in this point have the same inclination. Motion along the circular path is in its first point as if made on the corresponding inclined plane. It is for this reason that the force to descend along the circular path ("momento d'andare al basso") is the same as the force to descend along a corresponding inclined plane.[82]

---

[82]In *Le Meccaniche*, instead of "gravitas in plano," Galileo uses the term mechanical moment to refer to the force along the inclined plane by which quite generally he expressed the varying effect

Superficially, Galileo's argument seems to be essentially the same as the one in *De Motu Antiquiora* as it hinges on the geometrical relation between arc and tangent as defined by Euclidian geometry. However, in *Le Meccaniche*, motion along the circular path is no longer the abstractly conceived motion of an imagined body attached to a bent lever arm. Rather what is being compared are two concrete situations. In both, a body is placed on a surface, concave in the first case and straight in the second case. That the same force is required to hold the body steady in both cases, given that the slope is the same at the bodies' position, is immediately plausible based on mechanical intuition. To argue for it, Galileo resorts to, what could anachronistically be termed, virtual displacements. If the body at rest on the concave surface and on the corresponding inclined plane tangent were to move insensibly, motion would not be any different in both cases, and it is thus plausible to assume that the static forces required to resist the tendency to motion are likewise the same.

Upon closer inspection, Galileo's bent lever proof is thus revealed to rest on tangible considerations regarding concrete mechanical constellations rather than on rigorous theoretical inferences. That the bent lever should allow for an inference regarding the force required to hold a body on an appropriate concave surface can plausibly be surmised because it is intuitively obvious that a real bent lever could indeed be used to hold a real body on an appropriate real concave surface. That the same force is required to hold a body at a given position on a concave surface as to hold it on an inclined plane tangent to the surface can plausibly be surmised because it is intuitively obvious that the force required to hold a body on a real surface should not depend on the slope of the surface, other than at the position where the body is placed. Against the backdrop of the conceptual framework of mechanics available to Galileo, these intuitions cannot stringently be expressed, let alone established theoretically.

It is in this respect that Galileo's argument differs so decisively from the majority of the proofs of a law of the inclined plane that had been proposed before him.[83]

---

of a weight depending on the "arrangement which different bodies have among themselves (EN II, 159, transl. Galilei 1960, 151)." Galluzzi (1979), in his eminent study on the subject, refers to "momento" as the link between statics and dynamics. For a clarification of the relation between mechanical moment and "momento" as in "momentum velocitatis," see Chap. 10, Sect. 10.1. With regard to the bent lever proof, the substitution of "gravitas in plano" by "momento" appears to me to be less significant than has been assumed by, for instance, Festa and Roux (2008, 217), who state that "*De motu* brought together under the name of gravity two distinct quantities, weight and the effectiveness of weight; the Mecaniche distinguishes them by calling gravità the weight, and momento the effectiveness of the weight, that is to say the force required to support it or move it." It can be objected that in *De Motu Antiquiora* Galileo was well able to distinguish weight and the effectiveness of weight not only conceptually but also on a linguistic level, when he uses, for instance, "gravitas" as clearly distinguished and distinct from "gravitas super plano." With regard to this conceptual nexus, Clavelin (1983, 26) states succinctly that through the inclined plane "the tendency to descend, that is gravity as a motive force, is actually modified. A distinction thus becomes necessary between the weight and the natural motive force or, better still, between a properly gravific function and a motor function of weight."

[83] Being based on *model equivalence* rather than a physical instantiation of the balance-lever model, Galileo's argument seems to have been too abstract for some of his contemporaries. Davide Imperiali, for instance, devised his own version of a proof of the law of the inclined plane in which

2.2 Motion Along Inclined Planes and Dynamics

Even though his argument had clearly been inspired by contemplating upon very concrete mechanical constellations involving among others a bent lever, with regard to the argument itself, the lever no longer had to be conceived of as physically instantiated. Galileo had turned the bent lever into a device of theoretical abstraction, allowing him to deal with an aspect that could not appropriately be dealt with based on the extant conceptual frameworks and which, from our perspective, is recognized as the composition and decomposition of forces.

The idea of equating the force on the bent lever arm with the force along an inclined plane orthogonal to it may have been difficult if not impossible to justify based on the extant frameworks. Yet once it is accepted, solving the problem of determining the force on the inclined plane becomes trivial. For the bent lever it had been demonstrated, and in his proof Galileo presupposes this as known, that the weight of a body on a bent lever arm is compensated by the weight of a body on a horizontal lever arm of equal length if the ratio of the weights is in indirect proportion to the ratio of the length of the horizontal arm and the length of the projection of the bent lever arm onto the horizontal. With regard to Galileo's diagram, it thus held, for instance, that the body "is heavier at point d than at s, by as much as line da is longer than line ap (EN I, 298)."[84] According to the fundamental idea of Galileo's approach, it follows that the ratio of a body's heaviness on the inclined plane gh to its proper heaviness is as ap to ad.

By a trivial geometrical transformation, this ratio can be expressed in terms of length and vertical height of the plane and:

> it is hence certain that the same weight is pulled with a smaller force upward along an inclined ascent than along a right one, by as much as the right ascent is smaller than the oblique; and, consequently, the same heavy thing goes down with a greater force along a right descent than along an inclined one, by as much as the inclined descent is greater than the right one. (EN I, 298, tranl. Fredette)[85]

Hence the natural force of a body has to the force of a same mobile body on an inclined plane the same ratio as the length of the inclined plane has to its vertical height. In different contexts in which Galileo stated the law of the inclined plane, he used, in particular, different expressions to denote the force on the plane. Yet on an abstract level, all these statements articulate the same knowledge structure, which from a modern perspective is identified as specifying the relation between

---

he blended Galileo's approach with that of Pappus. Imperiali identified the abstract bent lever of Galileo's construction within a body of circular cross section on the inclined plane, thus essentially annihilating the advantages of Galileo's proof in favor of being able to provide an argument based on the idea of a physical instantiation of the balance-lever model. Imperiali himself claimed that his proof in no way fell short of Galileo's. Cf. Gatto et al. (1996, 38 and 88).

[84]...est autem tanto gravius in puncto d quam in s, quanto longior est linea da quam linea ap;

[85]constat igitur, tanto minori vi trahi sursum idem pondus per inclinatum ascensum quam per rectum, quanto rectus ascensus minor est obliquo; et, consequenter, tanto maiori vi descendere idem grave per rectum descensum quam per inclinatum, quanto maior est inclinatus descensus quam rectus.

the gravitational force and the component of the force along an inclined plane as determined by the sine of the angle of inclination with respect to the horizontal.

Galileo's statement is easily generalized to two planes of different inclination but equal height, in which case it holds that the mechanical moments on the planes are in the inverse ratio of the lengths:

$$H_1 = H_2 \Rightarrow \mathrm{mom}_1/\mathrm{mom}_2 \sim L_2/L_1.$$

Alternatively, for planes of equal length but different height and inclination, it holds that the mechanical moments are in the same ratio as the heights of the planes:

$$L_1 = L_2 \Rightarrow \mathrm{mom}_1/\mathrm{mom}_2 \sim H_1/H_2.^{86}$$

Dynamics of the inclined plane and the velocity theorem Determining how the force required to hold a body on an inclined plane depends on inclination was not, we remember, what Galileo had set out to do. What he had embarked on was demonstrating that, and why it was that, a heavy body moves faster along a more steeply inclined plane and, more specifically, to show how much faster exactly the free motion of a heavy body down along a plane of given inclination would be than motion of the same body down along another plane of different inclination. However, once the law of the inclined plane had been established by means of the bent lever proof, Galileo's goal could trivially be accomplished by invoking what in the following will be referred to as his *basic principle of dynamics* and by applying it to the case of free motion of heavy bodies along inclined planes.

In one of the preceding chapters, headed "[t]hat by which is caused the swiftness and slowness of natural motion," Galileo had stated regarding the "true cause of the slowness and the swiftness of motion (EN I, 260, transl. Fredette)" that[87]:

> care must be taken that swiftness is not separated from motion: for he who assumes motion, necessarily assumes swiftness; and slowness is nothing other than lesser swiftness. Consequently, swiftness comes from the same thing as does motion: and so, since motion comes from heaviness and lightness, it is necessary that slowness or swiftness come from the same thing; from a greater heaviness of the mobile comes a greater swiftness of that motion which happens because of the heaviness of the mobile, that is, motion downward. (EN I, 261, transl. Fredette)

---

[86] From a modern perspective, the restriction to planes of either equal height or length when stating the law of the inclined plane simply allows avoiding invoking the sine. The bent lever proof itself, in a sense, offered an alternative formulation of the principle in which indeed merely the inclinations, but not specific length or heights of the planes considered, need to be specified, and it was, as argued in Chap. 8, exactly for this reason that the argument and construction of the bent lever proof could provide the gateway for the construction of the so-called ex mechanicis proof of the law of chords.

[87] Unde causetur celeritas et tarditas motus naturalis.

2.2 Motion Along Inclined Planes and Dynamics

Fig. 2.4 Diagram associated to the velocity theorem in *De Motu Antiquiora* (EN I, 301)

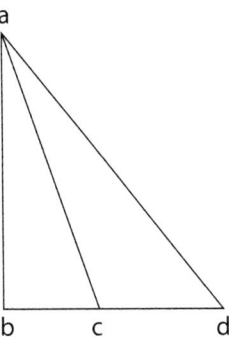

As Roux and Festa have phrased it for Galileo, "speed and motion cannot be distinguished from each other: gravity, the cause of motion, is also the cause of speed."[88] Not only does greater swiftness ensue from greater heaviness, but, in the case of natural downward motion, Galileo indeed assumes that cause and effect are directly proportional and thus "consequently, the ratios of the speeds follow the ratios of the heavinesses (EN I, 295–296, transl. Fredette)."[89] This assumption, namely, that the ratio of the heavinesses identified as the causes of motion is proportional to the ratio of the velocities of the ensuing motion as a measure of the effects, is what has above been referred to as Galileo's basic principle of dynamics. It "is nothing but a variant of the standard dynamical rule discussed by scholastic commentators, stating that the velocitas is proportional to the *potentia*...."[90]

In *On motion*, the basic principle of dynamics is applied to motion along inclined planes. Galileo identifies the force on the plane as the motive force and thus concludes that the velocities of motions along inclined planes have to be in proportion to the mechanical moments on these planes. This will be referred to as his *principle of inclined plane dynamics*. The ratio of the forces is given by the law

---

[88] See Festa and Roux (2008, 210). Souffrin (2001) similarly affirms: "The causes of the celeritates are the same as the causes of the motions, since motion and celeritas are one and the same thing."

[89] "... proportiones consequenter velocitatum, gravitatum proportiones, sequ[u]ntur." Wisan (1974, 151) comments: "this implicitly assumes the Aristotelian dynamic principle that a constant force generates a constant speed. Galileo initially assumes this relation, thinking of natural fall as uniform." The sentence quoted refers to motion through media and not to motion along inclined planes. Yet, as Souffrin (2001) has convincingly demonstrated in De Motu Antiquiora, exactly the same dynamical scheme forms the basis of Galileo's treatment of motion on inclined planes and his treatment of motion in media. For Aristotelian dynamics, see, for instance, Clagett (1979, 425–433). Aristotelian dynamics was subject to debate from the Middle Ages to Galileo's time, and alternatives were developed. These are discussed by Maier (1949). Van Dyck, 2006, Weighing falling bodies, Galileo's thought experiment in the development of his dynamical thinking, unpublished, available at http://www.sarton.ugent.be/index.php rightly comments that "the only model that he [Galileo] possesses for understanding forces is the balance which measures absolute weights; and all his dynamical thinking is based on the idea that speeds are caused by such forces."

[90] See Souffrin (2001) my emphasis. It needs to be added, however, that as regards natural motion, from an Aristotelian perspective, heaviness cannot be perceived as the potentia.

of the inclined plane which has just been established and proven, and Galileo can thus conclude that (for the associated diagram, see Fig. 2.4):

> just as the slowness on ad is to the slowness on ac, so line da is to line ac: hence as the swiftness on ac is to the swiftness on ad, so line da is to line ac. It is therefore certain that, the swiftnesses of the same mobile on different inclines are to each other inversely as the lengths of the oblique descents, provided that these hold equal right descents. (EN I, 301, transl. Fredette.)

Thus, Galileo has finally arrived at what has been referred to above as the velocity theorem and which we now recognize as but a particular expression of his principle of inclined plane dynamics.[91] As we will see, Galileo applied the theorem in the context of his theory of naturally accelerated motion. He did, however, not simply take for granted that the theorem would also apply in this case. Rather this was suggested to him by an analysis of early results regarding the pheno-kinematics of naturally accelerated motion.[92]

At this point, the goal set at the beginning had been reached. Yet before closing the chapter, Galileo adds the following remark:

> But since these things and others similar can easily be found by those who have understood the things said above, we deliberately omit them: only noting the following, that, just as has been said above concerning vertical motion, so also in the case of these motions on planes it happens that the ratios that we have put forward are not observed, sometimes for the reasons of the causes just now alleged, sometimes – and this is accidental – because in the beginning of its motion a lighter mobile goes down more swiftly than a heavier one: why this happens, we will make clear in its proper place; for this question depends on the one in which it is asked why the swiftness of natural motion is augmented. (EN I, 301–302, transl. Fredette)

It was not least this remark which has nourished the speculation that the law of chords, together with its so-called ex mechanicis proof, mentioned in a letter to Guidobaldo del Monte in 1602, might have been among the "things and others similar" that "can easily be found" based on Galileo's analysis of motion along inclined planes in *De Motu Antiquiora*.[93] The idea that the proposition was conceived and its proof established based on the velocity theorem while Galileo was still working under the framework of *De Motu Antiquiora*, i.e., particularly under the assumption that motion on inclined planes was uniform in principle, indeed by

---

[91] Festa and Roux (2008, 24) claim "that this result does not take into account the trivial observation that a body accelerates when it descends an inclined plane." However, at the end of the chapter, Galileo refers the reader forward to the chapter in which he goes on to give his account of accidental acceleration.

[92] Cf. Chap. 10.

[93] Vergara Caffarelli (2009, 101) referring to this remark by Galileo thus, for instance, claims that "Galileo wanted to tell us that he had found other applications of the theorem of the inclined plane (probably the theorem of chords)."

now seems more or less commonly accepted. Against this it will be demonstrated in Chap. 8 that the ex mechanicis proof proof was thought up by Galileo only after his conceptual shift to natural acceleration.

Secondly, Galileo's allusion to observation has been read as an indication that he may have been experimenting with bodies falling straight along inclined planes already, in the last decade of the sixteenth century. No direct evidence has, however, been put forward. As briefly mentioned above, indubitable evidence exists that Galileo at that time had indeed conducted an experiment involving motion along an inclined plane. Yet in this experiment it had not been motion straight down an inclined plane that was at stake. Galileo and Guidobaldo del Monte had rather used the inclined plane to make the trajectory of projectile motion directly accessible to observation. The conceptual foundation for this experiment, however, was laid, one can argue, in *De Motu Antiquiora*. Indeed, only once it is accepted that motion along an inclined plane behaves in essentially the same way as vertical free fall does it seem to make any sense at all to assume that the rolling of a ball along an inclined plane may characteristically show the same behavior and, in particular, assume the same trajectory as a projectile shot from a cannon. If this is indeed correct, then Galileo's treatment of motion along inclined planes in *De Motu Antiquiora* opened, if indirectly, the path toward the law of fall and thus more generally to his new science of motion. That Galileo had begun to venture this path in 1602 is evidenced by a letter to Guidobaldo del Monte written in this year. This letter will be discussed in quite some detail in the following chapter.

References

Archimedes, & Heath, T.L. (2002). *The works of Archimedes*. Mineola: Dover Publications.
Aristotle, & Hett, W.S. (1980). *Aristotle: In twenty-three volumes. Volume 14: Minor works* (Reprinted edn. Loeb classical library, Vol. 307). Cambridge: Harvard University Press.
Boccaletti, D. (2016). *Galileo and the equations of motion*. Heidelberg/New York: Springer International Publishing.
Bodnar, I. (2016). Aristotle's natural philosophy. In E.N. Zalta (Ed.), *The Stanford encyclopedia of philosophy* (Winter 2016 ed.). Metaphysics Research Lab, Stanford University, Stanford.
Burton, C. (1890). *An introduction to dynamics including kinematics, kinetics and statics*. London: Longmans, Green & Co.
Camerota, M. (1992). *Gli Scritti De Motu Antiquiora di Galileo Galilei: Il Ms. Gal. 71. Un'analisi storico-critica*. Cagliari: Cooperativa Universitaria.
Camerota, M., & Helbing, M. (2000). Galileo and Pisan Aristotelianism: Galileo's De Motu Antiquiora and the Quaestiones De Motu Elementorum of the Pisan Professors. *Early Science and Medicine, 5*(4), 319–365.
Clagett, M. (1941). *Giovanni Marliani and late medieval physics*. New York/London: Columbia University Press/P.S. King & Son.
Clagett, M. (1979). *The science of mechanics in the middle ages* (Publications in Medieval Science, Vol. 4). Madison: University of Wisconsin Press.
Clavelin, M. (1983). *Conceptual and technical aspects of the Galilean geometrization of the motion of heavy bodies* (pp. 23–50). Dordrecht: Springer Netherlands.

Damerow, P. (2006). Mentale Modelle als kognitive Instrumente der Transformation von technischem Wissen. In J. Renn, P. Damerow, M.D. Hyman, & M. Valleriani (Eds.), *Weight, motion and force: Conceptual structural changes in ancient knowledge as a result of its transmission* (Preprint 320). Berlin: Max Planck Institute for the History of Science.

Damerow, P., & Renn, J. (2012). *The equilibrium controversy: Guidobaldo del Monte's critical notes on the mechanics of Jordanus and Benedetti and their historical and conceptual backgrounds*, Vol. 2, Ed. Open Access). Berlin: Max Planck research library for the history and development of knowledge sources.

Damerow, P., Renn, J., & Rieger, S. (2001). Hunting the white elephant: When and how did Galileo discover the law of fall? In J. Renn (Ed.), *Galileo in context* (pp. 29–150). Cambridge: Cambridge University Press.

Damerow, P., Freudenthal, G., McLaughlin, P., & Renn, J. (2004). *Exploring the limits of preclassical mechanics*. New York: Springer.

Del Monte, G. (1581). *Le Mechaniche*. Venice: Francesco di Franceschi Sanese.

Del Monte, G. (1969). Mechanicorum liber. In S. Drake & I.E.T. Drabkin (Eds.), *Mechanics in sixteenth century Italy. Selections from Tartaglia, Benedetti, Guido Ubaldi and Galileo*. Madison: University of Wisconsin Press.

Del Monte, G., Renn, J., & Damerow, P. (2010). *Guidobaldo del Monte's mechanicorum liber*. (Max Planck research library for the history and development of knowledge sources, Vol. 1, Ed. Open Access). Berlin.

Dijksterhuis, E.J. (1943). *Simon Stevin*. M. Nijhoff, 's-Gravenhage.

Drake, S., & Drabkin, I.E. (1969). *Mechanics in sixteenth-century Italy*. Madison: The University of Wisconsin Press.

Drake, S., & Swerdlow, N.M., Levere, T.H. (1999). *Essays on Galileo and the history and philosophy of science*. Toronto/Buffalo: University of Toronto Press.

Festa, E., & Roux, S. (2008). The enigma of the inclined plane from Heron to Galileo. In R. Laird & S. Roux (Eds.), *Mechanics and natural philosophy before the scientific revolution*. Dordrecht: Springer.

Fredette, R. (1969). *Les De Motu "plus anciens" de Galileo Galilei: Prolégomènes*. Dissertation, University of Montreal.

Fredette, R. (1972). Galileo's De motu antiquiora. *Physis, 14*, 321–348.

Fredette, R. (2001). Galileo's De motu antiquiora: Notes for a reappraisal. In Montesinos, J., & Solís, C. (Eds.), *Largo campo di filosofare: Eurosymposium Galileo 2001* (pp. 269–280). La Orotava: Fundación Canaria Orotava de Historia de la Ciencia.

Galilei, G. (1960). *On motion and on mechanics: Comprising De Motu (ca. 1590). Translated with introduction and notes by I. E. Drabkin and Le Meccaniche (ca. 1600). Translated with introduction and notes by Stillman Drake*. Madison: The University of Madison.

Galilei, G. (2002). *Le mecaniche. Edizione critica e gaggio introduttivo*. Florence: Immagini della ragione, Olschki.

Galilei, G., & Wallace, W.A. (1992). *Galileo's logical treatises: A translation, with notes and commentary, of his appropriated Latin questions on Aristotle's Posterior analytics* (Boston Studies in the Philosophy of Science, Vol. 138). Dordrecht: Kluwer.

Galilei, G., Crew, H., & Salvio, A.D. (1954). *Dialogues concerning two new sciences*. New York: Dover Publications.

Galluzzi, P. (1979). *Momento*. Rome: Ateneo e Bizzarri.

Gatto, R., Stigliola, C., & Imperiali, D. (1996). *La meccanica a Napoli ai tempi di Galileo*. Napoli: Testi e documenti per la storia della scienza nel mezzogiorno, La Città del Sole.

Gentner, D. (1998). Analogy. In W. Bechtel & G. Graham (Eds.), *A companion to cognitive science* (pp. 107–113). Oxford: Blackwell.

Gentner, D., Holyoak, K., & Kokinov, B. (Eds.) (2001). *The analogical mind: Perspectives from cognitive science*. Cambridge: MIT Press.

Gillispie, C.C. (Ed.) (1981). *Dictionary of scientific biography*. New York: Charles Scribner's Sons.

Giusti, E. (1986). Ricerche galileiane: Il trattato de motu aequabili come modello della teoria delle proporzioni. *Bollettino di Storia delle Scienze Matematiche, 6*(2), 89–108.

References

Giusti, E. (1992). La teoria galileiana delle proporzioni. In *La matematizzazione dell'universo*, Lino Conti (pp. 207–222).
Giusti, E. (1993). *Euclides reformatus: La teoria delle proporzioni nella scuola galileiana*. Turin: Bollati Boringhieri.
Giusti, E. (1998). Elements for the relative chronology of Galilei's De motu Antiquiora. *Nuncius Ann Storia Sci, 2*, 427–460.
Grant, E. (1965). Bradwardine and Galileo: Equality of velocities in the void. *Archive for History of Exact Sciences, 2*(4), 344–364.
Grant, E. (1966). Aristotle, Philoponus, Avempace, and Galileo's Pisan dynamics. *Centaurus, 11*(2), 79–93.
Grant, E. (1974). *A source book in medieval science* (Source books in the history of the sciences). Cambridge: Harvard University Press.
Grant, E. (2010). *The nature of natural philosophy in the late middle ages* (Studies in philosophy and the history of philosophy, Vol. 52). Washington, D.C.: Catholic University of America Press.
Gregory, A. (2001). Aristotle, dynamics and proportionality. *Early Science and Medicine, 6*(1), 1–21.
Guerrini, L. (2014). Pereira and Galileo: acceleration in free fall and impetus theory. *Bruniana e campanelliana: Ricerche filosofiche e materiali storico-testuali, XX*(2), 513–530.
Halbwachs, F., & Torunczyk, A. (1985). On Galileo's writings on mechanics: An attempt at a semantic analysis of Viviani's scholium. *Synthese, 62*(3), 459–484.
Henry, J. (2011). Galileo and the scientific revolution: The importance of his kinematics. *Galilaeana, 8*, 3–36.
Heron (1976). *Herons von Alexandria Mechanik und Katoptrik*. Stuttgart: B. G. Teubner.
Hooper, W.E. (1992). *Galileo and the problems of motion*. Dissertation, Indiana University.
Jung, E. (2011). Intension and remission of forms. In H. Lagerlund (Ed.), *Encyclopedia of medieval philosophy* (pp. 551–555). Dordrecht: Springer Netherlands.
Laird, W.R. (2001). Renaissance mechanics and the new science of motion. In J. Montesinos & C. Solís (Eds.), *Largo campo di filosofare: Eurosymposium Galileo 2001* (pp. 255–267). La Orotava: Fundación Canaria Orotava de Historia de la Ciencia.
Lewis, C. (1980). *The Merton tradition and kinematics in late sixteenth and early seventeenth century Italy* (Vol. 15. Saggi e testi/Università di Padova, Centro per la storia della tradizione aristotelica nel Veneto). Padua: Editrice Antenore.
Machamer, P.K. (1978). Aristotle on natural place and natural motion. *ISIS, 69*(3), 377–387.
Maier, A. (1949). *Studien zur Naturphilosophie der Spätscholastik Bd. 01: Die Vorläufer Galileis im 14. Jahrhundert* (Storia e letteratura, Vol. 22). Roma: Edizioni di Storia e Letteratura.
Mersenne, M., & Galilei, G. (1635). *Qvestions physico-mathematiqves: Et Les mechaniqves du sieur Galilee: Avec Les Prelvdes de l' harmonie vniuerselle vtiles aux philosophes, aux medecins, aux astrologues, aux ingenieurs, & aux musiciens Les questions theologiques, physiques, morales, et mathematiques*. Paris: H. Gvenon.
Miller, D.M. (2014). *Representing space in the scientific revolution*. Cambridge: Cambridge University Press.
Oresme, N., & Clagett, M. (1968). *[Tractatus de configurationibus qualitatum et motuum.] Nicole Oresme and the medieval geometry of qualities and motions. A treatise on the uniformity and difformity of intensities known as Tractatus de configurationibus qualitatum et motuum*. Edited with an introduction, English translation, and commentary by Marshall Clagett. Madison, WI: University of Wisconsin Press.
Palmieri, P. (2003). Mental models in Galileo's early mathematization of nature. *Studies in History and Philosophy of Science, 34*, 229–264.
Palmieri, P. (2005a). The cognitive development of Galileo's theory of buoyancy. *Archive for History of Exact Sciences, LIX*, 189–222.
Palmieri, P. (2005b). Galileo's construction of idealized fall in the void. *History of Science, 43*, 343–390.

Palmieri, P. (2017). On *Scientia* and *Regressus*. In H. Lagerlund & B. Hill (Eds.), *Routledge companion to sixteenth century philosophy*. New York/Routledge: Taylor & Francis.

Pasnau, R., & Trifogli, C. (2014). Change, time, and place. In *The Cambridge history of medieval philosophy* (pp. 267–278). Cambridge: Cambridge University Press.

Renn, J., Damerow, P., & McLaughlin, P. (2003). Aristotle, Archimedes, Euclid, and the origin of mechanics: The perspective of historical epistemology. In J.L. Montesinos (Ed.), *Symposium Arquímedes Fundación Canaria Orotava de Historia de la Ciencia* (Preprint 239). Berlin: Max Planck Institute for the History of Science.

Salvia, S. (2017). From Archimedean hydrostatics to post-Aristotelian mechanics: Galileo's early manuscripts De motu antiquiora (ca. 1590). *Physics in Perspective, 19*(2), 105–150.

Schemmel, M. (2014). Medieval representations of change and their early modern application. *Foundations of Science, 19*(1), 11–34.

Schiefsky, M.J. (2008). Theory and practice in Heron's mechanics. In *Mechanics and natural philosophy before the scientific revolution* (pp. 15–49). Dordrecht: Springer.

Settle, T.B. (1966). *Galilean science: Essays in the mechanics and dynamics of the Discorsi*. PhD thesis.

Settle, T.B. (1983). Galileo and early experimentation. In *Springs of scientific creativity: Essays on founders of modern science* (NED – new edition edn., pp. 3–20). University of Minnesota Press.

Souffrin, P. (1990). Galilée et la tradition cinématique pré-classique. La proportionnalité momentum-velocitas revisitée. *Cahier du Séminaire d'Epistémologie et d'Histoire des Sciences, 22*, 89–104.

Souffrin, P. (1992). Sur l'histoire du concept de vitesse d'aristote à galilée. *Revue d'histoire des sciences, 45*(2/3), 231–267.

Souffrin, P. (2001). Motion on inclined planes and in liquids in Galileo's earlier De Motu. In P.D. Napolitani & P. Souffrin (Eds.), *Medieval and classical traditions* (Vol. 50). Turnhout: Brepols.

Stevin, S. (1586). *De Beghinselen der Weegconst*. Leiden: Druckerye van Christoffel Plantijn. By Francoys van Raphelinghen.

Sylla, E. (1971). Medieval quantifications of qualities: The "Merton school". *Archive for History of Exact Sciences, 8*(1/2), 9–39.

Sylla, E. (1973). Medieval concepts of the latitude of forms. The Oxford calculators. *Archives d'Histoire Doctrinale et Littéraire du Moyen Âge, 40*: 223–283.

Tartaglia, N. (Ed.) (1565). *Jordani opusculum de ponderositate*. Venice: Apud C. Troianum.

Valleriani, M. (2010). *Galileo engineer* (Boston Studies in the Philosophy of Science, Vol. 269). Dordrecht/London/New York: Springer.

Vergara Caffarelli, R. (2009). *Galileo Galilei and motion: A reconstruction of 50 years of experiments and discoveries*. Berlin/New York/Bologna: Springer/Società italiana di Fisica.

Wallace, W.A. (1981). *Prelude to Galileo: Essays on medieval and sixteenth-century sources of Galileo's thought*. Dordrecht: Reidel.

Wallace, W.A. (1984). *Galileo and his sources: The heritage of the Collegio Romano in Galileo's science*. Princeton: Princeton University Press.

Wallace, W. (1990). Duhem and Koyré on Domingo de Soto. *Synthese, 83*(2), 239–260.

Wisan, W.L. (1974). The new science of motion: A study of Galileo's De motu locali. *Archive for History of Exact Sciences, 13*, 103–306.

Wisan, W.L. (1978). Galileo's scientific method: a reexamination. In R.E. Butts (Ed.), *New perspectives on Galileo: Papers deriving from and related to a workshop on Galileo held at Virginia Polytechnic Institute and State University, 1975* (The University of Western Ontario series in philosophy of science). Dordrecht: Reidel.

Chapter 3
A Glimpse at a Challenging Research Agenda: Galileo to Guidobaldo del Monte in 1602

In November 1602, Galileo wrote a letter to his friend and patron Guidobaldo del Monte.[1] In the letter he outlined his current work. This letter has received particular attention as it provides the first explicit evidence that Galileo had returned to the question of the fall of heavy bodies on inclined planes which, as we have seen, he had already addressed in the 1590s in his *De Motu Antiquiora* treatise. Moreover, upon writing the letter, as Galileo himself claims, he had already been able to prove some propositions concerning the behavior of such motions. It is commonly held, however, that this letter antedates Galileo's conceptual shift toward the assumption that motion of fall along inclined planes is naturally accelerated, the very assumption that characterizes his new science of motion. As a consequence, it is commonly assumed that his considerations in 1602 must have been grounded in the belief, characteristic of his pondering on the problems of motion in *De Motu Antiquiora*, that motion of fall on inclined planes is essentially uniform. It has furthermore been surmised on this account that these considerations can be at best indirectly relevant for an understanding of the emergence of Galileo's new science of naturally accelerated motion.[2]

This being the predominant opinion, to the present day only a very limited number of entries from the *Notes on Motion* have been rather tentatively brought into connection with the letter written to Guidobaldo del Monte in November 1602. Consequently, beyond those he explicitly mentioned, Galileo's considerations and

---

[1] For Guidobaldo del Monte, see Gamba et al. (2013).

[2] Stilmann Drake, for instance, after having identified a number of folios of the *Notes on Motion* as containing early considerations by Galileo, claims that "these probably all belong to 1602, when it appears that Galileo, having revised and expanded his *Mechanics*, decided to write a new treatise on motion. Several propositions for this were neatly written out before he fully realized the importance of acceleration, and the sheets bearing these were mutilated in the course of the subsequent revisions of the projected treatise (Drake 1978, 67)."

achievements in the context of the intellectual work alluded to in the letter have largely remained the subject of speculation.[3]

In contrast, I will argue that the considerations referred to in the letter were in fact part of a much broader research agenda which Galileo was following at the time and which has left abundant traces in the *Notes on Motion*. The reconstruction and interpretation of Galileo's research agenda from this manuscript evidence will show that it was in fact conducted under his new governing assumption that motion on inclined planes is naturally accelerated and that it was exactly this program which—by providing a pivotal, though eventually insuperable challenge—conditioned the establishment of the new insights, which were successively turned into the core of a new science of motion.

Before venturing a reconstruction and interpretation of the challenging research agenda that Galileo was engaged in toward the end of 1602, based on the preserved manuscript evidence, a fresh exegesis of the letter written to Guidobaldo is befitting. Special attention will thereby be paid to those aspects which are relevant for my new interpretation of Galileo's early work on the problems of naturally accelerated motion to be advanced in this book. In what follows, the letter will merely be paraphrased. The full document together with a translation is given in the appendix.[4]

Galileo first came into contact with Guidobaldo del Monte in 1588 when he sent him a copy of his treatise on the centers of gravity of parabolic solids for comment. Over the course of the following years, the two men developed a close relation, which was to have significant impact on Galileo's further path. As a patron, Guidobaldo del Monte decisively supported Galileo's career; as an intellectual peer he provided him with crucial stimulus.

Most of the correspondence that was sent back and forth between Galileo and Guidobaldo over the course of their personal relation has been lost. Of the limited number of surviving letters, the one of 1602 is the most recent. The next antecedent letter preserved was sent by Guidobaldo del Monte to Galileo as early as December 1597. The letter of 1602, however, as becomes clear from its first paragraph, must have been preceded within a rather short period of time by at least two more letters which are now lost—one from Galileo and a response from Guidobaldo del Monte.

In the first letter, Galileo had evidently informed Guidobaldo about his view that the period of a pendulum was independent of its amplitude, i.e., of the alleged isochronism of the pendulum ("la proposizione de i moti fatti in tempi uguali nella medesima quarta del cerchio"), and had explicated an experiment on the basis of which this assumed isochronism of the pendulum could be demonstrated.

---

[3]Humphreys (1967), for instance, speculates about Galileo's work in 1602 without considering any of the material in the *Notes on Motion* at all. Galluzzi (1979, 268) more carefully restrained from speculation when he claimed concerning Galileo's work in 1602: "È tuttavia, impossibile riconstruire il modo in cui Galileo intendeva raggiungere la dimostrazione lavarando sulla proporzione dei *momenti*."

[4]The original letter is not extant. A copy in a later hand has been preserved. This copy was, however, not made directly from the original but from an earlier copy by Viviani, which is likewise lost today. Cf. EN X, letter 88, 97–100.

Taking into account the time it would have taken a letter to be delivered from Padua, Galileo's residence at the time, to Monte Baroccio, Guidobaldo del Monte's permanent residence in exile to which Galileo's first letter was most likely addressed (as was the one of November 1602), and from where Guidobaldo del Monte's answer was presumably sent, and taking into account that both protagonists did not necessarily compose their respective answers immediately, we can conclude that Galileo must have turned his attention to the problems discussed in the letter weeks, if not months before November 1602.[5]

With respect to the much debated question of how Galileo first arrived at his conviction that pendulum motion is isochronous, the letter thus confronts us with a certain conundrum. On the one hand, Galileo is eager to emphasize that this proposition (the isochronism of the pendulum) had always seemed admirable to him ("essendomi parsa sempre mirabile"), implying a dating of his insight to a rather indeterminate past while at the same time claiming that the truth of the isochronism of the pendulum became clear principally by means of the very experiment he discussed in the letter ("l'esperienza, con che mi sono principalmente chiarito di tal verità"), thus suggesting his insight may have been contemporaneous to the time of writing.[6]

In his answer to Galileo's first, now lost letter, Guidobaldo del Monte had obviously expressed doubts and even considered the assumption of the isochronism of the pendulum to be false. Galileo ascribed Guidobaldo del Monte's negative reaction, at least in part, to the fact that he himself had given a confused account of the experiment that should have demonstrated the isochronism of the pendulum ("da me confusamente stata esplicata"). By giving a new explanation of the experiment, Galileo wanted to make up for this deficit and in particular enable Guidobaldo del Monte to repeat it for himself so that he would eventually "also be able to ascertain this truth," i.e., the truth of the isochronism of the pendulum.

The experiment Galileo proposed is well-known, in part because he later presented variants of it in the *Dialogue* as well as in the *Discorsi*.[7] It had a particularly simple design and its performance was straightforward. Two pendulums of equal length with equal pendulum bobs were pulled back through different

---

[5]It will be demonstrated in Chap. 5 that Galileo must have been working on the problems he communicated to Guidobaldo in November 1602, since at least September. For Guidobaldo del Monte's exile, see Frank (2011, 512–519).

[6]For Viviani's account of how Galileo allegedly discovered the isochronism of the pendulum as a student when observing the swinging of a lamp in the dome of Pisa, see Viviani (1890–1909). For a contextualization of Galileo's engagement, theoretical as well as practical, with pendulum motion, see Büttner (2008). Drake (1970) has suggested that Galileo's interest in pendulum motion might well have been triggered by his father's musical experiments. Galileo's statements in the letter can be interpreted as based on a distinction between mere assumption and certitude. If this is the case, Galileo would be implying that it was his experiment conducted shortly before the antecedent letter that was sent that ultimately convinced him of the truth (verità) of the isochronism of the pendulum, which by then he had already held as an assumption for some time.

[7]See EN, VII, 474–476 and EN, VIII, 277–278. For a recent detailed assessment of these passages and Galileo's pendulum experiments in general, see Palmieri (2009).

angles from their equilibrium positions and then released from these positions at the same time. If an observer counted a large number of oscillations of one of these pendulums, e.g., 100, a second observer who observed the oscillations of the other pendulum would not have counted, according to Galileo, "even a single one more," this being "a most evident sign" for the isochronism of the pendulum.

Two aspects of Galileo's description of the experiment deserve particular attention here. Firstly, a bracketed side remark, in which Galileo ascertained that the isochronism of the pendulum would still hold even if unequal pendulum bobs were chosen for the experiment ("se ben niente importa se fussero [i palle] disuguali").[8] If unequal here is indeed to be understood as referring to the size, as well as the material of the bobs, this would imply that Galileo assumes the period of a pendulum to be independent, not only of the weight but also of the specific weight of the bob.

The period of a pendulum, indeed, depends neither on the mass nor on the bob's specific mass. Both claims are thus recognized as empirically justified from a modern perspective. As will be seen, in view of the way in which Galileo tried to link pendulum motion and motion along inclined planes, his remark receives particular significance. Whatever holds for pendulum motion in this respect should, based on Galileo's conceptualization of the relation between swinging and rolling, by implication also hold for naturally accelerated motion on inclined planes. Thus, more concretely, if pendulum motion is independent of weight, bodies of different weight should, all other things being equal, be characterized by the same kinematical behavior, and if pendulum motion is independent of specific weight, the kinematics of naturally accelerated motion should not depend on specific weight either.[9]

It has been noted in the literature that Galileo's instruction to count a large number of oscillations points to his understanding, in view of an unalterable absolute error of measurement, of how to reduce relative error by increasing the number of events observed. This is certainly true, but it is also rather trivial. A far more interesting aspect of why Galileo, in his desire to portray the experiment as confirmation of the isochronism of the pendulum, may have suggested comparing a large instead of a small number of oscillations has been revealed by analysis based on replication of Galileo's pendulum experiment.

Identical pendulums, oscillating with different amplitudes, do not swing isochronally. In fact, as physics informs us, sizable differences in the amplitude of the oscillations of pendulums of equal length correspond to significant differences

---

[8]Galileo had been experimenting with pendulums at least since the time of writing *De Motu Antiquiora* and pendulum bobs of different material figure in his very first allusion to such an experiment. In a passage of *De Motu Antiquiora*, Galileo writes: "The same thing is evident if two weights, one of wood, the other of lead, are suspended from two equal threads and, when they have received an impetus from an equal distance from the perpendicular, they are released; of the two, the lead will certainly be moved back and forth for a longer interval of time (EN I, 335, transl. Fredette)." See also Settle (1966, 96).

[9]Van Dyck (2006) convincingly shows that Galileo never genuinely got to grips with the problem of the relationship between weight, specific weight, and the naturally accelerated motion of heavy bodies.

between their periods. Such differences should be easily observable.[10] Moreover, if instead of just one or a small number, more oscillations are observed, it may be expected that the differences between the periods of the two pendulums would increase with every oscillation and thus should become even more manifest. Based on this type of argument, Galileo's descriptions of pendulum experiments, or at least of their results, have in the past occasionally been dismissed as fictitious.[11] The underlying argument is, however, deceptive.

Friction reduces the amplitudes of both pendulums over time and moreover reduces those of the pendulum that has been pulled further away from equilibrium at a greater rate. Yet the smaller the amplitudes of oscillation, the closer the pendulum's behavior is to actually being isochronous. Hence the deviation from a synchronous behavior per swing of the two pendulums in the experiment, in fact, decreases with the number of oscillations observed, and after a number of oscillations, the two pendulums swing almost synchronously.[12] Modern replications of Galileo's experiment have indeed yielded that his statement that after 100 oscillations the two pendulums are out of sync by less than one beat has to be considered as realistic.[13]

When describing the experiment, Galileo had stated the isochronism of the pendulum for half swings or half periods, i.e., he had affirmed that motion from one turning point of the pendulum's motion to the other is made in equal time, regardless of the amplitude, that is of the angle of the arc swung through. For his further arguments Galileo made a transition to stating the isochronism of the pendulum for quarter oscillations, i.e., to the statement that motion made from one of the turning points to the pendulum's equilibrium position and, vice versa, from the equilibrium position to the turning point, is completed in equal time regardless of the amplitude. As will be shown, this transition corresponds well to the theoretical approach Galileo was following at the time, which indeed focused on analyzing such quarter swings.

Instead of replicating the experiment in the manner that Galileo had proposed it, Guidobaldo del Monte altered the setup. However, the experiment which he conducted, rather than convincing him, apparently confirmed his doubts concerning

---

[10] The reader may explore the deviation of large-amplitude pendulums from isochronous behavior virtually at http://hyperphysics.phy-astr.gsu.edu/Hbase/pendl.html. Accessed 12 Jan 2015. Lima and Arun (2006) provide a penetrating discussion of the physics of large-amplitude oscillations of pendulums.

[11] In his translation of the *Discorsi*, Drake in Galilei (1974, 226, footnote 11), for instance, claimed that Galileo's portrayal of the results of the experiment could not correspond to actually observable facts.

[12] For angular amplitudes of less than 45 degrees, the period of a real pendulum deviates from that of an idealized pendulum in which the restoring force is assumed to depend linearly on the angle of elongation, which would swing truly isochronally, by less than one percent. The damping of a pendulum due to air drag forces is comprehensively treated in Nelson (1986). For large amplitudes, the rate of decay of amplitude due to damping is rather high.

[13] See MacLachlan (1976), Naylor (1974, 1976), and more recently Palmieri (2009).

the validity of the isochronism. In contrast to Galileo, in his version of the experiment, Guidobaldo replaced the swinging of a pendulum with the rolling of a ball in a circular bowl. This substitution of the motion of a pendulum supported from above by the rolling of a ball on a concave surface supported from below appears problematic from the perspective of modern physics.[14]

In the letter, however, Galileo expressed no objection to Guidobaldo's approach in principle. The substitution of pendulum swing by rolling along a concave surface indeed corresponded to the research agenda Galileo was pursuing at the time which, as will be seen, rested precisely on the assumption that swing through a given arc would be kinematically equivalent to rolling along this same arc.[15] Galileo, however, made two other objections to explain why Guidobaldo del Monte's version of the experiment may not have brought about the result he would have expected. On the one hand, he argued that the bowl may not have been perfectly circular, and on the other hand, he criticized Guidobaldo del Monte for trying to verify the isochronism of the pendulum (or better that of a ball rolling in a bowl) on the basis of a single transit of the rolling object, an error-prone procedure due to the problems of determining the exact time of the transit through the equilibrium position and/or turning points, respectively.

Apart from mentioning his failed experiment, in his preceding letter Guidobaldo del Monte had apparently also raised a theoretical counterargument against the validity of Galileo's claim of the isochronism of the pendulum. He had tried to counter Galileo's claim that pendulums swing isochronously with a concrete example which, not least through the use of large numbers, accentuated an especially counterintuitive consequence of Galileo's assumption. If the isochronism of the pendulum held, according to Guidobaldo's argument, which Galileo repeated in the letter, it would follow that "given a quadrant 100 miles in length, two equal mobiles might pass along it, one the whole length, and the other only a span, in equal times." Galileo's strategy in countering Guidobaldo del Monte's critique was twofold: first he argued for the physical and then for the mathematical possibility of the seemingly absurd consequence that the isochronism of the pendulum entailed for Guidobaldo.

Galileo's physical argument, instead of considering motion along arcs of the same circle as Guidobaldo del Monte had done, reverts to motion along inclined planes:

> ...... but if we consider that there can be a plane which is inclined so slightly, which would [for instance] be that of the surface of a river which flows extremely slowly, that on it a mobile will not have traveled naturally more than a span in the time in which another will have traveled 100 miles over a much inclined plane (or else if it is adjoined with a very big impetus, even over a small inclination).[16]

---

[14]Cf. Chap. 6.

[15]Much later in the *Dialogue*, Galileo indeed proposes to his readers to have a ball roll along a concave surface with the curvature of an upright quarter circular arc to observe that a ball's transit to the lowest point will always be made in the same time, regardless of the starting point on the arc (EN VII, 476), just as Guidobaldo del Monte had done.

[16]All English quotations follow the translation given in the appendix (see Chap. 14).

Superficially, the argument is reminiscent of Galileo's account of motion on inclined planes in the *De Motu Antiquiora*, in particular, of the velocity theorem. According to the velocity theorem, by choosing ever smaller inclinations, one could indeed make the velocity of natural descent on an inclined plane and thus the distance traveled in a given time arbitrarily small. Thus, if compared to the motion on an inclined plane of given inclination, it would always be possible to find another, less inclined plane such that the seemingly absurd effect which Guidobaldo del Monte had rejected could be produced, namely, that a body travels 100 miles in the same time that another body moves only a span.[17]

To this Galileo added that a body which travels 100 miles on a steeply inclined plane would potentially traverse the same distance in the same time on a less inclined plane if it had received an appropriate "grandissimo impeto" beforehand. Galileo is alluding to a situation where motion does not start from rest but, to speak in modern terms, with an initial velocity resulting from a preceding motion of fall.[18] The question of how motion along chords proceeded after an initial motion of fall played, as we will see, an important role in his considerations regarding swinging and rolling and appears to flare up in this remark, which is otherwise startlingly disconnected from Galileo's argument.

Galileo's argument can be justified based on the velocity theorem and could thus potentially be based on the assumption characteristic of the *De Motu Antiquiora* that natural motion on inclined planes is essentially uniform. Yet it would likewise be valid if the motions considered were assumed to proceed in a naturally accelerated manner. As will become clear, when writing the letter, Galileo was indeed already convinced that motion on inclined planes was naturally accelerated. It is hence conceivable that he consciously formulated his argument noncommittally, i.e., in a way as to leave the question regarding the underlying conception of motion open. In this way, the argument would be acceptable for Guidobaldo who, as far as we know, would not necessarily have shared Galileo's conviction regarding naturally accelerated motion.[19]

Galileo's second, his mathematical argument, was mainly directed against the apparent contradiction which the isochronism of the pendulum seemed to imply for Guidobaldo del Monte and which he had particularly emphasized by the use of large

---

[17] Galileo, anticipating what he would state in the next paragraph, is obviously not thinking of two arbitrary inclined planes, one measuring 100 miles and the other a span, but indeed about two chords which have these lengths and which are inscribed in the same circle sharing its nadir. Motion along such chords is completed in equal time as he will state and claim to have demonstrated a bit further down in the letter. Thus, where Guidobaldo had talked about motion along arcs, Galileo made his argument for motion along chords spanning these arcs.

[18] Alternatively, the "received impetus" may have been the result of an impact event, but this clearly does not seem to be what Galileo is alluding to here.

[19] The openness of the letter with regard to the question of how Galileo at the time conceptualized motion on inclined planes, i.e., as uniform or else as naturally accelerated, has certainly helped to nourish the incorrect understanding that in 1602 Galileo was still working under his old paradigm of uniform motion.

numbers in his example. Galileo's argument resorted to a geometrical construction that displayed a structural similarity to Guidobaldo del Monte's counterexample and entailed a similar consequence, which likewise seemed unreasonable, at least superficially. In Galileo's construction, the ratio of two sides of a triangle was allowed to increase above all limits, just as the ratio of the distances traversed in Guidobaldo del Monte's counterexample, and yet a related magnitude remained constant: the area of the triangle in Galileo's geometrical example and the ratio of times of the two motions in the motion situation under scrutiny, respectively. By showing that in the case of the geometrical example the consequence, despite being seemingly unreasonable, was well-founded in Euclidean geometry, Galileo's argument thus at least made it plausible that a similar conclusion regarding times of motion along different paths, where the ratio of the distances traversed could increase above all limits, was not as absurd as Guidobaldo had implied.

At this point Galileo, who had thus far merely commented on and complemented his previous exchange with Guidobaldo, turned to outlining his own current investigations. He commenced by putting forward two propositions he stated to have already proven. These two propositions are easily recognized as the law of chords, i.e., the statement that motion on inclined planes inscribed as chords in the same circle sharing either its apex or its nadir is made in equal time, and the *law of the broken chord*, i.e., the proposition according to which motion from rest over two conjugate chords is completed in less time than over a single chord spanning the same arc of 90 degrees or less of a circle and ending at the circle's nadir.

With respect to the diagram Galileo had included in the letter (see Fig. 3.1), the law of chords thus states that motion over the chords BA, CA, DA EA, FA, and SA is made in the same time, and the law of the broken chord states that less time is needed to traverse the conjugate chord SIA than the single chord SA spanning the same arc. Galileo introduced these two propositions as "no less unthinkable than the other," i.e., as no less conceivable than the isochronism of the pendulum,

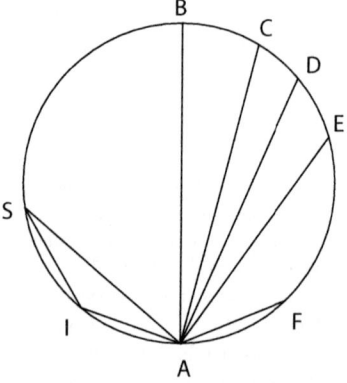

Fig. 3.1 Second diagram in Galileo's letter to Guidobaldo del Monte sent on 29 November. Cf. EN. X, pp. 97–100, letter 88

anticipating that Guidobaldo del Monte's earlier objection to the isochronism of the pendulum could, in a similar fashion, also be applied to the law of chords and the law of the broken chord.

Aside from remarking that he had demonstrated everything thus far "without transgressing the terms of mechanics" ("senza trasgredire i termini mecanici"), Galileo did not provide the slightest clue as to how he had proceeded in proving both propositions. As concerns the law of chords, it is generally agreed, based upon Galileo's remark, that he must have been alluding to a proof along the line of argument of the second of the total three proofs he provided for the law of chords in the *Discorsi*, the so-called ex mechanicis proof. With respect to the proof of the law of the broken chord that Galileo was alluding to in the letter, no such agreement exists on the other hand. Indeed, no proof for the proposition exists that in a similarly apparent fashion corresponds to Galileo's attribution as proven "without transgressing the terms of mechanics." In consequence, it has been assumed by a number of authors that the proof that Galileo was referring to in the letter has not come down to us.[20] Whereas the first assumption, namely, that Galileo was referring to the ex mechanicis proof of the law of chords, is essentially correct, it will be shown that the proof of the law of the broken chord which Galileo was alluding to in the letter is actually preserved in the *Notes on Motion* and that this proof furthermore corresponds to the proof which he provided for the proposition some 35 years later in the *Discorsi*.

From Galileo's wording, it is rather obvious that in the letter he was mentioning both propositions to Guidobaldo del Monte for the first time. In particular, in view of the fact that the law of chords was conceived by him as adding plausibility to the claim he had already made in his earlier letter—that pendulum motion is isochronous—this suggests that Galileo had conceived of the two propositions and constructed the respective proofs in the limited period between the composition of his earlier and the present letter to Guidobaldo.

As concerns the actual aim of Galileo's research agenda, the clause by which he ended this part of the letter is most revealing:

> ... but I cannot finish demonstrating how the arcs SIA and IA have been passed through in equal times and this is what I am searching for.

Here it becomes finally manifest that Galileo's considerations regarding heavy objects rolling down inclined planes were related to and directly aimed at demonstrating properties of the swinging motion of pendulums.

---

[20]There has been a tendency to assume that the proof of the law of the broken chord alluded to, just like the ex mechanicis proof of the law of chords, must have been based on Galileo's principle of inclined plane dynamics. Damerow et al. (2001, 58) have suggested that Galileo's remark should be understood as owing to a rhetoric strategy adopted to appease Guidobaldo del Monte, who was "skeptical with regard to studies involving motion" and who would have been reassured if "in spite of the novelty of the subject for traditional mechanics, he [Galileo] is still adhering to the principles of this mechanics." The proof of the law of the broken chord Galileo mentioned in the letter, as will be demonstrated here, was indeed entirely kinematical.

Galileo's assumption that swinging and rolling motion were related and that this relation could be exploited theoretically is by no means obvious.[21] Indeed, objects falling along inclined planes are supported from below and move over straight paths, whereas pendulum bobs are supported from above and move over continuously curved paths. Other than the fact that the two motions are anachronistically speaking brought about solely by gravitational force, i.e., that they proceed naturally, according to Galileo's understanding, there seems to be no communality among them. To expound the research agenda on swinging and rolling Galileo was following in 1602, and to explain how it had been engendered, is the task of the following chapters. Yet even before this has been accomplished, it is entirely clear from the letter that in the fall of 1602 Galileo was deeply immersed in the challenging problem of relating swinging and rolling. Although he said nothing more in this respect in the letter, he had said just enough to allow Guidobaldo and—due to the fortunate circumstance that this particular letter has survived—also the modern readers a cursory glimpse at the ambitious and challenging problem he was pursuing at the time.

The next and last paragraph, even though not directly related to Galileo's research agenda, still conveys information of interest for the ensuing discussion. In this paragraph, Galileo asked Guidobaldo del Monte to inform Lord Francesco that an experiment concerning the force of percussion had occurred to him and that he would personally inform Francesco about it in another letter. Galileo does not reveal any details of the experiment he had conceived, and the announced letter to Lord Francesco is not extant. It cannot even be made out unambiguously from Galileo's wording whether he had merely come up with the idea of how such an experiment could be designed or whether he had already conducted it. Whatever the case, Galileo seems not to have had any doubts about the feasibility of the experiment.[22]

All that is known in this respect is that Galileo conducted at least two experiments concerning the force of percussion. In one of these experiments, which is described in the *Discorsi*, Galileo basically attempted to measure the impact of a water jet by means of a balance.[23] Information about a second experiment concerning the force

---

[21] In a draft of a letter composed after the publication of the *Discorsi*, Galileo pointed out to Giovanni Battista Baliani, who likewise had assumed a relation between pendulum motion and motion on inclined planes and founded his own approach at a new science on that assumption, that the latter had proceeded haphazardly and had not appropriately accounted for the essential differences between the two types of motions. Cf. Chap. 9.

[22] Settle (1996, 18) has tentatively related the impact experiment mentioned in the letter to pendulum motion when he speculates that a "pendulum of sufficient length, with a standard weight, and swinging down from a fixed height, would deliver a standard percussive blow." There is, however, no concrete evidence that the experiment would have encompassed a pendulum.

[23] Some of the manuscripts of *Le Meccaniche* contain a chapter on the "force of percussion," which demonstrates that Galileo had turned his attention to the problem of impact by at least 1600, if not earlier. Galileo provided an account of the impact balance experiment in a short tract entitled "Della Forza della Percossa," which was appended to the *Discorsi* as an added day. See EN. VIII, 319–346. See also Favaro's *Avvertimento* to the *Discorsi*, ibid., 11–38. According to Settle (1996, 19), this experiment "seems to have taken place in Padua sometime between 1605 and 1610"; Drake

of impact has been handed down to us by Torricelli. In his recollections, Torricelli, who had become Galileo's disciple only a few months prior to the latter's death, describes an experiment which Galileo had told him he had conducted during his time in Padua. In this experiment, the amplitudes through which bow strings of different strength bows could be deviated from their equilibrium position by the impact of a falling body were taken as a measure for the force of percussion.[24]

Should Galileo, as appears to be rather likely, in fact have been alluding to one of these two experiments in the letter to Guidobaldo del Monte, this would underline that at the time of writing he was perceiving of conceptually complex experiments involving rather intricate setups and that he was, moreover, capable of carrying out such experiments, which may have required a specialized space such as a separate workshop.[25] As will be demonstrated here, at about the time of the letter to Guidobaldo del Monte, Galileo had indeed conducted a complex experiment in which he compared rolling along an inclined plane to the swinging of a pendulum, which in view of the length of the plane used, indeed would have called for such a special space.

Before the usual florid close, Galileo makes one last remark of a methodological nature. It regards the transition from geometry, a certain science, to a science dealing with matter, to which all kinds of contingencies are attached. His remark had apparently been triggered by a question on that subject which Guidobaldo had posed in the earlier letter and which is situated in the context of the contemporary debates regarding the applicability of mathematics to the physical world.[26] We can assume that when Galileo conceded to Guidobaldo that "when we begin to deal with matter, because of its contingency the propositions abstractly considered by the geometrician begin to change," this is meant to include matter in motion and thus in particular applies to exactly what Galileo was doing at the time.[27] Thus, it appears that Galileo is somewhat deceitful toward his friend and patron here. While claiming that a mathematical science of the motion of physical bodies is impossible

---

(1978, 126) dates it to 1608. Neither provides convincing evidence for their respective claims. For a more recent account, see Salvia (2014).

[24] See EN. II,191–192. Torricelli's description of the experiment is comprised in his second *Lezione Accademicae*. Drake (1978, 72) speculates that it may have been precisely this experiment that Galileo was alluding to in his letter to Guidobaldo del Monte in 1602.

[25] Damerow et al. (2001, 49) have argued that during his time in Padua, Galileo "through Guidobaldo del Monte ... came into close contact with the experimental techniques and the research interests of a leading engineer-scientist. . . ." This "practical turn," as the authors term it, included the establishment of a workshop in Galileo's house that offered facilities for the production of instruments but allegedly also for experimentation.

[26] For a discussion on the status of mathematics as a middle science in general, see Mancosu (1992), and in Galileo's work, in particular, see Biener (2004). See also Van Dyck (2006). For Galileo's defense of the application of mathematical reasoning in physics in his *Dialogue*, see Marshall (2013).

[27] Damerow et al. (2001) have likewise interpreted this remark as regarding matter in motion in particular.

("delle quali così perturbate siccome non si può assegnare certa scienza"), he is, should the interpretation presented in this book be correct, already heavily engaged in establishing what he took to be a new and certain science of motion.

Summing up: The letter discloses that by the end of the year 1602, Galileo was trying to find a demonstration for the alleged isochronism of the pendulum. Furthermore, it shows that he was attempting to construct such a demonstration by relating pendulum motion to motion on inclined planes and, last but not least, that he had already established at least two propositions concerning the latter type of motion and constructed proofs for them.

Galileo does not make explicit which conceptualization of natural downward motion of heavy objects he is adhering to, i.e., he does not explicitly state whether he assumes such motions to be characteristically uniform as he had still done in the *De Motu Antiquiora* or whether he was already conceiving of such motions as characteristically accelerated. As will be demonstrated here, the latter is the case, and Galileo would likely have imparted his position in this respect to Guidobaldo earlier. The letter, moreover, does not provide any more detailed information of how Galileo had gone about relating swinging and rolling, nor how he intended to proceed with his program, which he explicitly alleged had not yet been ultimately successful. Thus, if the letter had been the only evidence, we would have been doomed to speculate and could never have known for sure. Fortunately, however, not only the letter has survived to testify to the research agenda that Galileo was following at the time. As will be shown, Galileo's engagement with the challenging problem of the relation between swinging and rolling has left ample trace in the *Notes on Motion*. To reveal and interpret these traces and to show how this research eventually resulted in the establishment of the new science of motion are what this book sets out to do. I will begin in the succeeding chapter with the reconstruction and interpretation of a landmark experiment concerning the relation between swinging and rolling, which as it turns out, despite not communicating about it, Galileo must have already conducted by the time he wrote to Guidobaldo del Monte.

References

Biener, Z. (2004). Galileo's first new science: The science of matter. *Perspectives on Science, 12*(3), 262–287.

Büttner, J. (2008). The pendulum as a challenging object in early-modern mechanics. In R. Laird & S. Roux (Eds.), *Mechanics and natural philosophy before the scientific revolution* (pp. 223–237). Dordrecht: Springer

Damerow, P., Renn, J., & Rieger, S. (2001). Hunting the white elephant: When and how did Galileo discover the law of fall? In J. Renn (Ed.), *Galileo in context* (pp. 29–150). Cambridge: Cambridge University Press.

Drake, S. (1970). Renaissance music and experimental science. *Journal of the History of Ideas, 31*, 483–500.

Drake, S. (1978). *Galileo at work: His scientific biography*. Chicago: University of Chicago Press.

Frank, M. (2011). *Guidobaldo dal monte's mechanics in context*. Ph.D. thesis, University of Pisa.

References

Galilei, G. (1974). *Two new sciences*. Madison: The University of Wisconsin Press.

Galluzzi, P. (1979). *Momento*. Rome: Ateneo e Bizzarri.

Gamba, E., Bertoloni Meli, D., & Becchi, A. (2013). *Guidobaldo del Monte (1545–1607): Theory and practice of the mathematical disciplines from Urbino to Europe*. Max Planck research library for the history and development of knowledge sources, 868 Ed. Open Access, Berlin

Humphreys, W. (1967). Inclined planes an attempt at reconstructing Galileo's discovery of the law of squares. *British Journal for the History of Science, 3*(11), 225–244.

Lima, F.M.S., & Arun, P. (2006). An accurate formula for the period of a simple pendulum oscillating beyond the small angle regime. *American Journal of Physics, 74*, 892–895.

MacLachlan, J. (1976). Galileo's experiments with pendulums: Real and imaginary. *Annals of Science, 33*, 173–185.

Mancosu, P. (1992). Aristotelian logic and Euclidean mathematics: Seventeenth-century developments of the quaestio de certitudine mathematicarum. *Studies in History and Philosophy of Science Part A, 23*(2), 241–265.

Marshall, D.B. (2013). Galileo's defense of the application of geometry to physics in the Dialogue. *Studies in History and Philosophy of Science Part A, 44*(2), 178–187.

Naylor, R.H. (1974). Galileo's simple pendulum. *Physics, 16*, 23–46.

Naylor, R.H. (1976). Galileo: Real experiment and didactic demonstration. *ISIS, 67*, 398–419.

Nelson, R.A. (1986). The pendulum—Rich physics from a simple system. *American Journal of Physics, 54*, 112–121.

Palmieri, P. (2009). A phenomenology of Galileo's experiments with pendulums. *British Journal for the History of Science, 42*(4), 479–513.

Salvia, S. (2014). Galileo's machine: Late notes on free fall, projectile motion, and the force of percussion (ca. 1638–1639). *Physics in Perspective, 16*(4), 440–460.

Settle, T.B. (1966). *Galilean science: Essays in the mechanics and dynamics of the Discorsi*. Ph.D. thesis, Cornell University.

Settle, T.B. (1996). *Galileo's experimental research* (Preprint 52). Berlin: Max Planck Institute for the History of Science.

Van Dyck, M. (2006). *An archaeology of Galileo's science of motion*. Ph.D. thesis, Ghent University.

Viviani, V. (1890–1909). Racconto istorico di Vincenzo Viviani. In A. Favaro (Ed.), *Le opere di Galileo Galilei* (Vol. XIX, pp. 597–632). Florence: Tip. di G. Barbèra.

Chapter 4
Sparking the Investigation of Naturally Accelerated Motion: The Pendulum Plane Experiment

The search for the traces of the challenging research agenda—which according to his own account in the letter to Guidobaldo, Galileo was pursuing around the end of the year 1602—and thus more generally the quest for the origin of his new science of motion commence here with the reconstruction and interpretation of an experiment. As a matter of fact, it was not only Guidobaldo del Monte who, at that time, stimulated by Galileo's claim concerning the isochronism of the pendulum, had approached the problem of pendulum motion empirically. The *Notes on Motion* indeed comprises the full record of an experiment, in all likelihood conducted before the letter to Guidobaldo, though Galileo did not mention it there. The experiment involved the swinging of a pendulum and the rolling of a heavy object down along an inclined plane and indeed addressed exactly the challenging problem of the relation between swinging and rolling Galileo had hinted at in the letter.

The *pendulum plane experiment*, as it will henceforth be referred to, has almost totally been disregarded in the literature until now. In view of the attention that has been paid to Galileo's experiments including, in particular, those related to pendulum motion, this is rather surprising.[1] It is even more surprising in view of the fact that in the case of the pendulum plane experiment, the *Notes on Motion* actually contains the record of the experiment, which bequeaths far more concrete information about Galileo's experimentation with pendulums than the published sources such as the *Discorsi*.[2] It is finally most surprising once it is acknowledged

---

[1] Galileo's allusions to experiments involving pendulum motion in his published works have been scrutinized to the last detail. See, for instance, MacLachlan (1976). Drake (1975a), by changing the translation of merely a single word, for instance, revised his own earlier interpretation of a passage in *Discorsi* from a description of a thought experiment into the accurate description of an experiment Galileo actually conducted.

[2] Koyré prominently and influentially promoted the opinion that Galileo never conducted experiments with pendulums, at least not in the way he described them in the *Dialogue* and the *Discorsi*. Cf. in particular Koyré (1966). The following quote is exemplary: "It is obvious that the Galilean experiments are completely worthless: the very perfection of their results is a rigorous proof of

that the experiment is in fact conceptually one of the most rich and, at the same time, the best-documented of all of Galileo's experiments.[3]

Most importantly, however, the experiment grants us insight into the very origin of Galileo's pondering on a relation between swinging and rolling and thus takes us to the very origin of his new science of motion. When Galileo initially designed the experiment, he speculated that a particularly simple relation might hold between pendulum motion and motion along inclined planes. This initial understanding was soon to be replaced by a more elaborate conceptualization of the relation between swinging and rolling, and it is this already revised conceptualization that is lurking between the lines of the 1602 letter to Guidobaldo del Monte in which Galileo but cursorily outlined the underpinnings of his approach.

Correspondingly, the experiment involves two distinct and clearly distinguishable stages of work. The first stage comprises the design and execution of the experiment as well as Galileo's initial evaluation of the measurement data based on his particularly simple assumption concerning the relation between swinging and rolling. It is this first stage of work on the experiment, which is reconstructed in the present chapter. After substantial further inquiry based on his new conceptualization, triggered not least by the failure of the experimental results to comply with his initial expectations, Galileo returned to his original data and reevaluated it. In this second stage of evaluation, instead of comparing the motion of the pendulum to the motion along a single inclined plane spanning the arc of pendulum swing as he had done in the first stage, Galileo considered the relation of pendulum motion to motion along polygonal paths made up of adjacent inclined planes, inscribed into the arc of pendulum swing.

In the succeeding chapter, Galileo's considerations regarding motion along series of adjacent inclined planes inscribed into an arc, which provided the prerequisite for his revaluation of the experimental data, will be discussed. The reevaluation of the experimental findings on the basis of his new conceptualization of the problem

---

their incorrection. (Koyré 1968, 94)" The last decades of the twentieth century saw an experimental turn in Galileo scholarship sparked not least by an influential article by Drake (1975a). Many of the experiments described by Galileo have since been recreated, and it has become commonly accepted that most of them were indeed carried out by Galileo in the way he described. A survey of Galileo's experiments with pendulums is given in Palmieri (2009). Segre (1980) outlines the historically changing views on Galileo's experimental work.

[3]Drake (1987), later republished in Drake (1990), discusses some of the entries in the *Notes on Motion* related to the pendulum plane experiment. Like much of Drake's later work, the article strikes as being rather imaginative and is in parts notoriously difficult to follow. As David K. Hill put it: "Drake's analysis suffers from specific deficiencies to numerous to list. Its general defect is a Byzantine complexity which anyone familiar with the focused simplicity of Galileo's analytical and empirical procedures will find astonishing. (Hill 1994)" Hill himself has comprehensively discussed the experiment and my interpretation owes to Hill's and corresponds to it in several main points. However, I share neither of its main conclusions, which Hill summarized as follows: "The analysis also demonstrates that Galileo was well aware of the non-isochronism of the pendulum (despite his published claims asserting the opposite). He was also aware of rolling inertia, and had a close estimate of its effects. (Hill 1994, 515)"

in a second stage of work is reconstructed and discussed in Chap. 6. Some of the assertions made tentatively in the present chapter will be confirmed only there, and the crucial role of the experiment for the emergence and early development of Galileo's new science of motion will be discussed.

4.1 Experimenting with Swinging and Rolling

Manuscript evidence It is a distinctive feature of the entries in the *Notes on Motion* pertaining to the pendulum plane experiment that they contain numbers representing the dimensions of the experimental setup used or the time intervals measured. Based on the unique nature of these numbers, the pages containing entries in which experimental data was recorded and evaluated can be easily identified.

The record of the pendulum plane experiment and Galileo's considerations related to it are distributed over three folios of the *Notes on Motion* (see Fig. 4.1). The bulk of these entries, including, in particular, the complete record of Galileo's actual measurements, are preserved on folio 189 verso. Notably, all entries relating to the first stage of the experiment, which will be discussed in the present chapter, can be found on this page, and thus Galileo's two-stage proceeding is indeed reflected in the distribution of the relevant material over different folios of the manuscript. Results from this folio are partly restated and elaborated on three further pages, pages 189 recto, 90 verso, and 115 verso. Results Galileo obtained in calculations on folio page 90 verso were copied back to folio page 189 recto.[4]

What stands out is that the recto sides of folio 115 and folio 90, although devoted to an entirely different subject, are likewise closely related with respect to their content. They, in fact, bear two of the three identifiable examples of a scalar addition of impetus for projectile motion, which have been preserved in the manuscript.[5] These folios must hence pertain to Galileo's earliest attempt to establish a theory of projectile motion based on the assumption of natural acceleration. The distribution of content over the folios in question implies that this attempt dates to roughly the same time as the experiment.[6] At some later stage Galileo abandoned the idea that

---

[4] A close relationship between the cut folio pasted onto folio 90 on the one hand and folios 115 and 189 on the other is further corroborated by an additional observation made from the original manuscript. Both folio 189 and folio 115 are folded in the middle. The cut folio pasted onto folio 90 bears the same fold. As the folio it was pasted onto is itself not folded, the fold must have been there before the cut piece was pasted, suggesting that it was kept folded together with folios 189 and 115.

[5] For Galileo's attempt to add the impetus in a scalar fashion in the case of projectile motion, see Renn in Damerow et al. (2004, 223–226) and Caverni (1972, Vol. IV, 540).

[6] As the folio on which Galileo had originally written down his considerations regarding the pendulum plane experiment was subsequently cut and pasted onto folio 90 to make more room for Galileo's notes concerning projectile motion, this must thus slightly antedate the jotting down of the latter considerations. Moreover, as detailed below, the pendulum plane experiment

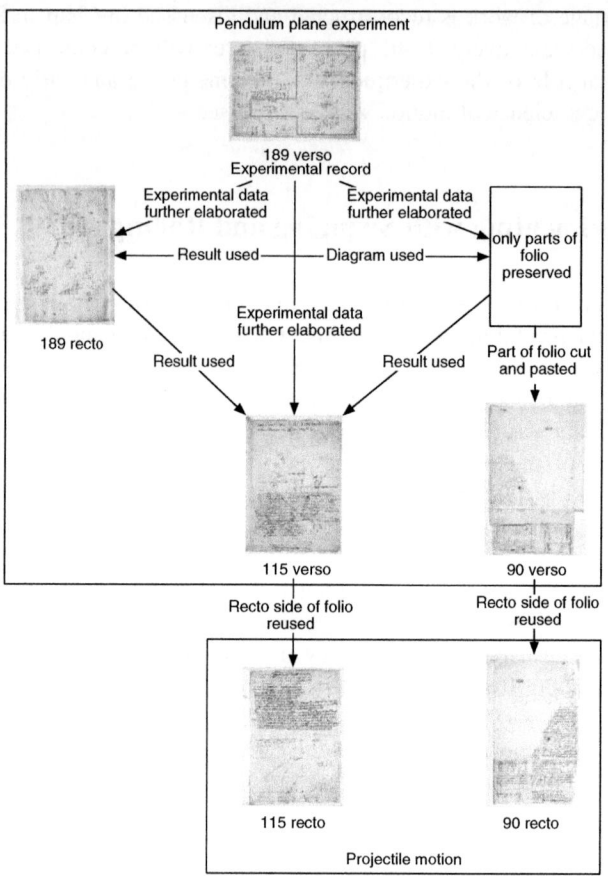

Fig. 4.1 Schematic representation of the relationships among the folios bearing entries pertaining to the pendulum plane experiment

the impetus of the horizontal and the vertical part of projectile motion could be added in a scalar fashion in favor of a semi-vectorial method of adding the impetus of the different components of the motions. When exactly this happened is, however, not known. The insinuated temporal proximity of the second stage of the evaluation of the pendulum plane experiment and the early stage of Galileo's considerations regarding projectile motion thus offers no decisive clue for dating the experiment.[7]

---

remained inconclusive, explaining Galileo's readiness to cut and thus, more or less, obliterate his corresponding notes. The verso side of the pasted part bears the number 90b in pencil.

[7]Evidence will be provided that dates the pendulum plane experiment to 1602, which suggests that Galileo drafted his first results concerning projectile motion at around this time, in contrast to the prevalent opinion, dating the onset of his considerations concerning projectile motion based on assuming natural acceleration to much later. Cf. for instance, Drake (1978).

4.1 Experimenting with Swinging and Rolling

Reconstruction of the design, execution and first stage of evaluation The experimental record preserved on folio page 189 verso reveals that Galileo timed two distinct physical processes: the swinging of a pendulum and the rolling of a heavy object down a long, rather slightly inclined plane.[8] A detailed reconstruction and interpretation of the content of the folio page (see the Appendix in Chap. 13), furthermore, reveal that Galileo, by means of theoretical calculations inferred from the time measured for motion along the long, gently inclined plane, the time it would take a body to roll down along another inclined plane related in a characteristic manner to the pendulum used in the experiment. The inclined plane for which Galileo inferred the time of descent did span the arc described by the pendulum bob in its motion from one turning point to the lowest, the equilibrium position of the pendulum as a chord. This configuration, i.e., the arc of pendulum swing in which the inclined plane is inscribed as a chord, was sketched by Galileo in a diagram close to the top of the page (see Fig. 4.2).

For the sake of brevity, the motion of the pendulum from one of the turning points to the lowest, the equilibrium position, will henceforth be referred to as the *quarter swing* and the path traversed by the bob in this motion as the *pendulum arc*. The motion of the heavy body falling along the chord spanning this arc will be referred to as the *single chord motion* and the chord itself as the *pendulum chord* (compare Fig. 4.2).

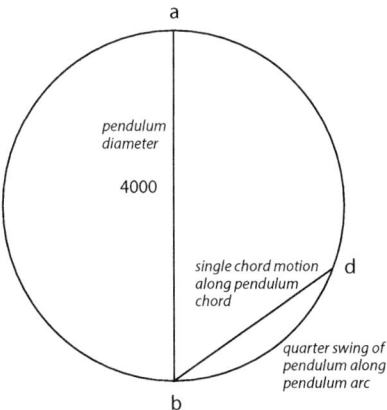

Fig. 4.2 Redrawing of Galileo's schematic diagram on folio page 189 verso. The terminology I have used in the text has been added for clarification

---

[8]The entries on folio page 189 verso discussed in the present chapter were written on the page first. At some later point Galileo reused this page for noting considerations unrelated to the experiment. These entries are written around the initial entries regarding the experiment. That Galileo's considerations pertaining to the second stage of evaluation of the experiment then had to be noted on fresh folios suggests that the second stage of evaluation succeeded the first, only after some temporal delay.

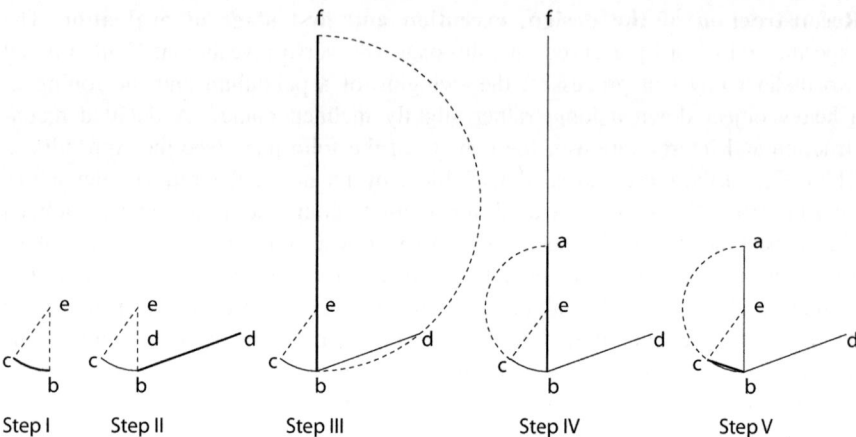

Fig. 4.3 Schematic representation of Galileo's measurements and calculations in the pendulum plane experiment. The paths of those motions whose times are measured or theoretically inferred by Galileo in each step are rendered as bold lines. The inferred time of motion along the chord *cb* in step V is compared to the time of the quarter pendulum swing measured in step I

The experiment was designed so as to compare the time of motion of the quarter swing with that of the motion along the pendulum chord. Timing the swinging of a pendulum evidently did not present Galileo with a major problem. With the help of some kind of timing device in which some measurable quantity was proportional to the time elapsed, Galileo measured the time for eight full swings of a pendulum whose length was slightly less than 2 m.[9] From this, he calculated, by simple division by 32, the time for a quarter swing of the pendulum, which amounted to 62 in the time unit imparted by the time measuring procedure Galileo had employed (step I in Fig. 4.3).[10]

---

[9] The experimental record on page 189 verso makes no explicit mention of the length unit used to measure the lengths of the pendulum and of the inclined plane. The length given in the text is based on the assumption that Galileo specified the lengths in *punti*, with the *punto* measuring approximately 0.96 mm. This, first of all, results in very plausible dimensions for the pendulum and the plane in the experiment. That the unit in which the dimensions of the experiment were specified is indeed the punto is vindicated by an entry on folio 115 verso, where a length derived from the dimension of the experiment is specified as measured in "p[unti]." For the *punto*, see the discussion of folio 166 recto in the Appendix in Chap. 13.

[10] As detailed in the Appendix 13 in Chap. 13, Galileo determined the overall time of motion measured for the pendulum swing by adding up a column of smaller numbers. Based on these numbers, Drake believed he could infer the flow rate of the water clock he supposed Galileo had used. Given how underdetermined the problem is, this was quite plainly an exercise in overinterpretation. Whereas Drake held that the individual numbers represented individual measurements of quantities of water flowed out, Hill (1994) assumed that Galileo only measured once and that the different numbers represent the "various small weights" that he had used to counterbalance the water that had flowed out. Drake (1987), moreover, sought to identify a schematic drawing of a water clock in the *Notes on Motion*. However, the diagram on folio page

4.1 Experimenting with Swinging and Rolling

Next, Galileo needed to determine the time consumed by naturally accelerated motion along the pendulum chord. An inclined plane inscribed as a chord into the pendulum arc would, however, have been relatively short and thus have been traversed in a rather short time. Measurement of such a small time interval would inevitably have resulted in a rather large relative error of measurement, taking into account an unavoidable absolute error of measurement depending on the timing procedure. Instead of timing the naturally accelerated motion on such a short plane, Galileo thus timed the motion of fall along a gently inclined plane of considerable length, namely, about 6.5 m. It goes without saying that, in order to allow for comparability of the measurements, the same timing device must have been used in both cases. For motion along an inclined plane of roughly 6.5 m in length and 8 degrees of inclination, Galileo's measurements yielded a time of descent of 280 time units (step II in Fig. 4.3).

Choosing such a long and gently inclined plane to refine measurement had the disadvantage, however, that Galileo was no longer able to compare his time measurements for pendulum motion and the motion on the inclined plane directly. Instead, he had to derive the time of motion along the appropriate pendulum chord by a theoretical argument. The way in which Galileo accomplished this is schematically depicted in Fig. 4.3.

In a first step, Galileo calculated the length of the vertical distance that would be traversed in free fall in the same time as the long inclined plane had been in his experiment. According to the law of chords—which, as we have seen in the preceding chapter, Galileo was not only aware of but had already proved when he wrote his letter to Guidobaldo del Monte in 1602—the length of this vertical was given by the vertical diameter of a circle in which the long inclined plane of his experiment could be inscribed as a chord ending at the lowest point (step III in Fig. 4.3). Next, from the time of motion through the long vertical, Galileo calculated the time of fall through a shorter vertical distance by applying the law of fall, whose length was given as the diameter of a circle described by the bob. This distance, equal to twice the pendulum's length, will be referred to as the *pendulum diameter* (step IV in Fig. 4.3).

The time of fall through pendulum diameter thus inferred was, by virtue of the law of chords, in turn identical to the time of motion along any chord that could be inscribed in the arc described by the pendulum, ending at its lowest point, that is, also the chord spanning the arc of the pendulum swing (step V in Fig. 4.3). This was the time that Galileo had sought to infer in order to compare it to the time he had measured for the quarter swing of his pendulum.[11]

---

107 verso he was referring to has since been identified by Damerow et al. (2001) as an attempt to construct the curve of a hanging chain.

[11] Step V remains implicit in Galileo's entries on page 189 verso where Galileo merely calculated the time of fall through the pendulum diameter, which, according to the law of chords, is also characteristic for motion along the pendulum chord. That Galileo perceived of the comparison as a comparison between two concrete motions connecting the same start and endpoints, the quarter swing of the pendulum and the motion along the corresponding pendulum chord, is suggested by his sketches of the experimental situation on 189 verso.

In the manner described above, from the results of his time measurement of motion on the long, gently inclined plane, Galileo inferred and calculated the time of motion along the pendulum chord to 80 2/3 of his experimental time units, which he compared to the time of 62 time units he had measured or rather inferred from his measurements for the quarter swing of the pendulum.

As will be detailed below, with all likelihood, Galileo expected these times to be equal and had in fact conceived and designed the experiment to test this hypothesis. Not only did the times differ by about 30 %, in addition, the time of the quarter swing, despite being made over the longer path, surprisingly turned out to be shorter than the time of motion along the single chord which Galileo had inferred from his measurement of the time of motion of a ball rolling down along a long, gently inclined plane.

Before discussing the rationale behind the experiment and the nature and origin of the underlying hypothesis, the experiment will first briefly be compared to other experiments by Galileo involving pendulums and/or motion on inclined planes. As it turns out, such comparison provides valuable additional clues for understanding the pendulum plane experiment but also, as will become clear, the famous inclined plane experiment Galileo described in the *Discorsi* for demonstrating the empirical adequacy of the law of fall.

4.2 The Experiment in Context

Galileo's experiments involving pendulums have left ample traces in the *Notes on Motion*, in his correspondence, and, as has already been mentioned, are alluded to in the *Dialogue* and in the *Discorsi*. As this material shows, making a pendulum such as was required for the pendulum plane experiment did not present a particular problem. The record of the experiment contains no information concerning material or the specific setup of the pendulum. It can plausibly be assumed, however, that the pendulum Galileo was using was somewhat similar to the one he described in the *Discorsi* as follows[12]:

> Imagine this page to represent a vertical wall, with a nail driven into it; and from the nail let there be suspended a lead bullet of one or two ounces by means of a fine vertical thread,

---

[12] The Florentine braccia measured between 58 and 59 cm and thus the pendulum described in the *Discorsi* measuring three braccia, about 175 cm, compares well in length to the one used in the pendulum plane experiment, whose length was somewhat less than 2 m. The frequency of such a rather long pendulum is sufficiently low, while at the same time the length of the pendulum is not so long as to become unmanageable.

4.2 The Experiment in Context

> AB, say from four to six feet long [due o tre braccia], on this wall draw a horizontal line DC, at right angles to the vertical thread AB, which hangs about two finger-breadths in front of the wall... (EN VIII, 206, trans. Galilei et al. 1954)

In his allusions to experiments with pendulums, Galileo repeatedly stressed that damping diminishes the arc swung through with every oscillation. This is a rather obvious observation, and we must assume that Galileo was already aware of this fact when he conducted the pendulum plane experiment. Given that amplitude diminishes with every swing, timing a number of successive swings of a pendulum to infer its period, as Galileo had done in the pendulum plane experiment, thus strictly speaking only makes sense if it is assumed that the period is independent of the amplitude, i.e., that pendulums swing isochronously. Thus vice versa, the method of time measurement employed suggests that when the pendulum plane experiment was conducted, Galileo must already have been convinced of the isochronism of the pendulum.

As concerns experiments involving motion along inclined planes, a strong resemblance of the pendulum plane experiment to Galileo's famous inclined plane experiment of the *Discorsi* cannot be missed. After he had introduced and proved the law of fall in the Third Day of the *Discorsi*, Galileo had his mouthpiece Salviati assert that the "Author," i.e., Galileo himself, had repeatedly tested ("farne la prova") that the acceleration of naturally falling bodies indeed followed the proportion that had been given ("accelerazione de i gravi naturalmente descendenti segua nella proporzione sopradetta"), i.e., the law of fall. Galileo described the principal setup of the experiment by which this was achieved in the following way:

> A piece of wooden moulding or scantling, about 12 cubits long, half a cubit wide, and three finger-breadths thick, was taken; on its edge was cut a channel a little more than one finger in breadth; having made this groove very straight, smooth, and polished, and having lined it with parchment, also as smooth and polished as possible, we rolled along it a hard, smooth, and very round bronze ball. Having placed this board in a sloping position, by lifting one end some one or two cubits above the other, we rolled the ball, as I was just saying, along the channel, noting, in a manner presently to be described, the time required to make the descent. We repeated this experiment more than once in order to measure the time with an accuracy such that the deviation between two observations never exceeded one-tenth of a pulse-beat. (EN VIII, 212–213, trans. Galilei et al. 1954)

Then an account of how the experiment was actually carried out follows. By varying the starting point appropriately, the times of descent over different distances along the inclined plane were measured and the ratios between the measured times compared to those the law of fall predicted. According to Galileo this procedure disclosed an exquisite correspondence between measured and predicted ratios.[13]

Not only is this beyond doubt the most famous of Galileo's experiments; it is arguably one of the most well-known experiments in the history of science.

---

[13]The inclined plane experiment has been alluded to in countless studies. A recent discussion is contained in Palmieri (2011), which provides a comprehensive bibliography on the subject.

At the same time, however, it ranks among the most debated.[14] Debates have centered on the question of whether Galileo actually conducted the experiment or whether he had made it up with the aspiration of convincing his readers. Whereas Mach, at the beginning of the twentieth century, portrayed Galileo as the father of experimental science and assumed as a matter of course that the experiment was actually conducted, Koyré somewhat later did not believe that Galileo had carried out the experiment in the way he described it and rejected the idea that if such an experiment were conducted in such a way, the results would be as accurate as Galileo claimed.[15]

The situation changed when in the early 1960s, Tom Settle replicated the experiment, showing not only that it could indeed be performed in the way described by Galileo but, moreover, that the results he had attained indeed compared in accuracy to Galileo's.[16] This did not show that Galileo had carried out the experiment but crucially undermined the premise based upon which Koyré had dismissed the experiment or at least its results as fictitious. Moreover, in wake of Settle's article, evidence for a number of experiments carried out by Galileo was unearthed, which had not been recognized until then.

The inclined plane that Galileo mentioned in the *Discorsi* measured 12 braccia, i.e., somewhat less than 7 m, and thus compares in length to the plane used in the pendulum plane experiment, which measured 6700 punti, i.e., approximately 6.5 m. The inclination of the plane was given as varying "between one and two braccia" in the *Discorsi* which, in light of its 12-braccia length, would correspond to angles between 4.8 and 9.6 degrees, comparing extremely well to the 8 degree inclination used in the pendulum plane experiment.

The time measuring device used in the *Discorsi* experiment was a water clock. The time unit Galileo employed in the pendulum plane experiment likewise, as will be shown, points to the use of a water clock. Furthermore, in the *Discorsi* it is said that the amount of water that flowed out was measured on a delicate balance. If indeed a water clock was used for time measurement in the pendulum plane experiment as well, the weighing described in the *Discorsi* can serve to explain the otherwise somewhat startling summation of the quantities representing time

---

[14]To this Naylor (1976, 153) remarked "[o]ne of the most controversial issues in the history of science has been the question of how far Galileo's achievement in mechanics was dependent on the use of experiment."

[15]Cf. Mach (1897, 122–126). Koyré strongly dismissed the idea that Galileo carried out experiments at all when he wrote that "experiments which Galileo, and others after him, appealed to, ... were not and could never be any more than thought experiments. (Koyré 1978, 37)" In later writings he took a somewhat more modest position.

[16]See Settle (1961). Naylor (1974) as well replicated the experiment. He came to the conclusion that from his "observations it does not appear that the precise ratios are obtainable as suggested by Galileo, particularly in the case of the smaller inclination. (Naylor 1974, 131)" He, in particular, points to the fact that lining the groove with parchment as described by Galileo, makes the outcome worse, which he takes as an indication that Galileo's description of the experiment cannot be taken at face value.

evidenced by the experimental record. What is being added would then be the counterweights that had to be put on, to bring the balance back to equilibrium after some kind of container had been filled with the water that had flown out during the measurement.

These correspondences strongly suggest that the *Discorsi* experiment and the pendulum plane experiment, as regards timing motion along an inclined plane, are, in principle, based on the same setup and that Galileo availed himself of the same hardware in both cases. This can be seen to lend additional support to the claim that the *Discorsi* experiment was actually conducted. It is, moreover, suggested that both experiments were probably not conducted at widely separate points in time, since it seems unlikely that Galileo would or could have kept the large apparatus used in both experiments, with its almost 7-meter-long plane, intact over a very long period. Rather, a temporal proximity between both experiments is suggested. It will be argued that the pendulum plane experiment must have been conducted around 1602. Dating the inclined plane experiment to about this time is in perfect agreement with a more overarching picture of the development of Galileo's new science.[17]

Information available with regard to either one of both experiments can, moreover, potentially be used to fill gaps in the record of the respective other, quite independently of the question of whether the inclined plane experiment of the *Discorsi* was actually conducted or only mentally conceived by Galileo. Should the latter, against all plausibility, indeed be the case, we can still assume that the description Galileo gave in the *Discorsi*, of what in this case would be an imagined experiment, was modeled on the experiment beyond all doubt actually conducted, namely, the pendulum plane experiment.[18] In particular, this suggests that in the pendulum plane experiment, just as described in the *Discorsi*, Galileo used a grooved board for an inclined plane. As argued in Chap. 6, this has a potential bearing on a modern assessment of Galileo's measurement data.

Yet another class of experiments involving naturally accelerated motions along inclined planes has left its traces in the *Notes on Motion*. In these experiments, quite generally, the motion along an inclined plane served to accelerate a ball that was then projected either obliquely or horizontally to observe features of the

---

[17] Whereas Galileo may have become aware of the quadratic relation between spaces traversed and times elapsed in free fall as a heuristic assumption as early as 1592 (cf. Damerow et al. 2001), he only started to avail himself of the law as a fundamental assumption in a systematic study in his research on swinging and rolling in 1602. This would thus seem to have been the right time to seek empirical confirmation of the underlying assumption by means of an experiment.

[18] Drake (1975b) thought he had identified direct evidence of an experiment involving naturally accelerated motion along an inclined plane in numbers recorded on 107 verso, which he took to be recordings of time measurements of motions along an inclined plane. Yet it has since been shown that these numbers are the result of Galileo's comparison of a hanging chain with a parabola and that, moreover, the diagram on that page, which Drake took to be a schematic representation of the water clock used in the experiment, is in fact Galileo's construction of the center of gravity of a constellation of weights suspended on a string or rope. See Damerow et al. (2001), in particular, footnote 119.

resulting trajectory. In these experiments, motion along the inclined plane was less the object of study than a means for Galileo to create a controlled projection. For all of these experiments save one, the attempts to unambiguously reconstruct the dimensions and inclinations of the inclined planes used have thus far failed. The one case in which the record of the experiment that has survived allows for a rather definite inference of the length and inclination of the plane used—the case of best documented and understood of these experiments, the record of which has survived on 116 verso—again shows a remarkable agreement to the pendulum plane experiment, and it is once more suggested that the very same inclined plane described in the *Discorsi*, and made use of in the pendulum plane experiment, was resorted to here as well.[19]

It is thus suggested that Galileo was using one and the same extraordinarily long inclined plane, specifically produced for this purpose, in a number of different experiments, likely all conducted within a rather confined period of time around 1602.[20] At that time a workshop formed part of Galileo's household, which would have provided a suitable space for experimentation with setups that, at least in some cases, reached considerable dimensions.[21]

4.3 The Single Chord Hypothesis as Starting Point

In the pendulum plane experiment Galileo compared two different motions, the quarter swing of a pendulum and the rolling of a heavy body along an inclined plane that could be inscribed into the arc of the quarter pendulum swing as a chord, where the lowest point of the pendulum arc also marked the end of the inclined plane. More specifically, he compared these two motions with re-

---

[19] For a comprehensive account of the horizontal projection experiment documented on folio 116 verso, see Hahn (2002). In the experimental record, the length of the plane used is not given. Yet the maximum vertical height above the point of projection from which the motion started is given as 1000 punti. If it is assumed that Galileo used a plane of 6700 punti length as in the pendulum plane experiment, this implies an inclination of 8.6 degrees, again in astounding accordance with the 8 degree inclination used in the pendulum plane experiment. A plausible scenario would be that Galileo simply started from the exact same setup used in the pendulum plane experiment and slightly lifted the end of his plane until as desirable for the projection experiment, the vertical height amounted to the round figure of 1000 punti. The assumption that a plane comparable in length and inclination to the one of the pendulum plane experiment was likewise used in the oblique projection documented on folios 114 verso and 81 recto may provide a way to promote the deadlocked discussion concerning these experiments.

[20] Vergara Caffarelli (2009, 121) has rightly emphasized that fabricating this plane would have been a difficult job that would have required a good craftsman and would certainly have been rather expensive.

[21] Apparently in late 1602, Galileo moved into a new house in Padua, much more spacious than his prior residence. See Brunelli Bonetti (1943). For the workshop Galileo ran in his house, see Valleriani (2010).

4.3 The Single Chord Hypothesis as Starting Point

spect to the times in which they were completed, times he had measured and inferred from measurement, respectively. It has already been anticipated above that in all likelihood Galileo hypothesized that these two motions would be completed in equal time, an assumption that will be referred to as the *single chord hypothesis*.

The approach Galileo took with the experiment is by no means obvious. After all, two entirely different motions are being compared, a swinging motion supported from above and a rolling motion supported from below, motions which share start and endpoints but otherwise proceed over entirely different paths. Why would Galileo engage in the single chord hypothesis and how can it be argued that the experiment was designed to probe the assumption that the two motions were completed in the same time?

A definitive answer can only be given once Galileo's further pondering on the relation between swinging and rolling has been discussed in Chap. 5 and the second stage of his evaluation of the experimental data has been reconstructed in Chap. 6. However, even if considered out of its wider context, the first stage of the pendulum plane experiment entails strong clues towards the answer. The assumptions concerning the two types of motion which Galileo explicitly or implicitly availed himself of in the design of the experiment and the evaluation of the data, and of which we can thus be sure that he held when the experiment was conceived, in fact directly suggest the single chord hypothesis by abduction.[22]

As argued above, the measuring routine chosen by Galileo to determine the quarter period of the pendulum is expedient only if it is assumed that the period of the pendulum does not depend on the angle of the arc swung through at all. Thus, Galileo was evidently convinced of the isochronism of the pendulum by the time he conceived the experiment. As we have seen, Galileo had communicated his view to Guidobaldo del Monte in 1602, validated by empirical test that pendulums swing isochronously.[23]

Moreover, to infer from the time measured for motion along the long gently inclined plane the hypothetical time of motion along a pendulum chord, Galileo

---

[22] The assumption of the single chord hypothesis is by no means implausible as the example of Marci von Kronland, who in fact gave the hypothesis as a theorem, shows: "Pendulum aequali tempore mouetur per arcum Circuli & chordam eidem subtensam." See http://archimedes.mpiwg-berlin.mpg.de/cgi-bin/toc/toc.cgi?dir=marci_figur_063_la_1648;step=thumb. Accessed 16 Feb 2017. Since Galileo never explicitly stated the single chord hypothesis in his published works, Marci must have arrived at the assumption independently, albeit based on the same conceptual prerequisites.

[23] According to Vincenzo Viviani, Galileo's last and most faithful disciple and his first biographer, Galileo discovered the isochronism of the pendulum as a student while observing the swinging of a lamp in the Cathedral of Pisa. Serious doubt has meanwhile been cast on Viviani's account. Cf., in particular, Ariotti (1971/1972). Segre (1998, 392) rightfully remarks that it "is the excellence of Viviani's essay that makes it so difficult to interpret; even today, historians have difficulties in distinguishing between reality and myth in its pages."

applied the law of fall and in particular also the law of chords.[24] That the latter proposition was known to Galileo and that he had even constructed a proof for it by the end of 1602 is likewise conferred to us by the letter to Guidobaldo.[25]

Both the isochronism of the pendulum and the law of chords define sets of motions as isochronous, i.e., they claim that all motions along different paths obeying a certain geometrical constraint are completed in equal time. According to the isochronism of the pendulum, given a fixed radius or pendulum length, each quarter swing of a pendulum is completed in equal time regardless of the angle of the arc swung through. According to the law of chords, given a circle of fixed radius, motion along any inclined plane that can be inscribed as a chord ending at the lowest point is completed in equal time, regardless of the angle of the arc spanned by the chord. Not only do both propositions make an isochrone statement for classes of motion, but the geometrical constraints defining the paths considered are similar and suggest relating them in the way Galileo does and to observe the case in which both motion situations are defined with respect to one and the same circle, i.e., in which the chords considered can be construed as chords inscribed in the circle bearing the arcs the pendulum is swinging through.

Together, the two assumptions, i.e., the isochronism of the pendulum and the law of chords, immediately imply that the relation between the time of a quarter swing of the pendulum and that of motion along the inclined plane inscribed as a chord neither depends on the angle of the arc the pendulum is swinging through nor on the angle of the arc spanned by the chord serving as an inclined plane. This will, in short, be referred to as *angle independence*. This implication is, however, not tested.[26] Rather it is presupposed in the way Galileo designed the experiment.

Thus the experiment must indeed have been aimed at questioning how concretely the isochronous time of motion to swing along an arc relates to the isochronous

---

[24] As will become clear, Galileo's insight into the law of fall must predate the earliest entries in the *Notes on Motion*. Damerow et al. (2001) have argued that Galileo most likely initially came to accept the law as a heuristic based on an experiment concerning the shape of the projectile trajectory conducted jointly with Guidobaldo del Monte, in 1592.

[25] Galileo's recourse to the law of chords in its theoretical evaluation of the data was strictly speaking not necessary. He could have determined the length of a chord with the same inclination as his experimental plane but inscribed into the arc of pendulum swing geometrically and from his measured time inferred the time of motion along this pendulum chord by application of the law of fall. That he employed the law of chords to gain essentially the same result indicates that he saw no reason to be skeptical of the law of chords and took it for granted by the time the experiment was conducted.

[26] In principle, such a test would have been possible, using the exact same setup and evaluation of the experimental data. Galileo would simply have needed to vary the angle and/or length of the plane or the initial angle of elongation of the pendulum. Apparently, however, he did neither and made just one-time measurement for each of the two types of motion. We will of course never know if the experimental record is complete and it is in fact plausible to assume that Galileo tinkered with the experimental setup before conducting the experimental run, whose results he ended up recording on 189 verso.

4.3 The Single Chord Hypothesis as Starting Point

time of motion to fall along an inclined plane inscribed into this very arc, and we are faced with two alternatives: either the experiment was explorative, i.e., Galileo held no particular expectation concerning the relation of the times he inquired into, or else he held an hypothesis that he sought to test experimentally.[27] In view of the effort Galileo invested in designing, setting up, and conducting the experiment and in further processing the measured data, the first alternative emerges as exceedingly unlikely, in particular in view of the fact that an empirically determined relation could hardly be expected to open a perspective for further exploration. In fact, as will be demonstrated, Galileo was never able to reconcile the result the experiment yielded, which ran counter to his initial hypothesis, with his theoretical speculations regarding the relation of swinging and rolling, despite immense further effort.

It is thus strongly insinuated that Galileo designed the experiment with regard to a particular hypothesis concerning the relation of the times in question. This hypothesis can hardly have been based on a logical inference as not even Galileo's matured conceptualization of the relation between swinging and rolling allowed for such an inference, as we will see. Much more plausibly the hypothesis had been arrived at by abductive inference.[28] The surprising fact that both swinging along any arc of the same circle and falling along any of the chords appropriately inscribed in the same circle are isochronous would more or less be a matter of course if it held that the motion of a pendulum from rest to the lowest point on the arc takes the same time as motion along an inclined plane connecting the start and endpoint of the quarter swing—the single chord hypothesis. Not only was there reason for Galileo to assume that this hypothesis held; moreover, its particular simplicity may have suggested that it should be possible to furnish it with a demonstration and eventually a physical explanation.[29]

The single chord hypothesis explains the similarity between the law of chords and the isochronism of the pendulum in the sense that, if it holds and can be demonstrated, the law of chords is revealed as a necessary consequence of the isochronism of the pendulum and vice versa. To explain this similarity, assuming angle independence of the relation of the times in question would suffice but as an assumption seems not prone to explanation. The single chord hypothesis, on the

---

[27] Drake (1987) referred to the ratio of these periods of the quarter swing and the time to fall through a vertical distance of twice the pendulums length as Galileo's constant.

[28] Peirce, who introduced the concept of abductive inferences into modern logic, illustrated them by the following famous schema: "The surprising fact, C, is observed; But if A were true, C would be a matter of course. Hence, there is reason to suspect that A is true. (Peirce et al. 1978, V, 189)" For abductive reasoning in general, see Aliseda (2006).

[29] Steinle, has introduced the label explorative experiments for experiments which serve first and foremost to gain orientation in a new field where the starting positions are unclear (see Steinle 2005). Even though according to the interpretation provided in the pendulum plane experiment Galileo tested a concrete hypothesis, the label arguably still applies as this hypothesis was not a strict consequence of a well-established theory but rather a heuristic, which, should it be confirmed, would guide further investigation.

other hand, potentially lends itself to an explanation, at the expense, however, of making a stronger claim. Given angle independence alone, the relation between the period of the pendulum and the time of motion along a chord inscribed into the arc of pendulum swings could in particular still vary with the length of the pendulum, i.e., the radius of the circle with regard to which both motions are defined. Yet nothing gives the circle specified by the pendulum chosen for the experiment precedence over any other. Clearly the single chord hypothesis was not conceived by Galileo as restricted to the concrete arbitrarily chosen case observed in the experiment but as being generally valid.

In other words, Galileo held that the quarter period of a pendulum was the same as the time of motion along the corresponding pendulum chord whatever the length of the pendulum, i.e., independent of the size of the circle considered. This will, in short, be referred to as the *size-independence assumption*, embraced by the single chord hypothesis. It is immediately apparent that just as angle independence is implied by and, vice versa, can be interpreted as explaining the similarity between the law of chords and isochronism of the pendulum, so size independence is implied by and can be interpreted as explaining the similarity between the law of fall and law of the pendulum.[30] According to the law of fall, the time to fall along a vertical distance is to the time to fall along another distance in the same ratio as the ratio of the roots of the distance traversed. For the times of motion along vertical distances to always be the same as the quarter periods of pendulums whose pendulum lengths correspond to half the distance, as claimed by the single chord hypothesis, it necessarily needs to hold that the period of a pendulum varies with the root of the pendulum length, i.e., the law of the pendulum.

This again leaves us with two interpretative alternatives. Either Galileo was aware of the law of the pendulum when he came up with the single chord hypothesis and designed the experiment, or not. In the former case, the challenging similarity between swinging and rolling would have been constituted by the similarity of the law of chords and the isochronism of the pendulum, on the one hand, and the law of fall and law of the pendulum, on the other. Angle independence as well as size independence would have been strictly implied and the leap from what was strictly implied to what was assumed in abduction to the single chord hypothesis would have been an extremely small one. In the latter case, abduction would have been based exclusively on the similarity between the law of chords and the isochronism of the pendulum. The leap between what was hypothesized and what was strictly implied

---

[30] By the law of the pendulum I refer to the assertion that the period of a pendulum varies in proportion to the square root of its lengths. From a modern perspective, it needs to be declared explicitly that the pendulum has to swing through the same angular amplitude, as the period, besides on the length of the pendulum, also depends on this angle of swing. For Galileo, who held that pendulum motion is isochronous, i.e., independent of the angle of the arc the pendulum is swinging through, the law of the pendulum held without such restriction.

4.3 The Single Chord Hypothesis as Starting Point

by the underlying assumptions would have been somewhat more sizeable.[31] Once the simple chord hypothesis is assumed, the law of the pendulum is, however, in turn strongly implied.

Galileo does not vary the pendulum's length in the experiment. In fact, the law of the pendulum figures neither in the design of the experiment nor in the theoretical processing of the measurement data. The experiment itself can thus not reveal whether or not Galileo was aware of the law of the pendulum when he conceived the single chord hypothesis and with it the pendulum plane experiment. Yet there are other indications to suggest that he was and that the single chord hypothesis was indeed based on his recognition of the strong and compelling similarity between swinging and rolling, encompassing the similarity of law of the pendulum and the isochronism of the pendulum, on the one hand, and of the law of fall and the law of chords, on the other.[32] Ultimately, it seems this question cannot be decided

---

[31] Should Galileo, against all odds, have been unaware of the law of the pendulum when he first conceived of the single chord hypothesis, the size-independence or scale invariance assumption it entailed would not have been directly implied by the underlying assumptions. That the conceived hypothesis embraced scale invariance in this case may have been the result of an unreflected choice, more or less without alternative. Indeed, based on the mathematical language of proportions, the relation between the quarter period of the pendulum and the time of motion along a corresponding pendulum chord, which can be conceived of as a ratio, cannot depend on the absolute size of the underlying geometrical configuration if it is in proportion to any of the characteristic distances defined by this configuration. Yet this is not necessarily the case, and Galileo was in fact able to deal with ratios which did change as the underlying geometry changed in size while remaining proportionally similar. In the *Discorsi*, Galileo thus indeed showed that the ratio between the moment of heaviness and the moment of resistance in a beam was not always the same for geometrically similar beams (EN VIII, 163–164). He was able to do so by analyzing the ratio of the ratios of the different kinds of moments for two similar cylinders instead of analyzing simply their ratio. Comparably, it was thus, at least in principle, conceivable that the relation of the period of a quarter swing to the time of motion along the corresponding pendulum arc may vary with the absolute size of the system. Galileo only commenced his investigations concerning the strength of materials in the summer of 1607, and it is not clear whether this would have been obvious to him earlier. Indeed, in the *Discorsi*, Galileo has Simplicio express surprise: "[t]his proposition strikes me as both new and surprising: at first glance it is very different from anything which I myself should have guessed: for since these figures are similar in all other respects, I should have certainly thought that the forces [momenti] and the resistances of these cylinders would have borne to each other the same ratio (EN VIII, 164, trans. Galilei et al. 1954, 125)." This may well portray Galileo's position with regard to the question of the relation of the times in question in the pendulum plane experiment. Galileo, albeit at a much later time, indeed explicitly stated that one may "come principio noto" suppose that the relation of the time of motion along arc and along the chord spanning the arc must be similar for similar geometrical constellations. Cf. Chap. 9.

[32] Galileo never made a claim to the discovery of the law of the pendulum. Since Galileo is generally not modest about his achievements, this can be taken as an indication that he did indeed not consider himself to be the person who discovered it. As concerns his published writings, Galileo stated a qualitative pendulum law in the *Dialogue*, i.e., he asserted that the period of a pendulum increased monotonously with its length. Ariotti (1968, 416–417) takes the exposition in the *Dialogue* as an indication that Galileo is "yet unclear on the exact relationship of the length and the period." In the *Discorsi*, Galileo finally introduced the law of the pendulum but gave no proof. Cf. (EN VIII, 139–140). Cf. Ariotti (1971/1972, 351–354). According to Drake (1970, 497–498), Galileo most likely witnessed his father Vincenzo Galilei's musical experiments in 1588–1589.

with absolute certainty based on the available evidence. However, whether Galileo's insight into the law of the pendulum preceded the single chord hypothesis or whether it was promoted by the latter is revealed as being ultimately inconsequential for the further course of events. Even if the law of the pendulum should initially have been implied to Galileo by his commitment to the single chord hypothesis, itself an unconfirmed heuristic, it seems clear that he accepted it as a true physical statement for reasons (certainly not the least empirical ones) other than the law being implied by a conceptualization between swinging and rolling, which did not stand up to experimental scrutiny.

The experiment was thus designed to test a particular conceptualization of the relation between swinging and rolling, which had been kindled by the realization of a challenging similarity between the two types of motion and found its expression in the single chord hypothesis. Yet when the first stage of the evaluation of the experimental data had been completed it had been shown that the time for the quarter swing of the pendulum measured differed from the time of motion along the appropriate inclined plane inferred by means of a theoretical argument from Galileo's measurements, by almost 30%.

As will be seen, Galileo kept and later even reused his measurement data as well as the results he had calculated from it. Hence, he clearly did not attribute the observed discrepancy with regard to his hypothesis to the poor quality of his measurement or to a flaw in his theoretically mediated processing of the data. The experiment had thus disproved the single chord hypothesis, indicating, moreover, a surprising tendency. The time of the quarter swing, despite being made over the long path, had turned out to be shorter than the time of motion

---

Drake claims that "an observer [of these experiments] can hardly escape the phenomena of the pendulum" and thus seems to imply that Galileo in this way first became aware of the law of the pendulum. In his *Methodi vitandorum errorum omnium qui in arte medica contingunt* published in 1602, S. Santorio introduced a device called a pulsilogium in which, by varying the length of a pendulum, the period of this pendulum is synchronized with a patients pulse. The resulting pendulum length was then interpreted as a measure of the pulse rate, which could then easily be compared, for instance, to the pulse rate at another time. Cf. Levett and Agarwal (1979). The pulsilogium does implicitly presuppose isochrone of pendulum motion. To conceive and build it, a qualitative understanding of the relation between length and period of a pendulum arguably suffices. Yet once constructed it obviously allows for qualitative observations that may suggest the law of the pendulum. In fact, in a marginal note in his private copy of the *Discorsi* (Gal. 79), the passage in which the law of the pendulum is introduced is annotated by the remark that "replicate esperienze" led to the law of the pendulum. Santorio and Galileo were part of the same learned circle in Venice. Cf. Recht (1931). Galileo may not necessarily have played a crucial role in the invention of the pulsilogium as commonly assumed, but he would certainly have been familiar with the device in 1602. Cf. Büttner (2008, 227–229). In the *Notes on Motion*, direct evidence of Galileo being aware of the law of the pendulum is provided by his considerations on folio 154 recto, where based on the law of the pendulum Galileo compared the observed time of a long pendulum to the hypothetical time calculated for a pendulum whose radius corresponded to the radius of the earth. Cf. the discussion of this folio in the Appendix in Chap. 13, where it is argued that other content on the page is related to Galileo's construction of the ex mechanicis proof and thus most likely dates to around 1602. It is thus strongly indicated that Galileo was familiar with the law of the pendulum at that time.

along the single chord spanning the arc of the swing. Was this the end of Galileo's aspiration to get to the bottom of the relation between swinging and rolling?

The single chord hypothesis, by which Galileo's assumptions concerning the two types of motion were networked, was after all a heuristic based on an abductive inference. Its empirical rejection did thus not force Galileo to reject any of his underlying assumptions regarding swinging and rolling. Chord and arc are, after all, different paths of motion and Galileo, as will be demonstrated in the succeeding chapters, modified his understanding of the relation between swinging and rolling appropriately.

4.4 Summarizing Conclusion

In this chapter, the so-called pendulum plane experiment has been reconstructed from the record preserved on folio page 189 verso. Based on the record, the setup and execution of the experiment can be reconstructed seamlessly. By means of the experiment, Galileo compared the quarter period of a pendulum to the time of motion along an inclined plane inscribable in the arc of pendulum swing as a chord, ending at the lowest, the equilibrium position. The latter time was not measured by Galileo directly. Rather he had timed motion along a long, gently inclined plane and from this calculated the time he sought by applying the law of fall and the law of chords.

The dimension and inclination of the plane Galileo used in the pendulum plane experiment, as can unambiguously be reconstructed from the record, correspond well to those of the inclined plane he used, according to his famous description in the *Discorsi*, in an experimental confirmation of the law of fall. This lends additional plausibility to the, by now, commonly accepted belief that the experiment recounted in the *Discorsi* had indeed been executed. Even if this should not be the case, it is suggested that the experimental setup Galileo described in the *Discorsi* was modeled on the setup used in the pendulum plane experiment. There is, furthermore, evidence to suggest that the very same inclined plane was likewise used in Galileo's early experiments involving oblique and horizontal projections.

It has been argued that Galileo had designed the experiment to test whether the quarter period of the pendulum was equal to the time of fall along an inclined plane inscribed into the arc of pendulum swing as a chord ending at the lowest point, which he had inferred from his measurement of the time of motion to roll along a long plane by means of a theoretical argument. This assumption has been referred to as the single chord hypothesis. It has been argued that the single chord hypothesis had its origin in an abduction based on the recognition of a challenging similarity between Galileo's assumptions regarding swinging and rolling, constituted by the structural similarity of the law of chords and isochronism of the pendulum and, most likely, also that of the law of fall and the law of the pendulum. That the law of chords, the law of fall, and the isochronism of the pendulum were known to

Galileo is entirely clear as they were invoked, implicitly or explicitly, in the design of the experiment and the further theoretical processing of the measured data. It can plausibly be assumed that Galileo was also aware of the law of the pendulum when the experiment was conceived. Should this not have been the case, the law of the pendulum would have directly been entailed by the single chord hypothesis together with the law of fall. The experiment thus reveals the very motive that turned Galileo's attention to investigating the relation between swinging and rolling in the first place and therewith, as will become clear, the true origin of his new science of motion.

If the single chord hypothesis held, the law of chords and isochronism of the pendulum, as well as the law of fall and the law of the pendulum, would directly imply each other. The particular straightforward nature of the hypothesis must, moreover, have made it seem possible to advance an explanation where the laws governing the swinging of pendulums could be demonstrated from those governing fall along inclined planes and vice versa.

Yet the experiment did not confirm the hypothesis it had been devised to test. Not only did the quarter period of the pendulum differ by about 30 % from the time measured and inferred for fall along a corresponding inclined plane, moreover, the former motion, despite being made over the longer path, turned out to be completed in less time than the latter. As the single chord hypothesis was not a strict logical consequence of Galileo's assumptions regarding the two types of motion but a heuristic assumption based on an abductive inference, its empirical rejection did not force Galileo to renounce any of the underlying assumptions regarding swinging and rolling. Rather Galileo reconceptualized his understanding of the relation between swinging and rolling. As will be demonstrated in the succeeding chapter, Galileo's new conceptualization was based on the idea of exhausting motion along the pendulum's arc by motion along polygonal paths of an increasing number of sides, ever more closely approximating the arc. The initial outcome of the experiment thus prompted Galileo to attempt to, somewhat metaphorically speaking, square motion along the pendulum's arc.

References

Aliseda, A. (2006). *Abductive reasoning: Logical investigations into discovery and explanation* (Synthese library). Dordrecht: Springer Netherlands.

Ariotti, P.E. (1968). Galileo on the isochrony of the pendulum. *ISIS, 59*(4), 414–426.

Ariotti, P.E. (1971/1972). Aspects of the conception and development of the pendulum in the 17th century. *Archive for History of Exact Sciences, 8*, 329–410.

Brunelli Bonetti, B. (1943). Nuove ricerche intorno alla casa abitata da Galileo Galilei in via vignali. In *Pubblicazioni Liviane e Galileiane a Ricordo delle Celebrazioni dell'Anno 1942. Supplemento al vol. LIX delle "Memorie"* (pp. 91–102). Padua: Tip. Penads.

Büttner, J. (2008). The pendulum as a challenging object in early-modern mechanics. In R. Laird & S. Roux (Eds.), *Mechanics and natural philosophy before the scientific revolution* (pp. 223–237). Dordrecht: Springer.

Caverni, R. (1972). *Storia del metodo sperimentale in Italia*. New York: Johnson Reprint.

References

Damerow, P., Freudenthal, G., McLaughlin, P., & Renn, J. (1992). *Exploring the limits of preclassical mechanics*. New York: Springer.
Damerow, P., Renn, J., & Rieger, S. (2001). Hunting the white elephant: When and how did Galileo discover the law of fall? In J. Renn (Ed.), *Galileo in context* (pp. 29–150). Cambridge: Cambridge University Press.
Damerow, P., Freudenthal, G., McLaughlin, P., & Renn, J. (2004). *Exploring the limits of preclassical mechanics*. New York: Springer.
Drake, S. (1970). Renaissance music and experimental science. *Journal of the History of Ideas, 31*, 483–500.
Drake, S. (1975a). New light on a Galilean claim about pendulums. *ISIS, 66*(233), 92–95.
Drake, S. (1975b). The role of music in Galileo's experiments. *Scientific American, 232*(6), 98–104.
Drake, S. (1978). *Galileo at work: His scientific biography*. Chicago: University of Chicago Press.
Drake, S. (1987). Galileo's constant. *Nuncius Ann Storia Sci, 2*(2), 41–54.
Drake, S. (1990). *Galileo: Pioneer scientist*. Toronto: University of Toronto Press.
Galilei, G., Crew, H., & Salvio, A.D. (1954). *Dialogues concerning two new sciences*. New York: Dover Publications.
Hahn, A.J. (2002). The pendulum swings again: A mathematical reassessment of Galileo's experiments with inclined planes. *Archive for History of Exact Sciences, 56*, 339–361.
Hill, D.K. (1994). Pendulums and planes: What Galileo didn't publish. *Nuncius Ann Storia Sci, 2*(9), 499–515.
Koyré, A. (1966). *Etudes Galiléennes*. Paris: Hermann.
Koyré, A. (1968). *Metaphysics and measurement*. London: Chapman & Hall.
Koyré, A. (1978). *Galileo studies* (trans: Mepham J). Harvester: Brighton.
Levett, J., & Agarwal, G. (1979). The first man/machine interaction in medicine: The pulsilogium of Sanctorius. *Medical Instrumentation, 13*(1), 61–63.
Mach, E. (1897). *Die Mechanik und ihre Entwicklung historisch-kritisch dargestellt*. Leipzig: F. A. Brockhaus.
MacLachlan, J. (1976). Galileo's experiments with pendulums: Real and imaginary. *Annals of Science, 33*, 173–185.
Naylor, R.H. (1974). Galileo and the problem of free fall. *The British Journal for the History of Science, 7*(2), 105–134.
Naylor, R.H. (1976). Galileo: The search for the parabolic trajectory. *Annales of Science, 33*, 153–172.
Palmieri, P. (2009). A phenomenology of Galileo's experiments with pendulums. *The British Journal for the History of Science, 42*(4), 479–513.
Palmieri, P. (2011). *A history of Galileo's inclined plane experiment and its philosophical implications*. Lewiston: Edwin Mellen Press.
Peirce, C.S., Hartshorne, C., Weiss, P., & Peirce, C.S. (1978). *Scientific metaphysics* (Collected Papers of Charles Sanders Peirce, 4th edn.). Cambridge: Belknap Press of Harvard University Press.
Recht, E. (1931). Life and work of Sanctorius. *Medical Life, 38*, 729–785.
Segre, M. (1980). The role of experiment in Galileo's physics. *Archive for History of Exact Sciences, 23*(3), 227–252.
Segre, M. (1998). The never-ending Galileo story. In P.K. Machamer (Ed.), *The Cambridge companion to Galileo*. Cambridge/New York: Cambridge University Press.
Settle, T.B. (1961). An experiment in the history of science. *Science, 133*(3445), 19–23.
Steinle, F. (2005). *Explorative Experimente: Ampère, Faraday und die Ursprünge der Elektrodynamik* (Boethius, Vol. 50). Stuttgart: Franz Steiner Verlag.
Valleriani, M. (2010). *Galileo engineer*. (Boston Studies in the Philosophy of Science, Vol. 269). Dordrecht/London/New York: Springer.
Vergara Caffarelli, R. (2009). *Galileo Galilei and motion: A reconstruction of 50 years of experiments and discoveries*. Berlin/New York/Bologna: Springer, Società italiana di Fisica.

Chapter 5
Squaring the Pendulum's Arc: Motion Along Broken Chords

The pendulum plane experiment had revealed to Galileo that his initial conceptualization of a relation between swinging and rolling based on the heuristic assumption that the quarter swing of a pendulum along any arc took the same time as the falling of heavy object along a chord spanning this arc was not tenable. Where could Galileo go from here? As it turns out, the similarity between swinging and rolling which ensued from his assumptions concerning the two types of motion was too challenging to give up on the problem. After all, swinging along an arc was entirely different from rolling along a chord, and, as we will see, Galileo modified his initial understanding of the relation between swinging and rolling in such a way as to accommodate this fact.

Galileo's new conceptualization, which shines through in the 1602 letter to Guidobaldo del Monte, rested on two pillars. The first is the assumption that the motion of a pendulum with the pendulum bob supported from above is kinematically equivalent to the motion of a heavy body on a concave surface supported from below as long as the two motions are made along the same arc.[1] This assumption, even though problematic, was not contentious for Guidobaldo del Monte and neither was it for Galileo.[2]

---

[1]Galileo did not explicate his assumption that rolling and swinging along the same arc were kinematically equivalent. The closest he came to doing so was in a letter sent to Lorenzo Realio in June 1637. In the letter, he stated that rolling along circular channels ("canali") was isochronous, and immediately after this, he stated that the same held true for pendulum motion, insinuating but not explicating a relation between both statements. Cf. EN XVII, letter 3496.

[2]As further detailed below, the swinging of a pendulum along an arc and the falling along the same arc are equivalent only if friction effects are completely ignored, i.e., in particular, when the body falling along the arc is sliding frictionlessly. In reality, due to friction, a body falling along the concave surface will usually not be slipping but rolling. As will be argued in the following chapter, Galileo was unaware of the difference between rolling and sliding with regard to the kinematics of the ensuing motion. Ultimately, this prevented him from successfully exploiting the results of his pendulum plane experiment.

The second pillar of Galileo's new conceptualization of the relation between swinging and rolling was the idea to model the rolling along an upright arc by rolling along a series of appropriately chosen, interconnected inclined planes.[3] The most simple such case, where the quarter arc is spanned by two interconnected chords, is already alluded to in the letter to Guidobaldo del Monte. However, Galileo's considerations did not remain restricted to the consideration of motion along such simple broken chords. His underlying idea was, as will become apparent, to exhaust motion along the arc by motion along polygonal lines of an increasing number of sides where each side was understood to be an inclined plane upon which motion took place. Should he be able to infer the time of motion along a polygonal path with infinitely many sides, it should, based on his assumption of the equivalence of falling and swinging along the arc, hence also be possible to demonstrate the properties of pendulum motion, in particular, the alleged isochronism of the pendulum, based on his assumption of the equivalence of falling and swinging along the arc. Figuratively speaking Galileo's new approach can thus be referred to as the attempt to square pendulum motion.

In the following, any series of inclined planes will be quite generally referred to as a broken chord path, and Galileo's investigation of the properties of motion along such broken chord paths or simply broken chords will be referred to as his broken chord approach. Galileo's considerations related to the broken chord approach are reconstructed in the present chapter. That these considerations eventually served the aim of relating swinging and rolling based on Galileo's new conceptualization is presupposed in the present chapter. It will become manifest, however, in the succeeding chapter, where it will be shown how, based on his new conceptualization and in particular based upon the concrete results achieved in the context of the broken chord approach, Galileo reevaluated the data his pendulum plane experiment had provided him with.

A considerable number of entries in the *Notes on Motion* are related to Galileo's attempt to represent or approximate the motion of fall along an arc by the motion of a body falling along a series of inclined planes inscribed into an arc.[4] These entries exhibit a rich number of mutual internal dependencies and relations at various levels, indicating that they indeed pertain to and represent a distinct stage of Galileo's work on the problems of motion. Moreover, many of the entries related to the broken

---

[3] Here and in the following, I refer to an upright arc as one that has a horizontal tangent in its lower endpoint.

[4] If I use the word "approximate" here and in the following, I do not mean to imply that Galileo held anything that would compare to a modern concept of approximation. In particular, for him, the idea that describing a physical process as a simplified model could be used to reach a sufficiently accurate solution would be rather meaningless, not least because he had no way of demonstrating how, given a certain model, a magnitude thus calculable deviated from the magnitude approximated only within a certain limited margin of error. For Galileo, considering motion along a polygonal path was merely a step toward the solution, preferably to be established by an argument based on the method of exhaustion.

chord approach are also externally related, i.e., independently of their content, such as, for instance, when different considerations are noted on the same folio page, suggesting that they had been drafted within a limited period of time. The overall picture which thus emerges is indeed that Galileo pursued the different strands of his investigations related to the broken chord approach in a rather restricted period of intense work. It will be shown that this work must in fact date to the year 1602, and the entries discussed in the present chapter are thus seen to document the very research program which Galileo had alluded to in his letter to Guidobaldo in this year.

In Sect. 5.1, Galileo's considerations pertaining to motion along a plain broken chord, i.e., a path made up of two adjoined chords, will be discussed. With regard to this configuration, as mentioned in the letter to Guidobaldo del Monte, Galileo had constructed a proof to show that motion along such a broken chord was always completed in less time than motion along a straight chord connecting starting and endpoints if the arc considered subtends an angle of 90 degrees or less and motion ends at the nadir. Section 5.3 is concerned with Galileo's considerations centered on paths composed of more than just two chords. The distinction made between paths composed of two and those composed of a higher number is not meant to imply that these represented separate problems for Galileo. Its purpose is merely to structure the discussion.

5.1 Toward a Proof of the Law of the Broken Chord

In November 1602 in the letter to Guidobaldo del Monte, Galileo explicitly mentioned that he had proven that motion along a broken chord was completed in less time than motion along the single chord spanning the same arc. This proposition has already been referred to as the law of the broken chord. A rather large number of folios that bear traces of Galileo's attempt to come up with a proof for the law of the broken chord are preserved in the *Notes on Motion*. These allow for a thorough reconstruction of the way in which Galileo constructed the proof and to work out how this endeavor was related to his ambitious program of integrating pendulum motion and naturally accelerated motion on inclined planes.

In the following, the proposition and its proof as published in the *Discorsi* are analyzed. This will serve as a reference point for the reconstruction of the path by which Galileo first arrived at the proof. Compared to most of the proofs that Galileo included in the Second Book of the Third Day of the *Discorsi*, the proof of the law of the broken chord shows some peculiarities. As will become apparent, these peculiarities find an explanation in the fact that the elaboration of the proposition and its proof happened at a particularly early stage of Galileo's inquiry into naturally accelerated motion and that the proof was subsequently included in the *Discorsi* without substantial alterations. When Galileo first drafted the proof, some of the methods and assumptions he used regularly later on were not yet available to him or at least had not yet been sufficiently corroborated to be formally employed as premises in a proof.[5]

Based on the analysis of the proof as published in the *Discorsi*, all entries in the *Notes on Motion* pertaining to Galileo's initial construction of the proof will be identified, and the developmental sequence that led to its establishment will be reconstructed. It will thus turn out that, in addition to merely demonstrating that motion along a broken chord took less time than motion along a single chord spanning the same arc, Galileo attempted to determine the ratio of these times of motion exactly. Finding a way to determine this ratio would, as will be demonstrated, have allowed him to advance his overall program. In his search for a solution, Galileo came up with an alternative approach on which he hoped to be able to base a new proposition concerning motion along a broken chord. This approach, apparently never completed, will likewise be reconstructed and discussed in this section.

The law of the broken chord in the *Discorsi* More than 35 years after Galileo had communicated to Guidobaldo del Monte that he had constructed a proof for the law of the broken chord, he published the proposition as Proposition XXXVI of the Second Book of the Third Day of the *Discorsi*:

> If from the lowest point of a vertical circle, a chord is drawn subtending an arc not greater than a quadrant, and if from the two ends of this chord two other chords be drawn to any point on the arc, the time of descent along the two latter chords will be shorter than along the first, and shorter also, by the same amount, than along the lower of these two latter chords. (EN VIII, 262–263, transl. Galilei 1914, 237)

The proof which Galileo gave is somewhat gruelling. It is one of the longest proofs in the *Discorsi* and rests on three additional geometrical corollaries. What makes the proof cumbersome to follow, above all, is that Galileo used a proof technique related to the way in which times of motion are geometrically treated and represented in a separate diagram that is alien to the other propositions of the Third Day.

With reference to the motion diagram associated with the proof (see Fig. 5.1), Galileo's argument can be paraphrased as follows:

Statement of the law of the broken chord:

$$t(DBC) < t(DC)$$

By law of chords:

$$t(DF) = t(DB)$$

---

[5] Wisan (1974, 179) refers to the proof as being "unusually clumsy" without, however, further explicating her opinion.

5.1 Toward a Proof of the Law of the Broken Chord

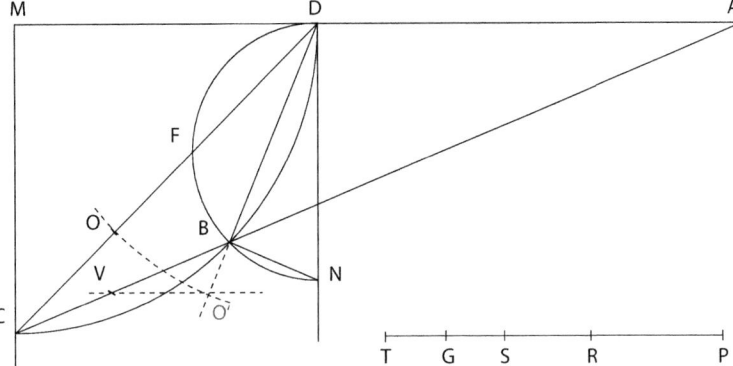

Fig. 5.1 Schematic rendering of the diagram associated with the proof of the law of the broken chord, Theorem XXII, Proposition XXXVI of the Second Book of the Third Day of the *Discorsi*. The dotted lines and the point O' are not part of Galileo's original diagram but have been inserted for the analysis. Their construction and meaning are explained in the text

Hence:

$$t(BC|D) < t(FC|D)$$

By drawing a circle around D with radius DB and by subsequently applying the law of chords to the motion along DF and DB, Galileo has reduced his problem to showing that the time of fall along FC from rest at D has to be longer than the time of fall along the lower chord BC, likewise after motion from rest at D.

Here Galileo states:

> At eadem temporis celeritate conficit mobile veniens ex D per DB ipsam BC, ac si venerit ex A per AB, cum ex utroque casu DG, AB aequalia accipiat velocitatis momenta

Thus, a reduced statement follows:

$$t(BC|A) < t(FC|D)$$

For Galileo, the kinematical behavior of naturally accelerated motion not starting from rest depended only on the height of the previous motion of fall and not on the concrete path taken. In the following, this assumption will be referred to as *path invariance*. In the case at hand, path invariance in fact implied that the time to fall along the lower chord BC from rest at D had to be as if the fall was made from rest at A instead of D.

A reader of the *Discorsi* who has reached Proposition XXXVI is long familiar with path invariance and of the use Galileo makes of this assumption in his arguments concerning naturally accelerated motion. Indeed, the assumption is first employed in Proposition X. Galileo argued by the same token in many of the following propositions. Here then, we have the first peculiarity of the proof of the law of the broken chord as published in the *Discorsi*. It is in this proof and only this one that Galileo does not simply presuppose path invariance but attempts to justify it further by emphasizing that, having fallen through the same height, the mobile arriving at B after fall from D and A, respectively, "aequalia accipiat velocitatis momenta" (receives equal moments of velocity).[6]

Galileo issued the law of the broken chord as one of the last proofs of the Second Book of the Third Day, where vindicating path invariance no longer made much sense. Genetically, however, the proof was, according to all evidence, the first or at least among the first proofs drafted where Galileo made an argument concerning naturally accelerated motion not starting from rest and in which he availed himself of path invariance. That the assumption is further elucidated in its first use as a premise in a proof is of course absolutely understandable. That Galileo took his statement in this respect over into the published version of the proof instead of moving it into the proof of Proposition X where path invariance is first made use of can then be considered an editorial oversight.[7]

The core part of Galileo's proof consists in demonstrating that the reduced statement holds. With respect to the way Galileo argues, we encounter a second peculiarity in the proof. Galileo constructs a timeline TP and upon this timeline marks distances which represent times of motion for the motions under consideration in the argument. The laws of motion applied show how, given the spacial geometry, the corresponding distances on the timeline have to be constructed. Galileo's way of proceeding is well founded yet it is unfamiliar to the reader of the *Discorsi*. The proof of the law of the broken chord indeed is the only proof in which Galileo constructed an argument that made recourse to a separate timeline. More commonly, in the proofs advanced in *Discorsi*, Galileo identifies and constructs, respectively, lines that represent the times of motion in question within the diagram representing the spacial configuration of the motion situation. This will be referred to as *integrated time representation*. This technique is usually paired with distance time coordination, i.e., with the choice to assume, for one particular motion, that the

---

[6]In the *Discorsi*, that bodies falling from the same height, here from A and D, respectively, to B acquire the same moments of velocity ("velocitatis momenta") had been formulated as one of the fundamental principles based upon which the new science of naturally accelerated motion on inclined was founded: "Accipio, gradus velocitatis eiusdem mobilis super diversas planorum inclinationes acquisitos tunc esse aequales, cum eorumdem planorum elevationes aequales sint. (EN VIII, 205)" Clavelin (1983, 45) has suggested that initially the assumption that equal degrees of velocity were reached after fall through the same height, independent of the path taken, was based on a "general intuition that ... is also the basis of statics and of the theory of simple machines." Strictly speaking path invariance furthermore presupposes what Wisan (1974) terms the "assumption of continuity of speeds at corners." This is, however, never made explicit by Galileo.

[7]In view of the fact that the majority of the proofs of the Second Book of the Third Day rely on path invariance, it is somewhat surprising that Galileo never explicitly formulated as a principle that, if naturally accelerated motion along an inclined plane is deflected to a second plane, the time elapsed in traversing this second plane is the same regardless of whether the preceding fall is made through the first plane or through any other plane of the same vertical height as the first plane.

time of this motion is measured by the same number of parts as the very distance along which motion proceeds.[8]

Based on integrated time representation and distance time coordination, arguments about naturally accelerated motion can be formulated more economically. It will be shown in Chap. 7 that integrated time representation indeed emerged as a spin-off of Galileo's attempt to construct a proof for the law of the broken chord. The technique was thus not available when he first drafted the proof, a draft taken over into the *Discorsi* without any substantial alteration, i.e., in particular, without reformulating the argument based on his new technique. If reformulated, the proof becomes more straightforward, however, and is easier to follow. For the analysis which follows, I have hence rendered Galileo's statements to assume the form more commonly used in his later arguments. Galileo's original statements are given in square brackets for ease of comparison and control.

Galileo demonstrated his reduced statement as follows.
By initial choice:

$$t(DC) = DC \text{ [equivalent to } PS].^9 \tag{5.1}$$

---

[8] One can distinguish between numerical and symbolic distance time coordination. Whereas numerical distance time coordination primarily renders calculations more convenient, symbolic distance time coordination has more far-reaching consequences. Galileo's numerical calculations of times of motion usually start by assigning a numerical value to the length of a distance traversed and a numerical value to a time elapsed during motion, usually motion along this distance. From a modern perspective, this amounts to assigning (arbitrary) units of distance and time. It is often convenient to identify the numerical value representing the measure of the length of a distance with the numerical value representing the measure of the time to traverse this distance, as this facilitates calculation. In the case of symbolic distance time coordination, two magnitudes, which are represented symbolically, that is, geometrically, with no particular numerical value being assigned to them, are identified with one another. Symbolic distance time coordination is typically established by statements of the kind: "[s]i itaque intelligatur, tempus per AC esse ipsamet AC …(EN VIII,229)" The time through AC in the example can of course not be the distance AC itself ("esse ipsamet") in a literal sense, but AC represents the time through AC quantitatively. In diagrams making use of integrated time representation, distance time coordination amounts to the identification of one and the same line as representing a distance covered and simultaneously a time elapsed. That Galileo had a precise idea of how distance time coordination functioned as a conventional and convenient assignment of an arbitrary measure is revealed by statements such as the following taken from discussion on the Third Day: "[s]tabilite ad arbitrio nostro sotto una sola grandezza *ab* queste 3 misure di generi di quantità diversissimi, cioè di spazii, di tempi e di impeti …(EN VIII, 227–228)."

[9] In the published proof, the times of motion are represented by distances on the separate external timeline TP. Galileo's original proof, for ease of comprehension, has been re-rendered into a form corresponding to his later argumentative technique based on integrated time representation. The original passages, whose re-interpretations are presented in the text, are provided as footnotes for means of reference and control. The sentence in Galileo's proof that corresponds to the statement transcribed in the symbolic notation as (5.1) in the text reads: "Sit autem PS tempus quo peragitur tota."

By (5.1) and law of fall:

$$t(DF) = mp(DF|DC) = DO \text{ [equivalent to } PR\text{]}.^{10} \tag{5.2}$$

By (5.1) and (5.2):

$$t(FC|D) = DC - DO = OC \text{ [equivalent to } RS\text{]}.^{11} \tag{5.3}$$

According to a schema of argument familiar from many other proofs of the Second Book of the Third Day, if the time of motion along DC is represented by the line DC, it straightforwardly follows that the line OC represents the time along the remainder FC after fall from D.[12]

The argument proceeds by showing that, given the choices already made, the time of motion along AC is represented by AC (equivalent to PT).

By (5.1) above, the law of chords and the law of fall:

$$t(AC) = AC \text{ [equivalent to } PT\text{]}.^{13} \tag{5.4}$$

This is simply a consequence of Proposition III of the Second Book of the Third Day of the *Discorsi*, i.e., the so-called length time proportionality.[14] However, and this then is the third peculiarity we encounter in the proof of the law of the broken chord, Galileo does not justify step (5.4) by the length time proportionality. I will in fact argue that Galileo was not familiar with the length time proportionality when

---

[10] In the proof: "erit tempus PR id, in quo mobile ex D peragit."

[11] In the proof: "RS vero id [tempus], in quo reliquum."

[12] It is worth noting that Galileo could have established the same by employing Proposition XI of the Second Book of the Third Day, the so-called generalized law of fall. As will be argued in Chap. 6, Galileo formulated this proposition only after he had drafted his proof of the law of the broken chord. Thus, the straightforward two steps argument, which leads from the law of fall to the generalized law of fall, had to be included in the proof of the law of the broken chord when it was originally devised and it remained part of the proof which, once completed was, as will be seen, never reworked but included almost verbatim in the *Discorsi*.

[13] In the proof: "erit PT tempus casus ex A."

[14] According to the length time proportionality, the times of naturally accelerated fall from rest along two inclined planes of equal height are to each other as the lengths of these planes. Theorem III proposition III of the Second Book of the Third Day of the Discorsi is actually restricted to the case where one of the two inclined planes is a vertical. The more general case of motion along two inclined planes of equal height referred to here as the length time proportionality is treated in a corollary to the proposition.

5.1 Toward a Proof of the Law of the Broken Chord

he originally devised his proof. Indeed, according to all evidence, he only became aware of this proposition in the context of the construction of the proof of the law of the broken chord.[15]

The way in which Galileo instead established that if the time along DC is represented by DC, then the time along AC is given by AC is as follows: According to the law of chords, it must hold that the times along BC and DC are equal. BC and AC are distances on one and the same inclined plane, yet fallen through from rest at B and A, respectively. Hence, the law of fall can be applied to these motions, albeit in a somewhat unfamiliar manner. What is sought is a line XY which represents the time $t(AC)$, given that the time $t(BC)$ is already determined to be represented by the line DC.

According to the law of fall, it must hold for the times of motion along BC and AC that $t(BC)/t(AC) \sim mp(AC|BC)/AC$.[16] Hence it must hold that DC is to the sought distance XY just as the mean proportional of AC and DC is to AC, i.e., it must hold that: $DC/XY \sim mp(AC|BC)/AC$.

From this last proportion, it is not immediately clear how the sought distance XY can be determined. However, for the given construction, DC just happens to be the mean proportional between AC and BC as Galileo established in the first lemma preceding Proposition XXXVI of the Second Book of the Third Day of the *Discorsi* (for the corresponding diagram, see Fig. 5.2).

> Let DC be drawn perpendicular to the diameter BA; from the extremity B draw the line BED at random; draw the line FB. Then, I say, FB is a mean proportional between DB and BE. ... (Galilei 1914, p. 235).

It hence follows that the sought distance XY must be equal to AC, exactly as concluded in (5.4).

Fig. 5.2 Diagram associated with the first lemma preceding the proof of proposition XXXVI of the second book of the third day of the *Discorsi*

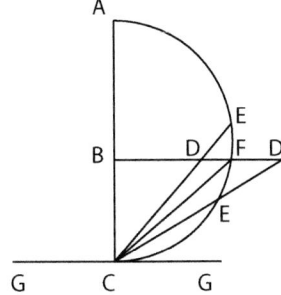

---

[15] Cf. Chap. 7, Sect. 7.2.

[16] In the *Discorsi*, Galileo provided a number of equivalent formulations of the law of fall, which he introduced as Proposition II of the Third Book of the Third Day of the *Discorsi*. The one that is used here and indeed the one that almost exclusively is brought to bear in the considerations documented by the *Notes on Motion* is the one Galileo formulated as the second corollary: "Colligitur, secundo, quod si a principio lationis sumantur duo spatia quaelibet, quibuslibet temporibus peracta, tempora ipsorum erunt inter se ut alterum eorum ad spatium medium proportionale inter ipsa. (EN VIII, 214)"

The argument by means of which Galileo established step (5.4) is complicated by the fact that the motions over the inclined planes BC and AC do not start from rest at the same point, as was usually the case when Galileo applied the law of fall. As will be seen, in an earlier version of his proof, Galileo had added an intermediate step at this point of the argument: the distance BC was relocated to a new position $B'C'$ on the inclined plane AC such that in this new position point B' coincided with point A and the two motions to which the law of fall was applied thus in fact started from the same point. In the published proof, this intermediate step, not strictly necessary for the argument, was dropped at the expense of rendering it somewhat more opaque. The proof proceeds:

By above and law of fall:

$$t(AB) = mp(AC|AB) = AV \text{ [equivalent to } PG\text{]}.^{17} \tag{5.5}$$

By (5.2) and (5.5), and path invariance[18]:

$$t(BC|D) = t(BC|A) = AC - AV = CV \text{ [equivalent to } GT\text{]}.^{19} \tag{5.6}$$

At this point it has been established that the time of motion along the lower of the broken chords from rest at A is measured by the distance CV, and that along the lower part of the single chord FC, also from rest at A, by the distance CO, and the problem has been reduced to having to show that:

Reduced problem:

$$CO > CV \text{ [equivalent to } RS > GT\text{]}.^{20} \tag{5.7}$$

This is a purely geometrical statement on the relationship between the lengths of the two lines CO and CV. In the remaining part of the proof, this statement is validated using the second and the third lemmata, which preceded the proof in the *Discorsi*. In the second lemma, Galileo had established that if two lines were given and both were divided in such a way that the ratio of the longer to the smaller part of the longer line was greater than the ratio of the respective parts of the shorter line, it

---

[17] In the proof: "erit PG tempus quo mobile ex A venit."

[18] This conclusion could have been arrived at before by invoking the generalized law of fall which, however, as argued, had not been formulated as an independent proposition when Galileo wrote the draft of the law of the broken chord, a draft he ended up publishing almost unaltered in the *Discorsi*.

[19] In the proof: "GT vero tempus residuum motus BC consequentis post motum ex A."

[20] In the proof: "ostendendum itaque est, RS maius esse quam GT."

5.1 Toward a Proof of the Law of the Broken Chord

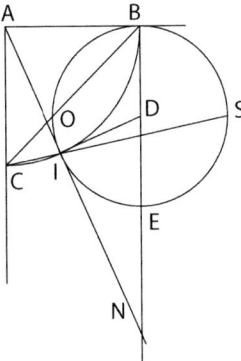

Fig. 5.3 One of the two diagrams associated with the third lemma preceding the proof of Proposition XXXVI of the Second Day of the Third Book of the *Discorsi*. With regard to the construction Galileo shows that "CI is always less than CO ...(Galilei 1914, 236–237)"

followed that the greater part of the longer line would exceed the greater part of the shorter line.[21] Galileo applied this lemma to the two lines FC and BC, divided by the points O and V, respectively (cf. Fig. 5.3).

By the third lemma, it held that FC had to be longer than BC. Hence, in order for the second lemma to become applicable, it remained to be shown that these two distances were divided by the points O and V in such a way that the ratio of the parts CO and CF was greater than that of the parts CV and VB, i.e., that $CO/OF > CV/VB$. In the remaining part of the proof, Galileo demonstrated that this inequality between the ratios in fact held; once this was done, the reduced statement, namely, that $CO > CV$, had been proven.

To sum up: The proof of the law of the broken chord in the *Discorsi* is distinguished from other proofs in that the times are represented by an external timeline rather than by lines that are part of the diagram representing the spatial geometry of the motion problem. Furthermore, only the law of fall and the law of chords enter the proof as premises. Even though a partial problem emerges within the proof, which could be solved immediately by applying the length time proportionality, this proposition is not used as a premise in the argument, and the partial problem is solved in a more complicated manner. Last, Galileo's proof is based on the assumption of path invariance, i.e., on the assumption that the time of motion on an inclined plane, which has been preceded by a motion of fall, depends only on the vertical height of the preceding motion of fall. Even though Galileo had invoked the assumption of path invariance in many of the preceding propositions, in the proof, the path invariance was not merely stated but further justified by claiming that bodies falling from the same height acquired equal moments of velocity. These peculiarities of the proof of the law of the broken chord find an explanation in the fact that, as will be demonstrated, the proof was among the earliest proofs concerning naturally accelerated motion that Galileo formulated.

---

[21]Lemma 2 to Theorem XXII, Proposition XXXVI of the Second Day of the Third Book of the *Discorsi* reads: "Let AC be a line which is longer than DF; and let the ratio of AB to BC be greater than that of DE to EF. Then, I say, AB is greater than DE. (Galilei 1914, 235–236)"

Comparing motion along a chord and broken chord spanning the same arc

Manuscript evidence In the following, those folios of the *Notes on Motion* will be considered that pertain to the motion situation in which one motion proceeds along a single chord and another along a broken chord spanning the same upright arc of a quarter circle or less. Diagrams which depict this particular motion situation are reasonably unique in the manuscript and easily discernible. Some of these diagrams, as will be seen, were part of Galileo's effort to calculate the times of motion along a polygonal path of an increasing number of sides, which will be discussed in the following section of this chapter. In the present section, only the content of those folios will be discussed, which, in addition to the pertinent diagram, contain a timeline characteristic of the proof of the law of the broken chord and/or some form of notation of proportions, which hold with regard to the motion situation under consideration.

If all the folios that meet the above criteria are identified, the following picture emerges, graphically rendered as Fig. 5.4: On a *core group* of folios comprising folios 186, 185, 149, and 150, the latter two of which are connected to form a double folio, Galileo elaborated all constituents, which pieced together, allowed the construction of a full proof of the law of the broken chord, including two of the three geometrical lemmata required for the proof as it was published in the *Discorsi*.[22] Galileo started to draft the law of the broken chord and its proof on folio 163, continuing to write on folio 172. The latter two folios are part of a group of folios referred to here as the star group folios. The content of this group is discussed in detail in Chap. 10. Sometime after Galileo's move to Florence, almost the entire contents of the star group folios were copied by Galileo's disciples Niccolò Arrighetti and Mario Guiducci onto fresh folios.[23] As part of this copying,

---

[22] Apart from the final draft on folio 172, no further notes can be identified which are related to the first lemma associated with the law of the broken chord. The lemma is so trivial that it was conceivable that Galileo drafted it directly on folio 172 without any preparatory notes.

[23] Of the 163 folios that make up the *Notes on Motion* 35 bear the hand of Niccolò Arrighetti or Mario Guiducci, who were both Galileo's disciples. For most of the entries in the hands of the disciples, an original in the hand of Galileo still exists and Favaro noted: "when we have a given fragment only in the hand of the disciples, we are allowed to surmise that they limited themselves to transcribing from a Galilean original which now is lost. (EN VIII, 34. My transl.)" All copies were made on the same type of paper not occurring elsewhere in the manuscript, i.e., no copy was made on a paper of different type, and neither does any folio of this type exist which does not contain disciple copy. The watermark is a spread-eagle with a crown above, the paper quality is thin, and white and all folios are folded in the middle. Niccolò Arrighetti and Mario Guiducci sometimes copied from the same original sheet without duplicating the work of the other, suggesting that they worked together at the same time under the supervision of Galileo. Two folios, folios 58 and 68, bear entries by both, Arrighetti and Guiducci. On folio 68 recto, the entry in the hand of Guiducci was written before the entry by Arrighetti. On folio 58, the situation is reversed. Here, the entries by Guiducci must have been made earlier, since Arrighetti added a remark referring to the content of the page that had been written by Guiducci. The entries by both disciples must hence have been made concurrently. Galileo first met Arrighetti long before his first encounter with Guiducci in 1614, which thus provides a terminus post quem for the copying. Galileo stayed in contact with both of his disciples more or less for the rest of his life and hence no terminus ante quem can be determined. Drake (1970) asserted that Guiducci must have made his copies as early as 1614 or,

5.1 Toward a Proof of the Law of the Broken Chord

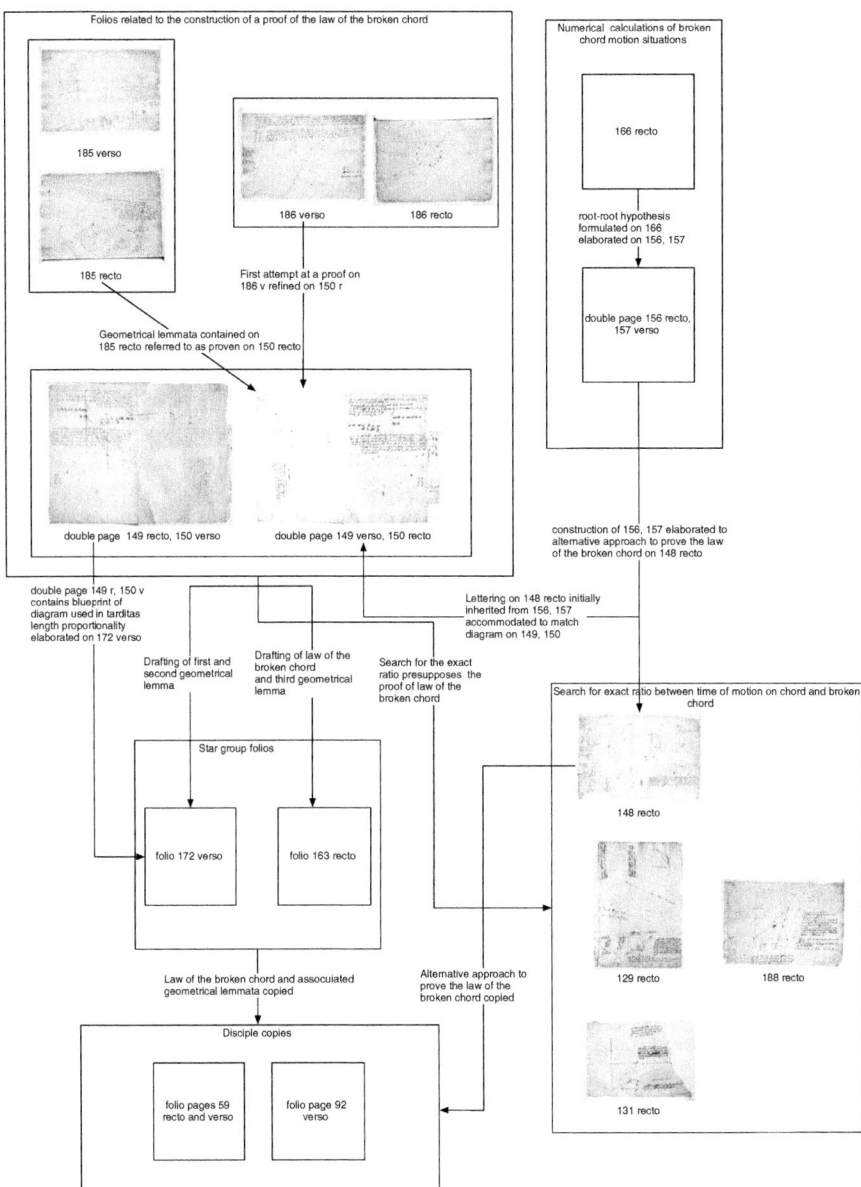

Fig. 5.4 Folios that can be related to Galileo's attempt to construct a proof for the law of the broken chord and their mutual relations. Folios represented by thumbnail images are discussed in detail in the current section

the draft of the law of the broken chord proof from folios 163 and 172 was copied by Mario Guiducci to folio 59.

An additional four folios, 148, 131, 129, and 188, also contain entries that match the criteria established above. These folios document Galileo's attempt to arrive at a stronger version of the law of the broken chord in which, instead of merely proving that descent was made in less time along the broken chord than along the respective single chord, he tried to determine the exact ratio of the two times of motion depending on the position of the junction point of the two chords making up the broken chord. This group of folios will be referred to as the *exact ratio group*.

Constructing a proof for the law of the broken chord

Page 186 verso The first in the progression of steps which led Galileo to establish a proof for the law of the broken chord can be identified with great certainty. Of the extant folios documenting his proof attempts, his entries on folio 186 verso certainly must be his earliest concerning this subject. Galileo opened the page with an explicit statement of what he wanted to prove (for the associated diagram, see Fig. 13.24), i.e., with a statement of the law of the broken chord:

Dicimus, tempus quo mobile permeat lineas *db*, *bc* brevius esse tempore quo permeat solam *bc*. Sit *ae* aequalis *bc*

As documented by the first paragraph on the page, Galileo initially approached the problem by constructing the external timeline xn, in which the line nm represented the time of fall over the lower of the broken chords bc from rest at b and xn represented the time of fall over the inclined plane ac.[24] However, in accordance with this initial choice, his attempt to construct other distances representing times on the timeline that were needed for the argument led him to a dead end. As a result of this first approach, Galileo was merely able to formulate a necessary, but not sufficient, condition for the law of the broken chord.[25]

In order to overcome the problems which had thus become manifest, Galileo developed a different approach. He introduced, as a new geometrical element, a

---

alternatively, between the years 1616 and 1620. Later he specified this dating to the year 1618, based on comparison of the watermark of the paper used for copying and the watermarks of letters written at this time. Cf. Drake (1978, 67, 262–263).

[24] 186 verso, entry T1A.

[25] The condition which Galileo arrived at in his first attempt to solve the problem was that the line xr had to be smaller than line nm. A detailed discussion is contained in the Appendix in Chap. 13.

5.1 Toward a Proof of the Law of the Broken Chord

Fig. 5.5 Diagram on page 186 verso, reduced to the elements necessary to reconstruct Galileo's approach documented by the second paragraph on the page

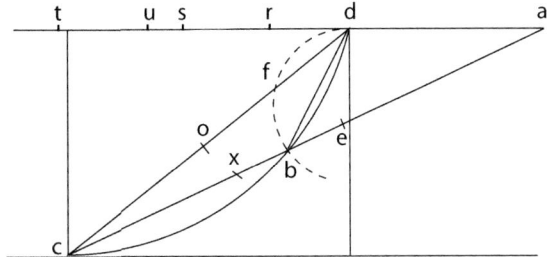

circle with its apex in d and with a radius chosen such that db, the first of his two conjugate chords, would be inscribed as a chord in that circle (compare Fig. 5.5). This circle intersected the single chord dc at point f. By application of the law of chords, the time along df and along the first of the conjugate chords, chord db, consequently had to be equal, and Galileo was thus able to reduce his original problem to showing that the time along fc, the remainder of the single chord, had to be longer than the time along bc, the second of the broken chords, both motions being made from rest at d. We recognize that Galileo here first introduced the argumentative strategy which he still used in the published proof. This particular construction was employed on every other single folio related to the proof of the law of the broken chord, and consequently, from then on Galileo started all of his considerations from the reduced statement, which clearly shows that of the extant folios on which Galileo approached the problem of constructing a proof of the law of the broken chord, folio 186 must have been the first one he worked on.

Based on his new construction, Galileo commenced a fresh attempt to construct the sought-after proof. This is documented by the second paragraph on folio 186 verso.[26] For this new attempt, Galileo constructed a new external timeline td. It needs to be noted for the following discussion that this timeline was integrated with the main motion diagram in a peculiar way.[27] One of the endpoints of the timeline, point d, coincided with the starting point of the motions considered, i.e., a point belonging to the spatial geometry of the motion situation.

In his entry, Galileo fully established the relations that must hold between the various lines representing times of motion in the timeline in terms of given geometrical relations between distances in the spatial geometry of the problem. Galileo determined that the line vt represented the time of fall over the chord bc from rest at a or, according to the assumption of a path invariance, from rest at d. This line vt had to be shorter than line rs, which represented the time of fall over fc from rest at d. Hence what remained to be proven at this point was that vt was larger than rs. On the remaining folios to be discussed in the present section, Galileo is indeed concerned with finding an argument to prove this reduced statement.

---

[26] 186 verso, entry T1B.

[27] 186 verso, diagram D01A.

It should be noted here that the timeline *td* in the diagram on folio 186 verso is, in contrast to the initial timeline *xn*, actually geometrically constructed, i.e., it preserves the proportions which, according to the applied laws of motion, must hold between the times of motion represented by lines in the timeline. This will in the following be referred to as a *conformably constructed timeline*. A conformably constructed timeline, besides serving the argument to be constructed, provided a means of testing the validity of the statement to be proven on paper as the distances representing the times in question could be measured off this timeline and be compared directly.

The statement of an inequality of two distances representing times in the timeline that Galileo had arrived at on 186 verso was not directly apt for completing the proof. To find a geometrical construction which would allow to construe an argument showing that the condition was fulfilled, Galileo needed to transform the condition formulated for lines representing times in the timeline back into a condition regarding distances in the spatial geometry of the problem.

As the space on folio 186 verso had more or less been exhausted by his previous entries, Galileo repeated the construction on a new folio, folio 150 recto, and, apart from the lettering of the points, he didn't even bother to ink the new diagram. Before proceeding to a discussion of folio 150 recto, however, we briefly need to turn attention to a last short note which Galileo wrote on the left margin of 186 verso.[28]

> Nota. Sit in circumferentia utcumque ducta *do*, et iungatur *co*: dico dico, citius moveri ex *d* in *o* quam ex *o* in *c*. Ostensum enim est aequali tempore moveri ex *o* in *c*, atque ex ex [*d*]in *c*; verum ex *d* in *o* patet celerius fieri motum quam ex *d* in *c*.

Galileo observed that motion was made faster, i.e., completed in less time over the distance *do* than over the distance *oc*, both motions being made from rest. In other words, for the junction point of a broken chord arbitrarily chosen on the arc, the time of fall along the first chord must be less than along the second chord, both motions being made from rest. This is, of course, a rather trivial consequence of the law of the broken chord, and it is interesting that Galileo deemed this noteworthy.[29]

---

[28] 186 verso, entry T2.

[29] In the central diagram on folio 186 verso, there are two points that are lettered *o*. As the first sentence shows, in his note Galileo is referring to the junction point *o* on the arc and not the point *o* marking the mean proportional on the chord *dc*. The time of fall along the second chord *oc* from rest at the junction point *o* was, by virtue of the law of chords, equal to the time of fall over the single chord spanning the arc of the broken chord, i.e., *dc*. The law of the broken chord in turn stipulated that the time of motion along the single chord was longer than the time of motion along the broken chord, and this necessarily also longer than the time of motion along the first part of the broken chord *do* as stated in the note.

5.1 Toward a Proof of the Law of the Broken Chord

The note can be understood as sharpening the counterintuitive nature of the law of the broken chord as the lower of the conjugate chords could be made arbitrarily small and yet the time to traverse it from rest would always be longer than the time to traverse the first of the broken chords. As I will show in Chap. 10, Galileo subsequently indeed elaborated the note to an independent proposition on folio 164 verso.

On the remaining three folios to be discussed in the following, Galileo elaborated all of the elements required by the published proof.

Page 150 recto On 150 recto Galileo proceeded with the construction of the argument begun on 186 verso. First he noted what he intended to prove[30]:

ostende *co* maiorem esse *cv*.[31]

Whereas Galileo had left off his considerations on folio 186 verso with a statement regarding the inequality of two lines representing times, on the new folio, he started his considerations from a statement about the inequality of two lines which were part of the spatial geometry. In the remaining entries on the folio, he shows that the former and the latter statements are in fact equivalent.

Galileo first established the proportions necessary for the construction of the new timeline td.[32] The construction of this new timeline differed somewhat from the one on folio 186 verso as Galileo chose to start his construction from the line representing the time along the single chord rather than from the line representing the time along the first of the broken chords, as he had done on folio 186 verso. Consequently, Galileo constructed the various other lines representing times in a different order, using moreover, slightly different labeling.

The condition which Galileo had formulated as $vt > rs$ on 186 verso, thus became $gt > sr$ with respect to the new labels on his new timeline. In a last entry, Galileo reversed the considerations by which the relations of the distances had been established. In this way he retranslated his statement about the inequality

---

[30] It appears that this entry, together with entry T3, which are both written in a conspicuously black ink, had been noted on the page first, as the other content is manifestly composed around these two entries. On 131 recto, likewise, two entries are made in an ink that is noticeably darker than the ink of the other entries on the page. On 131 recto the entries in the darker ink were likewise probably the ones composed earlier. As argued in Chap. 8, the entries in darker ink on 150 and 131 were probably made at more or less exactly the same time.

[31] 150 recto, entry T2A.

[32] 150 recto, entries T1A to T1D.

between two lines in the timeline into a statement about the inequality between the corresponding distances in the spatial geometry of the problem and thus demonstrated that indeed[33]:

$$gt > sr \Leftrightarrow co > cv.$$

The condition that co had to be larger than cv now indeed offered itself for finding a construction based on which it could be shown to hold. In fact, Galileo concluded his final entry on the page with the following remark:

> Ostenditur autem, per lemmata, co maior quam cv; ergo tempus rs maius est tempore gt: est autem rs tempus quo peragitur fc post df, gt vero tempus quo peragitur bc post ab: ergo patet propositum.[34]

185 recto On folio 185 recto, Galileo elaborated two of the lemmata he had referred to on 150 recto. These are lemmata two and three of the published proof. In addition to the elaboration of these lemmata, folio 185 recto contains a motion diagram of the broken chord motion situation including a conformably constructed timeline. No text is associated with this diagram; however, the page bears a numerical example, i.e., for the particular configuration of the broken chord depicted in the diagram, Galileo calculated the times of motion along the chord and the broken chord. Thus, he could have used the timeline to check the results of his numerical calculations and vice versa.

Compared to the motion situation depicted on folio 186 recto, the angle of the arc between starting and endpoint of motion considered on folio 185 recto is smaller. Galileo may hence have constructed this supplementary geometrical as well as numerical example to test whether the law of the broken chord actually held independently of the size of the arc considered, as long as this arc spanned an angle of 90 degrees or less. This may indicate that, this being the earliest stage of his considerations regarding naturally accelerated motion, Galileo was not overly confident regarding his inferences and deemed it appropriate to check them by calculating concrete examples.

Page 149 verso As we have seen in the analysis of the published version of the proof of the law of the broken chord, the only part of the argument which was not attributed to a lemma but remained within the proof was to show that, with respect to Fig. 5.1, the ratio of co to of was greater than the ratio of cv to vb. Galileo outlined how this partial statement could be demonstrated on folio 149 verso.

163 recto, 172 verso and 59 recto With the exception of one lemma, elaborations all of the constituents from which the final proof is composed have been identified above. Based on these preliminary considerations, Galileo

---

[33] 150 recto, entry T4.
[34] 150 recto, entry T4.

5.1 Toward a Proof of the Law of the Broken Chord

finally drafted a complete proof, and this draft can be identified on folios 163 recto and 172 verso. This draft corresponds almost verbatim to the version which Galileo finally published in the *Discorsi*.[35] Years later this draft was copied by Mario Guiducci to 59 recto. Apart from minor changes in punctation and one obvious copying error, the copy is identical to the original.

Constructing the law of the broken chord, a provisional summary Galileo's attempt to construct a proof of the law of the broken chord is documented by a number of folios. The proof is complicated and lengthy and required the solution of a number of partial problems. The way in which Galileo arrived at the proof of the law of the broken chord as it was later almost verbatim published in the *Discorsi* can consistently be reconstructed based on the manuscript evidence.

Galileo's construction of the argument was impeded by his use of an external timeline. Against the backdrop of his later way of arguing, based on integrated time representation, this entails a redundancy. As part of the argument, relations that hold between distances that are part of the spatial geometry are first translated into relations between lines in the timeline mediated by the laws of motion applied. These relations are in turn translated back into relations that hold between distances that are part of the spatial geometry. Furthermore, the fact that the length time proportionality was not used as a premise complicated the argument further, as this required introducing a number of argumentative steps which would not have been required had the length time proportionality been used.

Galileo announced he had found a proof for the law of the broken chord in the letter to Guidobaldo del Monte in 1602, and there is little reason to doubt that the proof which we have witnessed him construct and which was later published without substantial modification is indeed the one he alluded to in the letter. Yet this has been disputed, and independent evidence will be produced in Sect. 5.3 to corroborate that Galileo's construction of the proof, as documented by the manuscript, indeed happened in 1602.

---

[35] Most of the changes between the manuscript and the published version of the proposition are to be found in the first sentences. These changes do not concern the line of argument but for the most part serve to render more precise the geometrical construction required for the proof. The most relevant changes are the following: "mobile ex termino ferri ... tempore breviori" (draft version) vs. "tempus descensus ... brevius esse" (final version); "citius conficere" (draft version) vs. "citius permeare" (final version); "Quod sic ostenditur:" (final version) vs. "Quod sic demonstratur:" (draft version); "Constat insuper" (final version) vs. "Constat igitur" (draft version); and "veniens ex db ac si veniret ex ab" (draft version) vs. "ex D per DB ipsam BC, ac si venerit ex A per AB" (final version).

5.2 Expanding on the Law of the Broken Chord

Search for the exact ratio between the times of motion along a chord and broken chord Besides the folios discussed above, four additional folios, folios 129, 131, 148, and 188, contain diagrams, calculations, and texts that pertain to the motion situation of a chord and a broken chord spanning the same arc characteristic of the law of the broken chord, but which are not directly connected to Galileo's construction of a proof for the proposition. Despite differences in detail, the content of all these folios is unified by Galileo's search for a way to determine the ratio between the times of motion along the second part of the single chord and along the lower of the broken chords, times for which the law of the broken chord merely states an inequality. That this search was directly related to the construction of the law of the broken chord is indicated by an entry on folio 149 verso, on which Galileo elaborated a part of the proof of the law of the broken chord. Above these entries, however, he explicitly noted:

quaeritur ratio *co* ad *cv*.

This ratio, which as we have seen represents the ratio of the times in question, can be conceived of as a measure of the relative saving of time if motion along the broken is compared to the motion along the single chord. The ratio is not fixed but varies with the choice of the position of the junction point on the arc. Clearly, the closer the junction point is located to the starting point or the endpoint of motion, the closer this ratio must be to one. This, furthermore, immediately implies the existence of a position for the junction point that minimizes this ratio, i.e., that defines the configuration of the broken chord traversed in least time.[36] As will be discussed in Chap. 7, at some point Galileo in fact specifically turned his attention to the question under which geometrical configuration a bent plane would be traversed in least time. Finding a way to exactly determine the ratio of the times of motion in question would thus not only have served the formulation of a generalized version of the law of the broken chord, but was moreover needed to overcome a difficulty Galileo had encountered in the context of his work on the broken chord approach.[37]

---

[36] An analytic solution for the calculation of the time of motion along a broken chord depending on the position of the junction point is given in Erlichson (1998). The more general solution for the time of motion along a polygonal path made up of an arbitrary number of chords of equal length and inscribed into an upright quarter circle is given in the discussion of folio 166 in the Appendix in Chap. 13.

[37] See Sect. 7.3

5.2 Expanding on the Law of the Broken Chord

Page 148 recto The approach Galileo availed himself of on folio 148 recto to determine the ratio of the times along a chord and broken chord spanning the same arc had originated in the general context of his broken chord approach. The central diagram on folio 148 recto was indeed copied from double folio 156 and 157, including the original lettering, as well as the construction of point r which, as will be demonstrated below, embraced a particular hypothesis by Galileo regarding the sought ratio of times. As further discussed in Sect. 5.3, Galileo had originally advanced this hypothesis on folio 166 recto.[38] This latter folio, as will be detailed below, in turn, is the central document regarding Galileo's considerations pertaining to the broken chord approach. On double folio 156 157, Galileo was not only trying to show that the broken chord is traversed in less time than the respective single chord spanning the same upright arc but stated explicitly that he was searching to determine the ratio of these times precisely: [i]nveniendum sit tempus quo conficiuntur [due] *acb* [broken chord] in rationem ad tempus quo conficitur sola *ab* [chord]."

After the diagram had been copied, its original keying was changed by Galileo to match the lettering of the proof diagram associated with the complete final proof of the law of the broken chord on 163 recto.[39] Moreover, Galileo constructed two timelines on 148 recto of which the one underneath the diagram was constructed conformably. The construction principle of the timeline to the right of the main diagram is elucidated in a short textual entry below.[40] Folio 148 recto thus indicates that Galileo was endeavoring to reuse insights that had emerged from his explorative attempts to approximate motion on an arc by motion on a series of conjugate chords to formulate a proposition equipped with a proof, as he had done already in the case of the law of the broken chord.

The *Notes on Motion* contain no indication that in this respect Galileo further advanced the consideration begun on 148 recto, putting him in a position to formulate a stronger version of the law of the broken chord. Yet apparently Galileo deemed his incomplete considerations important enough to retain folio 148, and he even had its content copied by Mario Guiducci to folio 92. This is the only extant case of a copy by a disciple containing an unfinished proof. We are thus led to suspect that Galileo intended to eventually return to his approach, to complete his argument, and to thus arrive at a new proposition concerning the ratio of the times of motion along a chord and broken chord spanning the same arc.

---

[38]Point r is lettered in the diagram on double folio 156 157; it is constructed but not lettered in the diagram on 148.

[39]In the main diagram on folio 148 recto, point b was changed to point c. Originally, the endpoints of the vertical timeline had been a and b, where the distance ab simply represented the time of fall through the distance ab. Most likely to avoid confusion, Galileo changed the lettering of these points to q and p, respectively, and made the appropriate changes in T1, which he had already written.

[40]148 recto, entry T1.

118   5 Squaring the Pendulum's Arc: Motion Along Broken Chords

The content of three further folios can be related to his search for an argument to establish the ratio of times in question and to thus arrive at a new, stronger proposition.

Pages 129 recto and 188 recto Folios 129 recto and 188 recto document Galileo's attempt to exploit a particular geometrical property apparent in the construction underlying his proof of the law of the broken chord. With respect to Fig. 5.1, the distances CO and CV represent the times of fall over the second part of the single chord and over the lower of the broken chords, respectively, both motions being made from rest at D. If a new point O' is chosen on the prolonged chord DB such that DO' is equal in length to DO, O' lies at the same height as point V. With respect to the labeling of the diagram on 129 recto, point e is projected onto ds as b such that de is equal in length to db. Then b has the same height over the horizontal through s as point i. On 188 recto, point i is projected onto bl as point q such that bi has the same length as bq. With this labeling, point s and q are positioned at the same height.

On both folios, no explicit mention is made of motion or times of motion, and Galileo's considerations are, at least superficially, purely geometrical in nature. In both cases Galileo further investigated the geometrical property described above and its implications. Given the meaning of the points labeled O and V in the original proof diagram of the law of the broken chord, the conclusion that the geometrical considerations on 129 recto and 188 recto were meant to serve the construal of an argument regarding the relation of the time of motion along the lower part of the broken chord to that along the second half of the single chord is inescapable. Yet Galileo was apparently not able to exploit the geometrical property he was further investigating on both folios for the determination of the exact ratio of times he sought. At least the two extant pages, which bear traces of his analysis, show no indication of a successful advancement.

Page 131 recto The last folio that can be related to Galileo's considerations regarding the ratio of the time of motion along a chord and broken chord spanning the arc is folio 131 recto. The central diagram on the page corresponds almost exactly, including the labeling of the points, to the diagram associated with the final draft of the proof of the law of the broken chord on folio 163 recto. Hence the content of folio 131 recto must have been composed either at a late stage of Galileo's construction of a proof for the law of the broken chord or, more likely, after the construction of the proof had been completed.

Noticeably, different inks were used for the entries on the page. The central motion diagram, the first, deleted entry and the short sentence to the right of the diagram are characterized by a lighter brown ink. The remaining two textual entries on the page were written using a much heavier ink, as is particularly obvious when inspecting the verso side of the folio, where these two entries clearly shine through.

The latter two entries will be discussed in detail in Chap. 8, where it is argued that they were likely added to the page somewhat later than the remaining content.[41]

In an entry which Galileo later deleted by striking it through, he hypothesized that the mean proportional between the vertical heights rc and tb might have the same length as do, the sought after mean proportional between df and dc. This hypothesis may have been suggested to Galileo by his analysis on folio 188 recto, which had shown that with regard to the keying of the diagram on 131 recto the diameter ds of the auxiliary circle was the mean proportional between the vertical heights rc and tb.[42] Whereas ds is in fact the mean proportional between rc and tb, its length is, however, not the same as that of do. Galileo apparently realized this and canceled his entry. False or not, the hypothesis Galileo had expressed in the entry demonstrates that his entries on the page were initially intended to serve the further investigation of the ratio of the times of motion along a chord and broken chord spanning the same arc.

Expanding on the law of the broken chord—provisional summary A number of folios document Galileo's attempt to examine the ratio of the times of motion along a chord and broken chord spanning the same arc more closely, for which the law of the broken chord merely makes an inequality statement. As will be seen in the succeeding section, calculating this ratio for a given geometrical constellation, i.e., for an arc of a given size and a given junction point, was not a problem for Galileo. What distinguishes Galileo's consideration discussed above from such calculations is that they are concerned with finding a generally valid argument to determine this ratio which, if found, could of course be used to formulate a new proposition with proof. That Galileo indeed intended to do so is manifested above all by the content of 148 recto, where despite not having found an argument yet, Galileo had already constructed a timeline and spelled out the principles of its construction, which would necessarily be part of the proof to be formulated. The order in which the content of the four folios discussed was composed cannot be reconstructed with certainty. Though it must have been produced more or less contemporaneously to Galileo's construction, the proof of the law of the broken chord, his construction of the ex mechanicis proof and, moreover, with his attempt to exhaust motion on an arc by motion on a series of conjugate chords of increasing number to be discussed below.

---

[41] In her discussion of the considerations documented on folio 131 recto, Wisan (1974, 176–177) acknowledges that these considerations may have been "directed at a search for a proof of the isochronism, rather than the brachistochrone."

[42] The line tb in the diagram on 131 recto corresponds to nr marked on the radius of the big circle, na in the diagram on folio 188 recto. The mean proportional ds in the diagram on 131 recto reoccurs as no in the diagram on 188 recto.

5.3 Tackling the Problem of Motion on an Arc: The Broken Chord Approach

Galileo's considerations concerning the relation between motion along a chord and broken chord discussed in the previous section were embedded in his advanced approach toward investigating the relation of swinging and rolling, based on the idea of exhausting motion on an arc by motion along a polygonal path of an ever-increasing number of sides, ever more closely approaching the arc. The principal idea upon which this approach is based is straightforward, namely, that the more closely the path considered mimics the actual arc, the more closely the kinematical behavior of motion along such a path should resemble that of motion along the arc. Yet the program of exhaustion needs to be carried out. Preferably, the method of exhaustion, as is familiar from Greek mathematics, presupposes a solution, and it is then shown by reduction to absurdity that assuming that this is not the solution would lead to a contradiction.

In order to find such a solution, Galileo made the effort to represent motion along an arc by motion along polygonal paths composed of an ever higher number of inclined planes. Ideally, this would allow him to figure out a scheme relating motion along a path of a given number of sides to motion along a path with a higher number of sides. Such a scheme could serve as a heuristic for finding the general solution, which could then be proven by exhaustion. As we will see, Galileo calculated the times of motion for an impressive number of chords belonging to different polygonal approximations of the arc. This brought about valuable insights. Yet it did not indicate how an inference to the limiting case of motion on the arc could be achieved. Eventually, as will be seen, the broken chord approach indeed yielded more new questions than answers.

The computational expansiveness of the considerations and his remarkable self-imposed drive for numerical accuracy have a beneficial side effect for the reconstruction and interpretation. Many folios contain auxiliary calculations, and associating them with the broken chord approach is facilitated by the fact that they contain very accurate numbers, unique with regard to the rest of the manuscript.

Manuscript evidence The folios that bear entries pertaining to the broken chord approach and the relations among these folios at various levels have schematically been represented in Fig. 5.6. For ease of reference and for structuring the discussion, I have further subdivided the folios into a *core group* and an *exact ratio group*.

The core group comprises the five folios: 166, 167, 183, 184, and 192. Of these folios, 166 and 167 most likely formed a double page at the time Galileo wrote on them. The considerations on folio 167 verso can indeed be interpreted as a preliminary inquiry for the much more complex investigations begun on folio 166 recto. The content of all the remaining folios in the group is closely related to that on folio 166 recto. Folios 184 and 192 comprise calculations of times of motion along the various inclined planes inscribed as conjugate chords in a circle, which are diagrammatically represented in the central motion diagram on folio 166

5.3 The Chord Approximation Approach

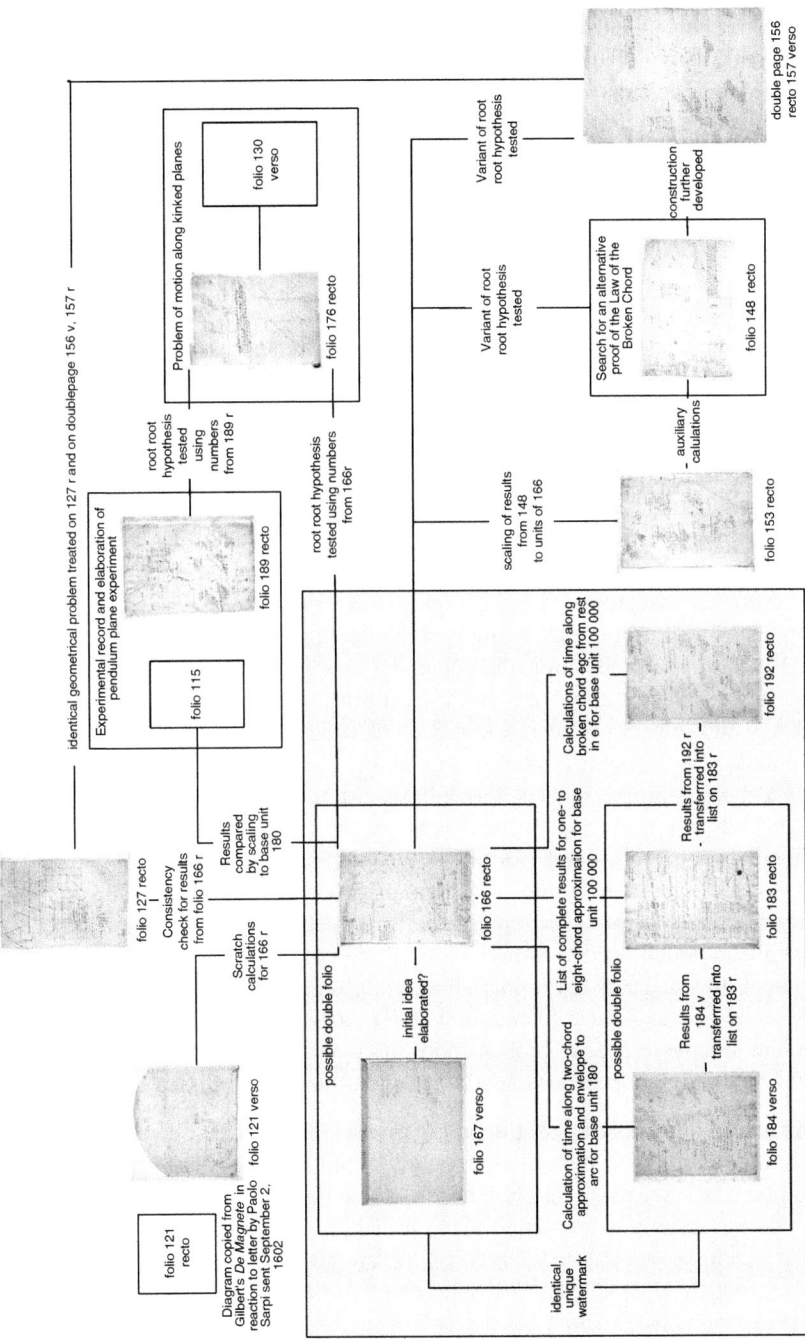

Fig. 5.6 Folios related to Galileo's attempt to represent motion on an arc by motion along a series of conjugate chords. The group of folios in the light gray area is referred to in the text as the *core group*. Folios represented by thumbnail images are discussed in detail in the current section

recto. Furthermore, folio 183 comprises a long list of results of these and similar calculations that are no longer extant but can be reconstructed on the basis of the calculational schema Galileo employed on folios 184 and 192.

The three remaining folios of the core group, 121, 127, and 189, contain auxiliary calculations for Galileo's considerations documented on the other folios of the group. Primarily, these three folios were used to record considerations not directly related to the broken chord approach and sometimes contain but a single auxiliary calculation pertaining to the latter. The impression is thus conveyed that Galileo was working frenetically on the broken chord approach and that, in doing so, he used every sheet of paper within reach for quick sketches and auxiliary calculations pertaining to this problem. This, in particular, suggests that the primary entries on the three folios bearing these auxiliary calculations must have been noted more or less contemporaneously to the work on the broken chord approach.

On folio 121 recto, Galileo retraced a construction for the determination of the declination of a magnetic needle depending on the position on the earth, which William Gilbert had introduced in *De Magnete*. In a letter from September 1602, Paolo Sarpi had asked Galileo to explain to him this particular construction from Gilbert's book. I will argue that the construction on 121 recto was indeed drafted in response to Sarpi's request. Folio 127 is part of the double folio 126–127, which besides auxiliary calculations pertaining to the broken chord approach contains drafts for two propositions belonging to the class of the so-called least time propositions as well as a sketch of the generalized law of fall. It will be argued in Chap. 7 that all three propositions had emerged as spin-offs of a specific problem that had grown out of the broken chord approach. The auxiliary calculations on folio 189, which otherwise bears the record of the pendulum plane experiment, finally indicated a temporal proximity to Galileo's explorations in the context of the broken chord approach to be confirmed in the succeeding chapter.

The second group of folios to be discussed, the exact ratio group, comprises five folios: folios 176, 148, and 153 and the double folio 156–157. On all of these, Galileo devoted himself to further investigating a problem that emerged from his elaborations on the folios of the core group. In order to make headway with his investigations, Galileo needed to find a way to determine the ratio of the time of motion along the single chord to that along a multi-chord spanning the same arc. In a first step, he thus needed to determine the ratio of the time of motion along a single chord and the respective broken chord. As will be seen, Galileo's attempts to determine this ratio far exceeded the considerations already discussed in Sect. 5.2.

The core group—exhausting motion along an arc The central folio documenting Galileo's attempt to exhaust motion on an arc by motion along paths of conjugate inclined planes of increasing number inscribed into this arc is folio 166 recto. The folio is dominated by a large motion diagram, whose central element is the arc of the lower quadrant of a circle. Galileo first examined the motion along a single chord spanning the complete arc. Next by dividing the arc into two equal halves, Galileo determined the junction point of a broken chord made up of two conjugate chords of equal length inscribed into the upright quarter arc. Dividing the arc in a

5.3 The Chord Approximation Approach

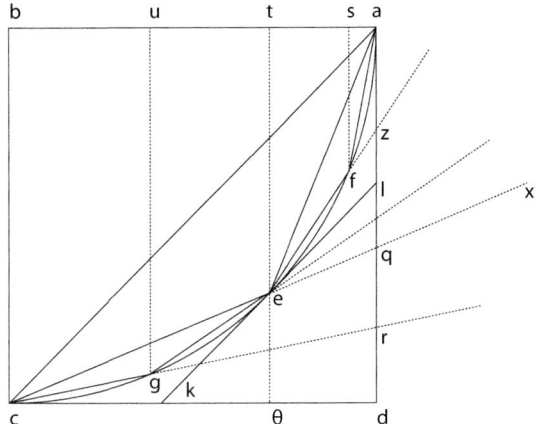

Fig. 5.7 Schematic representation of the central motion diagram on page 166 recto. The diagram has been simplified to only show the chord, two-chord and four-chord approximation of the arc

similar fashion into four and eight parts, respectively, and spanning these by chords, Galileo produced two more paths of motion that approached the arc more and more closely. I will refer to these paths as the *four-chord* and *eight-chord approximations*, respectively (cf. Fig. 5.7).

Galileo began his considerations by calculating the times of motion along each of the individual chords that made up the polygonal paths, where motion was always conceived as proceeding from rest at the apex *a* of the arc. The overall time of motion for the different approximations was then obtained by simply adding up the times of motion over all the individual conjugate chords of the respective approximation.

Initially, Galileo assumed the vertical radius to measure 180 distance units, a choice that was basically dictated to him by the width of his page. Indeed, measured in Galileo's standard distance unit, the punto, the edge length of the square in which the quarter arc was inscribed and which almost completely fills the width of the page, amounts to 180 punti. In other words, the diagram is a scale drawing with respect to the data used in Galileo's calculations. Proceeding in this manner had the apparent advantage that the results of calculations, such as those of the lengths of lines, could be controlled directly by means of the diagram. Next, by distance time coordination, Galileo chose the time of fall along this vertical distance to measure 180 time units. Galileo arbitrarily called this time unit "minute," to avoid confusing those numbers representing times and those representing distances. His calculations and considerations based on this particular initial choice of the basic distance and time unit will be referred to as the *base unit 180 approach*.

With the choice of a time and a distance unit, the problem was determined, i.e., all other times of motion along the different paths considered needed to be derived and calculated on the basis of this initial choice, by applying the laws of naturally accelerated motion. Galileo recorded the results of the calculations of the times of motions along the various paths up to the four-chord approximations in a table underneath the main diagram, reproduced here and rendered in a slightly

Table 5.1 List of results for one-chord to four-chord approximations for the base unit 180 approach on folio 166 recto

	longa puncta						
ad	180	sit tempus casus per ipsam m[inutum]	180	et per ambas *adc* m[inutum]	270		
ac	254 3/5	m[inuti]	254 3/5				
ae	138	tempus casus per eam post *ac* m[inutum]	75	et per ambas *aec* m[inutum]	239		
ec	138	tempus casus per illam m[inutum]	164				
af recta	70 1/2	tempus	113 1/2				
fe	70 1/2	tempus casus post *af*	48	et per ambas *afe* m[inutum]	161 1/2	et per 4 *afegc*	236 1/2
eg	70 1/2	tempus	39	et per ambas *egc*	75		
gc	70 1/2		36				

more legible form as Table 5.1.[43] This table not only listed the times of motion along the individual chords, but these times were also added up to yield the overall times along the various polygonal approximations considered. In addition, Galileo also listed the time of motion along the path adc, i.e., a vertical fall succeeded by a horizontal motion, which resulted from an application of his double distance rule.[44]

The calculations required to establish the times listed in the table are mostly no longer extant. However, one single folio has been preserved, folio 184 recto, which contains some of the calculations required and sufficient information to reconstruct the general method according to which Galileo could, and in all likelihood did, arrive at the other results listed. With reference to Fig. 5.8, in which a reduced version of the central diagram on folio 166 recto is rendered, the time of motion along a chord gc from rest at point a, $t(eg|a)$ could universally be determined in the following manner:

Determine the vertical distance of the highest point of the chord from the horizontal ba, i.e., for the chord ge determine by geometry:

$$te.$$

Extend the chord to intersect with the horizontal ba at x. Determine the length of the extended part, i.e., for the chord ge determine by geometry:

$$xe.$$

Assume according to distance time coordination that:

$$t(bc) = bc.$$

---

[43] Folio 166 recto, table C01.
[44] For Galileo's double distance rule, see Chap. 10.

5.3 The Chord Approximation Approach

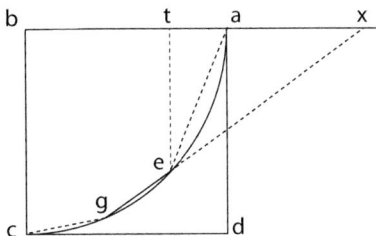

Fig. 5.8 Schematic rendering of Galileo's central motion diagram on folio 166 recto. Only those elements of the construction that are necessary for the calculation of time of motion along chord *eg* from rest at *a* have been included in the diagram

By law of fall:

$$t(te) = mp(bc|te).$$

By length time proportionality and above:

$$t(xe)/t(te) \sim xe/te \Rightarrow t(xe) = xe/te * mp(bc|te).$$

By above and generalized law of fall:

$$t(eg|x)/t(xe) \sim (mp(xg|xe)-xe)/xe \Rightarrow t(eg|x) = (mp(xg|xe)-xe)/xe * t(xe).$$

By path invariance:

$$t(eg|x) = t(eg|a).$$

In the central motion diagram on folio 166 recto, a path made of eight conjugate chords with the junction points on the arc is constructed, but no results of numerical calculations pertaining to motion along this path are listed in the table underneath the diagram. Indeed, the numbers in the table reveal the probable reason why Galileo did not simply continue his calculations. Looking at the table, we first of all realize, as Galileo will likely have done, that the time elapsed during the motion from *a* to *c* monotonously decreases with an increasing number of conjugate chords that make up the path considered. Furthermore, the time difference between motions along two successive approximations of the arc got smaller with the increasing number of conjugate chords. This behavior was in accordance with the assumption that just as the path of motion converged toward the arc, so should the time of motion converge, namely, to the actual time of motion along the arc, and the behavior observed in these trial calculations would indeed further the expectation that an argument by exhaustion should eventually be possible.

The overall times of motion Galileo had calculated for the two-chord and the four-chord approximations, 239 and 236 1/2, differed from each other by only about one percent. As Galileo did not use decimal fractions to express parts of a magnitude smaller than unity, but instead rounded the numbers to even multiples of fractions, mostly one half to one ninth, he had reason to change to a smaller base unit to guarantee the appropriate accuracy for his still outstanding calculation for the eight-chord approximation.[45]

---

[45] In rare cases, Galileo would also use multiples of fractions smaller than one tenth to express magnitudes smaller than the unit. Thus he had no problem in principle with expressing small

In fact, in a second approach, that I refer to here as the base unit 100,000 approach, Galileo chose the vertical radius to measure 100,000 distance units and, by distance time coordination, the time to traverse this distance by a freely falling body to be 100,000 time units.[46] Galileo laboriously recalculated the times of motion for the one-chord to the four-chord approximations he had already calculated using the base unit 180 approach and then proceeded to the calculation of the times of motion for the eight-chord approximation. None of these calculations, which must have followed the schema presented above and filled page after page, are preserved. Yet a list of their results has survived on folio 183 recto. Given the calculational effort to even calculate the time of motion along an individual conjugate chord, this is an indication of the importance Galileo attached to those calculations.

For now, it suffices to note that the results Galileo drafted on folio 183 recto show no regularity that would have indicated how he could move on to the next higher approximation, i.e., the 16-chord approximation, without having to calculate the times of motion along each of the 16 individual conjugate chords, let alone how he could have made an inference about the limiting case. On the other hand, the times of motion for the four-chord and the eight-chord approximations, respectively, differed by less than two in a thousand. Hence, Galileo could reasonably have expected that the actual time of motion on the arc would not deviate substantially from the time of motion calculated for the eight-chord approximation.

Nowhere on folio 166 recto nor, in fact, on any other page in the manuscript does Galileo explicitly spell out the program behind his broken chord approach. Nowhere does he state explicitly that it was his aim to show that motion along the arc and thus, ultimately, pendulum motion was isochronous. In fact, he makes no explicit allusion at all in the manuscript to pendulum motion or even to motion along the arc. Yet the remaining entries on the page provide a first indication that Galileo was actually seeking a way to determine the time of motion along the arc. First of all, in an entry on top of the page, Galileo noted the result of a calculation of the time that would elapse during naturally accelerated motion through a vertical distance equal to the arc's length.[47] This time represented the absolute lower limit for the time elapsed traversing the arc. On the other hand, Galileo may have thus checked whether the time of motion along the arc is in any meaningful relation to the time of vertical free fall along the same distance. Secondly, on 166 recto Galileo formulated a hypothesis regarding the ratio of the times of descent along a chord and its respective broken chord, which I will refer to as the *root-root hypothesis*.[48] This hypothesis, as will be seen, could potentially have provided the basis for a general method to infer from the time of motion for a given approximation the time of motion for the next higher

---

fractions. Yet going over to a smaller base unit to avoid having to deal with fractions, as in the example at hand, seemed to have offered the more convenient option.

[46]It is striking, yet likely coincidental, that on 166 recto Galileo specified the length of the radius of the circle to measure 100,000 units, while when illustrating the law of chords in his letter to Guidobaldo del Monte, he used the example of an arc whose radius measured 100,000 miles.

[47]166 recto, entry T1C.

[48]166 recto, entry T3A.

5.3 The Chord Approximation Approach

approximation. In fact, Galileo elaborated and checked as many as three different variants of the hypothesis. These considerations, which are dispersed on various folios of the exact ratio group, will be discussed in detail below. Yet none of the variants of his hypothesis proved tenable.

A last and rather decisive clue that the considerations on folio 166 recto aimed at establishing a relation between the time of motion along the chord and along the arc spanned by this chord is provided by the remaining two entries on the page, as well as from related calculations preserved on folio 192 recto. In line with his announcement in the letter to Guidobaldo del Monte that he was seeking a way to demonstrate the isochronism of the pendulum, Galileo did not merely consider approximations of a 90-degree arc by different polygonal paths. In addition to calculating the times of motion over these paths, he also calculated the times of motion over the broken chords *egc* from rest at point *e*, and 89*c* from rest at point 8, i.e., the times of motion along paths made up from conjugate chords of equal length inscribed in arcs spanning 45 and 22 1/2 degrees, respectively.

Just as the motions along all chords inscribed in the same circle ending at the nadir proceed in the same time, it could be expected that if instead of motion along such chords, motion along the corresponding broken chord was considered, then these motions would likewise be isochronous—at least for comparable broken chords, for instance, such broken chords where the junction point bisected the arc spanned in two equal halves. Yet calculations of the times of motions along such broken chords from rest at points positioned lower than the original starting point *a* revealed that these times of motion were not equal but in fact dependent on the choice of starting point. Thus, if motion on a chord was compared to motion along the respective bisecting broken chord, also the relative reduction in time turned out to differ with the angle of the arc considered. In the last two entries on 166 recto, Galileo started to further analyze this feature numerically.

The first of his entries shows that Galileo calculated the time of motion along the broken chord *egc* based on assuming that the time of motion along the chord *ec* from rest at *e* was measured by the length of *ec*, i.e., assuming that $t(ec) = ec = 76,536$. The time along the broken chord *egc* had resulted to $66,326$.[49] Next, Galileo assumed that the time saving with respect to motion on the single chord was the same as that when going over from the single chord *ac* to the broken chord *aec* and, under this assumption, calculated a hypothetical time of motion along *egc* of $71,757$.[50] The time calculated based on this hypothesis thus turned out to be longer than the actual time he had calculated by applying the laws of naturally accelerated motion. Galileo concluded that "movetur ergo citius per *egc* quam per *aec*."

The word faster ("citius") was then used here by Galileo in a somewhat uncommon manner.[51] Motion along *egc* is completed in but a little less time than

---

[49] The calculations are not preserved.

[50] 166 recto, entry T3B.

[51] In the *Notes on Motion*, Galileo does not usually use *citius*, the comparative of *citus*, synonymously with *velox* or *velocior*, respectively. Rather citius is almost always used as meaning in "in

that over *aec*; the distance covered is, however, almost half of that of the latter motion. Motion along *egc* is hence not faster in the kinematical sense of velocity as defined by the kinematic propositions. It is really faster only in the sense literally implied by Galileo's statement, namely, that the relative saving of time compared to motion on the single chord is greater in the first case than in the second one. Galileo repeated the same calculation for motions along broken chords from points *e* and 8. The result could not be compared directly to motion along *aec*, however, because Galileo had again used new distance and time units in these calculations, probably for reasons of convenience.[52]

This may have resulted in the idea of referring the times of motion along broken chords spanning different arcs to a standard time of 1000 time units for motion along the single chord. The times which Galileo thus calculated were noted by him as the last entry in the upper left corner of the folio.[53] These numbers are a direct measure for the relative saving in time which, as the numbers showed, increases as the angle subtended by the arc considered decreases.

From a modern perspective, the observed tendency reflects the fact that fall along the arc is not isochronous but takes longer the larger the angle of the arc. From Galileo's perspective, this result may have been unexpected and even puzzling. It did, however, not firmly threaten his conceptualization of the relation between swinging and rolling. The observed tendency for motion along broken chords by no means implied that motion along the arc could not be isochronous. Indeed, given the degree of freedom in the choice of the position of the junction point, Galileo's results did not even rule out finding broken chords along which motion would be isochronous for any two given arcs spanning different angles. Thus, from Galileo's perspective, the observed behavior had at best opened a new dimension of the problem under investigation.

The exact ratio group—the search for the ratio between the time of motion along a chord and broken chord revisited As indicated above, construing an argument based on exhaustion would have required finding a general way of inferring the time of motion for the respective next higher approximation from the time of motion for a given approximation. Search naturally begins by determining how the times of motions over the single chord and the broken chord spanning

---

less time." If the motions compared take place over equal distances the meaning of *citius* of course falls together with that of *velocior*. In his draft of Proposition XXXII of the Second Book of the Third Day of the *Discorsi* on folio 33 recto, Galileo, for instance, states: "Si in horizonte sumantur duo puncta, et ab altero ipsorum quaelibet linea versus alterum inclinetur, ex quo ad inclinatam recta linea ducatur, ex ea partem abscindens aequalem ei quae inter puncta horizontis intercipitur, casus per hanc ductam citius absolvetur quam per quascunque alias rectas ex eodem puncto ad eandem inclinatam protractas. (EN VIII, p. 253)" What is subsequently proven is that the motion claimed to be "citius absolvetur" takes less time than the motion which it is compared to without it being implied that it is also faster (velocior) according to Galileo's understanding. On 163 recto Galileo uses "duplo citius" to mean in half the time.

[52] 166 recto, entry T3B.
[53] 166 recto, entry T1D.

5.3 The Chord Approximation Approach

the same arc are related. Galileo's search for the exact ratio of these times was hampered by the fact that it was not a fixed magnitude but depended on an additional constraint, namely, the choice of position of the junction point on the arc, and Galileo thus needed to find a geometrical construction which would reflect this dependence and allow him to infer the ratio of times.

In the following, subsumed under the heading exact ratio group, folios 148, 153, and 176, and in addition the double folio 156–157, will be examined. On this group of folios, in what comes into view as a heuristically guided trial and error procedure, Galileo explored different ways of determining the ratio of the times of motion over a chord and the broken chord spanning the same arc. The different ways of determining the ratio he probed were based on varying a hypothesis he had initially noted on folio 166 recto. There, this root-root hypothesis, as it has been addressed above, was expressed in the following manner:

> Considera num tempus per *ac* ad tempus per [du]as *aec* sit ut radix radicis lineae quae a centro *b* super *ac* cadit perpendiculariter, ad radicem radicis perpendicularis ex eodem centro super *ae*.[54]

In the diagram on 166 recto, Galileo did not represent the distances of which he intended to take the roots of roots. In Fig. 5.9, Galileo's "lineae quae a centro *b* super *ac* cadit perpendiculariter" has been rendered as line *by* and his "perpendicularis ex eodem centro super *ae*" as line *bz*. A complication regarding the interpretation of Galileo's statement is immediately obvious. The time of motion along the single chord *ac* is greater than the time of motion along the broken chord *aec*. Hence the ratio $t(ac)/t(aec)$ must be greater than one. Yet the line *by* is shorter than the line *bz* and hence the ratio by/bz must be smaller than one. In consequence, also the root of the root of this ratio must be smaller than one, and superficially Galileo's statement does not seem to make sense at all.

If it is assumed, however, that Galileo was merely referring to the two magnitudes from which he wanted to form the ratio of the "roots of the roots," without implying the order in which the ratio had to be formed, his statement allows for a meaningful interpretation which can be transcribed as:

$$t(ac)/t(aec) \sim \sqrt{\sqrt{by}}/\sqrt{\sqrt{bz}}$$

What, if any, was the rationale behind Galileo's root-root hypothesis? When Galileo devised this hypothesis, he had already calculated, more than once, the time of motion along a single chord and the time of motion along the corresponding broken chord. In these numerical examples, the difference between these times had shown to be small and clearly not in proportion to the difference between the length of any lines in the geometry of the problem, which could uniquely be related to either of the two motions, such as, for instance, the lines *by* and the *bz*. Within Galileo's mathematical framework, a way of expressing such disproportionately

---

[54] 166 recto, entry T3A.

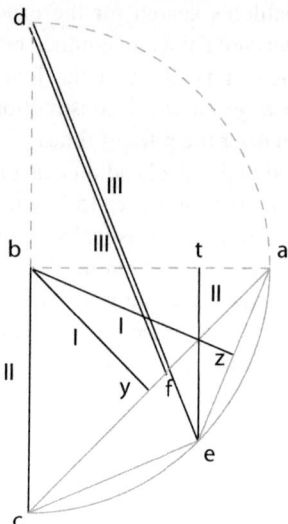

Fig. 5.9 Diagram illustrating the different variants of Galileo's root-root hypothesis. The respective pairs of distances from which Galileo calculated the ratio of the roots of roots have been numbered I to III according to the alternative versions of the hypothesis. The diagram has been composed to clarify the different variants of the root-root hypothesis and is not contained in the manuscript in this form. The constructions required by the different variants of the hypothesis are embodied in different diagrams contained on different folios of the manuscript

small change was in fact to take the root. Conceivably, Galileo first tried to take the root of the ratio between *by* and *bz*.

Indeed, next to his statement of the root-root hypothesis the number 229 1/2 is written. If the ratio of the root *bz* and the root of *by* is calculated and multiplied with the time of motion to fall along the chord *ac*, which in Galileo's base unit 180 approach measured 2543/5 time units, one arrives at a value of 230. It thus seems very likely that the number 229 1/2 was indeed the result of a corresponding calculation by Galileo that is no longer extant. Given that the calculations would have required extracting a root of two numbers, the small difference to 230 can be interpreted as due to rounding error. The number 229 1/2 differed from the exact value of 236 1/2 time units, which he had calculated for the time of motion along the broken chord *aec* by applying his laws of naturally accelerated motion, and if it was indeed calculated based on the ratio of the roots, this would have shown Galileo that and how his assumption would have to be altered. Instead of in proportion to roots of the lines *by* and *bz*, the ratio of the times could be in proportion to the ratio of the roots of the roots of these lines.

Quite possibly, by the root-root hypothesis, Galileo not only sought a way to relate motion along the single chord to motion along the broken chord spanning the same arc but may also have been striving to find a general method that would enable him to infer the times of motion for higher approximations and thus possibly even that of motion on the arc itself, from knowledge of the time of motion along

5.3 The Chord Approximation Approach

the single chord. Galileo's calculations of the times of motion up to the eight-chord approximation of the arc provided a good basis for extrapolating the way in which the time of motion along a regular polygonal path would change in moving to ever higher approximations. Until now, the more closely the polygonal path considered resembled the arc, the smaller the time of motion along it had become. At the same time, the difference of the times of motion for a given and the next higher approximation also decreased, and there could have been little reason to doubt that this tendency would prevail in going to ever higher approximations and that thus, as Galileo expected, the time of motion along a polygonal path of an increasing number of sides would converge to the actual time of motion along the arc.

If the root-root hypothesis was indeed intended to be applicable to polygonal paths representing approximations of higher order, it would qualitatively conform with this expected behavior. The line bz, falling perpendicularly on what is then interpreted to be the first chord of the approximation, would get longer with the order of approximation, the change of length diminishing with the order of the approximation to finally converge to the length of the radius of the arc.[55] Hence the time of motion would monotonously decrease with an increase of the number of conjugate chords considered and converge to a fixed and easily determinable value for the case of an infinite number of sides. Heuristically, the root-root hypothesis thus made absolute sense in that it qualitatively reproduced the behavior Galileo could expect of the correct solution based on his prior calculations and his conceptualization of the problem.

Yet the root-root hypothesis as Galileo had formulated it on folio 166 recto, and most likely also tested, was not only numerically untenable.[56] It also failed to correctly reflect the way in which the time of motion changes when the position of the junction point of the conjugate chords is changed. If the junction point of a broken chord approaches the starting point of motion, time of motion along the broken chord should approach that of the single chord as the paths traversed in both cases become more similar. When the junction point falls together with the start or endpoint of motion and the broken chord degenerates into a chord, the times of motion should finally be equal. According to Galileo's root-root hypothesis,

---

[55] Had Galileo indeed calculated the time of motion along the arc as the limiting case in the way described in the text, in the units of the base unit 100,000 approach, this would have yielded a time of 12,968 for motion on the arc. In comparison, he calculated a time of 131,078 time units for motion along the path made up of eight conjugate chords approximating the arc. From a modern perspective, the correct solution would be 130,622 time units. That the root-root hypothesis potentially applied not only to the broken chord but also to paths composed of a higher number of chords, thus allowing an inference to the limiting case of motion on the arc, however, remains merely plausible speculation.

[56] The hypothetical value that is calculated according to the root-root hypothesis for the time of motion along the broken chord aec is 242 time units. This overshoots the value of 236 1/2 arrived at by doing the calculation based on the laws of motion by about as much as it is undershot if instead of the ratio of the roots of the roots, only the ratio of the roots is used for the calculation.

however, the time of motion along the broken chord becomes smaller the closer the junction point is to one of the terminal points of motion.[57]

There is indeed some evidence that Galileo engaged in considerations of this kind and in consequence tested two more variants of his root-root hypothesis. Both variants are marked by the fact that the small change in the time of motion in going over from one approximation to the next higher approximation is still represented by the change of the roots of the roots of appropriately chosen distances in the geometry of the problem. These alternative variants of the original hypothesis, however, better reflect the dependence of the ratio of the times of motion on the position of the junction point.

On folio 176 recto, calculations can be identified which document that Galileo checked whether the ratio of times of motion he sought was possibly given by the ratio of the roots and, after this failed, by the ratio of the roots of the roots of the radius of the arc and the line that marked the height of the junction point beneath the horizontal. With respect to Fig. 5.9, he thus first tested whether:

$$t(ac)/t(aec) = \sqrt{bc}/\sqrt{te}$$

and after this failed to reproduce the correct behavior, whether:

$$t(ac)/t(aec) = \sqrt{\sqrt{bc}}/\sqrt{\sqrt{te}}$$

In his calculations on 176 recto, Galileo not only made use of the times of motion he had calculated and noted on folio 166 recto but also employed the results of corresponding calculations on folio 189 recto, the folio bearing the record of the pendulum plane experiment. According to the results, the hypothesis tested decently conformed with the times of motion he had inferred based on applying the laws of naturally accelerated motion. Yet the numbers calculated as roots of the roots were small, and thus the calculations were afflicted with a rather large rounding error. Moreover, the variant of the hypothesis tested on 176 recto only qualitatively shows the right behavior if the junction point is positioned between c and e with regard to the diagram in Fig. 5.9. In consequence, Galileo's calculational results are not too compelling. This may indeed have discouraged Galileo from further investigations along this line. In any case, this variant of the root-root hypothesis does not reoccur on any other folio of the manuscript. It is noteworthy that on the very same folio page, 176 recto, Galileo explored a geometrical property of a construction which, as

---

[57] According to the hypothesis, moving the junction point of a broken chord closer to the starting point of motions has the same effect as going over to an approximation of higher order, as it changes, the first incline of which the polygonal path is composed in the same manner and on which the construction of the hypothesis hinges. However, whereas the time of motion along a broken chord should get longer as the junction point moves toward the start point of motion, the time of motion along a path more closely resembling the arc should decrease. The construction underlying Galileo's hypothesis was not capable of correctly reproducing both inverse tendencies at the same time.

5.3 The Chord Approximation Approach

will be argued in Chap. 7, had resulted from his consideration of the configuration under which a bent plane was traversed in least time.

The third and most articulate variant of the root-root hypothesis can be identified on double folio 156–157. With respect to the central diagram (see Fig. 13.15), Galileo calculated the number 135,118 from the ratio of the roots of the roots of the lines ao and as (corresponding to de and df in Fig. 5.9). This represented the time of motion along the broken chord acb. The lines Galileo now considered in the ratio of the roots of the roots started from the high point of the circle and no longer fell perpendicularly onto the single chord and on the first of the broken chords, respectively, as they had done on his first construction. Rather both lines had the same direction determined by the connection of the apex of the circle and the junction point of the broken chord.

One advantage of this new construction is immediately obvious. This third variant of a root-root hypothesis qualitatively reproduced the desired behavior of times of motion along the broken chord if the position of the junction point was moved either toward the starting point or endpoint of the motion. In both cases, the lines considered converge with each other, and hence, according to this variant of the hypothesis, the time of motion along the broken chord will converge toward the time of motion along the single chord spanning the same arc.

The rationale behind this third variant of the root-root hypothesis can be reconstructed. The broken chord path motion along which Galileo centrally considered on the double folio was not, as might be expected, the path acb, but the path bsg. Both paths are identical save for their orientation, and hence with regard to kinematics, motions along these paths can be treated as interchangeable. Assuming the ratio of the times of motion along chord bg and the broken chord bsg to be given by the ratio of the roots of the roots of the distances ao and as is, as can be shown, equivalent to the assumption that the inclined plane ar in the diagram is traversed in the same time as the broken chord bsg itself. Thereby, Galileo had defined point r as the mean proportional between ao and as.

What would be known about the position of point r on as such that the motions along bsg and along ar were completed in the same time? It would first of all have been clear to Galileo that this point had to be positioned between o and s, which marked a lower and an upper limit for the length of the line ar. The point o marks a lower limit for the line ar because, according to Galileo's construction, co is equal in length to the lower of the broken chords cb. Plane co, however, is more steeply inclined than cb. Hence motion over aco must take less time than motion over the broken chord abc (or, what is the same over bsg). The line cs represents an upper limit for the sought plane ar, because according to the law of chords this cs is traversed in the same time as the single chord ab, which, according to the law of the broken chord, is traversed in a longer time than the broken chords acb or bsg.

Not only must point r thus be positioned between the o and s, but Galileo's calculations had furthermore shown that it should lie somewhat closer to o than to s. Thus, rather than the arithmetic mean, it could be assumed that the point r marked the geometric mean between the distances ao and as. Thus ar to as is as the root of ao to the root of as, and consequently, the times of motion along ar and as are as the root of the root of *ao* to the root of the root of *as*. However, the time of 135,118 units, which Galileo in effect calculated from the time of motion along the chord and the root of the root of the ratio of *ao* and *as*, did not correspond well to the time of 132,593, which he had calculated and noted on 183 recto for motion along the broken chord.

As already discussed, Galileo adapted his construction from double page 156, 157 to the geometrical construction based on which he had constructed his law of the broken chord. He did this on folio 148 recto. The diagram on this page, in particular, comprises the arc *dcsg*, the chords *ds* and *cg*, the point *o* marking the intersection of the latter two chords, as well as the mean proportional between the lines *do* and *ds*, which is marked but remained unlettered. Underneath the diagram Galileo had conformably constructed a timeline on which he introduced a point t' such that the distance qt' would represent the time of motion along the broken chord *dbc*.[58] This time could be compared directly to the hypothetical time which resulted according to the root-root hypothesis by transferring the appropriate distance to the timeline, albeit, with negative result.

For the time being, Galileo was apparently not able to determine the position of point *r* on the extended chord *db* such that the broken chord and the inclined plane *ar* thus defined would be traversed in equal time. In the *Discorsi*, a solution to this particular problem is presented as Proposition XV. The folios that can be identified in the manuscript on which Galileo elaborated this proposition with certainty all date, however, from the Florentine period. Obviously the solution to this particular problem, which had first cropped up in the 1602, was not found until much later.[59]

In summary, Galileo's work on the root-root hypothesis presents a particularly interesting example of how he developed a hypothesis as to the general solution of a problem in a situation where he knew a number of the conditions that such a solution had to fulfill from his prior considerations. The hypothesis was subsequently numerically tested based on comparing what it predicted for a concrete case, with what Galileo could calculate for this concrete case based on his laws of motion. Failure to correctly predict the case, but also to meet the constraints of the general solution, fed back and led to a suitable modification of the hypothesis twice. Galileo's heuristic attempt to find, along this line, a general method for determining the ratio of the time of motion along a chord and the corresponding broken chord spanning the same arc relates a number of folios of the manuscript, thus adding another tile to the mosaic of the overall interpretation.

---

[58] The point in question is marked with a dotted line and labeled *t* by Galileo. To distinguish it from the point labeled *t* on the timeline, it is referred to here as t'.

[59] In Proposition XV of the Second Book of the Third Day of the *Discorsi*, the first motion proceeds over a vertical and the successive one over an inclined plane. However, the proposition can easily be generalized to apply to the case of a broken chord in which the first motion likewise proceeds along an inclined plane as well. Folios 61 and 76, on which Galileo elaborated this proposition, both bear a radiant sun as a watermark, and the paper is clearly of Florentine origin. It thus appears that Galileo was only much later able to solve this particular problem that he first encountered in the context of the broken chord approach.

5.3 The Chord Approximation Approach

The chord-approximation approach in context—dating Galileo's research program to 1602 Folios 121 and 127 contain auxiliary calculations for Galileo's considerations in the context of the broken chord approach discussed above, and analysis of their content hence allows a contextualization of the broken chord approach. The considerations laid out on folio 121, in this way, eventually allow a dating of the broken chord approach to the end of 1602. It is thus established beyond doubt that Galileo's investigations in the context of the broken chord approach, as they are preserved in the manuscript, indeed correspond to the challenging research program Galileo had alluded to in his letter to Guidobaldo del Monte in the fall of the same year. Before turning to a discussion of folio 121, I will, however, briefly review folio 127. The content of this folio also provides an important piece of information for contextualizing the broken chord approach.

Folio 127 recto, which is part of the double folio 126, 127, bears but a single auxiliary calculation related to the broken chord approach. By means of this calculation, Galileo checked the consistency of the results he had gained from his base unit 180 approach and from his base unit 100,000 approach.[60] Besides this auxiliary calculation, the double folio contains some geometrical sketches of unidentified purpose as well as elaborations of three propositions, including a proof. Two of these propositions belong to the group of least time propositions; the third is a variant of the so-called generalized law of fall. According to the interpretation that will be advanced in Sect. 7.3, all three propositions drafted on the folio had been formulated by Galileo as the direct outcome of his elaboration of the problem of motion along bent planes under a varying position of the junction point. Moreover, some of the geometrical sketches comprised on double folio 126, 127 reappear in identical form on double folio 156, 157, implying that both double sheets were in use at approximately the same time. The overall impression conveyed by the multiple relations between the folios in question is that of a number of folios lying on the desk and being concurrently used in a limited period of time during an intense phase of work, characterizing Galileo's work of the research program on the relation between of swinging and rolling. This work will, in the following, be dated to 1602, independently of information bequeathed by the letter sent to Guidobaldo del Monte in the fall of that year.

The purpose of most of Galileo's entries on folio 121 verso cannot be reconstructed unambiguously. Two calculations on the page, however, can clearly be identified as auxiliary calculations pertaining to Galileo's elaborations documented on the folios of the core group. Galileo computed the length of the line $e\theta$ in the central motion diagram on 166 recto (compare Fig. 5.7). The same calculation is in fact carried out twice on 121 verso, once in the units of the base unit 180 approach

---

[60]The calculation 127 recto D01A, $141{,}422 * 239/254 \ 3/5 = 132{,}75[6]$, can be interpreted as a consistency check of the result calculated for the times of motion along the broken chord aec with respect to a base unit of 100,000 and a base unit of 180 on folio 166 recto. It must hold that: $t(ac)_{base\ 100{,}000} * t(ac)_{base\ 180} / t(aec)_{base\ 180} = t(aec)_{base\ 100{,}000}$. Aside from rounding errors, the calculated result should hence correspond to the time of motion along aec with respect to the base unit of 100,000 as it is listed on folio 183 recto, namely, 132,593.

and a second time in the units of the base unit 100,000 approach. Thus folio 121 must have been on Galileo's desk, readily available for jotting down these auxiliary calculations, when he was pursuing his problem of approximating motion on an arc by motion along a series of conjugate chords, centrally documented on folio 166.

Otherwise, 121 recto is completely covered by a geometrical construction. Upon closer inspection this diagram turns out to be a near identical copy of a diagram contained in Gilbert's *De Magnete*.[61]

In 1600, Gilbert published what was quickly to become the period's most important work on magnetism. It was in this book that Gilbert linked the polarity of a magnet to the polarity of the earth and investigated the effects that the magnetism of the earth had in orienting a magnetic needle. In Chapter V of Book V of *De Magnete*, Gilbert introduced his primary empirical tool for the study of this phenomenon, the *terrella*. The terrella consisted of a perfectly spherical magnetic stone placed in a wooden compartment paired with a magnetic needle, a so-called *versarium* which was fixed at the center of a quadrant. The magnetic stone was allowed to turn beneath this arrangement such that, by means of the quadrant, the declination of the *versarium* could be measured for any latitude.[62] With this arrangement Gilbert intended to mimic the declination of a magnetic needle by the earth's magnetic field, which could in this way be studied.

In the succeeding chapter, Chapter VI, Gilbert introduced the problem of finding the correct "proportion of declination," i.e., of determining the degree of deflection of the needle toward the earth depending on the latitude. In hindsight, the construction Gilbert provided as a solution to this problem appears to have emerged as the result of an interpolation between two known limiting cases—the orientation of the magnetic needle on the pole and on the equator. At the pole, the needle should point to the center of the earth, whereas at the equator it should be tangential to the surface of the earth. In order to determine the declination of the needle for latitudes between these two limiting cases, Gilbert basically constructed a spiral line such that the inclination tangent to this spiral at any given latitude would represent the declination (i.e., inclination) of the magnetic needle.

After an exemplary construction for the determination of the declination at a latitude of 45 degrees, schematically reproduced here as Fig. 5.10, Gilbert issued a more complete diagram in which the declination of the needle was constructed for every fifth degree between the equator and the pole.[63] It was precisely this

---

[61] That the diagram on 121 recto is a modified copy of a diagram from Gilbert's book was realized by Drake. The way in which he relates the remaining content of the folio to Galileo's work program outlined in the letter to Guidobaldo del Monte in 1602, is tentative. Cf. Drake (1978, 67).

[62] Gilbert's book is in the inventory list of Galileo's library (Favaro 1886, 261, nr. 200). According to Drake, Galileo's copy of the book was most likely given to him by Cremoni, who "seemed afraid to keep it on his shelves lest it infect his other books" (Drake 1978, 62–63). The discussion in the text is based on Gilbert (1958).

[63] The diagram is reproduced from the English edition, *On the magnet, magnetick bodies also, and on the great magnet the earth: A new physiology, demonstrated by many arguments & experiments* from 1900. It is reproduced from the Project Gutenberg EBook (http://www.gutenberg.org/files/

5.3 The Chord Approximation Approach

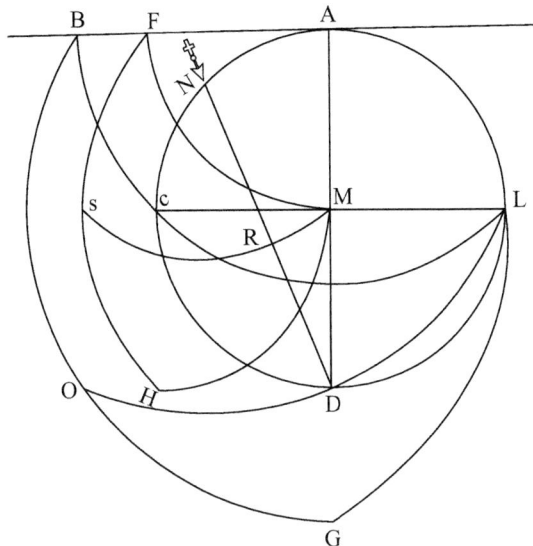

Fig. 5.10 Gilbert's construction for the determination of the declination of a magnetic needle depending on the latitude of the observer's position on the earth. The circle $ACDL$ represents the *terrella* and the earth, respectively. AD is the equator; C and L represent the poles of the earth. This particular diagram serves the determination of the declination of the magnetic needle at point N at 45 degrees latitude. The connection between N and D gives the declination of the magnetic needle

construction that Galileo copied onto folio 121 recto, whereby he, with one exception, did not consider every fifth degree as Gilbert had done but only every tenth degree between 0 and 90 degrees. In addition, Galileo introduced a slight but practical modification (Fig. 5.11).

In addition to the degree scale on the outer circle, Galileo added another degree scale starting at zero on the inner circle (representing the circumference of the earth), ran counterclockwise, and ended at the three o'clock position.[64] This second scale was not given by Gilbert and is only implicit in his construction. In addition, Galileo marked the degrees of latitude by ten small circles starting from 0 at the equator, up to 90 degrees at one of the poles; the center of an additional, 11th circle marked a latitude of 45 degrees. These circles represent a convenient way of dividing the arc in order to mark the latitudes and were perhaps meant to symbolize small versaria on the surface of the earth. Galileo only needed to mark the latitudes on a quarter of the circle because by definition the problem was symmetrical with respect to the

---

33810/33810-h/33810-h.htm. Accessed 10 Dec 2016) and reused under the terms of the Project Gutenberg License.

[64] With regard to the degree scale, Galileo's version differs from the original diagram by Gilbert, which has otherwise been copied more or less accurately. Galileo drew the diagram with the page turned 180 degrees with respect to the modern orientation. With regard to modern orientation of the folio, what I refer to as the apex of the circle, thus occurs as the nadir.

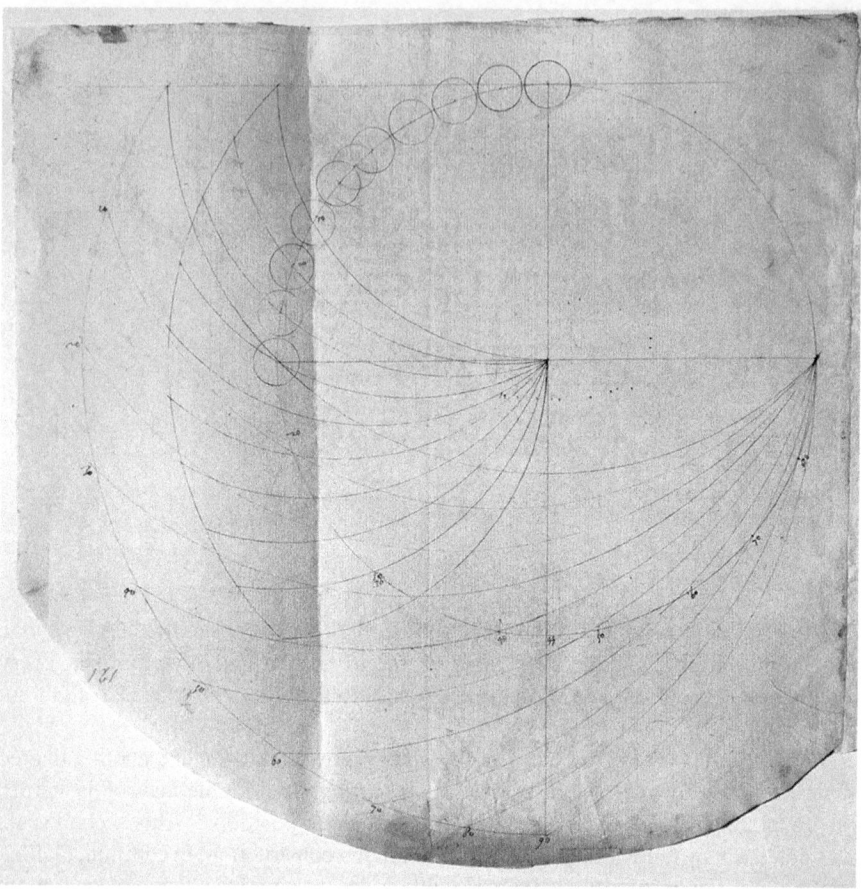

Fig. 5.11 Folio 121 recto, bearing a copy with modification of a diagram from Gilbert's *De Magnete*. (Courtesy of Ministero dei Beni e le Attività Culturali – Biblioteca Nazionale Centrale di Firenze. Reproduction or duplication by any means prohibited)

plane of the equator, as well as to the polar axis. Galileo's new construction made it particularly easy to find the declination of a magnetic needle for a given degree of latitude. One simply needed to connect the position on the earth/terrella specified by the given latitude, with the point marking the same degree on the newly added inhomogeneous scale. The inclination of the line thus produced corresponded to the inclination of the needle at that latitude.

Galileo's correspondence provides a decisive clue about why and when he copied and elaborated the diagram from Gilbert's book. As becomes clear from a letter Paolo Sarpi had sent to Galileo in September 1602, the two men had discussed Gilbert's book on the occasion of their last personal meeting and in particular

5.3 The Chord Approximation Approach

the question of magnetic declination.[65] That Paolo Sarpi, who merely alludes to "il nostro autore" in the letter, was indeed referring to Gilbert and his book *De Magnete* becomes clear in the second paragraph of the letter, where Sarpi discusses a statement that Gilbert had argued for at length in the fifth chapter of his book, namely, that it is not attraction but a "disposing and rotating influence" that causes the *versarium* to take on its orientation with respect to the magnetic stone. Sarpi concurs with Gilbert's opinion.

Paolo Sarpi quite understandably had difficulties with Gilbert's construction serving to determine the declination of the magnetic needle. Having outlined these difficulties, he asked Galileo:

> I beg Your Lordship to have a bit of consideration concerning my difficulties, and to make up for the lack of our author, who was silent on the causes of the most obscure things: he could at least have said how they have come to his attention. (EN X, letter 83)

It was almost certainly upon Sarpi's request that Galileo copied and elaborated the diagram from Gilbert's book. Indeed, the way in which Galileo had modified Gilbert's original diagram on folio 121 recto provides rather clear indications that his construction was a direct response to Sarpi's questions. Among other things, in the letter Sarpi had asked Galileo to explain the construction of the various different scales in Gilbert's original diagram.

> Then, why must they be divided into as many parts as a quadrant, both the big ones and the small ones?

As we have seen, Galileo modified Gilbert's diagram in such a way as to reduce the different scales used in constructing the declination to just one scale, serving essentially the same function. It is thus strongly suggested that Galileo indeed introduced the modification in an attempt to answer the question his friend Sarpi had posed him. The diagram on folio 121 may have been the blueprint for a diagram Galileo sent to Paolo Sarpi along with an answer. However, no letter containing a reply by Galileo has been preserved. Hence, it seems we cannot be certain that folio 121 recto was in fact drafted by Galileo in direct response to Sarpi's request. Galileo may have copied the diagram from Gilbert's book to 121 recto earlier or much later than September 1602. Strictly speaking, all the diagram on folio 121 recto provides us with is the year 1600 as a terminus *post quem*.

It is, however, not the presence of the diagram copied from Gilbert's book on folio 121 recto alone, but the convergence of different, independent arguments which so compel me to date the composition of the diagram to the fall of 1602. On the one hand, Galileo's work on the broken chord approach, as documented in the manuscript, perfectly ties in with the cursory description of a challenging research program Galileo gave in his letter to Guidobaldo del Monte of November 1602. By an inference to the most plausible explanation, we are led to assume that the broken chord approach indeed corresponds to the program outlined by Galileo in the letter

---

[65] For the context of Galileo's and Sarpi's exchange on magnetism and Gilbert in particular, see Heilbron (2010, 97).

and this in turn implies that the auxiliary calculation on 121 verso related to the broken chord approach should date roughly to the fall of 1602. On the other hand, with regard to the diagram on the page, the most plausible hypothesis is that it was copied by Galileo in reaction to the request made by Sarpi in September 1602, i.e., just about 1 month before the letter to Guidobaldo del Monte. Last, but not least, the way the auxiliary calculation was jotted down suggests that this did happen at approximately the same time that the diagram was copied to the page and that hence the diagram was drawn while Galileo was working on the broken chord approach.

None of the three assumptions listed above is in itself absolutely mandatory, yet together they mutually support each other. Indeed, negating any of the three assumptions almost necessarily forces us to likewise reject at least one or even both of the others. If, for instance, one were to maintain, as has been done in the literature, that the content of the folios documenting the broken chord approach was not composed in 1602, but only much later when Galileo allegedly returned to the very problem he had communicated to Guidobaldo del Monte in 1602, one would be forced to assume that Gilbert's diagram was copied to folio 121 not in September 1602 but likewise at a much later time, which coincidentally would have been the time when Galileo was working on the broken chord approach, despite the fact that this work is in no way related to the copying of the diagram. If, in addition, it were upheld that the diagram was indeed copied in 1602, one would even more implausibly have to assume that at a much later point Galileo took the folio out of his drawer to scribble just one auxiliary calculation on it. The simultaneous occurrence of an auxiliary calculation and Gilbert's diagram on the same folio page indicates beyond reasonable doubt that Galileo used this folio to write upon in the fall of 1602 and thus unambiguously allows us to date Galileo's investigations into the relation between swinging and rolling, as they are documented in the manuscript, to that time.

5.4 Summarizing Conclusion

The present chapter has dealt with Galileo's considerations regarding motion along polygonal paths inscribed into an arc. The fundamental idea of his approach, as has been argued, was that swinging and falling along the same arc could be assumed to be kinematically equivalent. Galileo's assumptions about naturally accelerated motion did not allow for an inference regarding the motion of fall on the arc directly. Instead, he represented or approximated motion on the arc by motion along a series of adjoined inclined planes inscribed as chords into the arc. This has been referred to here as the broken chord approach. If the number of inclined planes is increased above all limits, the polygonal path exhausts the arc and Galileo, in this way, hoped to eventually be able to construe an argument regarding falling, and thus swinging, along the arc.

Galileo's new, refined approach toward investigating the relation between swinging and rolling had, as has been argued, emerged after the pendulum plane

5.4 Summarizing Conclusion

experiment had revealed that his less elaborate assumption—the single chord hypothesis, according to which swing along an arc and fall along a chord spanning this arc might be completed in the same time—was not tenable. As will be demonstrated in the succeeding chapter, not only was the new conceptualization stimulated by the experiment but Galileo indeed reinterpreted his experiment in light of the results yielded from his new understanding.

As a first substantial result of his new approach, Galileo was able to demonstrate that motion along a broken chord always proceeded in less time than motion along a single chord spanning the same quarter arc or less. In 1602, in the letter to Guidobaldo del Monte, Galileo mentioned this proposition, the law of the broken chord, and claimed to have found a proof for it. Galileo's construction of such a proof is indeed documented on a number of folios in the manuscript. The proof Galileo thus finally came up with was drafted on 172 verso and 163 recto. It was published, essentially unaltered, as the proof of Proposition XXXV of the Second Book of the Third Day of the *Discorsi*. That the proof was originally drafted in 1602 is demonstrated. It is thus shown, not only that it must correspond to the proof Galileo alluded to in the letter to Guidobaldo del Monte, but, moreover, that it was among the first, if not the first, argument that Galileo wrote down concerning naturally accelerated motion as they were finally published more than 35 years later.

Situating the genesis of the proof of the law of the broken chord in the earliest phase of Galileo's pondering regarding naturally accelerated motion has served to explain a number of oddities that become apparent if the proof is compared to the other proof Galileo included in the Second Book of the Third Day. Galileo managed to establish the proof based on just the law of fall and the law of chords, that is, on exactly the same presumption about naturally accelerated motion on which his pendulum plane experiment had been based. In particular, he did not invoke the length time proportionality as a premise even though this would have greatly simplified his argument. As has been argued, the reason for this is that Galileo was simply not yet aware of the length time proportionality when he constructed his proof of the law of the broken chord. It will be argued in Chap. 7 that Galileo first realized and successively devised a proof for the length time proportionality as a direct outcome of his considerations regarding the law of the broken chord.

The proof required constructing an argument regarding motion of fall not starting from rest but being preceded by a prior motion of fall made along a plane of different inclination. To make such an argument Galileo resorted to path invariance, i.e., the assumption that naturally accelerated motion along a given path would proceed in the same manner after different preceding motions of fall if only the preceding motions had covered the same vertical height difference. The assumption was invoked for the first time in a written out argument, and Galileo therefore further explained it by stating that equal degrees or moments of velocity were reached by fall over the same height regardless of the path taken. This explanation survived into the published proof, where it is, due to the order of exposition where the law of the broken chord is one of the last propositions, however, misplaced. Last, but not least, the proof is the only proof in the *Discorsi* in which times of motion are diagrammatically represented as lines in a timeline separated from the diagram representing the spacial geometry of the motion situation. This unusual technique, from the perspective of Galileo's mature science, complicated his argument. It was in fact his construction of the proof of the law of the broken chord that, as will be argued in Chap. 7, gave rise to a new technique of representing distances covered and times elapsed diagrammatically, which Galileo from then on would use exclusively, as it substantially facilitated arguing about the pheno-kinematics of naturally accelerated motion.

The statement Galileo had thus been able to infer by a well-formed argument based on his assumptions about naturally accelerated motion was, despite its surprising and somewhat contra intuitive nature, supported by a tendency that had been observed in the experiment. Galileo's experimental data had shown him that swinging along an arc took less time than rolling along the chord spanning this same arc. This suggested that if motion along a single chord was compared to motion along a path made up of two adjoined chords more closely approximating the arc, the motion along the latter path, should be completed in less time despite being made over a longer distance. This is of course exactly what the law of the broken chord stated and what Galileo had proven.

This encouraged Galileo to proceed to investigate motion along paths composed of more than just two chords. A great number of folios could be related directly or indirectly to his broken chord approach, which in this case has come into view as a coherent research agenda pursued in an intensive phase of work over a limited period of time. His investigations culminated in calculating the time of motion along a path made up of as many as eight conjugate chords inscribed into a quarter arc. From a modern perspective, this differs only about one in 1000 from the time it actually takes to fall along the arc. Yet Galileo did not know the time to fall along the arc; indeed this is what he was trying to infer. His calculations suggested that the time of motion along the eight-chord path should not be too far off from the solution. However, finding a way to rigidly infer the solution quite generally required finding a way to make an inference from motion along a given path composed of a given number of inclined planes to the motion along another path composed of the higher number of inclined planes, which proved difficult.

Galileo engaged, as was shown, in a heuristic problem-solving strategy. With the root-root hypothesis, he devised a hypothesis regarding the ratio of the times of motion along different paths approximating the arc that was chosen so as to correctly reproduce conditions which, as Galileo knew, the correct solution had to fulfill. Failure to reproduce the results Galileo had concretely calculated applying his laws of motion led him to modify his hypothesis twice, where each of the two further variants of the hypothesis thus produced specifically answered a shortcoming of the previous variant. Yet Galileo was not able to accomplish what he was aspiring to, namely, to come up with an argument allowing him to infer the pheno-kinematics of fall along the arc.

Despite considerable partial accomplishments, overall his approach had bestowed Galileo with more questions than answers with little leads on how to proceed. Galileo's considerations had in fact even yielded indications of a problem in the underpinning conceptualization. Comparison of the motions made along broken chords spanning arcs of the same radius but different angles showed that the times of these motions were no longer equal, as was the case for motion along the respective single chords spanning these arcs, but in fact depended on angle of the arc. From a modern perspective, this is not surprising and reflects the fact that motion along the arc opposed to motion along the corresponding chords is not isochronous. For Galileo, who was firmly convinced and wanted to demonstrate that pendulum motion and thus fall along the arc was isochronous, the tendency that had become evident through his calculations must have represented a serious complication. For motion along the arc to be isochronous, isochroneity lost when going over from motion along the single chord to motion along the broken chord would have had to be resurrected when motion along paths made up of an ever higher number of chords were considered. Based on Galileo's results, this would not have seemed likely but could not be excluded with logical necessity. Expressed the other way around, nothing in his considerations thus far made the conclusion that falling, and thus also swinging along the arc, was not isochronous mandatory.

In a nutshell, Galileo's theoretical considerations regarding the relation between swinging and rolling did not so much fail as lead him into a blind alley. His aspiration to carry his investigation on the relation of swinging and rolling further was brought to an end when he revisited the empirical data produced in the pendulum plane experiment and interpreted, in light of the results achieved, based on the broken chord approach, as will be discussed in detail the succeeding chapter.

References

Clavelin, M. (1983). *Conceptual and technical aspects of the Galilean geometrization of the motion of heavy bodies* (pp. 23–50). Dordrecht: Springer.
Drake, S. (1970). *Galileo studies: Personality, tradition, and revolution*. Ann Arbor: The University of Michigan Press.
Drake, S. (1978). *Galileo at work: His scientific biography*. Chicago: University of Chicago Press.
Erlichson, H. (1998). Galileo's work on swiftest descent from a circle and how he almost proved the circle itself was the minimum time path. *The American Mathematical Monthly, 105*(4), 338–347.
Favaro, A. (1886). La libreria di Galileo Galilei. *Bullettino di bibliografia e di storia delle scienze matematiche e fisiche, 19*, 249–293.
Galilei, G. (1914). *Dialogues concerning two new sciences*. New York: Macmillan.
Gilbert, W. (1958). *On the magnet: [Magnetick bodies also, and on the great magnet the earth; a new physiology, demonstrated by many arguments & experiments]* (Nachdr. der Ausg. London 1900 edn.). New York: Basic Books.
Heilbron, J.L. (2010). *Galileo*. Oxford: Oxford University Press.
Wisan, W.L. (1974). The new science of motion: A study of Galileo's De motu locali. *Archive for History of Exact Sciences, 13*, 103–306.

Chapter 6
Swinging and Rolling Revisited: Motion Along Broken Chords and the Pendulum Plane Experiment

In Chap. 4, we have witnessed Galileo experimentally comparing the swinging motion of a pendulum to that of a heavy body rolling along an inclined plane. The pendulum plane experiment, as has been argued, had initially been designed to test a particular simple hypothesis Galileo held concerning a relation between the two motions, namely, the assumption that the swing of a pendulum from one of its turning points to its equilibrium position is completed in the same time as motion along an inclined plane connecting these two points, thus spanning the arc of the pendulum swing as a chord. Galileo's single chord hypothesis was not confirmed by the experimental data.

In Chap. 5 we surveyed Galileo's extensive and arduous investigations of falling motion along polygonal paths made up of adjacent inclined planes. The polygonal paths Galileo considered were chosen so as to, with an increasing number of inclined planes considered, more and more closely approach an arc. The objective of Galileo's approach was to thus eventually be able to make an inference about the kinematic behavior of rolling motion along a circular arc. Galileo further assumed that rolling motion along an arc supported from below would characteristically show the same behavior as a swinging motion along the same arc, supported, however, from above, i.e., pendulum motion—an assumption which will be referred to in the following as *kinematic equivalence under path correspondence*. A potential inference regarding rolling motion along a circular arc would thus likewise hold with respect to the swinging motion of a pendulum through the same arc, and demonstrating the properties of the latter motion in particular was, as we have seen, what Galileo was striving for.

Galileo reasonably supposed that with an ever-increasing number of inscribed chords making up the polygonal path, just as the path traversed would ever more closely resemble the arc, the time of motion would converge to the actual time of motion on an arc and thus the time of pendulum motion. This assumption will be referred to in the following as *polygonal approximability*. The assumptions of polygonal approximability and kinematic equivalence under path correspondence were the two pillars of Galileo's new, refined conceptualization of the relation between swinging and rolling, replacing his earlier understanding, which had found its expression in the single chord hypothesis. That Galileo's investigations of motion along paths made up of multiple conjugate inclined planes were motivated by his refined conceptualization of the relation between swinging and rolling has thus far been plausibly supposed. It is eventually confirmed beyond doubt by the reconstruction of the second stage of Galileo's evaluation of the data that the pendulum plane experiment had provided him with. Galileo's reevaluation of the experimental data, reconstructed and discussed in the present chapter, is indeed revealed as the exploration of the empirical tenacity of his new conceptualization.

To this end, Galileo did not repeat the experiment or otherwise produce any new experimental data. He rather accommodated the theoretical results achieved in the context of his broken chord approach to his original measurement data, together with the results of the first stage of their evaluation. In the first stage of evaluation, from his measurement of the time of motion along a long plane, Galileo had inferred the time of motion along a shorter plane spanning the arc of the swing of the pendulum used in the experiment as a chord. In the second stage of evaluation, based on the results of his broken chord approach, Galileo accordingly inferred from the same measured time along a long inclined plane, the time along polygonal paths more closely resembling or approximating the arc the pendulum had swung through in the experiment.

Based on Galileo's refined conceptualization of the relation between swinging and rolling, it could indeed be expected that the closer the polygonal path considered matched the arc of the pendulum swing in the experiment, the closer the inferred time of motion along this path would be to that of the quarter pendulum swing measured. In the first section of this chapter, Galileo's second stage of evaluation of the experimental data is reconstructed based on the manuscript evidence. The results Galileo achieved did not comply with his expectations. Rather the measured time for the quarter period of a pendulum differed considerably from the times inferred for motions along polygonal paths closely resembling the arc of pendulum swing. In the following section, the pendulum plane experiment is analyzed from a modern perspective. This reveals the origin of the discrepancy to reside in the difference between rolling and sliding, unacknowledged by Galileo, as well as in the fact that pendulum motion, as opposed to Galileo's firm conviction, is not isochronous. In the last section, a synthesizing interpretation will reveal the perplexing situation Galileo was thrown into with his attempt to penetrate his experimental findings based on his new theoretical understanding of the relation between swinging and rolling.

6.1 Reevaluating the Experimental Data Under a New Conceptualization

Manuscript evidence Traces of Galileo's efforts to combine the results of his investigations of motion along polygonal paths approximating an arc with his experimental measurement data and data inferred in the first stage of evaluation of the experiment can be identified on three folio pages, 115 verso, 189 recto as well as on a sheet that was cut and pasted to folio 90. The relation between the relevant pages is schematically rendered in Fig. 4.1. The entries on the three pages concerned with a reevaluation of the experimental data are cursory and in the case of the cut sheet, in addition, severely mutilated. They suffice, however, to reconstruct Galileo's general approach.

The results of Galileo's calculations of the times of motion along polygonal paths were not directly applicable to the experimental data, where a plane of given length measured in a given unit (the punto) was traversed in a particular time, measured in a unit dictated by the time measuring procedure he employed. To exploit the results of his calculations Galileo, expressed in modern terms, assigned the (implicit) unit of his time measurements in the experiment to the arbitrary time units in which his calculations of the times of motions along broken chords had been carried out. He thus inferred the times in which motions along paths assimilating the arc described by the pendulum in the experiment would take place. As the times thus inferred were represented using the same unit as that used to measure the time for the quarter pendulum swing, these times could then be directly compared and such comparison in turn directly probed Galileo's new conceptualization of the relation between swinging and rolling.

Galileo did this twice. Once on the basis of his calculations of a broken chord motion situations documented on 189 recto, i.e., on the back of the very folio page which carries the experimental record. The necessary calculations are begun on folio 90 verso, and, using the result achieved, the time of motion along a broken chord in appropriate units is calculated directly on 189 recto. Essentially the same calculation, albeit in a somewhat more straightforward fashion, is again carried out on 115 verso, this time, however, based on Galileo's much more elaborate calculations of motions along polygonal paths comprising a higher number of inclined planes documented on folio 166 recto.

The distribution of the relevant content over the folios in question, the sketchy character of the notes, and the fact that two of the folios on which these notes were taken were, as discussed in more detail in Chap. 4, later reused for different purposes imply the following scenario. The second stage of evaluation was clearly distinct from the first. It succeeded the latter and was begun only after Galileo had engaged in additional considerations in the context of the broken chord approach. Moreover, the two separate sets of calculations constituting the second stage of evaluation were apparently not carried out in one consolidated effort but successively.

More concretely, after the single chord hypothesis had proven untenable in the first stage of evaluation, Galileo must have first engaged in his considerations concerning motion along a broken chord path documented on 189 recto and next, as documented on 90 verso and 189 recto, have adjusted the results of these considerations to the experimentally measured data. Finally, motivated most likely, as will be argued, by the inconclusive result of this first step, Galileo proceeded to a more thorough investigation of motion along polygonal paths composed of higher numbers of chords chiefly documented on 166 recto and, in a final step, documented on 115 verso, again adjusted the new results thus achieved to the experimental data.[1] The outcome was unsatisfactory and after further work on 166 Galileo apparently abandoned, for the time being, his program on swinging and rolling for reasons to be discussed in the present chapter.

Reconstruction In the pendulum plane experiment Galileo had measured the time it took a body to roll down a long inclined plane to be 280 time units. From this he had inferred, using the law of chords and the law of fall, that the time it would take a body to fall along the diameter of the pendulum used in the experiment, and thus again by virtue of the law of chords also the time along any inclined plane spanning a quarter swing of the pendulum, to be 80 2/3 time units. This latter time was to be directly compared to the time he had measured for the quarter swing of this pendulum of 62 time units. Essentially, the question Galileo tackled in the second stage of evaluation of the pendulum plane experiment was what time it would take a body to move, instead of over a single chord, over broken chord paths more closely assimilating the actual arc of pendulum swing. Like before, this had to be inferred from the measured time for rolling along the long inclined plane by means of a theoretical argument.

In order to do so, on the sheet pasted onto folio 90, Galileo started from the time it would take a body to fall through a distance equal to the pendulum's diameter of 4000 punti that he had already calculated to be 80 2/3 experimental time units. From this time he inferred the time it would take the body to fall through the pendulum's radius, i.e., a vertical distance equal to the pendulum's length. He did so, however, not as might be expected by applying the law of fall directly, i.e., he did not derive the sought time to fall along the radius based on knowledge of the concrete distances fallen through, that is, 4000 and 2000 punti, respectively.

---

[1]Galileo's calculations on folio 166 recto do also comprise the calculation of the time of motion over a simple broken chord, i.e., a path made up by two conjugate chords. That Galileo referred to the corresponding calculations on 189 recto instead of to these suggests that the in-depth analysis on folio 166 recto had not yet been carried out when he started to reevaluate his experimental data.

6.1 Reevaluating the Experimental Data Under a New Conceptualization

Rather Galileo referred back to his calculations on 189 in which the ratio of the time of motion through a diameter and that of motion through a corresponding radius had already been calculated in arbitrary units. In his calculations on 189 recto, with the time through the radius measured by 100 parts, the time for motion along the diameter had resulted to 141 and 1/2 of the same parts. Assuming, as it is indeed directly and trivially implied by the law of fall, that the ratio of these two numbers uniquely characterizes the relation between the times of two vertical motions, one being made over a distance twice as long as the other, starting from the time of motion through the diameter known to him in experimental units, Galileo calculated the time through the radius. This amounted to 57 of the experimental time units.

The scheme underlying this first calculation is also characteristic of Galileo's subsequent calculations in the context of the second stage of evaluation of the pendulum plane experiment and, for this reason, deserves some closer scrutiny. It is based on two assumptions. The first is that a ratio of times that is inferred for two naturally accelerated motions by applying one (or more) law(s) of motion applies in all geometrically similar motion constellations, i.e., to use a modern term, it is scale invariant. From a modern perspective, this is somewhat problematic and indeed does not generally hold true.[2] From Galileo's perspective, however, it was a direct consequence of the fact that all of his propositions concerning naturally accelerated motion along inclined planes are themselves scale invariant, i.e., quite generally establish relations between times of motions whose paths are specified only with respect to each other and not to some absolute dimension. This does not change if a ratio of two times of motion is inferred or calculated not only by one but by the successive application of a number of such propositions, as in the case of Galileo's calculations of the times of motion along polygonal paths. The second assumption on which Galileo's inference was based is that a ratio of two magnitudes expressed as a ratio of numbers is not altered if the numbers are replaced by same equimultiples. This is a fundamental assertion of the Euclidean theory of proportions and allowed Galileo to appropriately rescale, to use a modern expression, the time unit.[3]

If, as in Galileo's calculations of times of motion, a magnitude is expressed by a number, be it a time or a distance, this implicitly invokes a part. If a ratio of two magnitudes of the same kind is expressed as, or more precisely said, expressed to be in proportion to a ratio of numbers, this implies that the two magnitudes are quantified with respect to the same common part. Replacing the two numbers forming a ratio by their same equimultiples as Galileo does to allow for comparability with his experimental data amounts to replacing one arbitrary

---

[2]The ratio of the time to fall through a certain vertical distance to that to fall through half that distance is, according to modern understanding, not scale invariant, for instance, due to the inhomogeneity of the earth's gravitational field or the dependence of friction forces on the speed of motion. To be fair, such effects are irrelevant to all empirical observations that Galileo's means would have allowed for.

[3]Cf. definition V of Book V of Euclid's *Elements*. For Galileo's understanding of some of the fundamental concepts of Euclidean proportional theory, see Palmieri (2001).

common part serving as a measure by another and can thus, from a modern perspective, be addressed as a change of the underlying unit. Galileo exploits this and alters the numbers in such a way that with respect to the new common part, a given time, i.e., in the case at hand the time of fall through the vertical diameter, has as many of these new parts as it has parts with respect to the unit of the experimental time measurement.

In his subsequent considerations, Galileo exploited the same general scheme with regard to a more complex case. He assumed that the ratio of time to fall along the vertical radius of a circle to the time to fall along a polygonal path approximating an arc of this circle to be independent of the absolute length of the radius. Thus the ratio between times of motion of fall along the radius of a circle and that of fall along a particular broken chord inscribed in this circle, which had amounted to 100 to 132 in the calculation on 189 recto for motion along bf and cef, respectively, was characteristic of all similar motion situations and thus, in particular, also applied to the concrete case where the radius of the arc measured 2000 punti, as had been the case in the experiment. As Galileo had already determined on the sheet pasted to folio 90 that the time of motion through a vertical radius of 2000 punti would amount to 57 measured in experimental time units, he could immediately infer the time of motion along a broken chord inscribed as cef in the arc of pendulum swing by again simply rescaling the unit. In experimental units, this time of motion along a broken chord inscribed into the quarter arc of the pendulum used in the experiment turned out to be 75 1/4.

This time, derived from the experimentally measured data by a theoretical inference, was to be compared to the measured time of 62 time units for the quarter swing of a pendulum. It thus turned out that with a, compared to the single chord, closer fit of the broken chord to the arc, the time of motion along the broken chord was indeed closer to the time Galileo had measured for the quarter swing of the pendulum. However, the times of motion along a broken chord and that of the quarter swing of the pendulum still differed by about 20%. From Galileo's perspective this result must have been inconclusive. On the one hand, it confirmed that, just as Galileo would have expected, if a path more closely resembling the arc of pendulum swing was considered, the time of motion along that path indeed more closely matched the actual time of quarter swing. At the same time, the difference of 20% was still rather large. There was, moreover, no way to foresee, without actually analyzing and calculating the times of motion, if it would vanish when considering motion along paths more closely resembling the arc, i.e., paths composed of more conjugate inclined planes.

Galileo's extensive calculations of the time of motion along paths made up from conjugate chords inscribed into an arc are contained on folio 166 recto, and related pages in the *Notes on Motion* were discussed in the previous chapter.[4] We need to recall that in the context of this broken chord approach, Galileo calculated the times of motion for the single-chord up to the four-chord approximation of an arc

---

[4] Cf. Chap. 5, Sect. 5.3.

6.1 Reevaluating the Experimental Data Under a New Conceptualization

for a base length of the radius of 180 distance units. Moreover, he repeated the same calculations once again and in addition calculated the times of motion for an eight-chord approximation of the arc, now, however, more accurately, i.e., with respect to a different unit in which the radius of the arc measured 100,000 units.[5]

Pushing his investigation of the relation between swinging and rolling further Galileo referred back to his calculation of the time along a path composed of four conjugate chords. Based on these calculations, he assumed that if the vertical radius was fallen through in 180 time units, the actual time to fall through the arc would be very near to 235 units.[6] Proceeding in essentially the same fashion as before to accommodate this result to his experimental data, the required rescaling of the time unit was carried out by Galileo on folio page 115 verso. He thus inferred that the time to roll along the arc in the units of the experiment would be very near 74 time units. Thus instead of using a two-chord approximation, for the time being, the best estimate derived from a four-chord approximation had yielded a decrease of the time of one and a half units or less than two percent, and the expected correspondence to the measured time of a quarter swing was far from materializing.

Galileo made no attempt to accommodate the results of his most accurate approximation of motion on an arc by motion on a series of inclined planes, the eight-chord approximation, to his experimental data. At least no such attempt can be identified in the manuscript. However, after Galileo had calculated the time of motion for the eight-chord approximation on folio page 166 recto, he would have had little reason to attempt an accommodation of these results to his experimental data. In fact, his calculations showed that the time of motion along the path of the eight-chord approximation differed only very minimally from the time calculated for motion along the path composed of the four-chord, and thus when converted into the suitable unit, it differed only minimally from the time of 235 units he had already probed.

As the difference between the times of motion in going over from one approximation to the next was rapidly decreasing, it was, moreover, virtually excluded by his results that considering even better approximations of motion on an arc, i.e., considering paths made up of an even higher number of conjugate chords, could, when accommodated to the experimental units, yield time of motion that would even remotely compare to the time of 62 units measured for the quarter swing

---

[5] It is strongly suggested that Galileo's accurate calculations of the times of motion along polygonal paths made up of four and eight inclined planes on 166 recto were triggered by the inconclusive results brought about by the accommodation of a two-chord approximation to the experimental data.

[6] The time calculated for motion along a four-chord approximation had been 236 1/2 time units. Galileo's use of the slightly smaller time of 235 as an estimate for the actual time of motion along the arc can be seen as reflecting his awareness that the time for the limiting case of motion along a polygonal path of infinitely many sides would be smaller but not much smaller than the time calculated for the four-chord approximation. Cf. the detailed discussion of folio 115 in the Appendix in Chap. 13. Amazingly, Galileo's value of 235 time units for motion along a quarter arc whose radius is fallen through in 180 time units is off only two parts per thousand from the value predicted by modern physics.

of a pendulum. Paired with the insight that motion along conjugate chords was no longer isochronous indicating that the same might hold for motion along the arc, the result of Galileo's attempt to reevaluate the experimental data under his new conceptualization must have been discouraging. Indeed, as will be discussed in Sect. 6.3, in view of the problems he saw himself confronted with Galileo discontinued not only the reevaluation of the pendulum plane experiment but indeed his entire broken chord approach. Before moving on to this discussion, however, an analysis of the experiment from the perspective of classical mechanics is provided allowing for a clear understanding of why Galileo did not, and in fact, could not, recover the correspondence between swinging and rolling he was searching for and expected to find and for an assessment of what Galileo did or in fact could have concluded from the outcome of his efforts.

6.2 The Pendulum Plane Experiment from a Modern Perspective

With regard to pendulum motion, in the experiment Galileo timed the quarter period of a pendulum of a given length. For an ideal pendulum, whose equation of motion has been linearized in the angle of elongation, this time is given by:

$$t = \frac{\pi}{2} * \sqrt{\frac{l}{g}} \qquad (6.1)$$

Given the length of the pendulum Galileo used, we can thus first of all give an estimate of its period and thus of the time for a quarter swing. Converting the length given as 2000 punti into metric units, we can thus get an estimate of the time interval we expect Galileo to have measured when timing the quarter swing of his pendulum[7]:

$$t = 0.69\,\text{s} \qquad (6.2)$$

Comparing this with the time Galileo measured in his own units, namely, 62 units, it follows that Galileo's unit corresponds to a time interval within the order of magnitude of 10^{-2} s. This a priori excludes a harmonic process whose oscillations are counted as the method of time measurement.

In the *Discorsi*, Galileo mentioned the use of a water clock to time the descent along an inclined plane. Consequently, a water clock has also been proposed as a means of time measurement in the case of the pendulum plane experiment. From

---

[7]For this and the following calculations, I have assumed the punto to measure 0.95 mm. Cf. the discussion of folio 166 in the Appendix in Chap. 13.

the sparse information in the experimental record, it is not unambiguously possible to infer which technique of time measurement it was that Galileo applied. The use of a water clock is, however, certainly compliant with the concrete data.[8]

Equation 6.1 provides a good approximation for the quarter period of the pendulum only for small angles, and a more accurate formula must be used if the pendulum is oscillating beyond the small-angle regime.[9] Galileo did not note the angle through which the pendulum was swinging. Indeed, this angle is not well defined based on his measuring routine, which involved the counting of a number of oscillations during which, due to the damping, the arc of pendulum swing would have diminished. As Galileo apparently did not call into question the validity of the isochronism of the pendulum at the time he conducted the experiment, recording the initial angle would have made but little sense because from his perspective, the period of the pendulum should be independent of this angle. The fact that due to damping the amplitude of the pendulum was diminishing during the time of the measurement wouldn't have mattered from his perspective. Vice versa, as will be detailed below, the results of the pendulum plane experiment become, if not meaningless, then at least problematic, once it is accepted that pendulum motion is not isochronous.

Galileo will likely have had his pendulum swing through larger rather than smaller amplitudes as the former are more easily observable. Moreover, as detailed in Chap. 4, he referred to larger angles of oscillation in his description of other experiments involving the swinging of pendulums. At the same time, in view of the practical difficulties connected with it, it is rather unlikely that Galileo would have used initial elongations of his pendulum greater than 90 degrees. Without a concrete estimate of the angle Galileo used, we can, for now at least, specify a lower and an upper limit of the time we expect him to have measured. The lower limit is given by Eq. 6.1. The upper limit is given by the period of a pendulum swinging through a 90-degree arc, which is about 17.6% longer than that given by the equation of motion of the linearized pendulum. It thus holds for the time we expect Galileo to have measured that:

$$0.69\,\text{s} \le t \le 0.81\,\text{s} \tag{6.3}$$

Concerning the time of motion on the inclined plane, the theoretical value for the time it takes an object to fall freely, i.e., to slide frictionlessly along an inclined plane of length L and inclination α, is given by:

---

[8] If the unit in which Galileo recorded the weight of the water that had flowed out of the water clock during measurement corresponded to a rather small weight, for instance, one gram (in the order of magnitude of the Florentine Scrupulum, cf. Zupko 1981), this would correspond to a flow rate in the order of magnitude of 100 grams per second corresponding, as a rough estimate based on Poiseuille's Law shows, to plausible dimensions of a water clock, which Galileo thus could have been using.

[9] Determining the period of a pendulum exactly requires solving an elliptical integral. For the approximation of the solution by a power series, see Lima and Arun (2006).

$$t_{\text{falling}} = \sqrt{\frac{2*L}{g*\sin\alpha}} \qquad (6.4)$$

In view of the small angle of inclination of the plane Galileo was using, we are forced to assume that Galileo was timing the motion of a rolling object, most likely some kind of metal ball.[10] However, rolling, as compared to free falling and frictionless sliding, respectively, diminishes the translational acceleration of the rolling object, because some part of the potential energy is transformed into rotational energy. This diminishing in fact depends on the type of object rolling, specifically on its geometric form. For a homogeneous ball, the factor by which the translational acceleration is diminished is 5/7. Consequently, the time it takes a ball to roll over a certain distance is prolonged compared to sliding over the same distance by a factor of $\sqrt{7/5}$, which will be referred to here as the *rolling factor*. Hence the expected time for a ball rolling on Galileo's long inclined plane is[11]:

$$t_{\text{rolling}} = \sqrt{\frac{2*L}{g*\sin\alpha}\frac{7}{5}} \qquad (6.5)$$

In Chap. 4 it was argued that Galileo mostly likely used some kind of grooved board as an inclined plane just as the one described in the *Discorsi* where Galileo, however, did not specify the shape of the groove. As Hahn has noticed, this leaves open the possibility that Galileo was using a V-shaped groove in which the ball would have had two contact points, effectually diminishing the rolling radius and thus even increasing the rolling acceleration of the ball as compared to that of the same ball rolling on a flat surface. In this case the expected time for rolling along the inclined plane would be even greater.[12]

With Eq. 6.5 the expected time for rolling along the 6700 punti long plane amounts to 3.4 s. This again allows for an estimate of the magnitude of the unit of time measurement employed by Galileo. The time Galileo measured to be 280 of his units corresponds to 3.4 s. His unit of time measurement thus corresponds to 0.0122 s plus/minus an error margin corresponding to the error margin of Galileo's

---

[10] In 1599, Galileo ordered two lead balls from a foundry. From entries in Galileo's bookkeeping accounts, it can be inferred that they weighed approximately 1.6 kg, having a diameter of approximately 6.5 cm. Cf. Damerow et al. (2001). It may have been one of these balls that Galileo used in the pendulum plane experiment but this assumption is not crucial for the analysis of the experiment.

[11] The friction required in order that the point of contact of the lead ball with the surface of the plane will not slip does not work against the motion of the ball, and thus no energy is lost. Other influences such as, for instance, the friction forces due to the surrounding air or the flexing of the ball, which would dissipate some energy, are absolutely negligible for the present purpose.

[12] See Hahn (2002). However, in view of the fact that Galileo mentioned in the *Discorsi* that the groove was polished and lined with parchment, a V-shaped groove is unlikely and it can plausibly be assumed that Galileo was using a concave groove to roll the ball in. By the same token of reasoning, Naylor (1974, 133) also concludes that the sphere must have been rolling in the groove.

6.2 The Pendulum Plane Experiment from a Modern Perspective

measurement. With this estimate of the magnitude of Galileo's time unit, the time he measured for the quarter period amounts to 0.76 s. If the equation for the period of a pendulum is solved for the angle of elongation, it is found that this corresponds to an angle of 69 degrees. Of course the error margin of this estimate of the angle of elongation is rather large as it depends on the error in two of Galileo's measurements.[13] Yet that Galileo should have used an initial elongation of about 70 degrees seems to be very plausible.

From Galileo's point of view, the experiment would have been a success if the time of motion brought about by the accommodation of the results of his best approximation of motion on an arc to his experimental results had corresponded, within an acceptable limit, to the time measured for the quarter swing of a pendulum. It would have confirmed his idea that the limit of the chord approximation was in fact motion on the arc and the assumption of kinematic equivalence under path correspondence according to which that latter motion was essentially equivalent to pendulum motion made over the same arc. Both assumptions are essentially correct from a modern perspective, yet the expected correspondence did not show. The above analysis has put us in a position to identify the different factors that contributed to the deviation of what was observed from the expected behavior. These factors are listed below, and their interplay is schematically rendered in Fig. 6.1:

- Galileo's measuring error in the two time measurements. Galileo can be expected to have had a grasp of the approximate size of the error thus introduced, gained, for instance, by comparing successive time measurements of the same process.[14]
- The difference between his best approximation, i.e., motion made along a path of eight conjugate chords, and the exact time of motion to fall along a 90-degree arc. From a modern perspective, we recognize this difference to be less than 0.5%.[15] Galileo must have been aware that he had not inferred the time of motion on the arc precisely. However, he had good indications that the time of motion he had actually inferred by analyzing motions along polygonal paths closely approximating the arc could have differed but little from the former.
- The difference between rolling and sliding. Galileo had timed rolling descent along an inclined plane. Thus when he adjusted the results of his broken chord approach to this time measurement, he was implicitly still regarding a rolling motion. Hence, what he inferred was in fact the time to roll along rather than to fall along the paths made up of conjugate chords. This then was compared, however, not to the rolling but to the swinging of a pendulum bob. The difference

---

[13] As in view of the considerably longer time measured in the case of pendulum swing, assuming the same method of measurement, the relative error can be expected to be an order of magnitude smaller than in the case of the inclined plane.

[14] Regarding the accuracy that can be reached measuring a time interval with a water clock of the type Galileo presumably used, see Settle (1961).

[15] The difference between Galileo's best value of 235 time units for motion along a 90-degree arc if the radius is fallen through in 180 units and the correct, modern result is even smaller and amounts to only two in a thousand.

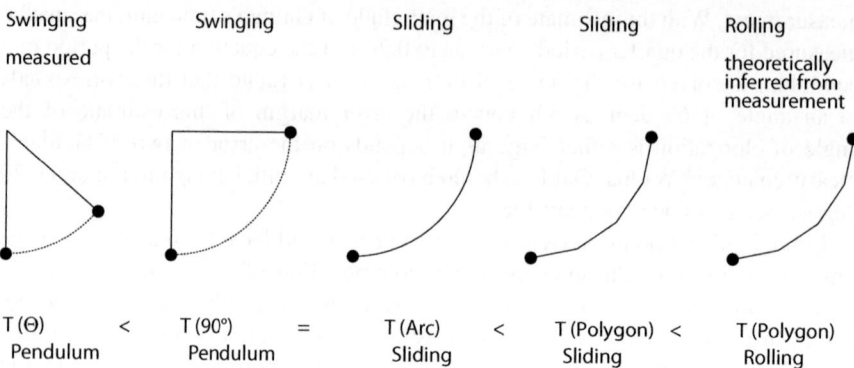

Fig. 6.1 Schematic representation of the factors accountable for the deviation between the time of 62 measured for the quarter period of the pendulum (left) and the time of 72 theoretically inferred from the time measurement of rolling along an inclined plane (right). From right to left: The time of rolling along a polygonal path Galileo had inferred is greater than the time to slide along the same path. The time to slide along the polygonal path is slightly longer than that to slide along the arc spanning 90 degrees. The time to slide along a 90-degree arc corresponds to the quarter period of a pendulum swinging through a 90-degree arc. The quarter period of a pendulum swinging through a 90-degree arc is longer than that of a pendulum swing through a smaller arc, which Galileo had measured in the experiment

between rolling and swinging along the same arc is given by the rolling factor of $\sqrt{7/5}$ and depends on no other parameter. In this way, Galileo overestimated the chord approximation time by 18.3% with regard to a corresponding sliding motion. Galileo, as will be argued, was unaware of the source of this deviation from his expected result.

- The difference in the period between swings through different angles, i.e., the non-isochronous behavior of real pendulums. Galileo compared the results of his chord approximations, calculated for an arc of 90 degrees, to the period of a pendulum which had swung through a smaller arc in the experiment. The deviation, thus introduced, depends on the size of the arc of pendulum swing actually used in the experiment. Assuming that Galileo's pendulum was elongated to 75%, this accounts for about 5% of the deviation with regard to expected behavior.

David K. Hill, the only other author to have considered the pendulum plane experiment in detail, has come to widely different conclusions. In his analysis of the experiment, he engaged in considerations similar to those outlined above using, however, apparently only the linearized equation for estimating the period of the pendulum.[16] For this to be warranted, it has to be presupposed that Galileo used a vanishingly small angle of initial elongation for his pendulum. Hill seems not to have recognized this implication.

---

[16] Hill (1994) does not provide the formulas he used to attain the numerical estimates he supplies. The estimates can, however, be reproduced using the linearized pendulum equation.

6.2 The Pendulum Plane Experiment from a Modern Perspective

As argued above, the difference between the time measured for the quarter period and that inferred for motion along a polygonal path closely approximating the arc is made up mainly of two contributions, the difference between the period of swing through a 90-degree arc to the period of a smaller swing Galileo actually measured and the difference between rolling and sliding along the same path. As further argued above, according to all evidence, Galileo used a rather large initial angle of elongation. Hence the empirically observed difference is mainly due to the difference between rolling and sliding. Hill's implicit assumption that Galileo used a vanishingly small angle for his pendulum swing entails that he would have committed an overall measuring error of about 15%. As it is indeed not reasonable to attribute such a large measuring error to Galileo, Hill sought a different explanation.[17]

Coincidentally, the difference of about 15% between the period of a pendulum swinging through insensibly small angles and the period of a pendulum swinging through a larger arc as in the experiment corresponds rather well to the difference between rolling and sliding along the same plane which, given by the rolling factor, is about 18%. Thus, if it were assumed that Galileo, in addition to using insensibly small amplitudes, timed a sliding instead of a rolling motion, his reckoned measuring error is reduced to a more realistic value. Hill interpreted this as an indication that Galileo must have been aware of the difference between rolling and sliding motion and had accounted for this difference in the design of the experiment and the evaluation of the data.

Hill, however, did not claim that Galileo was using a sliding object, as indeed in view of the small angle of inclination of the inclined plane, this is virtually excluded. Instead, he offered a different explanation, maintaining that Galileo was aware of the difference between rolling and sliding and that, in order to compensate for the difference, he had in the experimental record not given the real length of the plane he had used but an effective length. The effective length, according to Hill, is the length of a longer inclined plane that would have been traversed in the same time as the one in the experiment had the ball been sliding instead of rolling. In modern notation, where L_{real} represents the real length and $L_{\text{effective}}$ the corrected length, it must hence hold for the most simple case that[18]:

$$L_{\text{real}} = L_{\text{effective}} * 5/7 \qquad (6.6)$$

---

[17] As stated above, the major contribution to this error should stem from Galileo's timing of the rolling of a ball down the long and gently inclined plane. If the alleged error of 15% is, for the sake of the argument, attributed to the time measurement of motion on the inclined plane alone, Galileo's absolute measuring error would rather unrealistically range in the order of magnitude of half a second.

[18] Hill surmised that Galileo had actually used an inclined plane which differed in both length as in inclination from the values given and implied, respectively, by the experimental record. According to Hill's interpretation, the inclined plane had a length of 7500 punti and an inclination of just under 13 degrees. Cf. Hill (1994, footnote 12). Hill's argument has here been rephrased for the actual dimensions and an effective plane, which is merely altered in length. This change does not affect the implications with regard to the question as to whether Galileo was aware of the difference between rolling and sliding.

To support his hypothesis, Hill pointed to some of Galileo's elaborations on the back of the page containing the experimental record, i.e., on folio page 189 recto, which he interpreted as being Galileo's analysis of the problem of the comparison between rolling and sliding motion and, in particular, as an indication of Galileo's knowledge of the numerical value of the rolling factor mediating between the two. The elaborations by Galileo, which Hill had taken as evidence of Galileo's analysis of the effects of rolling if compared to sliding, can, however, be shown to have served an entirely different purpose.[19]

Even if Hill's interpretation rests on an unwarranted presupposition, Galileo may still have been aware of the difference between sliding and rolling. However, it is exceedingly unlikely that if Galileo was aware of this difference, so directly relevant to his new science, he would have failed to mention it even once in his entire œuvre or his private correspondence.

In a nutshell there is no reason to assume that Galileo was aware of the difference between rolling and sliding and thus of the effect this difference had on the times of motion observed in the experiment, on the contrary. In addition, Galileo's experiment rested crucially on the, from a modern perspective, unwarranted belief that pendulum motion was isochronous. Both assumptions, the implicit assumption of a kinematical identity of sliding and rolling and the explicit assumption of the pendulum's isochrony, not only are false but both contributed to the difference between observed and expected behavior in a different way. Whereas the first contributed by constant factor, the contribution due to the second assumption depended on a parameter of the experiment, the initial angle of elongation. In consequence, there was hardly any way that based on the results of the second stage of evaluation of the experiment, Galileo could have progressed to isolate the factors which contributed to the difference from the behavior he expected and thus to recognize one of his underlying assumptions as false.[20] The full scope of the epistemological conundrum Galileo had run into can only be appreciated if Galileo's attempt to reevaluate the experimental data is considered in the context of Galileo's new conceptualization of the relation between swinging and rolling on which it was based.

---

[19]The considerations alluded to by Hill as being an attempt by Galileo to relate the lengths and inclinations of two planes in such a way that sliding descent along one took the same time as rolling descent along the other are shown to pertain to the elaboration of a proposition later published by Galileo as Proposition V of the Second Book of the Third Day of the *Discorsi*. See the discussion of Sect. 13.20 in the Appendix in Chap. 13.

[20]There is no evidence to indicate that Galileo ever became aware of the difference between rolling and sliding with regard to kinematics. The unacknowledged difference also impacted on Galileo's experiment involving horizontal projection, whose experimental record is preserved on folio 116 verso, where the difference between rolling and sliding figures in the form of a deviation of the measured data from the theoretically expected values, as Crawford (1996) has demonstrated. Cf. also Shea and Wolf (1975). In the case of this projection experiment, Galileo had thus clearly not taken the provision, according to Hill allegedly taken in the case of the pendulum plane experiment, to account for the difference between rolling and sliding.

6.3 Swinging and Rolling: From Promising Start to Dead End

The two distinct stages of work that have been identified in which Galileo evaluated the measurement data of his pendulum plane experiment literally enclose his theoretical considerations concerning the relation between swinging and rolling. The negative outcome of the first stage of evaluation triggered a reconceptualization, which, after theoretical elaboration in the context of the broken chord approach, as we have seen, became the backbone of the second stage of evaluation of the experimental data. In the previous chapter, Galileo's work on the broken chord approach was unambiguously dated to 1602. Hence the pendulum plane experiment must have been conducted at about the same time as well, and the experiment is thus recognized as the empirical component which complemented the broken chord approach as the theoretical component of Galileo's research agenda on swinging and rolling. Both experiment and Galileo's theoretical investigations were conceptually closely interwoven, as has become particularly apparent from the way in which Galileo reevaluated the experimental data.

Both components have been treated more or less detached in the previous chapters. A synopsis as presented in the current section gives rise to an overarching account of the progression of Galileo's investigations into the relation of swinging and rolling. It will thus become clear that Galileo's systematic study of naturally accelerated motion along inclined planes first started out with these investigations. The problems, theoretical as well as empirical, which piled up rather quickly, however, forced Galileo to discontinue his exploration of the relation of swinging and rolling.

Galileo's considerations started from the recognition of a challenging similarity between the properties of pendulum motion and those of motion along inclined planes resulting in the assumption that, generally speaking, pendulum motion and naturally accelerated motion on inclined planes must be related in some specific manner that would serve to explain this similarity. On a phenomenological level, the isochronism of the pendulum and the law of the pendulum compare to the law of chords and the law of fall. Galileo was indubitably committed to the law of chords, the law of fall, and the isochronism of the pendulum when he commenced his research agenda some time around 1602. We can be absolutely sure of this because all three propositions are in fact encompassed by the way Galileo designed and evaluated the pendulum plane experiment. Furthermore, as argued in Chap. 4, Galileo was in all likelihood also aware of the law of the pendulum when the experiment was originally designed or else he must at least have become aware of it early on, as the law of the pendulum was indeed directly implied by his initial but also by his refined understanding of the relation of swinging and rolling.

When Galileo commenced his agenda on swinging and rolling, none of his claims concerning the pheno-kinematics of the two types of motion were well established and rigidly justified. Rather they must have had the status of more or less well-founded, heuristic assumptions. Galileo, as we know from the letter of 1602, had tested the isochronism of the pendulum empirically and, as is argued in Chap. 4, using the very inclined plane used in the pendulum plane experiment very likely also had submitted the law of fall to an experimental test, as he indeed claims to have done in the *Discorsi*. As will be indicated in Chap. 8, Galileo may even have probed the law of chords experimentally at about this time.

Furthermore, Galileo was certainly in possession of arguments which, even though not capable of establishing them beyond doubt, lent the law of fall and the law of chords additional credibility.[21] With regard to the law of chords, by the end of 1602, while working on the problem of swinging and rolling, Galileo had even been able to advance a proof for the law of chords as he likewise disclosed in his letter to Guidobaldo. Considering the nascent character of his consideration concerning naturally accelerated motion, this proof, however, did not incontestably establish the law of chords, and indeed doubt was thrown on the proof advanced in 1602 by Galileo's further considerations, as will be demonstrated in Chap. 8.

The similarity between the assumptions concerning swinging and those concerning rolling implied a relation between the two types of motion, and as he indicated in the letter to Guidobaldo del Monte, exploration of this relation opened Galileo the prospect of being able to infer or demonstrate his assumptions concerning swinging from those concerning rolling or vice versa. The pendulum plane experiment was originally designed to test the perceivably most simple assumption concerning a relation between swinging and rolling, the single chord hypothesis, which, if it held, would allow such an inference or demonstration. If the quarter swing along a pendulum arc and fall along a chord spanning the same arc were always completed in the same time, then indeed the isochronism of the pendulum and the law of chords would mutually imply each other and so would the law of the pendulum and the law of fall. The single chord hypothesis, itself a heuristic assumption, thus, in a sense, networked Galileo's assumptions about swinging and rolling. It was promising, as if it could be justified, it would open a road for further inquiry. Thus what would have remained to be done, had the experiment confirmed the single chord hypothesis, would have been to provide an explanation of why it held. Due to the particularly simple nature of the hypothesis, this may have appeared feasible to Galileo. Yet the pendulum plane experiment disproved the single chord hypothesis.

The hypothesis had to be refuted, not so, however, any of the underlying assumptions concerning swinging and rolling by which it had been engendered. This would only have been the case if the hypothesis had been logically implied

---

[21] Damerow et al. (2001) show that already in 1592, Galileo had come to accept the parabolic shape of the trajectory of projectile motion. They argue that the decomposition of the parabola "into a neutral, horizontal motion and a natural, vertical motion ... implies the proportionality of the vertical distances to the squares of the times represented by the horizontal distances, that is, the law of fall (Damerow et al. 2001, 53)." In consequence, in "1602 the law of fall as implication of the symmetrical parabolic shape of the projectile trajectory must have been so familiar to Galileo that he not even made a point of mentioning it (Damerow et al. 2001, 88)." Cf. also Chap. 2. An argument which may have led Galileo to come to accept the law of chords is reconstructed in Chap. 8.

6.3 Swinging and Rolling: From Promising Start to Dead End

by the underlying assumptions in a strict manner. Yet the single chord hypothesis was not a rigid logical consequence of Galileo's assumptions concerning the pheno-kinematics of swinging and rolling. Rather it was, as has been detailed, the result of abductive reasoning. Indeed, the single chord hypothesis makes a claim that is much stronger than what is actually implied by the underlying assumptions about swinging and rolling, whose relation it was supposed to explain.

Besides showing that the hypothesis could not be maintained empirically, the experiment thus did not disprove any of Galileo's assumptions. It merely indicated that a different approach to investigating and eventually explaining the relation between swinging and rolling needed to be sought. The difference observed in the experiment between the time of motion along a chord spanning an arc and that to swing along the arc was small enough to suggest a modification of the underlying hypothesis.

Galileo's new conceptualization of the relation between swinging and rolling was more complex and can, from a modern perspective, be described as based on two assumptions—kinematic equivalence under path correspondence and polygonal approximability. According to kinematic equivalence under path correspondence motion along an arc supported from above, i.e., pendulum motion, is kinematically equivalent to motion along the same path, supported, however, from below, i.e., falling along a concave surface. Polygonal approximability has been introduced here as a shorthand way of referring to Galileo's conjecture that, just as a polygonal path inscribed into an arc, approaches the arc with an increasing number of sides, so the kinematic behavior of fall along this path approaches the kinematic behavior of fall along the arc, which would result as the limiting case for motion along a polygonal path of infinitely many sides.

Instead of claiming a correspondence between motion along a chord and the corresponding arc as the single chord hypothesis did, kinematic equivalence under path correspondence claims correspondence between swinging and falling along the same arc. This is intuitively more plausible, and the prospect of being able to provide an explanation, for instance, along the line Galileo had argued in *Le Meccaniche*, was promising.[22] Whether this correspondence held, was, as regards Galileo's specific approach, not a theoretical question but something that had to show empirically.

How the fall along an arc relates to the fall along one or a series of conjugate inclined planes to which the law of fall and law of chords apply was, on the other hand, a theoretical question. The extensive theoretical investigations, referred to as the broken chord approach, by which Galileo tried to recover the kinematical behavior of a body falling along an arc based on the idea of polygonal approximability were reconstructed in the previous chapter.

---

[22] As discussed in Chap. 3, Galileo indeed granted that Guidobaldo del Monte, instead of observing the swing of a pendulum as Galileo had proposed, observed the rolling motion of a ball along a circular arc, to observe whether this motion was isochronous.

Galileo's approach is, in principle, correct from a modern perspective. Kinematic equivalence under path correspondence applies from a modern perspective if the net forces are same. The pull to the pendulum string and the push to the concave surface are passively counteracted by tensile forces in the material and thus in both cases the same force acts on the same body moving along the same circular path regardless of whether it is supported from above by a pendulum string or from below by a concave surface. Yet, as the real motions of heavy bodies are concerned, however, another type of force comes into play, friction force. Due, in particular, to the friction between the falling object and the spherical surface, a body moving on the arc, in contrast to a pendulum bob, starts to roll.

By the way the experiment was designed, Galileo implicitly compared a swinging to a rolling motion made over the same path. He was, as has been argued, completely unaware of the difference this made with regard to the kinematics and thus had absolutely no reason to scrutinize the assumption of kinematic equivalence under path correspondence. From a modern perspective, however, the assumption of kinematic equivalence under path correspondence served as the conduit for the implicit false assumption of an equivalence of rolling and sliding, all other things being equal.

The first result Galileo achieved based on his new conceptualization was to show that despite being made over the longer path than motion on the respective single chord, motion on a broken chord spanning the same arc and composed of two adjacent planes was completed in less time. This corresponded to a somewhat surprising tendency Galileo had observed in the experiment, which had indeed yielded that swinging along a quarter arc was completed in less time than falling along a chord spanning this arc, despite the fact that the path of the latter motion was shorter. Galileo's theoretical results were thus found to correspond with the empirically observed tendency, which must have reaffirmed Galileo regarding his approach. The proof he constructed for the law of the broken chord is mentioned in the letter to Guidobaldo del Monte.

Galileo's further proceeding was contingent on the fact that the revised conceptualization of the relation between swinging and rolling rested on weaker presumptions than his initial hypothesis. The single chord hypothesis directly entailed that the relation of falling along a chord and swinging along the corresponding arc was neither dependent on the angle nor the size of the arc considered. It thus also straightforwardly entailed that the law of chords and isochronism of the pendulum, as well as the law of fall and the law of the pendulum, mutually implied each other. Based on Galileo's new conceptualization, this could no longer be presumed but had to be theoretically recovered. Concretely, it had to be shown that what held for motions along appropriate chords according to the law of chords, namely, that these motions were isochronous, also held for motions along the corresponding arcs and that what according to the law of fall held for times of naturally accelerated motion along different distances over the same inclination also held for the periods of pendulums with different pendulum lengths, namely, that the times are in proportion to the roots of the corresponding distances.

As regards the relation between the law of fall and the law of the pendulum, it is rather obvious that if the two assumptions upon which Galileo's refined conceptualization of swinging and rolling rested held, it would follow with necessity that the law of fall implied the law of the pendulum and vice versa. According to polygonal approximability, motion along the arc can be considered as the limiting case of motion along a polygonal path made up of infinitely many planes. Even if Galileo was not able to devise an argument that would have allowed him to transgress to the case of an infinite number of sides for the polygonal paths considered, it was nevertheless implied that the time of motion for this limiting case had to scale like the time of motion along a radius or a chord, i.e., according to the law of fall with the root of the distance traversed.[23] Provided there is indeed no kinematical difference between the limiting case of falling along the arc and swinging along the same arc this immediately implies the law of the pendulum. However, nowhere in the manuscript does Galileo explicitly make this argument, and, as will be shown in Chap. 9, it took until almost 40 years after the publication of the *Discorsi* for Galileo to formulate a corresponding proof.

With regard to theoretically recovering the isochronous behavior of pendulum motion, the situation was different. Indeed the isochronism of the pendulum doesn't hold and thus cannot be recovered, at least not based on an argument that is recognized as essentially correct from a modern perspective. The question arises whether Galileo's investigation of swinging and rolling allowed him to conclude that the isochronism of the pendulum was unwarranted as has indeed been claimed, in particular by Hill.[24] If pendulum motion is isochronous, then by kinematic equivalence under path correspondence so is fall along the arc and, according to polygonal approximability, also motion along a polygonal path of infinitely many sides exhausting the arc. Given the law of chords, an obvious way to argue that the time of motion along a polygonal path of infinitely many sides does not depend on the angle of the arc considered would be by induction.

Yet Galileo's considerations on 166 recto had shown that the ratio of the time to fall along a chord and a corresponding broken chord with the junction point chosen in such a way that it halves the arc considered did depend on the angle of the arc considered. The smaller the angle of the arc, the bigger the relative saving in time for motion along a broken chord had shown to be. From a modern perspective, this reflects the fact that neither rolling nor swinging along an arc is isochronous but that it does indeed take less time to traverse an arc spanning a smaller angle than it takes

---

[23] As demonstrated above, Galileo's accommodation of his theoretical results concerning descent along polygonal paths to his measurement data was based precisely on the assumption that the ratio of the time to fall through a characteristic distance, for instance, the radius of the circle, to the time to fall along any given polygonal path inscribed into an arc of this circle was scale invariant and that this must also hold for the limiting case.

[24] "Galileo's work in this context plainly establishes that *circular pendulums are not isochronous* (Hill 1994, 507)."

to traverse an arc of the same radius but spanning a bigger angle, notwithstanding the fact that the corresponding chords spanning these arcs are indeed fallen through in the same time.

Hill's surmise that Galileo thus recognized that the isochronism of the pendulum did not hold is, however, haphazard. The loss of isochronous behavior when going over from motion over the single chord to a particularly defined two-chord approximation must certainly have puzzled Galileo. It may even have been interpreted by him as threatening the underlying conceptualization. Yet, as long as Galileo was not able to devise an argument regarding the kinematics of the limiting case of motion along the arc, it did not imply with necessity that fall or swing along an arc was not isochronous. Isochrone, somewhat inexplicably lost when going over from motion along a single chord to motion along two-chord conjugate chord, may after all be resurrected in the limiting case, despite the fact that Galileo's results strongly pointed in a different direction. It was thus apparently not only the theoretical complications that had cropped up that brought Galileo's research agenda on swinging and rolling to a halt.

Indeed, the second stage of evaluation of the experimental data had left Galileo with a devastatingly large difference of somewhat more than 20% between the time measured for a quarter swing and the time measured and inferred for his best polygonal approximation of motion along an arc, greatly at odds with his theoretical expectation that these times should be nearly equal.[25] The fact that Galileo had not been able to transgress to the limiting case of motion along a polygonal path with an infinite number of sides could clearly not be held accountable for this difference. Based on the results of his calculations, he must have realized that with an increasing number of adjacent inclined planes considered, the times of motion along the paths considered was convergingly rapid, which indicated that the actual time of motion along the arc should not differ too much from the time calculated for motion along a polygonal path of eight conjugate chords.

Based on the evaluation of the empirical data produced in the experiment could Galileo thus have inferred either that a difference prevailed between rolling and sliding or that pendulum motion was not isochronous or maybe both? This seems exceedingly unlikely. Both his assumptions that no difference prevailed between rolling and sliding (an unconscious assumption implicit in the design of his experiment) and that pendulum motion was isochronous (a conscious assumption) were built into the design and evaluation experiment. Both contributed to the unexpected discrepancy observed in a different manner. The unacknowledged difference between rolling and sliding contributed to the discrepancy by a constant factor, while the contribution due to the non-isochronous behavior of pendulum motion depended on a parameter of the experiment, the angle of the arc of the quarter

---

[25] What Galileo would have accepted as an agreeable correspondence between measured and derived, and theoretically expected results, is not at all clear. Damerow et al. (2001) have shown, for instance, that when Galileo compared the curve of a hanging chain with a parabola, despite having detected a considerable discrepancy between the two curves, he clung to the hypothesis that both curves were identical and even drafted a theoretical proof based on this.

6.3 Swinging and Rolling: From Promising Start to Dead End

pendulum swing, which had been timed, which Galileo had not noted, and which indeed was not well defined based on the way the experiment was designed and conducted.

When Galileo designed the experiment, he took the assumption that pendulums swing isochronously for granted. He indeed incorporated it into the design of the experiment so as to refine the experimental procedure. On the one hand, instead of timing a single quarter swing, Galileo had timed a number of consecutive quarter swings and calculated the time of the single swing by division. This was advantageous as it diminished the relative error but, given that pendulum swing is damped, was permissible only under the assumption of isochrone. Furthermore, instead of timing the quarter swing through the 90-degree arc for which he had carried out his theoretical calculations, Galileo used a smaller initial angle of elongation of the pendulum. Again, this was experimentally advantageous but permissible only under the assumption of isochrone.

The experimental data could thus meaningfully be exploited only as long as the isochronism of the pendulum was not to be called into question. However, as just argued, Galileo's theoretical considerations at some point furnished the isochronism of the pendulum with a problematic character, and the outcome of the second stage of evaluation may have nourished doubts in this respect. If, however, the isochronism of the pendulum was no longer taken for granted, the only way for Galileo to proceed would have been to redesign his experiment in such a way so as to allow swinging through an arc of a certain angle to be compared to falling through an arc of the same angle directly. In order to do this, Galileo would either have had to redo his calculations for the concrete angle of pendulum swing used in the experiment, or he would have had to retime the pendulum swing, this time using an initial angle of elongation of 90 degrees. Moreover, both alternatives would have required, instead of timing of a number of consecutive swings, timing only one swing through an arc whose angle would be clearly defined.

Only in this way would it have been possible to disentangle the contributions due to the non-isochronous behavior of pendulum swing and due to the difference between rolling and sliding. There is, however, no indication whatsoever that Galileo returned to and modified the experiment. Indeed, given the complications he had run into, he would have had little motivation to do so. Even if he had surmised that pendulum swing might not be isochronous, this would hardly have accounted for the observed difference.[26] In hindsight, it seems that the only way for Galileo to make headway and to eventually accept that the isochronism of the pendulum didn't hold would have been to realize that kinematic equivalence under path correspondence didn't hold, i.e., that there was a decisive difference between

---

[26] As argued above, Galileo must have observed rather large amplitudes of pendulum swing, i.e., an initial elongation not too different from 90 degrees. If the difference of 20% were due entirely to the difference of the quarter period of the pendulum swinging through 90 degrees to that of the quarter period of the pendulum swinging through approximately 75 degrees, the isochronism of the pendulum would be gravely violated.

swinging and rolling along the same arc and that it was this difference which mainly showed in the experiment. But this insight, as argued above, remained denied to him.

To sum up, Galileo's research agenda on swinging and rolling had bestowed him with serious complications, theoretical as well as empirical. On the theoretical end, he had not been able to complete the broken chord approach and to infer the time for the limiting case of motion along the arc. This was due primarily to the fact that in Galileo's time the appropriate mathematical means were lacking.[27] Furthermore, his theoretical considerations had indicated that motion along an arc might not be isochronous in the end but without being able to transgress to the limiting cases this remained a conjecture. On the empirical end, Galileo's evaluation of the experimental measurements had resulted in a large deviation with respect to his theoretical expectation. This deviation was due to the unacknowledged difference between rolling and sliding and the fact that pendulum motion is not isochronous. It could thus not be accounted for by rejecting the isochronism of the pendulum.

None of the complications Galileo had run into did, however, imply with necessity that his underlying conceptualization of the relation between swinging and rolling was flawed, as it rested on unwarranted assumptions. His research agenda thus was not overthrown. Rather it had been complicated to the point where pursuing it further did not appear promising. In fact, besides the material discussed in this and the previous chapters, no further entries can be identified in the *Notes on Motion* that would show that Galileo further engaged in considerations directly related to the broken chord approach or the pendulum plane experiment. Around 1602, Galileo apparently discontinued work on swinging and rolling as his research agenda had run into a dead end.

The twist between swinging and rolling is not limited to the history of pop. The above synopsis of Galileo's early research agenda on swinging and rolling from its inception in the realization of a challenging similarity to its running into undisentangable complications and thus a dead end manifests a perplexing almost paradoxical aspect of the early genesis of Galileo's new science of naturally accelerated motion. The conjecture from which Galileo's considerations started out, as obvious and plausible as it may seem, is bluntly speaking false. Pendulum motion is not isochronous, and the alleged similarity between pendulum motion and motion along inclined planes is a chimera.[28]

---

[27] Based on the mathematical means at the time it was not impossible, in principle, to devise valid arguments concerning properties of the limiting case of an iteratively defined geometric construction. Yet with the benefit of hindsight, we realize that the mathematical expression for the time of motion along a polygon with infinitely many sides is a very complex power series, which, as can safely be stated, could not adequately have been rendered with the mathematical means available to Galileo.

[28] Ariotti (1968, 429) maintains that the various observations Galileo alluded to in its favor actually "do not support the claim of isochrony or near isochrony. Yet he emphasizes that "[t]he thesis of isochrony has the appeal of an archetype, of the real essence of things. It is consonant with the principle that nature is simple, ordered, essentially mathematical, and understandable through reason-that nature 'does not multiply things unnecessarily ... she makes use of the easiest and simplest means for producing her effects ... '. (Ariotti 1968, 426)".

If it weren't for this chimera, however, Galileo's considerations wouldn't have gotten underway in the first place. Moreover, the match between his assumption and physical reality was close enough to not be easily debunked. Instead, Galileo was able to retain his basic idea of a similarity between swinging and rolling, implying a close relation between the two types of motion and to tackle the problems he encountered in investigating this relation by more than once modifying and adopting his approach. Somewhat pointedly speaking, it was Galileo's attempt to recover his false assumption of the isochronism of the pendulum that provided the motor for his ongoing efforts. Galileo would not give up his assumption easily as this would have deprived his research agenda on swinging and rolling of its basis and meaning.[29]

The false assumption at the same time was the germ of an epistemological dilemma. As productive as his investigation of the relation between swinging and rolling may have been, ultimately it was not and indeed could not be crowned with the success Galileo had hoped for. Yet it did bestow him with something else, a host of new insights concerning naturally accelerated motion that could stand on their own independently of the agenda on swinging and rolling. These new insights, spin-offs from the agenda on swinging and rolling, became the seeds of a new science of motion, as will be discussed in the next chapter.

References

Ariotti, P. (1968). Galileo on the isochrony of the pendulum. *ISIS, 59*(4), 414–426.
Crawford, F.S. (1996). Rolling and slipping down Galileo's inclined plane: Rhythms of the spheres. *American Journal of Physics, 64*(5), 541–546.
Damerow, P., Renn, J., & Rieger, S. (2001). Hunting the white elephant: When and how did Galileo discover the law of fall? In J. Renn (Ed.), *Galileo in context* (pp. 29–150). Cambridge: Cambridge University Press.
Hahn, A.J. (2002). The pendulum swings again: A mathematical reassessment of Galileo's experiments with inclined planes. *Archive for History of Exact Sciences, 56,* 339–361.
Hill, D.K. (1994). Pendulums and planes: What Galileo didn't publish. *Nuncius Ann Storia Sci, 2*(9), 499–515.
Lima, F.M.S., & Arun, P. (2006). An accurate formula for the period of a simple pendulum oscillating beyond the small angle regime. *American Journal of Physics, 74*(10), 892–895.

---

[29]Despite this lack of success Galileo apparently did not completely give up on his general idea, however. With regard to the *Notes on Motion* the overall impression is that the older the material, the more selected it appears to be. This is not very surprising and would correspond to a plausible work routine in which, during various revisions over the course of almost 40 years, Galileo preserved only those pages containing results that retained importance, even after further progress in particular pages containing completed drafts of propositions. The fact that Galileo retained the sheets documenting the experiment and its evaluation, and that furthermore an unusually large number of entries preserved in the *Notes on Motion* can be related to Galileo's work on the broken chord approach (which is after all the reason why it can be reconstructed so seamlessly), thus strongly implies that he intended to return to and complete the challenging problem of swinging and rolling.

Naylor, R.H. (1974). Galileo and the problem of free fall. *The British Journal for the History of Science, 7*(2), 105–134.
Palmieri, P. (2001). The obscurity of the equimultiples: Clavius' and Galileo's foundational studies of Euclid's theory of proportions. *Archive for History of Exact Sciences, 55*(6), 555.
Settle, T.B. (1961). An experiment in the history of science. *Science, 133*(3445), 19–23.
Shea, W.R., & Wolf, N.S. (1975). Stillman Drake and the Archimedean grandfather of experimental science. *ISIS, 66*(3), 397–400.
Zupko, R.E. (1981). *Italian weights and measures from the Middle Ages to the nineteenth century.* Philadelphia: American Philosophical Society.

Chapter 7
Accumulating Insights: The Problem of Motion Along Broken Chords Driving Conceptual Development

The research program regarding the relation of swinging and rolling, which Galileo had hinted at in the letter to Guidobaldo del Monte in 1602, can be identified in the *Notes on Motion*, as demonstrated in the previous chapters. This research, independently datable on the basis of manuscript evidence to the year 1602, marks the beginning of Galileo's work on problems of naturally accelerated motion. In particular, it was his attempt to approximate motion on the arc—and thus, as he believed, in due course pendulum motion—by naturally accelerated motion along a path composed of a series of inclined planes of increasing number that bestowed Galileo with a number of novel, complex problems. Addressing these problems, Galileo established new propositions, developed new methods, gained structural insights and encountered yet more new problems, which in turn stimulated further investigation. Some of the results, or spin-offs from the research on swinging and rolling as they in fact were, will be surveyed in the present chapter. Emphasis is put on those results which were to become particularly relevant for the further development of Galileo's new science of motion.

The first section will discuss the genesis of a new technique for the diagrammatical representation of motion that Galileo developed as a by-product of his attempts to construct a proof for the law of the broken chord. The new technique, referred to here as integrated time representation, allowed Galileo to represent times elapsed and distances covered by motions in just one diagram. This diagrammatical technique turned out to be better suited to elaborating the kind of problems Galileo was concerned with and soon replaced the earlier scheme of representing motion diagrammatically in which times elapsed and spaces traversed were represented separately.

In the next section, it will be argued that the so-called length time proportionality, which served a crucial function for the foundation of the new science of naturally accelerated motion as presented in the *Discorsi*, likewise emerged as a spin-off from Galileo's research program on swinging and rolling.

In the last section, it will be demonstrated how a particular type of problem, that of determining paths traversed in the least time, was suggested to Galileo by his considerations in the context of the broken chord approach, and how, by systematically elaborating this type of problem, he arrived at a new set of propositions that he later presented in the *Discorsi* and which will be referred to here as the *least time propositions*. It will, moreover, be argued in this section that it was precisely this type of problem which directed Galileo's attention to a more systematic examination of the behavior of motion made along bent planes ("plana inflexa"), resulting in the formulation and proof of two propositions, referred to in the following as the *generalized length time proportionality* and the *generalized law of fall*, which were to become essential ingredients of the new science as published in the *Discorsi*.

7.1 A New Technique for the Diagrammatical Representation of Motion

When representing and elaborating motion problems, one of the central tasks that early modern thinkers were confronted with was, most generally speaking, establishing a relation between the spatial and the temporal aspects of motion, i.e., determining its pheno-kinematics. Central to this endeavor was the graphic representation of motion by means of diagrams.[1] One fundamental benefit of the graphical representation of motion was that it allowed the application of the rich knowledge of Euclidean geometry in the investigation of the problems at hand. The graphical representations of motion, however, presented a problem. Whereas it was quite obvious by the seventeenth century how to represent the path of a moving body on paper, the question of how to render the kinematics of such a motion in diagram is not a trivial one. This requires not only representing the time elapsed during motion in addition to the path traversed but, moreover, uniquely relating spaces traversed and times elapsed, thus rendering the kinematics of the motion situation. A commonly adopted solution to this problem was to isolate the spatial from the temporal aspects of motion, to represent each of them in the form of a separate diagram and to express the relationship between the two diagrams by means of natural language. Solutions of this type will, in the following, generally be referred to as the *separate time representation*.[2]

---

[1] For the history of representing motion by tracing its path and the moving body on paper and for the role this representational form played in the establishment of preclassical external ballistics, see Büttner (2017).

[2] Separate time representation usually embraces three components: a diagram representing the spatial geometry of the motion situation, a second, separate diagram representing its temporal aspects, and, in addition, some form of representation of the relation between the elements of the two diagrams, usually expressed in language employing the theory of proportions. Quick sketches of motion problems in the *Notes on Motion* often lack an explicit description of the relations

7.1 A New Diagrammatical Technique

As we have seen in Chap. 5, in his various attempts to establish a proof for the law of the broken chord, Galileo did indeed draw separate timelines to represent the times in which the motions under consideration were completed, i.e., he employed a form of separate time representation. However, in later approaches to solving problems of accelerated motion on inclined planes, as documented by the majority of the propositions of the Third Day of the *Discorsi*, Galileo mainly used a different method for graphically representing motion, namely, the method referred to here as integrated time representation. That is, he abandoned the representation of times of motion in a separate diagram and instead identified, within the diagram representing the spatial geometry of a given motion problem, lines that served to represent times of the motions under consideration.[3] Internal time representation generally occurs in conjunction with distance time coordination, i.e., with choosing a scale for the representation of the temporal and the spatial aspects such that at least one line of the diagram represents and measures both a distance traversed and the time to traverse that distance.

How exactly the new technique of representing motion diagrammatically emerged as a by-product of Galileo's elaborations of a proof of the law of the broken chord, a proof which, as has been shown, had been engendered by his research program on swinging and rolling, will be demonstrated in the following. Once Galileo had familiarized himself with the new technique, he hardly ever again represented times of motion by means of an external timeline. Indeed in the *Discorsi*, the proof of the law of the broken chord is the only proof in which Galileo made use of separate time representation at all.[4]

As will be detailed below, the intermediate step that facilitated the transition from external to integrated time representation can be reconstructed by means of analyzing uninked construction lines contained on one of the sheets documenting

---

between the diagrams, since those relations were immediately clear to Galileo. The diagrams representing the times of motion are, as a rule but not always, directional and monotonous, that is, the points representing later moments in time are always located to the same side of points representing earlier moments in time. Alternatively, Galileo sometimes rendered lines representing intervals of time as parallel lines.

[3] Galileo's representation of times in diagrams based on integrated time representation is always conformable, i.e., it preserves the ratios that should hold between the respective times according to the laws of motion employed. It is indeed the crucial advantage of integrated time representation that ratios that hold between times according to the laws applied can be geometrically constructed based on the spacial geometry of the motion situation. Palmieri (2006) has recently emphasized the importance of integrated time representation, which he refers to as "embedding of times of descent in diagrams," without, however, providing an account of its origin, for Galileo's considerations.

[4] A limited number of propositions of the Second Book of Third Day of the *Discorsi*, do employ an external timeline. Not, however, to represent times of motion inferred as part of a proof argument, but rather when the ratio of times is given and a specific configuration is sought under which the times elapsed during two motions obey this given ratio. In such cases, separate lines are used to represent the given time intervals and, thus, also their ratio as, for instance, Proposition XXII of the Second Book of the Third Day of the *Discorsi*, the separate lines labeled A and B.

Galileo's attempts to construct a proof for the law of the broken chord.[5] There, Galileo still used an external timeline to represent times of motion. However, for reasons discussed below, he chose to geometrically construct this timeline conformably, i.e., in such a manner that it would preserve the ratios that should hold between the times of motion according to the respective laws of motion applied. Moreover, Galileo chose a special scale for this conformable timeline that corresponded, in a particular manner, to the scale of his spatial diagram. This choice had the advantage that lines constructed within the spatial diagram of the motion situation could, preserving their length, successively be simply transferred to the timeline where they then represented times of motion. The transition from this particular construction to a fully developed integrated time representation then corresponds to the realization that the transfer of lines constructed within the spacial geometry of the motion problem to a separate timeline is redundant for all practical purposes and, thus, is not strictly necessary.

The use of integrated time representation in making inferences about times of motion is, first of all somewhat more efficient as it repeatedly allows the avoidance of a, strictly speaking, redundant step. Its true strength, however, plays out, when consecutive inferences are made where in each step a yet undetermined time of motion is inferred from a known time of motion based on propositions relating times elapsed and distances traversed in such motions. In the combined space-time diagrams of integrated time representation such a chain of consecutive inferences simply corresponded to a number of consecutive geometrical operations within one and the same diagram. Each of these operations transforms a line representing a time into another line, again representing a time, albeit of a different motion. Thus, the line generated can then simply become the object of the next geometrical transformation irrespective of the geometrical construction by which it was generated. While it is not the case that integrated time representation allows for inferences that could not, in principle, be reached employing separate time representation, it greatly facilitates establishing complex chains of inferences. Arguably, integrated time representation thus allowed Galileo to achieve results he would not otherwise have been able to achieve.[6]

---

[5]Uninked construction lines scored into the paper and the incision marks of the compass on the pages of the *Notes on Motion* have thus far been virtually ignored in studies of the manuscript. The sole exception I am aware of are Damerow et al. (2001, 379), where resorting to information embraced by uninked construction marks remains restricted to the analysis of a single construction by Galileo. I owe my information concerning the uninked constructions to a group of scholars from the Max Planck Institute for the History of Science, who took photos of the manuscript under raking light in which these constructions become visible.

[6]Integrated time representation as argued in the text greatly facilitates chains of inferences as part of an argument and thus trivially also the putting of arguments into words. An example is provided by Proposition XIXXX of the Third Book of the Third Day of the *Discorsi*, in which the proposition is stated in terms of a proportion between ratios of distances and ratios of times, while in the proof itself integrated time representation is exploited and times are specified in reference to lines in the diagram.

7.1 A New Diagrammatical Technique

Galileo, indeed, quickly abandoned separate time representation in favor of integrated time representation and the emergence of the new technique thus provides an additional criterion for establishing a relative chronology. Entries in which Galileo still used the older technique of separate time representation, should, thus, be considered as being earlier than or at least roughly contemporaneous with Galileo's construction of a proof of the law of the broken chord, which resulted in the adoption of the new technique. The transition from separate time representation to integrated time representation, furthermore, gave rise to a terminological ambiguity, since now it was possible for one and the same line to represent a distance as well as time. For some time, this led Galileo to use transitional forms of graphical representations of motion that allowed for a disambiguation. The difficulty was finally resolved by an adaptation of the terminology by which Galileo referred to the different elements of his new diagrams, spaces traversed, and times elapsed. The use of the new versus the old terminology is, thus, likewise indicative of a relative ordering among entries in the *Notes on Motion*.

The genesis of internal time representation Manuscript evidence allows the reconstruction of how integrated time representation emerged in great detail. A geometrical construction on folio 186 verso, indeed, contains elements which link internal and external time representation and which thus facilitated the transition between them. More concretely, it will be demonstrated that Galileo, in constructing an external timeline in the course of his attempt to prove the law of the broken chord on this folio, used a particular geometrical construction that can be regarded as interim between separate time representation and integrated time representation. The relevant parts of the construction are not immediately apparent as they were carried out by Galileo with the help of uninked construction lines which, together with construction marks, are represented in Fig. 7.1.

As discussed in detail in Sect. 5.1, the central diagram on 186 verso depicts a motion situation characteristic of the law of the broken chord. The horizontal line td represents the timeline Galileo used in a second attempt at a proof, documented by the second paragraph on the page.[7] In contrast to the timeline xn—which Galileo drew on the same page and which is associated with his first attempt at constructing the sought proof documented in the first paragraph—this line and, in particular, the distances on it were not the result of a freehand sketch; rather it was constructed as a conformable timeline. The seemingly sparse information of the uninked construction lines and marks on the page suffices to exactly reconstruct the way in which Galileo constructed the points on the timeline td, in accordance with the requirements of a proof of the law of the broken chord, as they are formulated in the second paragraph on the folio.[8] In the following, Galileo's construction will be followed step by step.

---

[7] 186 verso, T1B.

[8] The argument at the basis of Galileo's diagram on folio 186 verso is discussed in detail in Chap. 5.

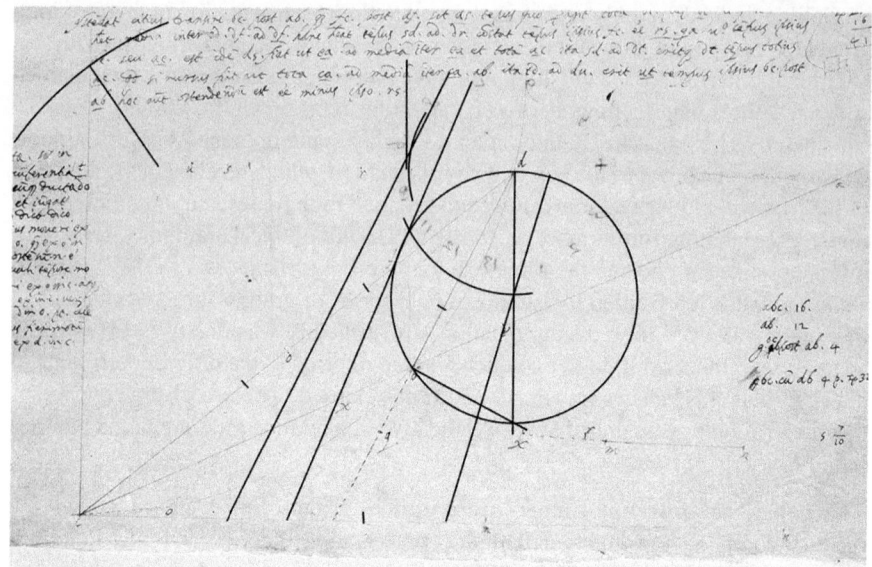

Fig. 7.1 Detail of folio 186 verso with all uninked details that could be identified on the page under raking light shown as bold black lines. (Courtesy of Ministero dei Beni e le Attività Culturali – Biblioteca Nazionale Centrale di Firenze. Reproduction or duplication by any means prohibited)

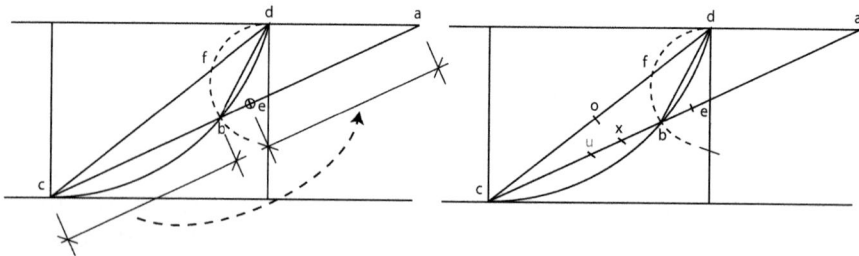

Fig. 7.2 Schematic rendering of the central diagram on 186 verso showing the construction of point e such that $ae = bc$ and points o, x, and u marking the mean proportionals $mp(df|dc)$, $mp(ae|ac)$, and $mp(ab|ac)$, respectively (point u is not lettered in Galileo's original drawing)

The points o, x, and u mark the mean proportionals do, ax, and au (the latter point was indicated but not lettered in the diagram. For ease of reference, I have lettered it u in accordance with the diagram published together with the law of the broken chord in the *Discorsi*). For the conformable construction of the timeline td which Galileo was aspiring to, the position of these points and, thus, the lengths of the corresponding mean proportionals had to be known. More concretely, Galileo had constructed point o such that $do = mp(df|dc)$, point x such that $ax = mp(ae|ac)$, and finally point u such that $au = mp(ab|ac)$ (see Fig. 7.2).

7.1 A New Diagrammatical Technique 175

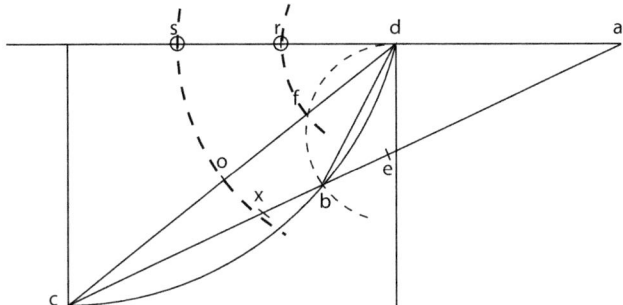

Fig. 7.3 Schematic rendering of the central diagram on 186 verso showing the construction of the points s and r such that $ds = do$ and $dr = df$

In the first step of the construction of the timeline, Galileo constructed the points s and r and hence the lines ds and dr representing the times of fall through dc and df, respectively, by transferring the distances do and df respectively to the timeline such that $|ds| = |do|$ and $|dr| = |df|$ (see Fig. 7.3). As the distances were transferred by means of a compass, the only remaining construction marks in this step are incisions of the compass at points d, s, and r. This construction trivially satisfied the requirement that followed from the law of fall and which Galileo expressed as:

> Sit ds tempus quo peragitur tota dc ... et quam rationem habet media inter cd, df ad df, hanc habeat tempus sd ad dr ...[9]

Galileo's construction not only guaranteed that the times represented by ds and dr, i.e., the times of motion along dc and df respectively, were in the same ratio as df and $mp(df|dc)$. Moreover, the line dr of the timeline representing the time of motion along df was chosen in such a way that its length equaled that of the distance fallen. This choice coupled the scale of the timeline to that of the spacial diagram. The time of fall through df is not yet represented by the line df itself as it would be in later cases when Galileo employed distance time coordination in combination with integrated time representation. Yet the central idea is already present, namely that one particular given time, here the time of motion along df represented by dr, is measured by the same number of parts with respect to the time unit as the corresponding distance traversed, here df, measured with respect to the unit of length measure.

Next Galileo constructed a point t on the timeline such that the line dt would represent the time of fall through the inclined plane ac. According to the law of fall, this line dt had to fulfill a requirement which Galileo's expressed as:

> fiat ut ea ad mediam inter ea et totam ac, ita sd ad dt, eritque dt tempus totius ac.[10]

---

[9] 186 verso, T1B.
[10] Ibid.

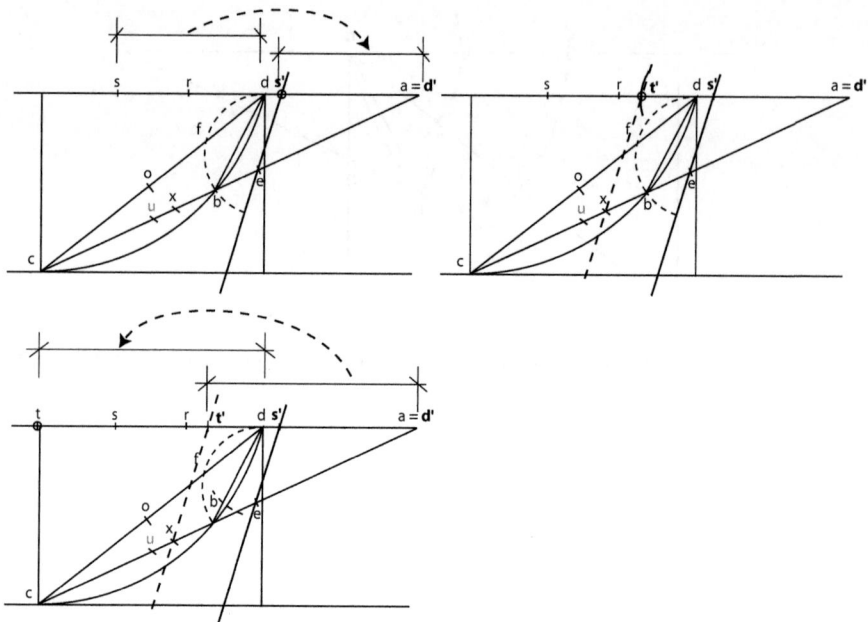

Fig. 7.4 Schematic rendering of the central diagram on 186 verso showing the construction of point t by means of the theorem on intersecting lines in such a fashion that $ae/ax \sim ds/dt$. To this end, the auxiliary points s' und t' are introduced such that $ae/ax \sim as'/at'$, and finally t is constructed such that $dt = at'$

Hence the point t had to be chosen such that $sd/dt \sim ae/ax$. Yet no simple geometrical construction lent itself to the realization of this proportion, thus allowing for a geometrical determination of point t. Finding an appropriate construction was hampered by the fact that the timeline comprising the lines sd and dt did not appropriately intersect with the inclined plane ac comprising the lines ae and ax. Galileo, however, found a relatively simple way to construct the line dt so that it would meet the given constraint (the single steps of Galileo's construction are rendered in Fig. 7.4).

Using the compass, he simply transferred the distance ds to a, thereby constructing a new point s' on the upper horizontal where $|ds| = |as'|$.[11] Next, Galileo drew a straight line through the points e and s' and a second line, parallel to this first line, through point x, cutting the upper horizontal at point t'. Both these lines are preserved as uninked construction lines. The theorem of intersecting lines can be applied to this construction and it thus holds that $as'/at' \sim ae/ax$. Since as' was equal in length to ds, at' had the length of the sought distance dt. To appropriately

---

[11] The points s', t' and v' were merely marked by Galileo but not lettered.

7.1 A New Diagrammatical Technique

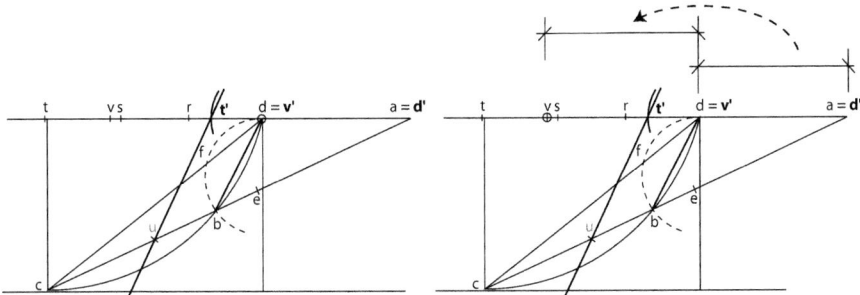

Fig. 7.5 Schematic rendering of the central diagram on 186 verso showing the construction of point v, where it holds that $ad = dv$

construct the point t on the timelime Galileo merely had to transfer the distance at' back to d in such a way that a coincided with d.

The last point that remained to be constructed on the timeline was the point v, such that line dv would represent the time of fall along ab. Then, finally, the difference between dt and dv, i.e., tv, would represent the sought time of motion along bc after fall from a. The constraint for this construction, again resulting from the law of fall, was the following:

> Quod si rursus fiat ut tota ca ad mediam inter ca, ab, ita td ad dv...[12]

In modern notation, the constraining condition for the construction of point v can hence be expressed as the proportion $dt/dv \sim ac/mp(ac|ab) \sim ac/au$. In order to satisfy this constraint, Galileo used the same strategy he had used in the construction of point t (see Fig. 7.5). He joined point t' and point u, which marked the mean proportional between ab and ac, by an uninked construction line. In order for the theorem of intersecting lines to become applicable, Galileo needed to draw a line parallel to $t'u$ through b. As it happened, with Galileo's particular choice of geometry for the problem, the chord db was coincidentally parallel to line $t'u$. Thus, the point v', whose distance from point a had to be the same as that of the sought point v from d, coincided with d, and all that remained to be done to complete the last step of the construction was to transfer the distance av' to d of the timeline, thus constructing point v.

What motivated the construction of a conformable timeline on 186 verso? The statement about naturally accelerated motion that the law of the broken chord expressed was by no means self-evident. On the contrary, the insight that motion along a broken chord was completed in less time than motion along a chord spanning the same arc was surprising and counterintuitive. It had been hinted at by the

---

[12] 186 verso, T1B.

pendulum plane experiment and proven valid in calculations. A generalization of the insight as expressed by the law of the broken chord must have been problematic, as long as no rigid proof had been constructed. In view of the immature state of Galileo's pondering on naturally accelerated motion, even a general proof would have left room for doubts.

Galileo's construction of the conformable timeline on 186 verso was thus, certainly to some extent, a means of testing the statement of the law of the broken chord and indirectly also the inferences upon which his attempt to establish a proof of the proposition rested, as the construction of a conformable timeline allowed the direct measurement and comparison of the distances representing times. A conformable timeline, in particular, allowed him to check whether the statement of the law of the broken chord indeed held for the concrete geometrical setup considered. In fact, a last element of the uninked construction, which has not yet been considered, shows that Galileo indeed used the timeline to test the validity of the statement of the law of the broken chord for the concrete situation under consideration.

A small uninked mark between the points r and s indicates that Galileo had transferred the distance ut to r to compare the lengths of the lines ut and rs directly. Since rs represents the time of motion along fc from rest at d, and ut the time of motion along the lower of the broken chords bc, also from rest at d, this comparison amounts to a test of the validity of the reduced statement of the law of the broken chord according to which the law of the broken chord is indeed equivalent to stating that it takes less time to move along bc when motion is made from rest at d than it does to move along fc, likewise from rest at d. This reduced statement, as detailed in Chap. 5, indeed marks the endpoint of Galileo's considerations documented on 186 verso, while at the same time providing the starting point for all of his further considerations in this context.

The construction of a conformable timeline thus had required the construction of lines of the desired length within the spatial geometry of the problem, which subsequently had to be transferred to the correct position on the one-dimensional timeline. However, in view of the likely purpose of the construction, i.e., to enable the direct comparison of two times to validate the assumption of the law of the broken chord, the transfer of those lines to a timeline was not essential. The insight that those steps are in fact unnecessary and can simply be dispensed with lead directly to the new method of integrated time representation.

The transition to internal time representation—early use of the new technique and its consequences The shift from separate time representation to integrated time representation occasioned, as will briefly be exemplified in the following, the emergence of transitional forms of diagrams, as well as a change in the terminology Galileo used in dealing with problems of motion. Both the transitional forms of diagrams and Galileo's adaptation of a new terminology ultimately rest on an ambiguity brought about by integrated time representation. If one and the same diagram serves to represent the spatial as well as the temporal aspects of a

7.1 A New Diagrammatical Technique

given motion situation, ambiguity arises as to whether a given line in the diagram represents a distance covered or a time elapsed during motion, or possibly both.

In the following, a number of entries pertaining to this transitional phase will be discussed. Independent evidence can be adduced, that the folios comprising these entries belong to the period of work during which Galileo pursued his challenging research program on the relation of swinging and rolling. The majority of folios showing the transitional phenomena in question can indeed be directly related to the broken chord approach, lending additional support to the interpretation presented above according to which integrated time representation first emerged as a consequence of Galileo's elaborations of the proof of the law of the broken chord.

A first motion diagram displaying a transitional form of representing times can be found on folio 147 recto. The diagram is part of a proof of the law of fall from the law of chords and the length time proportionality as premises. The details of this proof need not concern us here as we are primarily interested in the means with which Galileo countered the ambiguities that arose from the use of his new technique of representing times.[13]

In the considerations documented on folio 147 recto, Galileo actually employed two different, albeit related, strategies to counter potential ambiguities: when furnishing his diagram with the results of the numerical calculations of the times of motion for the depicted motion situation, he marked the numbers representing times by associating them with a dotted line to thus distinguish them from numbers representing distances. An example is the line *ab* in the diagram which, as the number 10 written next to it indicates, measures ten distance units. Galileo had inferred, that this distance was traversed in 12 time units, which he indicated by writing the number 12 preceded by a dotted line next to the same line *ab*.

On the other hand, Galileo drew a distance-time ratio diagram, in which he represented in abbreviated form the proportions which must hold between the distances traversed and the times of motion in the given example.[14] To avoid

---

[13] For a detailed discussion of the proof of the law of fall on 147 recto, see Sect. 10.1.

[14] Throughout Galileo's *Notes on Motion*, a number of diagrams can be found which embody a particular mnemonic technique for dealing with complex proportions between ratios, often, but not necessarily, between ratios of times and ratios of distances. For an example, see Fig. 7.6. In order to do so, these diagrams re-represent lines from the main diagram, ignoring their spatial orientation, focusing merely on their lengths as the relevant characteristic with regard to the proportions under investigation. Depending on the context, the appearance of an distance-time ratio diagram, can vary somewhat. These diagrams can be read like a table where usually one row contains distances covered, and the other times elapsed in traversing these distances in a given motion situation, all magnitudes being represented by lines. Times in one cell represent the times of motion along the distance represented in the cell in the same column. Neighboring columns represent valid proportions. Thus, in the example in Fig. 7.6, the first two columns represent the proportion $t(ab)\ t(ad) \sim ab\ ad$, as a result of the length time proportionality for the motion situation under consideration. The second and third columns represent the proportion $t(ad)\ t(ae) \sim ad\ as$, as a result of the law of fall. Such diagrams aid the construction of valid proportions. In the example provided it can, for instance, immediately be read off that the ratio between the time elapsed in

Fig. 7.6 Schematic rendering of the ratio diagram on folio 189 verso

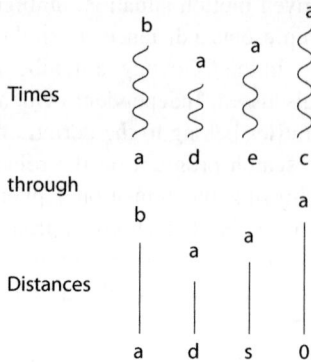

potential confusion in the distance-time ratio diagram Galileo adopted a convention comparable to the one used in the main diagram. He simply rendered those lines representing a time as dotted lines and those representing distances as a solid line.

At least two more instances can be identified where, when drawing distance-time ratio diagrams, Galileo introduced graphical conventions that allowed him to distinguish lines representing distances covered by motion from lines representing times elapsed. On double folio 156 recto 157 verso, beside the large central motion diagram, Galileo drew a small diagram serving as a mnemonic aid in the elaboration of the problem for which the large diagram was drawn. As argued in Chap. 5, the considerations documented on the double folio are part of Galileo's attempt to precisely determine the ratio of the times of motion along a chord and a corresponding broken chord, a problem which had arisen as part of the broken chord approach. The distance-time ratio diagram beneath the small diagram compares to that on folio 147 recto where, once again, lines representing times are disambiguated from those representing distances by the use of dotted lines.

Folio 189 verso, i.e., the very page comprising the record of the pendulum plane experiment, bears an attempt by Galileo to prove a proposition that was later published as Proposition V of the Second Book of the Third Day of the *Discorsi*. In this attempt, just as in the two cases already mentioned, he employed a distance-time ratio diagram in which, however, a somewhat altered convention served to disambiguate lines representing times from those representing distances. On folio 189 verso, Galileo did not use a dotted line but a wavy one to signal that a line represented a time of motion as opposed to a distance covered by motion.

---

traversing *ab* and the time elapsed in traversing *ac* must be in proportion to the ratio of the lines *ab* and *a*0.

7.1 A New Diagrammatical Technique

All three examples discussed above, in which special graphical conventions are used to disambiguate lines representing times from those representing distances traversed, are thus indeed directly linkable to Galileo's research program on swinging and rolling, lending additional support to my interpretation, according to which the new technique first emerged as a by-product of the program Galileo was pursuing toward the end of 1602.

Yet the transition from external to integrated time representation created an ambiguity, not only in the graphical representation of motion problems but also when referring to such diagrams with written language. Hence, as soon as integrated time representation started to be used in solving motion problems, the ambiguities it entailed also became apparent when, as part of formulating proofs, reference was made to integrated time representation diagrams.

Thus, for instance, on folio 164 recto, Galileo concluded an argument by stating:

> Patet insuper, tempora casuum per gb, fc, ed esse ut lineas gb, fc, ed; non tamen a magnitudinibus ipsarum linearum gb, fc, ed esse determinandas eorumdem temporum quantitates, si temporis mensura ponatur ab, in quo tempore conficiatur linea ab, sed desumendas esse a lineis rb, tc, vd.[15]

Again, details of Galileo's argument do not concern us here.[16] Central to our present interest is the added remark in which Galileo warned himself of a pitfall potentially arising from the new technique of integrated time representation.[17] Galileo, as it were, warned himself not to conclude from the proportionality between the times and the respective distances of motions made along inclined planes of equal height after initial fall from the same point that the lines representing these inclined planes could simply be thought of as representing and, in particular, as providing the measure of the times of motion on them as it might have seemed to indicate but which would indeed be wrong in the case at hand.

The new means of integrated time representation, in fact, required an adjustment of terminology. As long as Galileo was only using separate time representation, the expression time xy (tempus xy) was completely unambiguous. Time xy indubitably meant the time represented by the line xy, part of a timeline, and could by no means refer to the time needed to traverse a distance xy. Indeed, a distance labeled xy would, as a rule, not even be part of the spacial geometry of the problem. With the introduction of integrated time representation, the unqualified expression tempus xy had, however, become ambiguous, as it could now mean the time represented by the

---

[15] 164 recto, T1.

[16] A more detailed discussion of Galileo's entry on 164 recto is contained in Sect. 10.1.

[17] Wisan (1974, 192) was the first to recognize that Galileo's remark on 164 recto was directly linked to the use of integrated time representation and concluded that "[t]his suggests a chronological development in which folio 164 recto represents the transitional stage between earlier proofs using a separate time line and later proofs which do not."

line xy, or alternatively indeed the time needed to traverse the distance xy, thus, in fact facilitating false conclusions of the kind Galileo alluded to in the addendum on 164 recto.[18]

On 147 recto, the very folio page already discussed above, for instance, we find Galileo using the expression "tempus ae" to refer to the time needed to traverse the distance ae, which in the example at hand was not represented by the line ae but instead by the mean proportional as.[19] To avoid any possible confusion, Galileo adapted his terminology and would henceforth usually mention both the line representing the time and the line representing the distance covered by the motion under consideration, in expressions such as xy tempus casus per vw or simply xy tempus per vw. The majority of the very limited instances in the *Notes on Motion* where Galileo used the expression tempus xy to signify the time in which motion along xy was completed rather than the time represented by xy, without further qualification, can in fact be securely attributed to an early period of work before integrated time representation had emerged and with it the need for a more precise specification.

We have seen how, as a spin-off from his research on swinging and rolling, Galileo contrived and developed a new conceptual tool in integrated time representation and how he was able to overcome the complications which initially arose. Once those impediments were removed, the application of integrated time representation would decisively shape Galileo's further investigations into the nature of naturally accelerated motion. These investigations, as will be discussed in the following sections, also brought about some very tangible new insights about the pheno-kinematics of naturally accelerated motion.

7.2 The Length Time Proportionality as a Spin-Off

The length time proportionality, according to which the times elapsed during naturally accelerated motion along inclined planes of different inclinations but equal height are in direct proportion to the length of these planes, has hardly figured at all in Galileo's considerations analyzed, thus far. It is, in particular, strikingly absent from his attempts to construct a proof for the law of the broken chord and did not play any role in the design and initial evaluation of the pendulum plane experiment.

---

[18] Wisan observes and correctly links the change in terminology to Galileo's use of integrated time representation. She inaccurately claimed, however, that the entry on 164 recto provides the sole instance where tempus xy is used to refer to a time represented by line xy, instead of to the time to traverse the line xy. On folio 77 recto, for instance, Galileo states "...erit *bi* tempus per xa ...," and when he continues his argument, he refers to the line *bi* representing a time simply as "tempus *bi*." Other examples can be found on 87 verso or folio 96 recto. By the time Galileo wrote these entries, he would have been so familiar with the new technique that the ambiguity entailed would no longer have been a source of confusion for him. Cf. also the discussion of folio 164 in Chap. 10.

[19] 147 recto T5. Cf. Chap. 11

7.2 The Length Time Proportionality as a Spin-Off

This is in striking contrast to the prominent position of the proposition in the new science of naturally accelerated motion as laid down in the Second Book of the Third Day of the *Discorsi*. Not only does the length time proportionality, i.e., Proposition III, figure as a premise in the majority of the proofs of the other propositions of the Second Book of the Second Day; furthermore, it is one of only three propositions of the Second Book that are not proven based exclusively on other propositions making statements about times elapsed during naturally accelerated motions. Rather the proof given for the length time proportionality is based on more fundamental assumptions about such type of motions.

As will be demonstrated in Chap. 11, the length time proportionality achieved the status of a fundamental proposition early on, i.e., about the time Galileo was pursuing his research on swinging and rolling. It is not least for this reason that an understanding of the origin of the proposition and the way in which it came to take on its fundamental role for the new science are indispensable for an overall interpretation of the early conceptual development of Galileo's new science of naturally accelerated motion. The question concerning the origin of the length time proportionality has, indeed, previously attracted some attention in Galileo scholarship. Yet no persuasive interpretation has been put forward to the present day. Alternative interpretations that have been given range from the assumption that Galileo became aware of the proposition by means of experiment, to the hypothesis that it was initially deduced by him based on the conceptual framework of the *De Motu Antiquiora* and was accepted as valid even after the transition to the assumption that falling motion was, in principle, accelerated.

The majority of existing interpretations are unified by the assumption that the length time proportionality must in some way or another have been part of the repertoire of propositions concerning naturally accelerated motion from which the development of Galileo's new science took its outset. Yet none of the interpretations which have been brought forward in this respect are based on concrete manuscript evidence.[20]

---

[20] Renn holds that the length time proportionality was initially "merely a plausible assumption which Galileo at first also believed to be a consequence of this [*De Motu Antiquiora*] theory (Damerow et al. 2004, 205)." With his interpretation, Renn followed Wisan who has in fact provided an account of how Galileo, still working under the assumption that motion along inclined planes was in principle uniform, may have first arrived at the insight expressed by the length time proportionality. Her speculative account strikes me as not very convincing, as it presupposes that Galileo committed a trivial error when he applied "this rule [one of the kinematic propositions] while overlooking the necessary condition that the distances be equal (Wisan 1974, 188)." She, moreover, suggests that Galileo may have successfully tested the proposition in an experiment. Wisan (1974, 200–201) attempts to supported her interpretation by making reference to an entry on folio 177 recto in the hand of Niccolò Arrighetti with no original in Galileo's hand extant. The relevant entry is, furthermore, crossed out. Independent of the question of whether it documents a consideration by Galileo at all, the entry poses some serious interpretational difficulties. Caverni (1972) likewise located the origin of the length time proportionality in Galileo's *De Motu Antiquiora* theory, attributing a similar error to Galileo, as Wisan does.

In the following it will be argued that the length time proportionality emerged as a spin-off from Galileo's work on the broken chord approach and thus, more generally, his research program on swinging and rolling. More concretely, it will be maintained that the length time proportionality emerged as a consequence of Galileo's elaboration of a proof of the law of the broken chord. This first of all provides an explanation as to why the proposition was not used in Galileo's various attempts to prove the law of the broken chord. In contrast to what earlier interpretations have claimed, the length time proportionality was thus not a prerequisite but rather a consequence of Galileo's early considerations on naturally accelerated motion.

In the final draft of the law of the broken chord eventually published in the *Discorsi*, Galileo did not use the length time proportionality as a premise. This has already been observed by Wisan: "Galileo could have avoided the construction of the first Lemma by using Theorem III [the length time proportionality]". This theorem, however, is not employed in the proof; nor is it used as a heuristic device in any of the fragments exhibiting the search for a proof of Theorem XXII. This argues that Theorem III is not yet part of his deductive machinery (Wisan 1974, 183)."[21]

Wisan's conclusion that Galileo did not use the proposition because it was not part of his "deductive machinery" upon devising the proof of the law of the broken chord is plausible and her argument can be substantiated. As exemplified in Chap. 5 it was not only that Galileo could have saved himself the trouble of constructing the first geometrical lemma used in the argument by invoking the length time proportionality as a premise. As illustrated in Fig. 7.7 this would in fact have substantially simplified the entire argument. Indeed, the proof of the law of the broken chord contains a rather protracted partial argument which results in the statement of the length time proportionality and, thus, taken in itself, constitutes a proof for the proposition.

If the partial argument by which the length time proportionality is established, and which could have simply been replaced by invoking the proposition as a premise, had only been made once by Galileo, this could hardly be taken as evidence that the length time proportionality was not known to him at the time. Indeed, more often than not, a proof will be formulated in a less efficient manner than would, in principle, be possible given the available premises. However, as shown in Sect. 5.1, Galileo struggled to establish a proof of the law of the broken chord. In consequence, the somewhat overly complicated argument, besides in the final draft as it was later almost verbatim published in the *Discorsi*, is repeated twice in the *Notes on Motion*. This, indeed, strongly indicates that the length time proportionality was not part

---

[21] Wisan is somewhat inconsistent in her assessment as she on the one hand asserts that the length time proportionality was an early insight achieved before Galileo's conceptual shift to the assumption of natural acceleration, while on the other hand she maintains that the proposition was not "part of his deductive machinery" until much later.

7.2 The Length Time Proportionality as a Spin-Off

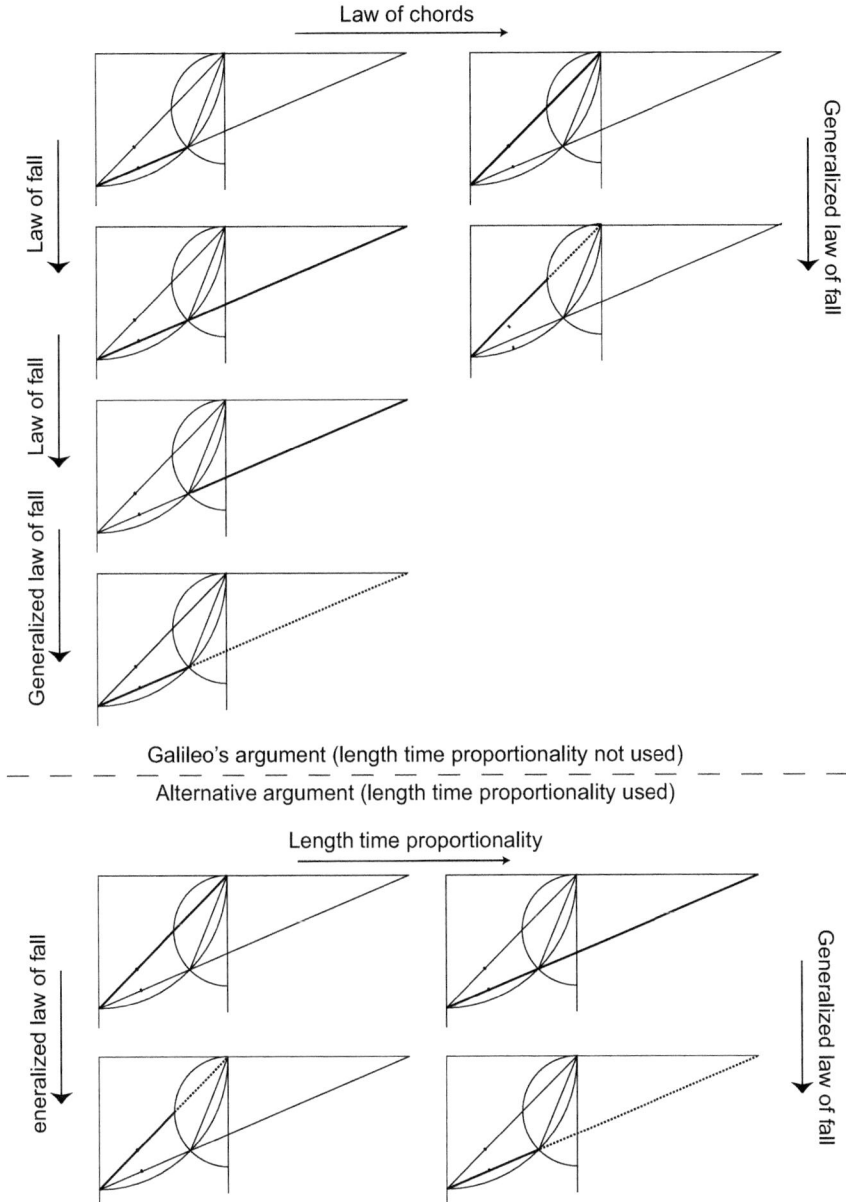

Fig. 7.7 Schematic representation of the line of argument of the proof of the law of the broken chord as published by Galileo in the *Discorsi* (above) and of an alternative proof that could be constructed by employing the length time proportionality as a premise in the argument (below). The transition from one diagram to the next represents an inference based on applying a law of motion, noted next to the arrow representing the respective inferential step. Bold lines represent a known time of motion on the basis of which another time of motion is inferred in a deductive step; dotted lines indicate that the motion considered has been proceeded by fall through the distance indicated by the dotted line

of Galileo's repertoire when the respective notes were taken and hence the partial argument in question could not simply be replaced to formulate a simpler, more elegant version of the proof.

With regard to the upper right of the four diagrams in Fig. 7.8, in his proof of the law of the broken chord, Galileo had established by the partial line of argument, that the time along the chord BA is to the time along a less inclined plane CA of the same height in the same ratio as the lengths of these planes are to each other, just as stated by the length time proportionality. The basic geometrical insight that Galileo made use of in this derivation was formulated by him as the first of the three geometrical lemmata preceding the law of the broken chord, i.e., Proposition XXXVI of the Second Book of the Third Day of the *Discorsi*.[22] A draft of this lemma can be found together with a draft of the law of the broken chord itself on 172 verso.

The assumption that Galileo, in principle, became aware of the length time proportionality as a consequence of his efforts to formulate a proof for the law of the broken chord receives compelling support by a consideration documented on folio 147 recto, already alluded to above. The content of this folio was, as will be argued in detail in Chap. 11, drafted contemporaneously or slightly later than Galileo's work on the broken chord approach and thus likely also only slightly after his composition of the proof of the law of the broken chord. Of particular interest for the present argument is a proof of the law of fall contained on the lower two thirds of the page. As illustrated in Fig. 7.8, this proof is based on a geometrical constellation that reproduces parts of the geometrical constellation underlying the proof of the law of the broken chord. However, not only are the underlying geometrical constructions similar, but both proofs indeed comprise similar lines of argument.

The proof of the law of fall on 147 recto, in particular, makes use of a geometrical lemma that Galileo referred to as "ex demonstratis." With regard to Fig. 7.8 it is concretely stated in the proof that "ex demonstratis" AB is the mean proportional of AD and AC. It has been claimed that the lemma, thus referred to, was a geometrical proposition noted by Galileo at that time but lost today. However, upon closer inspection, the geometrical statement is found to correspond precisely to what is proven in the first geometrical lemma of the law of the broken chord. It was, thus, in all likelihood exactly this lemma Galileo was referring to by the phrase "ex demonstratis."

Indeed, as is argued in detail in Chap. 11, folio 147 must belong to the same phase of work as folio 172, on which Galileo had drafted the lemma in clean form, as part of his effort to collect and systematize his results achieved thus far. The proof of the law of the broken chord and the proof of the law of fall are, thus, not only structurally similar. The proof of the law of fall was very likely formulated in close temporal proximity to the drafting of the final version of the law of the broken chord, perceivably with folio 172 still lying on the desk. When Galileo

---

[22] Cf. Chap. 5, Sect. 5.1.

7.2 The Length Time Proportionality as a Spin-Off

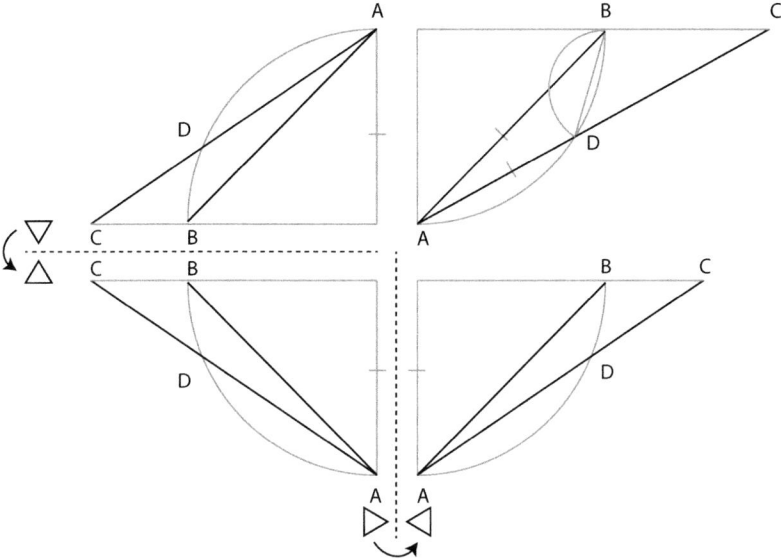

Fig. 7.8 Structural similarity between Galileo's proof of the law of the broken chord and a proof of the law of fall contained on folio 147 recto. The upper left diagram reproduces the proof diagram from 147 recto, reduced to the elements relevant to the argument to be made here. The diagram in the upper right corresponds to the proof diagram of the law of the broken chord as contained in the *Discorsi*. If the diagram from 147 recto is first mirrored on a horizontal (lower left), and successively on a vertical (lower right), the similarity of the diagrams corresponding to similar partial arguments within the respective proof immediately strikes the eye. In both proofs, the fact that AB is the mean proportional of AD and AC, which Galileo established in a lemma, plays the decisive role

elaborated the proof of the law of fall on 147 recto, he was thus very likely inspired by reflecting upon the proof of the law of the broken chord constructed before, and he explicitly referred to one of the geometrical lemmata formulated for the proof.

The argument on folio 147 recto does not identically reproduce the line of reasoning contained within the law of the broken chord by means of which a statement equivalent to the length time proportionality is inferred from the law of fall, the law of chords, and the geometrical lemma. Rather the reasoning is turned around, retaining, however, its structure. On folio 147 recto, the law of fall is proven based on the length time proportionality, the law of chords, and the lemma as premises. Thus, compared to the reasoning embraced within the law of the broken chord, merely the role of premise and consequence, or else *explanans* and *explanandum*, are exchanged, otherwise the arguments are identical. The reflection upon the proof of the law of the broken chord, thus, not only provided Galileo with the insight into the length time proportionality but furnished him with the line of argument by means of which the proposition could be proven, or, by exchanging the role of premise and consequence, by means of which the law of fall could be proven based on assuming the length time proportionality.[23] At about the same time, Galileo finally added a proof for the length time proportionality to folio 163 verso which, as one of the star group folios, will be discussed in detail in Chap. 10. The length time proportionality was by far not the only proposition which resulted as a direct consequence from Galileo's research program on swinging and rolling, as will become clear in the succeeding sections.

7.3 Bent Planes Traversed in Least Time

In the present section, it is demonstrated how, in the context of the broken chord approach, Galileo was attacking an extremely ambitious problem. He considered a specific variation of a bent plane, i.e., he investigated motion along a bent plane connecting two points where the position of the junction point of the two planes together forming the bent plane was allowed to vary in a specific manner. He sought the path of least time, i.e., that geometrical configuration of the bent plane where motion along it would be completed in least time. The question ultimately remained unanswerable with the means available to him. Yet Galileo's approach yielded some profound insights solving at least some of the partial problems he had encountered in his attack on the problem. A number of these insights, as will be argued in the successive section, were immediately drafted by Galileo as propositions, and these were published in the *Discorsi* more than 35 years later.

Manuscript evidence On folio 130 verso, Galileo centrally attacked the question of how the time of naturally accelerated motion along a bent plane connecting two points varies with a specific variation of the bent plane, i.e., of the position of the junction point of the two planes forming the bent plane. The content of this folio is related in multiple ways to the content of other folios, and these relations are depicted schematically in Fig. 7.9. A number of folios can directly be linked to the problem on 130 verso because they either contain variations of the geometrical construction under examination, calculations of the relevant distances on 130 verso, or attempts to solve partial problems which had emerged as part of Galileo's work documented on the page.

---

[23] As detailed in Chap. 5, the length time proportionality was employed by Galileo in all extant numerical calculations of broken chord motion situations in the context of the broken chord approach, which according to the interpretation presented, suggests that these calculations postdate Galileo's inception of the length time proportionality as a spin-off from his search for a proof of the law of the broken chord.

7.3 Bent Planes Traversed in Least Time

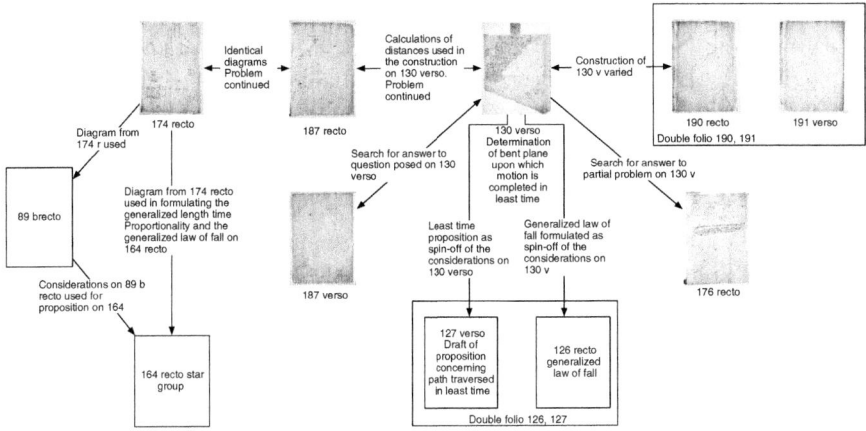

Fig. 7.9 Folios related to the problem of motion along bent planes elaborated on 130 verso. Folios represented by thumbnail images are discussed in detail in the current section

Thus, the double folio 190, 191, contains three diagrams, characterized by essentially the same construction as that of the central diagram on folio 130 verso, albeit for different geometrical constraints, determining the variation of the bent plane under consideration. Folio 176 recto contains an argument regarding a particular geometrical property of the construction on folio 130 verso, which had to be presupposed in order to solve a partial problem that had emerged from Galileo's approach on 130 verso. Despite being correct from a modern perspective, Galileo marked his argument as false. Folio 176 recto, moreover, bears auxiliary calculations for Galileo's considerations of motion along polygonal paths composed of a high number of inclined planes documented on folio 166. This links Galileo's considerations regarding variation of a bent plane directly to his broken chord approach. This link is not very surprising as we saw in Chap. 5 that as part of the broken chord approach Galileo had already probed certain variations of bent paths connecting two points, albeit not in the systematic manner he started to do on folio 130 verso and also not under the general question of finding the path of least time.

Folio 187 is related in a twofold manner to 130 verso. The recto of the folio, just as that of 130, contains the calculations of the length of some distances, which were successively transferred into the construction on 130 verso. On the verso of the folio Galileo explored whether a hypothesis explicitly formulated on 130 verso held. Likewise, on the verso of 187 Galileo, moreover, repeated the geometrical construction underlying his geometrical argument on 176 recto and this new diagram may have indicated to Galileo that his argument was correct after all.

One of the geometrical constructions from 187 verso reoccurs on 174 recto. Galileo's further elaboration of this construction, documented by entries on the same folio, as well as by some sparse entries on 89b recto, resulted directly, as will be shown, in the drafting of two entries on 164 recto. One of these entries embraces the statement of the generalized law of fall as well as of the generalized length time proportionality in a particular manner, that, as will be argued, is related to Galileo's new technique of integrated time representation; the other is a draft of the generalized length time proportionality as a proposition proper.

A proof of the generalized law of fall based on the very geometrical property investigated on 174 recto was drafted by Galileo on double folio 126, 127. On the same double folio, Galileo noted two other propositions which belong to the group of propositions collectively referred to here as the least time propositions. These least time propositions were, as will be argued, likewise, formulated as a direct spin-off from Galileo's work on folio 130 verso. Via the occurrence of auxiliary calculations, this double folio is again directly related to Galileo's investigation of motion along polygonal paths centrally documented on folio 166.

Galileo's considerations on 130 verso, thus, connect the content of a rather large group of folios and situate it firmly within Galileo's research program on the relation of swinging and rolling.

The problem on folio 130 verso—how does time of motion along bent planes depend on the configuration of the bent plane? Folio 130 verso has so far received virtually no attention in studies of Galileo's *Notes on Motion*. Merely one scholar, Stillman Drake, has referred to this folio at all. Drake claimed that on this particular folio Galileo had solved a "purely mathematical problem" and assumed that this happened before his conceptual shift toward the assumption of natural acceleration. According to Drake's interpretation, Galileo, after having "discovered" the law of fall, returned to the page and added some details to the construction, which were "related to the law of fall." Drake neither specified what he believed these additions were, nor did he give an interpretation of their purpose.[24]

At first glance the content of folio 130 verso indeed seems to exclusively pertain to the proofs of geometrical statements related to the complex construction of the central diagram on the page (see Fig. 7.10). Taking into consideration all the available evidence, including, in particular, the parts of the construction Galileo carried out by means of uninked lines and incisions of the compass, the page, however, reveals that it contains the elaboration of an intricate and ambitious problem concerning accelerated motion along bent planes which, as will be demonstrated, resulted in a number of insights that Galileo transformed into propositions and went on to publish more than 30 years later.

It will emerge from the reconstruction that the actual purpose of Galileo's considerations was to inquire which of the bent inclined planes *anc*, *abc*, *aec*, *afc*, and *amc* would be traversed in naturally accelerated motion in the least time. More

---

[24] See Drake (1978, 96).

7.3 Bent Planes Traversed in Least Time

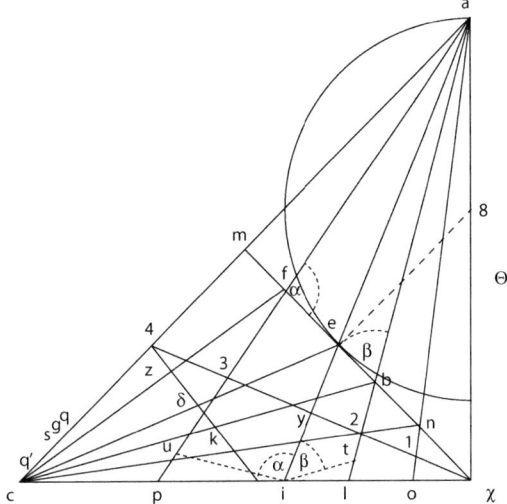

Fig. 7.10 Schematic rendering of the central diagram on folio 130 verso

generally, given two points *a* and *c* whose vertical distance equaled their horizontal distance, and further given a line *xm* that halves the line *ac* and is orthogonal to it, Galileo sought the junction point of a bent plane, under the constraint that this point had to be located on the line *xm*, such that a motion from *a* to this junction point succeeded by a motion from there to *c* would proceed in the least time.

In a first step, Galileo approached this question quasi-empirically, i.e., he geometrically constructed the times that would elapse during motion along the various bent planes and compared these times.[25] This showed that among the planes considered, the plane *aec*, i.e., the broken chord, was indeed the one traversed in the least time.[26] Only once this had been observed did Galileo set out to seek a theoretical explanation.[27]

The points q', *s*, *g*, *q*, located close to point *c* on the line *ac*, designate the results of Galileo's construction of the times of motion along the various bent planes.[28] Galileo's choice to assume, by distance time coordination, the time of

---

[25] Galileo's construction is explained in detail in the Appendix in Chap. 13. Another instance where in a first step Galileo approached testing the validity of his assumptions, in this case the so-called theorem of equivalence, by brute force calculation of relevant cases has been analyzed in detail by Renn (1990). As discussed in Chap. 5, Galileo explored and probed the potential validity of his root-root hypothesis concerning the time of descent along bent planes in a similar fashion.

[26] The bent plane *aec* is a broken chord as the junction point *e* is positioned on the quarter arc connecting the points *a* and *c*.

[27] The hypothesis test by calculating or rather constructing the times of motion for different concrete cases could of course only show that, of the cases considered, motion along the bent plane, *aec* was the one traversed in the least time. What Galileo attempted to prove, theoretically, was that this was the path of absolute least time under the given constraints. Yet as argued, he was not able to achieve this.

[28] In Galileo's original diagram two points are lettered *q*. To disambiguate, I here refer to the closest one of the two to point *c* as q'.

motion through the inclined plane ac to be measured by the length of this plane meant that aq' represented the time of motion along the bent plane anc, as the time of motion along the bent plane abc, aq the time of motion along the bent plane aec, and finally ag the time of motion along the bent plane afc. As a mnemonic device Galileo coded these relations in abbreviated form.[29] As the remaining content of the page is written around this entry, we can safely assume that Galileo had constructed these times of motion before he began with his further theoretical considerations.

Galileo's construction shows two things in particular. On the one hand, the line aq was revealed to be the shortest; in other words, of the bent planes concretely considered, the bent plane aec was shown to be traversed in the least time. On the other hand, the times of motion did not show a symmetry with respect to ae as Galileo may have expected. The path afc, for instance, was traversed in less time than the path abc, even though the angle fae was greater than the angle eab.

How did Galileo construct the times of motion along the various bent planes? By analyzing the uninked construction marks, Galileo's procedure can be reconstructed. A detailed discussion of Galileo's approach is included in the treatment of the folio in the Appendix in Chap. 13. Quite generally, Galileo proceeded in two steps. First he sought a construction which in all generality would allow the inference of the times of motion along the first parts of the bent planes, and then a similarly general construction which would allow an inference of the times of motion along the second halves of the various bent planes under consideration.

To infer in all generality the times of motion along the first parts of the bent planes, Galileo drew the line $x4$, whose intersection with the lines ao to ap, marked as points 1 to 3 in the diagram, represented the mean proportionals between the lines an and ao to am and ac, respectively. Since Galileo had chosen by distance time coordination to represent the time through ac by the length of the line ac, according to the length time proportionality, it followed that the lengths of the remaining inclined planes ap to ao likewise represented the times of motion along these inclined planes. Furthermore, by the law of fall, the mean proportionals $a1$ to $a4$ represented the times of motion over the first parts of the bent planes an to am.

From the vantage point of the developed new science of naturally accelerated motion, the construction described above is very straightforward and should not have presented any particular problems. Still Galileo had calculated the positions of the mean proportionals on the recto side of the folio first, transferred them to the diagram and apparently only then realized that these mean proportionals could simply be constructed as the intersection points of the inclined planes with one line, namely the line $x4$.[30] As uninked arcs show, by means of the compass Galileo

---

[29] 130 verso, T2.

[30] In the diagram, the positions of the points marking the mean proportionals show slight deviations with respect to their geometrically correct positions, indicating that Galileo had calculated the positions and based on the results of these calculations marked the points before. Only in a second step did he add the line $x4$, whose intersections with the inclined planes determined the exact positions.

7.3 Bent Planes Traversed in Least Time

transposed the lengths of the lines $a1$ to $a3$ onto the plane ac, where they could be compared directly. This first part of his construction showed him that, of the inclined planes an to am, the plane ae was traversed in the least time, a result which, as we will see, Galileo elaborated upon further in a next step.

Finding a similar construction which would have allowed Galileo to infer the time of motion over the second parts of the bent planes after motion from rest at a, posed a bigger problem. A diagram on folio 176 recto can be identified as auxiliary considerations for finding such a construction. Based on these additional considerations, Galileo eventually realized that by drawing a line from point 4 to the midpoint of the horizontal cx, he could accomplish what was sought.[31] In fact, the intersections of this line with the lower parts of the bent planes marked the mean proportionals between θn and θc to θf and θc. θ in each case marks the intersection between the extended respective lower parts of the bent plane with the horizontal through a, as indicated by the one θ Galileo wrote on the right margin of the diagram (see Fig. 7.10). These points θ are then conceived as the starting points of motions to c and, according to the logic of the generalized law of fall for straight paths—which as will be argued, for him, was merely a trivial consequence of the law of fall— the times of motion along the lower parts for motion begun from rest at θ were given by the lines cz to ck. Finally, by path invariance, these equaled the times of motion along the same lower part, where motion was conceived to have started, instead, from rest at Θ, from rest at a and then to have been deflected at points f, e, b, and n, respectively.

In a last step of his construction, again by means of the compass, Galileo added the lines representing the times of motion over the second parts of the bent planes to the lines representing the times of motion over the respective first parts, which he had already transposed to ac. This procedure finally yielded the sought points q', s, g, and q and, thus, the lines cq', cs, cg, and cq representing the times of motion along the various bent planes under consideration. Notably on folio 130 verso, Galileo did not yet exploit the insight that z and 3, δ and y, etc. are all positioned pairwise at the same vertical height as he would later do in the bent variant of the generalized law of fall published in the *Discorsi*. I will in fact argue in the following that Galileo read this property from a construction on 187 recto.

After Galileo had constructed the times of motion along the various bent planes he started to turn his attention to a theoretical investigation of why motion along aec was allegedly completed in the least time as his construction had suggested. Whether this ambitious question, which ultimately remained unsolvable for Galileo, was first

---

[31] The points z, δ, k (the mean proportional on nc remained unmarked) do not lie exactly on the line from 4 to the midpoint of the horizontal cx, indicating again that their positions were first calculated by Galileo and that the general construction was added only later. The deviations from the geometrically precise positions are, thus, due to rounding errors. Folio 176 recto contains a proof of the fact that all points marking the mean proportionals lie on the line from point 4 to the midpoint of the horizontal cx. The proof is correct, yet it was marked by Galileo as "[f]alsa est." The reason for this is not clear.

inspired by the very construction on 130 verso, or whether Galileo, intrigued by the implication of the law of the broken chord that in naturally accelerated motion the path of least time was not always the straight connection between two points, had already started his considerations on the page with the aim of finding a path of least time along bent planes, cannot be decided. As variations of the basic construction from 130 verso that are contained on folios 190 and 191 indicate, it is even conceivable that Galileo had already conceived the basic structure of his later brachistochrone argument and thus by means of the considerations on folio 130 verso wanted to show that the junction points of bent planes traversed in least time always had to be positioned on an arc connecting the starting point and the endpoint of motion.[32]

Whatever the initial motivation for the considerations on 130 verso may have been, in his search for a theoretical explanation, Galileo first turned his attention to demonstrating that of all the lines that could be drawn from point a to the line xm, it was in fact the inclined plane ae that was traversed in the least time. To this end Galileo proved that

> [rectangulum] iae esse omnium minimum lab, oan, paf etc.[33]

Since the mean proportional between the sides of a rectangle is proportional to the root of its area, this amounted to showing that ay was the shortest of all the mean proportionals, and thus, of all the inclined planes running from a to xm, the plane ae was the one traversed in the least time. Next, he went so far as to show that the time of motion would monotonously increase the more the angle of the inclined plane considered differed from that of ae.[34] From a modern perspective, this can be rephrased into the statement that besides the global minimum, no local minima for the time of motion along the inclined planes defined by the variation under consideration existed.

In a slightly different hand, which may indicate that the entry was made somewhat later than the other entries on the page, Galileo added an alternative argument for his statement that motion over the first part of the bent plane was completed in the least time if the junction point was positioned at e. It reads:

> Aliter brevius. Posito angulo ae8 aequali [angul]o eam, erit linea e8 parallela am; ergo perpendicularis ad mx, eritque aequalis 8a: quare, centro 8, intervalo 8e, circulus tanget mx in e: unde patet propositum.

This alternative proof added a new element to the central diagram. A circle around point 8 on ax was introduced such that it touched the line mx at point e (see Fig. 13.2). Galileo casually concluded by virtue of this construction that his statement concerning the rectangle of minimal area held.

---

[32]Cf. Chap. 9.
[33]139 verso, T1A and T1B.
[34]130 verso, T1D.

7.3 Bent Planes Traversed in Least Time

No geometrical reasoning is immediately apparent to justify Galileo's claim. Yet a look at a proposition which Galileo published as Proposition XXXII of the Second Book of the Third Day of the *Discorsi* clarifies why the introduction of the circle with radius 8*e* provided the basis of a proof for Galileo's statement, at least if interpreted as a statement about naturally accelerated motion.

According to the line of argument of Proposition XXXI, motion from *a* along any chord to the periphery of the circle is completed in equal time by the law of chords. Since all the lines emanating from *a* to the line *xm* surpass the periphery of the circle except for the line *ae*, it is thus proven that *ae* is traversed in the least time. This suggests that Proposition XXXI emerged as a spin-off of Galileo's bent plane problem on 130 verso, and evidence for this assumption will indeed be provided below.

What remained to be done at this point was to prove that not only the first part *ae* but actually the entire bent plane *aec* was traversed in the least time in naturally accelerated motion. Galileo's first attempt at solving this problem is evidenced by a final entry on 130 verso, which reads: "Vide num *ya* ad *a*3 sit ut δc ad *cz*."[35] For the motions along *afc* and *aec* Galileo aspired to check whether the times of the first parts of the motions were in the same ratio as the second parts of the motions. Should this be the case, it would of course immediately follow from what Galileo had already proven, namely, that *ae* was traversed in the least time, that *aec* was also the path of motion completed in the least time. Galileo actually tried to validate his hypothesis on folio 187 recto, however, without success.

Indeed Galileo's hypothesis is false and showing that the times of motion along the first parts of the bent planes cannot exhibit the same ratio as the times of motion along the respective second parts is rather elementary. We can assume that at some point Galileo must have realized that the assumption was false as this is in fact indicated by his diagram on 130 verso. In the diagram, *cz* falls almost vertically on the line 4*k* and is thus the shortest of the lines *c*4, *cz*, *c*δ, *ck*. Hence, of the motions along the second parts of the bent planes Galileo had concretely considered, it was motion along *fc* that was completed in the least time, not motion along *ec*.

The time of motion along the bent plane, thus, had to be conceived as resulting from two trends. With respect to the variation under consideration, the time along the first part assumed its minimal value under a different geometrical constellation than the time along the second part. Determining the constellation in which their sum, i.e., the overall time along the bent plane, would be minimal, turned out to be too complex a problem to be solvable based on Galileo's mathematical means. If the problem is analyzed from modern perspective, it turns out that the constellation of the bent plane along which motion is made from *a* to *c* is not given by *aec*. Rather the junction point *w* of the configuration of the bent plane *awc*, which indeed minimizes the time of motion, lies somewhat closer to *m* such that the angle *wa*χ measures almost exactly 26° (cf. Fig. 7.11). Given the observed behavior, Galileo could indeed no longer even be sure that it was the bent plane *aec* that was traversed

---

[35] 130 verso, T1F.

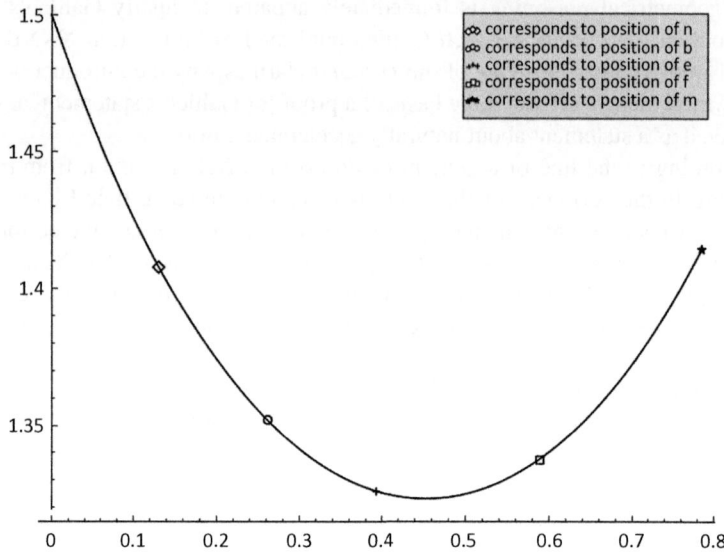

Fig. 7.11 Time of motion along a bent plane awc with w on mx over the angle of the first plane aw to the vertical. The time unit is chosen so that the time along the vertical ax measures 1. The time along the first part of the bent plane is given by: $\sqrt{\sqrt{1/2} * 1/(\cos(\alpha) * \cos(\pi/4 - \alpha))}$. The time along the second inclined plane is given by: $\sqrt{(1/\sin(\alpha) - \sqrt{1/2}/\cos(\pi/4 - \alpha)) * 1/\cos(\pi/2 - \alpha)}$. The overall time of motion along the bent plane is then simply the sum of the two terms. The discrete cases for which Galileo calculated the times of motion are individually marked. For an initial angle of zero degrees, the path is axc and the time of motion 1.5 times that of motion along ax alone corresponding to Galileo's double distance rule. For an angle of 45°, the bent plane degenerates to the straight chord amc, and the time of motion is that of ax times $\sqrt{2}$. The time is minimal for an angle of 26°

in the least time. Yet since the actual solution is close to aec, his calculations could not produce evidence to refute the assumption, which could hence be maintained. Yet Galileo's work on the problem was not to remain entirely fruitless. Rather, as argued in the following section, it resulted in the formulation of a number of propositions that would eventually be published in the *Discorsi*.

7.4 Galileo's Considerations on Motion Along Bent Planes as the Source of New Propositions

The propositions Galileo formulated as a direct consequence of his considerations documented on folio 130 verso are the generalized law of fall and the generalized length time proportionality, as well as a group of propositions collectively referred to here as the *least time propositions*.

7.4 New Propositions

The generalized law of fall and the generalized length time proportionality In the *Discorsi*, as Proposition X of the Second Book of the Third Day, Galileo issued a proposition in which he stated that the times along inclined planes of the same height but different inclination were in the ratio of the lengths of these planes, regardless of whether the motions started from rest or whether they had been preceded by a fall from the same height. This proposition generalizes the length time proportionality to the case of motion not made from rest and has, thus, already been referred to here as the generalized length time proportionality.

In the succeeding proposition in the *Discorsi*, Proposition XI, Galileo generalized the law of fall. Whereas the law of fall applies to motions made from rest along planes of the same inclination, Proposition XI applies to the case where a first motion is succeeded by a second motion which, thus, does not start from rest. The proposition allows the drawing of inferences about the times elapsed during arbitrary parts of naturally accelerated motions along inclined planes and has hence already been referred to here as the generalized law of fall. In the *Discorsi*, the proposition is stated as follows:

> If a plane be divided into any two parts and if motion along it starts from rest, then the time of descent along the first part is to the time of descent along the remainder as the length of this first part is to the excess of a mean proportional between this first part and the entire length over this first part. (EN VIII 229, trans. Galilei et al. 1954)

The statement follows trivially from the law of fall. This is reflected in the proof Galileo provided for the proposition in the *Discorsi*, which basically consists of only one argumentative step taking the law of fall as a premise. Galileo terminates his argument with the remark "quod est propositum." Yet the proposition carries and a modified motion situation is introduced:

> Quod si motus non fiat per continuatam ..., sed per inflexas ...

Thus, the requirement that the motions considered proceed over the same inclination is dropped, and Galileo, in the second part of the proposition, indeed considers the case of motion along a bent plane. Galileo proves that in this case "the time along AC will be to the time along CD as the length AC is to the length CE" where point E has the same height over the horizontal as point F, marking the mean proportional between AC and AB (see Fig. 7.12). The argument Galileo provides invokes Proposition X as a premise.

Despite being formulated as just one proposition, two distinct statements are thus made, and those statements are proven by different arguments. To distinguish and

Fig. 7.12 Diagrams associated with Proposition XI of the Second Book of the Third Day of the *Discorsi* (Cf. EN VIII. 229–230)

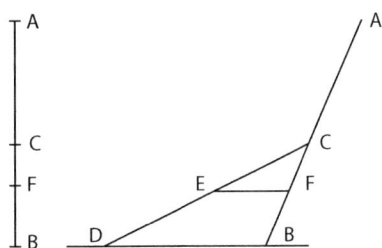

refer to the different parts, I allude to the first as the *generalized law of fall straight* and to the second as the *generalized law of fall bent plane*.

For Galileo, the generalized law of fall straight was a trivial consequence of the law of fall. This is indicated by the fact that many of the propositions of the Second Book of the Third Day of the *Discorsi* do not make use of the generalized law of fall in situations when it could directly be invoked as a premise in the proof. Instead the proofs are based on the law of fall as a premise. The straightforward one-step argument which leads from the law of fall to the generalized law of fall straight is then simply part of the respective proofs.

The generalized law of fall bent plane in contrast, as well as the generalized length time proportionality, were the direct results of Galileo's considerations regarding motions along bent planes centrally documented on folio 130 verso and discussed in detail above. Both build upon a geometrical property Galileo hit upon in his attempt to solve a partial problem brought about by his considerations on folio 130 verso and from which indeed both the generalized length time proportionality and the generalized law of fall bent plane immediately follow.

Central to the reconstruction are Galileo's considerations on folio 174 recto, which served as the direct blueprint for the first drafts he produced for both propositions. The content of folio 174 recto, in turn, is related to Galileo's investigations on folio 130 verso via a third folio, namely, 187 recto. Folio 187 recto contains two diagrams. The upper of the two picks up Galileo's construction from 176 recto. The diagram in the lower part of the folio essentially repeats the construction from 130 verso. It, however, contains one additional geometrical element not present in the diagram on 130 verso, namely, an inclined plane starting at a and running parallel to ec, the lower chord of the broken plane aec. This new inclined plane was to become central in Galileo's successive considerations, and the reason why it was introduced will become clear below.

The diagram on 187 recto remained unlabeled. Yet as the entries on the page show, Galileo conceived of it as being equipped with the same labels as the original diagram on 130 verso. The diagram was obviously drawn in order to investigate the hypothesis Galileo had formulated on 130 verso, namely, that $ya/a3 \sim \delta c/cz$. In search of an answer, Galileo hit upon a proportion he noted to the left of the diagram in dual form as a ratio diagram and above in writing as the short statement "ut ay ad yi ita $\Theta\delta$ [ad] δc." What this meant was essentially that the points δ and y marking the two mean proportionals relevant for the determination of the time of motion along the bent plane aec had the same height above the horizontal $c\chi$.

Galileo copied the diagram from 187 recto to the right-hand side of folio 174 recto in clean form and added new labels (Fig. 7.13).[36] Galileo first recalculated the time of motion along the broken chord designated afs with the new labels. The calculations are not preserved but their results were noted by Galileo scattered around the diagram.[37] Underneath, separated from the diagram and the initial

---

[36] Galileo wrote on 174 recto with the page turned 90° with respect to the modern orientation in the manuscript. Positions are specified with regard to the orientation of the page as it was used by Galileo.

[37] 187 recto, C2 to C5 and T2.

7.4 New Propositions

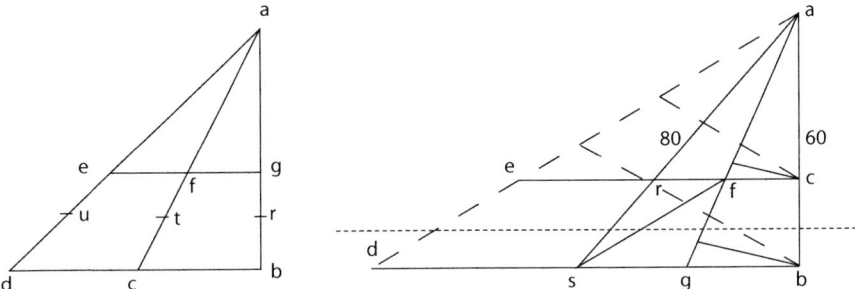

Fig. 7.13 Schematic representation of the two diagrams contained on 174 recto

result, is a table listing the lengths of virtually all relevant distances in the diagram together with the times needed to traverse these distances in naturally accelerated motion. Slight deviations with regard to the initial results indicate that Galileo had recalculated the example for the list.[38]

Even though the calculations themselves are not preserved, the way in which Galileo noted the results shows that the scheme according to which he must have calculated the times of motion under consideration, must have differed from the one employed on 130 verso. Central to Galileo's new approach is the one plane ad parallel to the lower of the two planes fs that make up the bent plane afs that had already been introduced on 187 recto. On this plane, ad, Galileo marked a point e at the same height above the horizontal as point f the junction point of the bent plane, and calculated the time of motion along ed for motion starting from rest at a.[39] This time was the same as motion along the lower plane fs with motion starting from a and Galileo exploited this fact when calculating the sought time along the bent plane afs, which in his first set of calculations amounted to 76.[40]

Thus, where in his earlier consideration of motion along bent planes, in particular those on 130 verso, Galileo had extended the lower inclined plane to reach the same height as the starting point of motion, he now, in addition, had transposed the extended plane by a parallel translation such that its starting point fell together with the starting point of the motion along the bent plane under consideration. This

---

[38]The time along af resulted to 54 instead of 53 time units. Thus, the time along afs amounted to 77 time units, calculated and noted by Galileo to the right of the list.

[39]174 recto, T2.

[40]That Galileo proceeded in the same way in the more extensive calculations whose results are put down underneath the diagram is indicated by the way he noted the results. In the table, the distances for ae, ad, and ed are given together with the times of motion along these distances. The respective values for fs are not contained in a separate table row, but Galileo simply wrote fs next to ed, which indicates that what he had calculated were the respective values for ed which, based on his construction, also applied to fs.

neither changed the physics nor the required calculations but produced a geometrical configuration that offered itself to a combined application of the law of fall and the length time proportionality, and it is this type of inference from which, as we will see, the generalized length time proportionality and the generalized law of fall bent plane immediately followed.

Indeed, in addition to e and f, Galileo also marked point r on as and c on ab both at the same height above the horizontal db as point f. Thus, the diagram comprised four inclined planes of equal height, ab, as, ag, and ad, on which shorter distances ac, af ar, and ae were marked such that the latter likewise represented planes of equal height. In Galileo's calculations, the law of fall had been applied to each pair of distances of the same inclination, for instance, ae and ad. On the other hand, the times of motion along the longer planes and the shorter planes of equal height could be related to each other via the length time proportionality.

By length time proportionality the times of motion along ac, af ar, and ae are to each other in the same ratios as the distances covered and so are the times of motions along the longer planes ab, ag, as, and ad. As the distances on the shorter planes are among each other in the same ratio as the distances covered on the longer planes, it, thus, follows immediately by division of ratios that the same must hold for the times along the remaining cb, fg, rs, and de starting from rest at a. This is what the generalized length time proportionality states. The law of fall then implied that points marking the mean proportionals between the longer and the shorter distances, likewise, had all to lie on the same heights. As argued above, this was the case in the construction on 187 recto, and it was even more obvious in the new construction where the inclined planes under consideration all started from the same point. Indeed, in the diagram, Galileo added an uninked horizontal line through the points marking the respective mean proportionals.

That Galileo engaged in a consideration along this very line is shown by the diagram he drew on the left of 174 recto (Fig. 7.13) together with the short entries associated to the diagram, some of which are contained on the cut leaf 89b. The diagram on the left-hand side reduced the diagram on the right to only show the planes of equal heights with shorter distances on them, likewise, of equal height. In addition, Galileo marked and labeled the endpoints of the mean proportionals. Finally, instead of four, as in the original diagram, Galileo only drew three inclined planes. As evidenced by the entries associated to the diagram, Galileo indeed related the times of motion along the inclined planes of equal height by the length time proportionality and applied the law of fall to pairs of motions along each plane determining, in particular, the ratio of the times along the first parts to the times of motion over the successive parts, not made from rest.[41]

---

[41] 147 recto, T1C (incorrectly transcribed in the electronic representation of Galileo's *Notes on Motion*) and 89b recto T1A and T1B.

7.4 New Propositions

The left-hand diagram from folio 174 recto was transferred by Galileo to folio 164 recto, and there the associated short entries were expanded into a full argument by means of which Galileo concluded:

Patet insuper, tempora casuum per gb, fc, ed esse ut lineas gb, fc, ed[42]

This is the generalized length time proportionality. As will be detailed in Chap. 10, Galileo's incentive for drafting this entry was not primarily to formulate the generalized length time proportionality but to reflect upon and clarify an ambiguity potentially arising from the application of the new method of integrated time representation in arguments about the kinematics of naturally accelerated motion. Yet on the very same folio, 164 recto, Galileo also recorded the generalized length time proportionality as a proposition complete with proof, where the argument provided is basically equivalent to the argument as published in the *Discorsi*.

Thus, Galileo's consideration of motion along bent planes under variation, which began on folio 130 via a number of intermediary steps, led directly to the generalized length time proportionality. Evidence that the generalized law of fall bent plane was engendered in essentially the same way is somewhat more indirect. What is essential for the generalized law of fall bent plane as published in the *Discorsi* is the horizontal line through the point marking the mean proportional labeled F in Fig. 7.12 and the point E defined as the intersection of this horizontal with the lower part of the bent plane CE. This line was first introduced by Galileo as shown on 174 recto as an uninked line running horizontally through the points marking the mean proportionals between ae and ad, ar and as, and finally between af and ag. Given this line and the generalized length time proportionality as a premise, the generalized law of fall bent plane followed trivially, and Galileo drafted the argument on folio 126 recto. Together with folio 127, folio 126 forms a still uncut double folio. As mentioned in Chap. 5, the double folio contains auxiliary calculations pertaining to Galileo's investigation of motion along polygonal paths centrally documented on folio 166. Folio 127, moreover, as discussed in the succeeding section, contains drafts of two of the least time propositions which, as will be argued, likewise resulted as spin-offs from the work which had begun on 130. The content of the double folio, thus, fits consistently into the overall picture that has been pieced together here.

Before proceeding to a discussion of the least time propositions, however, a last remark concerning the generalized law of fall is appropriate. The majority of propositions drafted during the Paduan period were transferred by one of Galileo's two disciples Mario Guiducci and Niccolò Arrighetti to fresh folios. Among these copies a proof of the generalized law of fall can be identified on folio 49 recto for which no original in Galileo's hand is extant. Such an original may have been discarded or perhaps Galileo dictated the proposition to Mario Guiducci directly.

---

[42] 164 recto, T1.

Opposed to the proof of the generalized law of fall bent plane on 126 recto, the argument on 47 recto does not take the generalized length time proportionality as a premise. Instead, the argument is based on the law of fall.

Many of the copies made by the disciples bear the trace of a later stage of work in which Galileo re-elaborated the copied material. Often, Galileo added the general statements to the propositions copied. In the case of the generalized law of fall on 49 recto, however, Galileo went beyond this. Before the entry he noted:

> Hic pr[a]emittenda videtur sequens propositio.[43]

What Galileo deemed necessary as a prefix to the generalized law of fall bent plane was the generalized law of fall straight which he, indeed, added underneath. Thus, it is only when reworking his collected material and when presenting his propositions as a deductively organized system in treatise that Galileo decided an argument for the generalized law of fall straight should precede the argument for generalized law of fall bent plane. Galileo, as we have seen, followed this decision when he published the generalized law of fall straight and the generalized law of fall bent plane together as just one proposition, Proposition XI in the *Discorsi*. Yet the argument provided for the generalized law of fall bent plane in the *Discorsi* was not the one drafted on 47 recto but the one he had drafted on folio 126 recto. This argument, in particular, required the generalized length time proportionality as a premise. In the *Discorsi*, Galileo paid tribute to this fact by issuing the generalized length time proportionality as Proposition X before the generalized law of fall published as Proposition XI.

The proof of the generalized length time proportionality, however, presupposes the generalized law of fall straight. To avoid a formal inconsistency in the proof of the generalized length time proportionality, Galileo did not introduce the generalized law of fall straight, proved only in the succeeding proposition as a premise, but instead repeated the one-step argument by which it is established as part of the proof of Proposition X.[44] Indeed, as the reconstruction above has shown, the generalized law of fall straight was a late addition to the new science of naturally accelerated motion owing to the intention to issue clear, comprehensible deductive structure. This explains why the generalized law of fall straight, despite being applicable not only in the proof of Proposition X but also in the majority of the other proofs of the Third Book of the Second Day, was hardly ever explicitly invoked as a premise. When Galileo originally drafted these proofs, he had simply not yet envisaged incorporating the generalized law of fall straight as an independent proposition into the new science of naturally accelerated motion.

---

[43] 47 recto, T2A.

[44] Galileo could have resolved this problem by splitting up what he issued as one proposition, Proposition XI, into two independent propositions. The generalized law of fall straight could then have been presented before the generalized length time proportionality from which then the generalized law of fall bent plane could have been derived.

7.4 New Propositions

The least time propositions As Propositions XXX, XXXI, and XXXII, Galileo published three propositions which are distinct from the other propositions of the Second Book of the Third Day in that they are concerned with a path of motion traversed in the least time and that the only premise employed in the respective proofs is the law of chords. These propositions have already been referred to above as the least time propositions.

The fact that all three least time propositions depended solely on the law of chords as a presupposition has resulted in the assumption that they must have been conceived and the proofs drafted before Galileo came to accept that descent along inclined planes is naturally accelerated, or at least before he gained insight into the law of fall as the law governing the behavior of this type of motion.[45] This assessment directly hinges on the assumption that Galileo first formulated the law of chords when he still regarded motion along inclined planes as essentially uniform. As will be argued in the succeeding chapter, Chap. 8, there is, however, nothing to support this assumption.

As demonstrated above, the general question regarding paths traversed in least time had its origin in Galileo's investigation of motion along bent planes under variation on folio 130 verso. Galileo's consideration documented on this folio had, furthermore, engendered the basic geometrical construction common and central to the proofs of all three least time propositions, a circle appropriately constructed so as to allow for the conclusion that a chord to the arc of the circle is the path of least time. What distinguish the three propositions are the elements of the underlying construction which are given and sought, respectively. As mentioned above, Galileo drafted Proposition XXXI and Proposition XXXII, on the same page, folio 127 verso.[46] Proposition XXX was drafted on folio 140 recto. Both folios are part of double folios that have remained uncut to the present day, and both double folios bear the same watermark/countermark combination of crown and crossbow, indicating that all three propositions were drafted at approximately the same time. In both cases, moreover, the content of the respective other half of the double folio

---

[45] Wisan and Drake both claimed that Galileo formulated the least time propositions prior to his insight into the law of fall. Drake (1978, 79), referring to elaborations of two of the least time propositions on folios 140 and 127, states that they "follow immediately from Galileo's theorem [the law of chords] without considering acceleration at all," that is, before Galileo started to assume that falling motion is naturally accelerated. Wisan (1974, 171), who provides a thorough analysis of the propositions of the least time group, is somewhat more careful and concludes her discussion of the relevant propositions with the remark: "They [the propositions of the least time group] depend only on the law of chords which appears to be the first of the published propositions on motion, and they may precede the times-squared law." Caverni (1972, Vol. 4, 361–363), in contrast, assumes that Galileo formulated the least time propositions after his conceptual shift toward the assumption of natural acceleration.

[46] A somewhat more elaborate variant of Proposition XXXII, including a lemma required for its proof, was drafted by Galileo on 168 recto, likewise, on a double folio bearing the watermark/countermark combination crown, crossbow. The facing page contains a draft of 2/12-th-12, which exactly reproduces the logic by which Galileo on 130 verso had determined the time of motion along the bent planes under consideration.

can be directly linked to Galileo's considerations concerning the variation of bent planes on 130 verso.

The assumption that the least time propositions originated before Galileo's conceptual shift toward the assumption that motion along inclined planes was naturally accelerated must, thus, be rejected. Rather these propositions surfaced as spin-offs from Galileo's investigations into the relation between swinging and rolling in 1602 and, more concretely, as the by-product of the far more complex problem concerning the paths of least time, which he had addressed but not been able to solve. Yet solutions of partial problems that had resulted from Galileo's attack on the complex problem, or, in some cases, even his mere attempt to solve them, led to new insights. From Galileo's perspective, these insights were valuable in their own right. By furnishing them with proofs, they were turned into propositions which could potentially be presented as part of the subject matter of a new science which would later be put into print. That Galileo aspired to do so already in this early phase of his work on the problem of naturally accelerated motion will be demonstrated in the second part of this book. In this manner, as demonstrated in the present chapter, at least five of the propositions Galileo later published as part of the Second Book of the Third Day of the *Discorsi*, the generalized law of fall, the generalized length time proportionality, and the three least time propositions, came into being, owing more or less directly to the problem Galileo attacked on folio 130 verso.

7.5 Summarizing Conclusion

Galileo's considerations discussed in the present chapter are part of his work on the challenging problem of the relation between swinging and rolling. In this sense they do not differ from his considerations discussed in the previous chapters. Yet in hindsight it can be recognized that some of the insights achieved, propositions formulated, and methods developed that ensued from his investigations into the relation between swinging and rolling were to have repercussions beyond these investigations. Some of the results Galileo achieved became core assets of the new science of motion as it was published more than 35 years later in the *Discorsi*, in which the original research program, as argued in Chap. 9, no longer played a role. They are, thus, spin-offs from the problem of swinging and rolling, not only in that they resulted from Galileo's attempt to solve the problem but also in that they emancipated themselves from the narrower context in which they were originally conceived and assumed a new meaning and function in the new science of motion.

It has thus been shown how Galileo's attempt to construct a proof for the law of the broken chord brought forth a new method of representing motion diagrammatically. With the new technique, that has been referred to as integrated time representation, the spacial, as well as the temporal, aspects of motion are represented as lines in one and the same diagram. The new technique substantially

7.5 Summarizing Conclusion

facilitates inferences regarding the kinematics of motion as successive applications of laws of motion are rendered by successive geometrical constructions in the diagram. The new method was immediately adopted by Galileo to become his almost exclusive means of the diagrammatic representation of motion. This, as has been argued, can be exploited to establish a relative ordering between considerations in which Galileo still employed the old and in which he already availed himself of the new technique.

Application of the new diagrammatic technique in considerations regarding naturally accelerated motion did bring about certain ambiguities. One and the same line in a diagram could now represent a time of motion or a distance covered by motion and in some cases both. It has been shown that Galileo initially employed certain graphical conventions for disambiguation and eventually even changed the terminology with which he referred to his constructions in natural language. These transitional phenomena have been shown to occur on folios which can all be independently related to the earliest phase of Galileo's work on naturally accelerated motion, resulting in consistent interpretation.

There is a difference between sketchily noting insights concomitant to the elaboration of a problem and the formulation of polished propositions. The former is characteristic of notes taken in the exploration of a problem meant to be of use and potentially understandable only to their author. The latter is characteristic of the formal presentation of results in a book or treatise addressed to a wider audience. In the present chapter, a number of cases have been surveyed in which Galileo turned the insights achieved into propositions drafted in a form conducive to their public dissemination. This indicates that, while still being deeply immersed in the elaboration of the problem of the relation of swinging and rolling, Galileo had already begun to perceive of at least the possibility of a new science to be presented as a treatise and had begun working toward it. That this was indeed the case will ultimately be corroborated in the second part of this book.

It has thus been argued that the length time proportionality, notably absent from Galileo's earliest considerations regarding swinging and rolling, and specifically not invoked as a premise in his various attempts to construct a proof of the law of the broken chord, indeed, emerged as a spin-off from these attempts. Appreciating that the length time proportionality emerged as an albeit early result of Galileo's pondering on naturally accelerated motion undermines the commonly shared assumption that it was a presupposition of Galileo's work, rooted, as has been claimed, in his older *De Motu Antiquiora* framework.

In the second half of the chapter, Galileo's work on a particularly challenging problem that had emerged directly from his research program on swinging and rolling has been reconstructed and discussed in detail. Galileo examined the question of how the time of motion along a bent plane connecting two given points varies with a particular variation of the junction point of the two planes which make up the bent plane. He aimed at finding the geometrical configuration of the bent plane which would be traversed in the least time. Trial calculations indicated that the bent plane traversed in the least time might be the broken chord, but Galileo was not able to demonstrate this let alone find a construction that would have allowed him to establish the relation between the position of the junction point and the time of motion along the bent plane in all generality. Yet his attempt to find a solution brought about insights which resulted in the drafting of a number of propositions.

Galileo's search for the path of least time, in particular, engendered a construction which implied the generalized law of fall as well as the generalized length time proportionality, issued by him as Propositions XI and X. These two propositions allow the treatment of naturally accelerated motion not starting from rest and thus play a particularly important role and are used as the premises in the majority of the proofs contained in the Second Book of the Third Day of the *Discorsi*. Both propositions, as has been shown, were drafted by Galileo with recourse to the geometrical construction and the insight into a particular geometrical property that resulted from his search for the bent plane traversed in least time.

Even though Galileo was not able to determine the configuration of the bent plane traversed in the least time, in his attempt to do so, he hit upon an argument which at least allowed him to infer the configuration of an individual plane obeying a particular geometrical constraint such that it is traversed in least time. By variation of the argument, as has been shown, Galileo formulated three propositions, referred to here as the least time propositions. This variation did not and, in fact, could not produce any further insights. It was, as can be argued, not explorative but aimed at a form in which the results could potentially be published in the future. Indeed, all three proposition were later included in the Second Book of the Third Day of the *Discorsi* as Propositions XXX to XXXII.

Thus, in general, Galileo's research regarding the relation of swinging and rolling resulted in the formulation of quite a number of propositions making statements about the spacio-temporal behavior of naturally accelerated motion along inclined planes. More than 35 years later, these propositions were to form the subject matter of the new science of motion. Even though around 1602 Galileo had not found all the propositions he would later publish, he certainly had come up with enough to endeavor to turn his new findings into a new science. Indeed, as later presented in the *Discorsi*, the propositions of the new science of naturally accelerated motion almost exclusively deal with only two basic configurations, the straight and the bent plane, and by 1602 the scaffolding for dealing with these cases had been erected.[47]

---

[47] Wisan (1981, 319) writes: "[m]ore results then quickly follow as Galileo uses these new tools to develop enough mathematical theorems on motions along inclined planes to make up a whole treatise. ...Galileo now has a genuinely new 'science' of motion. However, it has neither an experimental nor a theoretical basis. The times-squared law was most likely confirmed by an early experiment, but he has no proof from established principles, and there remain several fundamental puzzles concerning accelerated motion. From this time on, Galileo is more and more preoccupied with the problem of clarifying and explaining accelerated motion and finding properly evident principles on which to base the growing body of mathematical theorems." This is an excellent portrayal of Galileo's situation around 1602. Yet I do not share her interpretation on how Galileo arrived there.

How, what were originally by-products of the exploration of the relation between swinging and rolling, were turned into a new science by Galileo is discussed in the second part of this book. In this process, the traces of the research program which had originally given rise to these insights were almost completely erased. Indeed, only a few hidden traces remain of the challenging program in the *Discorsi*. Only once this program has been reconstructed, as has been done here, is it even possible to discern them. To unearth these traces and to show the program on swinging and rolling had an unexpected aftermath will be tackled in Chap. 9. Before, however, one last proof that resulted as a spin-off of Galileo's investigation into swinging and rolling will be analyzed, which, due to its importance for reconstructing the emergence of Galileo's new science, merits discussion in a separate chapter that follows here.

References

Büttner, J. (2017). Shooting with ink. In M. Valleriani (Ed.), *The structures of practical knowledge* (pp. 115–166). New York: Springer.

Caverni, R. (1972). *Storia del metodo sperimentale in Italia*. New York: Johnson Reprint.

Damerow, P., Renn, J., & Rieger, S. (2001). Hunting the white elephant: When and how did Galileo discover the law of fall? In J. Renn (Ed.), *Galileo in context* (pp. 29–150). Cambridge: Cambridge University Press.

Damerow, P., Freudenthal, G., McLaughlin, P., & Renn, J. (2004). *Exploring the limits of preclassical mechanics*. New York: Springer.

Drake, S. (1978). *Galileo at work: His scientific biography*. Chicago: University of Chicago Press.

Galilei, G., Crew, H., & Salvio, A.D. (1954). *Dialogues concerning two new sciences*. New York: Dover Publications.

Palmieri, P. (2006). A new look at Galileo's search for mathematical proofs. *Archive for History of Exact Sciences, 60*, 285–317.

Renn, J. (1990). Galileo's theorem of equivalence: The missing keystone of his theory of motion. In T.H. Levere & W.R. Shea (Eds.), *Nature, experiment, and the sciences: Essays on Galileo and the history of science in honour of Stillman Drake*. Dordrecht: Kluwer.

Wisan, W.L. (1974). The new science of motion: A study of Galileo's De motu locali. *Archive for History of Exact Sciences, 13*, 103–306.

Wisan, W.L. (1981). *Galileo and the emergence of a new scientific style* (pp. 311–339). Dordrecht: Springer.

Chapter 8
Toward a Foundation: The Ex Mechanicis Proof of the Law of Chords

This chapter is devoted to a particular argument by Galileo, the so-called ex mechanicis proof of the law of chords. Galileo's construction of the proof was, as will be demonstrated, a spin-off from his work on the relation of swinging and rolling. In this respect, it compares to his considerations discussed in the previous chapter. At the same time, however, the considerations, which eventually resulted in the formulation of the ex mechanicis proof, decisively differ from Galileo's work discussed thus far. The ex mechanicis proof, as will become clear, was the first proof devised in the context of the investigation into naturally accelerated motion in which Galileo fundamentally availed himself of concepts other than times elapsed and spaces traversed. The proof is based on regarding the moments of bodies on an inclined plane as the forces causing the motions. The velocities of the resulting motions are understood as measures of the effects in turn conceived to be proportional to the active causes. The ex mechanicis proof is thus based on a dynamical argument.

As will be argued, the reason why Galileo turned his attention to dynamics is not to be sought in the narrower context of his investigation into the relation between swinging and rolling any longer. It rather has to be understood as motivated by the attempt to found his new insights regarding naturally accelerated motion along inclined planes on basic concepts anchored in the established conceptual frameworks of mechanics. This first attempt to bear fruit from the results of his ultimately futile research program on swinging and rolling by turning them into a new science is discussed in the second part of this book. As will be demonstrated, the conceptual preconditions based on which the ex mechanicis proof was established had been engendered by this attempt. The present chapter could, thus, as well have been placed in the second part of this book. Yet, as will become clear, the proof was likewise a direct by-product of the research program on swinging and rolling and thus by conventional choice is discussed here.

8 Toward a Foundation: The Ex Mechanicis Proof of the Law of Chords

In the following, the content and relative chronology of a number of entries on different folios of the *Notes on Motion* pertaining to Galileo's construction of the ex mechanicis proof of the law of chords is reconstructed. New evidence is brought forward concerning the internal relations between these folios as well as their relations to other groups of folios in the manuscript. The interpretation of the genesis of the ex mechanicis proof which thus results is found to be in striking contradiction to existing interpretations.

Existing accounts of the development of Galileo's new science of motion have in fact assumed almost unanimously that the law of chords, mentioned by Galileo in his letter to Guidobaldo del Monte of 29 November 1602, was among his earliest propositions—if not the earliest—concerning motion on inclined planes. Furthermore, these accounts maintain that the proposition was initially derived and proved within the conceptual framework of the *De Motu Antiquiora* theory, i.e., in particular, under the premise that the motion of fall along inclined planes was essentially uniform. It is commonly assumed that Galileo's first derivation proceeded exactly along the line of the argument of the ex mechanicis proof. This will be referred to here as the *standard interpretation*.

The assumption that Galileo first came up with the ex mechanicis proof of the law of chords before his conceptual shift toward the assumption that fall along inclined planes was naturally accelerated was nurtured above all by the observation that a kinematic proposition invoked as a premise in the proof was assumed to apply to uniform motion only. This assumption has since been refuted as it has been shown that Galileo and his contemporaries assumed the proposition valid independent of the particular type of motion considered.[1] This, however, has not called into question the claim whose fundamental support it had undermined, namely, that Galileo devised the ex mechanicis proof and thus first became aware of the law of chords before his conceptual shift toward the assumption of natural acceleration. In contrast, in the present chapter, it will be demonstrated that all entries pertaining to Galileo's construction of the ex mechanicis proof that can be identified in the *Notes on Motion* in fact succeed his conceptual shift toward the assumption of natural acceleration, like indeed all entries in the manuscript. There is, moreover, no evidence to support that Galileo ever thought of the law of chords as being valid for motion along inclined planes conceived as uniform.

In Sect. 8.1, the ex mechanicis proof as published in the *Discorsi* is first analyzed. Based on this analysis, the entries in the manuscript pertaining to Galileo's construction of the proof are then identified, and the sequence of steps which led him to a draft of the proof is reconstructed. This sequence is in Sect. 8.2 contextualized within the more general picture of Galileo's early work on the

---

[1] In a series of articles together with different co-authors, Souffrin has analyzed Galileo's concept of velocity. Cf. Chap. 2. He comments upon Galileo's application of one of the kinematic propositions as a premise in the ex mechanicis proof as follows: "It is generally assumed that Galileo's *mechanical* demonstration of theorem VI on accelerated motion in the *Discorsi* relies on an illegitimate use of a theorem valid only for uniform motions. I will show that this view depends on an anachronistic interpretation of the terminology of theoretical kinematics, and propose an alternative interpretation which conforms more to kinematics at the time of Galileo and which leads to a demonstration that is devoid of the alleged error (Souffrin and Gautero 1992, 269)."

problem of natural acceleration established here. The standard interpretation is ruled out by the consistent interpretation which thus emerges and to which it will be juxtaposed. In particular, the claim that Galileo might first have achieved insight into the law of chords along a line of argument similar to the one found in the ex mechanicis proof is excluded. An alternative to how Galileo may have first come to accept the law of chords as a heuristic assumption is presented in Sect. 8.3.

8.1 The Formation of the Ex Mechanicis Proof of the Law of Chords

The law of chords, correct but insignificant in terms of classical mechanics, states that the time of fall along all chords inscribed in a circle, comprising either the highest or alternatively the lowest point of the circle, i.e., the apex or nadir, are equal. As we have seen, the proposition was mentioned for the first time in the letter to Guidobaldo del Monte written in late November 1602, where Galileo claimed to have proven it "without transgressing the boundaries of mechanics."[2]

In the *Discorsi*, the law of chords was articulated as Proposition VI of the Second Book of the Third Day where Galileo advanced a total of three different proofs for the proposition.[3] It has already been assumed in Galileo scholarship that the second of these proofs, which Galileo explicitly characterized as a demonstration "ex mechanicis," preserves the line of argument of the proof alluded to in the letter.[4] As this assumption is indeed vindicated here, the ex mechanicis proof as published in the *Discorsi* will first be discussed in some detail in the following. This provides a point of reference and will allow the identification of the entries in the manuscript from which the sequence of steps that led Galileo to the final proof can be reconstructed.

The ex mechanicis proof in the *Discorsi* For analysis purposes, the ex mechanicis proof as published in the *Discorsi*, has been rendered as a table. The left column comprises the text, and the right column comments the individual steps of the argument which are rendered in modern notation.[5] Rows individuate argumentative steps. Steps implicit in Galileo's proof but which, from a formal point of view, are

---

[2] "Sin qui ho dimostrato senza trasgredire i termini mecanici (EN X, 99)" Cf. Chap. 3.
[3] Theorem VI Proposition VI of the Second Book of the Third Day of the *Discorsi* compared motion along a chord to motion along the vertical diameter of the circle. The more general case stating that motion along two arbitrary chords inscribed in the same circle and comprising either the apex or nadir was made in the same time, i.e., the statement referred to here as the law of chords, is actually established in a corollary to the proposition.
[4] "Idem aliter demonstratur ex mechanicis (EN VIII, 221)".
[5] EN, VIII, 221–223.

Fig. 8.1 Schematic representation of the diagram associated with the ex mechanicis proof of the law of chords in the *Discorsi* (Cf. EN VIII, 222)

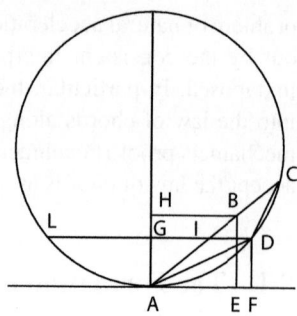

required to complete the reasoning, are represented as rows in which the left cell is left blank. The diagram associated with the proof is provided as Fig. 8.1.

Si a puncto sublimi vel imo circuli ad horizontem erecti ducantur quaelibet plana usque ad circumferentiam inclinata, tempora descensuum per ipsa erunt aequalia…	1. General statement
Idem aliter demonstratur ex mechanicis	2. Identification of the proof as based on the principles of mechanics (ex mechanicis proof)
nempe, in sequenti figura, mobile temporibus aequalibus pertransire CA, DA.	3. General statement geometrically instantiated: $$t(CA) = t(DA)$$
Sit enim BA aequalis ipsi DA, et ducantur perpendiculares BE, DF:	4. Definition of points B,F,E: $$|BA| = |DA|$$ $$BE, DF \perp AE$$
constat ex elementis mechanicis, momentum ponderis super plano secundum lineam ABC elevato ad momentum suum totale esse ut BE ad BA, eiusdemque ponderis momentum super elevatione AD ad totale suum momentum esse ut DF ad DA vel BA;	5. By the law of the inclined plane ("constat ex elementis mechanicis") and by above: $$\text{mom}_{AC}/\text{mom}_{\text{tot}} \sim BE/BA$$ $$\text{mom}_{AD}/\text{mom}_{\text{tot}} \sim DF/DA$$ $$\sim DF/BA$$
ergo eiusdem ponderis momentum super plano secundum DA inclinato ad momentum super inclinatione secundum ABC est ut linea DF ad lineam BE;	6. By above: $$\text{mom}_{DA}/\text{mom}_{AC} \sim DF/BE$$
	7. By above and the basic principle of inclined plane dynamics: $$v_{DA}/v_{AC} \sim DF/BE$$

8.1 The "Ex Mechanicis" Proof in the Sources

quare spatia, quae pertransibit idem pondus temporibus aequalibus super inclinationibus CA, DA, erunt inter se ut lineae BE, DF, ex propositione secunda primi libri.

8. By above and "ex propositione secunda primi libri":

$$\text{for all } X \subset CA, Y \subset DA$$

$$\text{if } t(CX) = t(DY) \Rightarrow CX/DY \sim DF/BE\ {}^6$$

Verum ut BE ad DF, ita demonstratur se habere AC ad DA; ergo idem mobile temporibus aequalibus pertransibit lineas CA, DA.

9. By above[7]:

$$\text{for } X = Y = A$$

$$\text{if } CA/DA \sim BE/DF \Rightarrow t(CA) = t(DA)$$

Esse autem ut BE ad DF, ita CA ad DA, ita demonstratur:

10. Reduced statement:

$$CA/DA \sim BE/DF$$

Iungatur CD, et per D et B, ipsi AF parallelae, agantur DGL, secans CA in puncto I, et BH:

11. Definition of points G,L,H,I:

$$DGL, BH \parallel AF$$

$$I = DGL \cap CA$$

eritque angulus ADI aequalis angulo DCA, cum circumferentiis LA, AD aequalibus insistant, estque angulus DAC communis. Ergo triangulorum aequiangulorum CAD, DAI latera circa aequales angulos proportionalia erunt,

12. By geometry:

$$\angle ADI = \angle DCA$$

$$\triangle CAD \sim \triangle DAI$$

et ut CA ad AD, ita DA ad AI, id est BA ad AI, seu HA ad AG, hoc est BE ad DF: quod erat probandum.

13. By above, geometry and theory of proportions:

$$CA/AD \sim DA/AI \sim BA/AI$$

$$\sim HA/AG \sim BE/DF \text{ q.e.d.}$$

---

[6] According to the wording, Galileo's sentence could also be interpreted as implying that if the time of motion along any distance on CA was the same as the time of motion along any other distance on DA then the spaces traversed must be in the ratio of BE to DF. This is, however, quite clearly not what was intended, and it is assumed here that Galileo was referring to motions from rest in C and D, respectively, only.

[7] In drawing this conclusion, with regard to the premise expressed in the previous step, Galileo seems to have reversed the direction of the inference. Yet, as we will see, the relation expressed in the previous step was for Galileo indeed a logical equivalence and thus an inference from equal times of motion to a ratio of distances traversed was for him as legitimate as the inference from a ratio of distances to the equality of the times of motion.

Besides the transition from Steps 6 to 9, Galileo's argument is rather straightforward. It receives a consistent interpretation if it is assumed that the transition is mediated by application of what has been referred to as Galileo's principle of inclined plane dynamics, now however applied to accelerated motion.[8] According to the principle, the velocities of the two motions should be in the same ratio as the forces causing these motions, i.e., the mechanical moments. The ratio of the mechanical moments on inclined planes of different inclination is given by the law of the inclined plane, i.e., "ex mechanicis" and thus the ratio of the velocities can be determined.[9] This line of reasoning is indicated by the implicit Step 7 that has been introduced in the table above.

Furthermore, according to the second kinematic proposition, for two motions completed in the same time, the distances traversed are in the same ratio as the velocities of the motions. Vice versa, if the spaces that have been traversed by two motions are in the same ratio as their velocities, it follows that these motions are completed in the same time. For the motions over the inclined planes, CA and DA, Galileo showed by geometry that the lengths of the inclined planes CA and DA were in the same ratio as the velocities of motion along them and could thus indeed conclude that the motions along these planes are completed in equal time.

In the proof, Galileo explicitly stated that his conclusion was justified by "propositione secunda primi libri." The second proposition of the First Book of the Second Day reads:

> Si mobile temporibus aequalibus duo pertranseat spatia, erunt ipsa spatia inter se ut velocitates. Et si spatia sint ut velocitates, tempora erunt aequalia. (EN VIII, 193)

This statement corresponds to the second Aristotelian kinematic proposition. However, the First Book of the Third Day of the *Discorsi* was exclusively devoted to treating uniform motion.[10] Accordingly, even though not explicitly expressed in the statement of the proposition nor actually required in the proof Galileo gave for it, it was thus presupposed that the motions considered had to be uniform, which the motions that Galileo considered in Proposition VI clearly were not.[11]

---

[8] Cf. Chap. 2.

[9] Interpretations of the argument of the ex mechanicis proof that are comparable to the one given here have been advanced by Souffrin and Gautero (1992), Wisan (1974, 162–165) and Giusti (2004, 119–124).

[10] Galileo introduced the propositions advanced in the First Book of the Second Day of the *Discorsi* as follows: "...in prima parte consideramus ea quae spectant ad motum aequabilem, seu uniformem (EN VIII, 19)."

[11] Renn has already remarked that neither the statement nor the proof of Proposition II of the First Day are directly restricted to the case of uniform motion and concluded that "[t]heorem II, corresponding to one of the Aristotelian proportions, is in fact also valid for non-uniform motion ...(Damerow et al. 2004, 260)."

That Galileo invoked as a premise a proposition whose validity had formally been sanctioned for the case of uniform motions only, arguably has crucially contributed to the creation of the standard interpretation. If the premise applied to uniform motion only then the argument in which it was invoked should be, or should have been, an argument about uniform motion. As the law of chords in fact holds regardless of whether the motions along the chords are considered accelerated or uniform, it would then indeed seem to suggest that the ex mechanicis proof argument must have been conceived by Galileo while he was still working under the conceptual framework of the *De Motu Antiquiora*. What is commonly overlooked, or at least concealed by those who have adopted such a view, is that, if this were the case, Galileo's argument would have lost its justification as soon as he started to regard motion along inclined planes to be characteristically accelerated. Thus, in particular, his inclusion of the proof in the *Discorsi* would have been unwarranted.[12]

Yet as will be demonstrated in the following, such a position cannot, in any case, be maintained as it will be shown that Galileo conceived of and first drafted the ex mechanicis proof certainly only after his conceptual shift toward assuming natural acceleration. Manuscript evidence, furthermore, unambiguously attests that the kinematic proposition Galileo invoked as a premise in his argument was indeed not conceived by him as restricted to the case of uniform motion. Rather it was assumed by Galileo and his contemporaries to be universally valid. Thus, when Galileo referred to Proposition II of the First Book as a premise in the ex mechanicis proof in the *Discorsi*, he had either overlooked or simply ignored the fact that via the deductive structure of the *Discorsi*, he had superimposed a restricted scope of applicability on a proposition conceived by him as more generally valid.[13] Compared to the, according to standard interpretation, unwarranted inclusion of the ex mechanicis proof in the *Discorsi*, this amounts to rather venial inconsistency.[14]

---

[12]If the ex mechanicis proof had indeed been conceived under the old conceptualization of motion of fall, it could be argued that Galileo may at some point have become convinced that it was likewise valid under his new conceptualization, not least because the premises employed were ultimately indifferent to the type of motion. Such a position was taken, for instance, by Souffrin. When issuing a version of the same proposition Torricelli, possibly because he realized the problem entailed by Galileo's reference to Proposition II of the First Book, explicitly stated that it was valid for accelerated motion. Cf. Torricelli (1919, 112).

[13]The same happens once more in the *Discorsi*, when in the proof of Proposition III of the Second Book of the Third Day of the *Discorsi*, i.e., the length time proportionality, Galileo shows that the velocities of the motions along two inclined planes of equal height are equal and then concludes "Sed demonstratum est, quod si duo spatia conficiantur a mobili quod iisdem velocitatis gradibus feratur, rationem habent ipsa spatia, eamdem habent tempora lationum (EN VIII, 216–217)." This is an explicit reference to Proposition I of the First Book, where thus again a proposition proved formally only for uniform motion is invoked in a conclusion about naturally accelerated motion.

[14]It is tempting to attribute the slight inconsistency in the deductive structure to the unfavorable circumstances under which the *Discorsi* were composed. Cf. Drake (1978), in particular Chap. 20.

Based on the above analysis, the argument of the ex mechanicis proof is recognized as crucially resting on three argumentative steps, which together uniquely identify the proof and distinguish it from other arguments contained in the manuscript. These steps can be summarized as follows:

- For inclined planes appropriately inscribed as chords in the same circle, the effect of the heaviness of a material body on these planes, the mechanical moment, can be determined by the law of the inclined plane. Together with some geometrical reasoning, it can be shown that the mechanical moments are in the same ratio as the lengths of the chords.
- The mechanical moments are subsequently interpreted as the causes of motion of fall along the chords. The basic principle of inclined plane dynamics then entails that the velocities of the motions are in the same ratio as the mechanical moments, i.e., in the ratio of the distances traversed.[15]
- If the velocities of motions are in proportion to the distances traversed, the times of fall are equal according to the second Aristotelian kinematic proposition.

In modern notation, with reference to the geometrical configuration rendered in Fig. 8.2, this can be expressed as:

- By law of the inclined plane and geometry

$$\text{mom}_{AB}/\text{mom}_{AC} \sim AB/AC$$

- By Galileo's basic principle of inclined plane dynamics

$$V_{AB}/V_{AC} \sim \text{mom}_{AB}/\text{mom}_{AC}$$

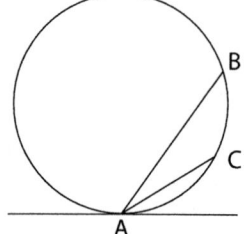

Fig. 8.2 The geometrical setup of the law of chords. Motion along the chords AB and AC is made in equal time

---

[15] For Galileo's recourse to his basic principle of dynamics in the ex mechanicis proof compare, in particular, Giusti (2004) as well as Souffrin (1990) and Souffrin and Gautero (1992).

8.1 The "Ex Mechanicis" Proof in the Sources

- By above and the second Aristotelian kinematic proposition

$$V_{AB}/V_{AC} \sim AB/AC \Rightarrow t(AB) = t(AC)$$

Reconstructing the developmental sequence that resulted in the formulation of the ex mechanicis proof

Manuscript evidence Three distinct blocks of content that pertain directly to Galileo's construction of the ex mechanicis proof can be identified in the *Notes on Motion*.[16] They are contained on folios 151 recto, 160 recto, and 172 recto.[17] In addition, a fourth folio, 131 recto, will be discussed here whose content is related to that of folio 151 in twofold manner. On the one hand, on folio 131 recto, in the context of his considerations regarding the law of the broken chord, Galileo had determined the ratio of mechanical moments on different chords inscribed in the same circle. This, as will be argued, directly inspired his first sketch of the ex mechanicis proof drafted on folio 151 recto. On the other hand, before noting this first sketch on 151 recto, Galileo had compiled a table of relations of distances and times of motions on the page pertaining directly to the motion situation depicted in the central motion-diagram on folio 131 recto, substantiating beyond doubt that the content of both folios is directly related and must have been drafted at the same time. The relations between the folios discussed in the present chapter have been rendered graphically in Fig. 8.3.

Folio 131 recto Folio 131 recto is dominated by a large motion diagram to which four short textual entries are associated. One of the entries was deleted by Galileo. Despite the sparse textual information, the motion diagram, reproduced here as Fig. 8.4, can be unambiguously identified as a diagram pertaining to Galileo's considerations in the context of the law of the broken chord. The diagram indeed corresponds almost exactly to the motion diagram drawn on 163 recto along with the proof of the proposition, including even the keying, which is nearly identical.

---

[16] In a number of additional entries Galileo argued for motions along inclined planes completed in equal time in a manner comparable to the ex mechanicis proof. However, in contrast to the law of chords, in these considerations Galileo did not specify the planes under consideration as chords inscribed in a circle but through a different geometrical constraint. These considerations will be discussed in Chap. 10 where it will be argued that these considerations must postdate the construction of the ex mechanicis proof.

[17] The ex mechanicis proof was eventually copied by Mario Guiducci from folio 172 recto to 47 recto introducing only some minor corrections that may have been directly dictated by Galileo to his disciple.

Fig. 8.3 Folios of the *Notes on Motion* bearing consideration pertaining to Galileo's formulation of the ex mechanicis proof of the law of chords

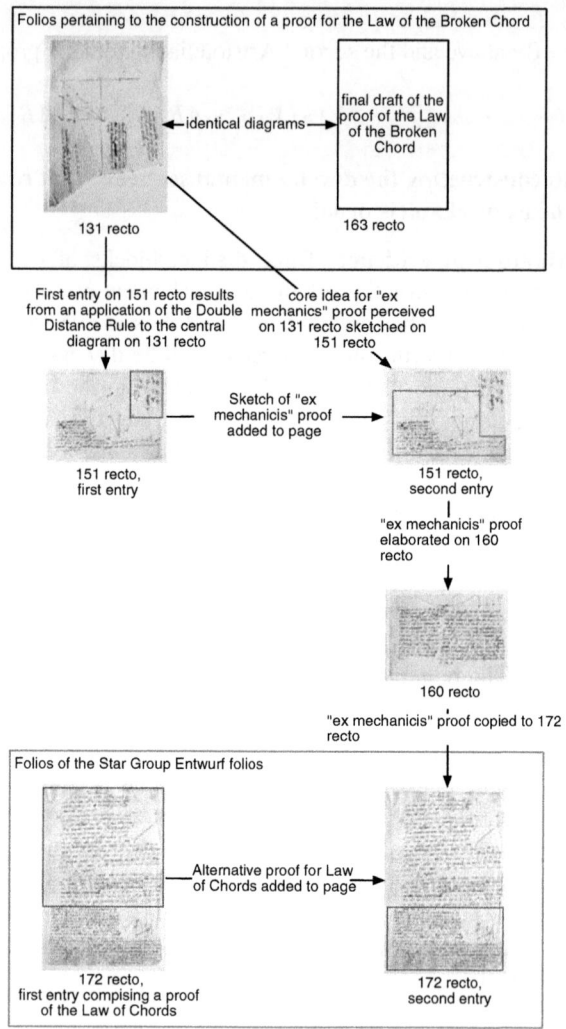

As discussed in detail in Chap. 5, Galileo's attempt to construct a proof for the law of the broken chord had passed through various stages, a development mirrored in an evolving motion diagram which only gradually stabilized in its final form, the form in which it was later published in the *Discorsi*. It is thus suggested that the motion diagram on folio 131 recto pertains either to a late stage of the development of Galileo's construction of the proof of the law of the broken chord or else that it was drafted after he had already completed the proof and set out to explore further consequences. In light of the detailed analysis in Chap. 5, the second alternative provides the more likely scenario. As argued there, after having found a proof for the law of the broken chord, Galileo proceeded to further analyze motion along a broken chord with the aim of finding a way to infer the exact ratio between the times

8.1 The "Ex Mechanicis" Proof in the Sources

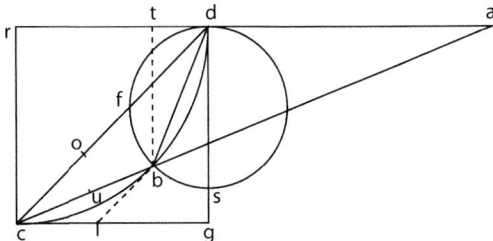

Fig. 8.4 Schematic representation of the diagram on 131 recto, reduced to the elements relevant for the discussion in the present chapter

for which the law of the broken chord merely stated an inequality. Indeed, the two entries on 131 recto, written in somewhat lighter ink than the other two, appear to have been directed toward this aim.[18]

In the two remaining entries on the folio, which were written in a noticeably different ink and allegedly later, Galileo as well compared motion along a single chord to motion along a broken chord spanning the same arc. However, he introduced the notion of mechanical moment into his analysis, alien to his prior investigations in this context.[19] The likely reason why Galileo here introduced the concept of mechanical moment is obvious.

If the motion diagram on folio 131 recto is compared to the geometrical configuration of Galileo's bent lever proof of the law of the inclined plane (see Fig. 2.3), a similarity is immediately realized. The former diagram, in particular, comprises almost all geometrical elements of the latter. The diagram on folio 131 recto incorporates, in particular, an inclined plane lb tangent to a circle and the distance rt, which can be interpreted as the projection of the bent lever arm rb, that is orthogonal to the inclined plane lb, onto the horizontal. Only a representation of the bent lever arm rb itself, connecting the fulcrum at r with the point of contact b between the inclined plane and the circle, is missing from the diagram on folio 131 recto but can easily be substituted in order to complete the geometrical configuration of the bent lever proof. This fortuitous resemblance of the geometrical configurations of the two diagrams was apparently noticed by Galileo, who, as discussed in detail in Chap. 11, once he had realized that the principle of inclined plane dynamics should also be applicable in the case of naturally accelerated motion, was certainly compelled to seek a possibility for a dynamical analysis of naturally accelerated motion.

---

[18]131 recto, T1 and T3.

[19]The different inks used on folio 131 recto are suspiciously similar to the different types of inks used on folio 150 recto. As argued in Chap. 5, the entries on the latter page should have been drafted at about the same time as those on folio 131 recto.

In the first of the two entries on folio 131 recto, Galileo avails himself of the concept of mechanical moment and repeats an argument concerning the diminishing of the mechanical moment of a body positioned on a quarter circle as the position approaches the lowest point of that circle, which he had already made in *De Motu Antiquiora* and later restated in *Le Meccaniche*:

> Consider that the moment in the single points of the circumference of the quarter circle diminishes according to the ratio of the perpendicular approach of the point, as t to the center.[20]

This statement was an immediate consequence of the bent lever proof. As is explained in detail in Chap. 2, this proof rested on the assumption that the motive force in any point of descent on an arc is the same as the motive force of descent along an inclined plane tangent to the arc in that very point. Wisan has claimed that in this entry Galileo was no longer referring to the static case but actually already had in mind a body moving on an arc, i.e., ultimately, pendulum motion. Whether this was the case, it seems, cannot be decided.[21] Be that as it may, at the time when he was working on folio 131 recto, Galileo was apparently no more able to exploit the insight expressed than he had been at the time of writing the *De Motu Antiquiora*. The insight is indeed neither picked up on folio page 131 recto nor anywhere else in the *Notes on Motion*.

The entry which turns out to be most revealing with regard to Galileo's construction of the ex mechanicis proof is the entry on the bottom of the page. It reads[22]:

> momentum super plano dc ad totale momentum est ut linea tr ad rd, ducta lb aequidistante cd.

This statement results from the bent lever proof paired with an additional assumption. As discussed in Chap. 2, the bent lever proof provided a way of determining the ratio of the mechanical moment of a heavy body on an inclined plane tangent to a circle to the total moment of the same body on the vertical. With reference to Galileo's motion diagram, according to the bent lever proof, it thus held that:

$$\text{mom}_{\text{tot}}/\text{mom}_{lb} = rd/tr.$$

---

[20] 131 recto, T2.

[21] Not only did Wisan claim that here Galileo was already referring to a body moving continuously but she has even proposed that Galileo may, in this way, have inferred that the descent of a body on an arc was accelerated. Cf. Wisan (1974, 161–162). Galluzzi, in contrast, emphasized that Galileo does not talk explicitly about a continuous descent, and concluded that "la sua analissi si mantiene, cioè, ancora sostanzialmente <<statica>>. (Galluzzi 1979, 217)" He consigns, however, that Galileo probably would have asked himself how the diminishing static moment on the arc could be reconciled with the accelerated motion of a descending pendulum. Galluzzi indeed suggested that it may have been due to precisely this problem that Galileo decided to present a purely static analysis and avoided discussing any dynamic implication in *Le Meccaniche*. Cf. Galluzzi (1979, 216–217 and 269).

[22] 131 recto, T4.

8.1 The "Ex Mechanicis" Proof in the Sources

Furthermore, as will be demonstrated below, in the context of the considerations for which the diagram on 131 recto had originally been drafted, Galileo had considered motion along the chord cd parallel to the inclined plane lb. As the moment along an inclined plane depends only on the inclination, the mechanical moments of one and the same mobile body on the parallel planes lb and cd are the same:

$$\text{mom}_{lb} = \text{mom}_{cd}.$$

Thus, it indeed immediately followed that:

$$\text{mom}_{\text{tot}}/\text{mom}_{cd} = rd/tr.$$

Hence, the ratio of the mechanical moments on two chords in a circle (one of them being the vertical diameter) had been determined. Thus, what is achieved by resorting to the law of the inclined plane in the published proof is achieved differently in the considerations documented on folio 131 recto, namely, by harking back to the construction of the bent lever proof and a parallel transposition of the plane in question, exploiting the fact that the mechanical moment solely depended on inclination.

As will be seen, in his first sketch of the proof idea of the ex mechanicis proof in an entry on folio 151 recto, Galileo inferred the sought-after ratio of mechanical moments in the very same manner as on 131 recto, i.e., by resorting to the bent lever proof and by exploiting the fact that the mechanical moments on parallel planes are equal, indicating that the proof idea sketched was indeed directly stimulated by Galileo's considerations on 131 recto.[23]

Folio 151 recto The content of folio 151 recto can be broken down into two clearly distinguishable blocks. In the upper left corner, we find a table in which various ratios of times of motions are coded. The remaining, main block of content, composed of a central motion diagram, a smaller diagram and a longer text entry were written with the page turned 90 degrees clockwise.[24] The layout of the folio implies that the table was put on it before the remaining content.

---

[23] A diagram on folio 154 recto compares to that on 131 recto in that it contains a plane tangent to a circle together with a construction that would allow for an immediate application of the argument of the bent lever proof, as well as a parallel transposed plane inscribed as a chord in a circle. Just as based on the construction on 131 recto, likewise based on the construction on 154 recto, the ratio of the mechanical moment on this chord and on the vertical chord could immediately be inferred. The diagram on 154 recto is unfortunately not associated with any textual entry, and it cannot, thus, unanimously be concluded that the diagram served such a consideration even though this appears to be rather likely. Folio 154 recto otherwise contains considerations concerning the period of a pendulum whose length corresponds to the radius of the earth. Cf. Büttner (2009).

[24] 151 recto, T1A, T1B and D01A as well as D02A.

Fig. 8.5 Schematic representation of the central diagram on folio 151 recto

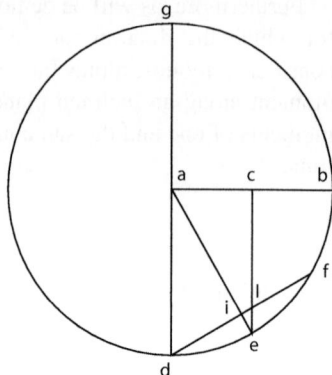

Galileo commenced his considerations pertaining to the construction of the ex mechanicis proof with the following statement (for the associated diagram see Fig. 8.5):

> Sit gd erecta ad orizontem, df vero inclinata: dico, eodem tempore fieri motum ex g in d et ex f in d.

Thus, Galileo considered the case where one motion proceeds on the vertical diameter and the other on a chord appropriately inscribed in a circle. What follows is a condensed argument, which, despite its cursory nature, is recognized as obeying the argumentative schema characteristic of the ex mechanicis proof:

> Momentum enim super fd est idem ac super contingente in e, quae ipsi fd esset parallela; ergo momentum super fd ad totale momentum erit ut ca ad ab, idest ae.

Galileo infers the ratio of the mechanical moments on the chord and on the vertical diameter, which he simply refers to as the total mechanical moment. In doing so, as already indicated, he avails himself of the very line of argument we saw him develop on folio 131 recto. The central motion diagram on folio 151 recto again comprises most of the geometrical elements required for the bent lever proof. Here, even a representation of the bent lever arm ae itself is included. A tangent to the circle at e is missing from the diagram but is explicitly alluded to in the text.

With this construction, according to the logic of the bent lever proof, the ratio of the mechanical moment on an inclined plane tangent to the circle at point e to the total mechanical moment was given as the ratio of ca to ab. Furthermore, just as on 131 recto, Galileo made use of the fact that the mechanical moment on the tangent is equal to that on the chord df parallel to it.[25]

---

[25] That the construction on folio 151 recto is closely related to Galileo's bent lever proof has already been noted by Galluzzi. He argued that the method and terminology used on folio 151 recto, in particular Galileo's use of the expression "momentum super" instead of "gravitas in plano", places the sketch of the ex mechanicis proof on the page conceptually closer to *Le Meccaniche* than to *De Motu Antiquiora*. Cf. Galluzzi (1979, 267).

8.1 The "Ex Mechanicis" Proof in the Sources

In the remaining part of the entry, Galileo demonstrated that as ca to ab (which is the same ca to ae) so is fd to dg. Thus, he showed that as the mechanical moment on the vertical diameter is to the mechanical moment on the chord, so is the length of this diameter to the length of this chord, completing the first argumentative step of the ex mechanicis proof:

> Verum ut ca ad ae, ita id ad da, et dupla fd ad duplam dg; ergo momentum super fd ad totale momentum, scilicet per gd, est ut fd ad gd:

Rendered into modern notation this can be expressed as:

$$\text{mom}_{\text{tot}} = \text{mom}_{gd}$$

$$\text{mom}_{fd}/\text{mom}_{gd} \sim fd/gd.$$

The entry ends somewhat abruptly:

> ergo eodem tempore fiet motus per fd et gd.

Galileo does not explicate the reasoning that leads him to this conclusion. There can, however, be little doubt that the reasoning was the same as that resulting in the same conclusion in the ex mechanicis proof as published in the *Discorsi*. By the basic principle of inclined plane dynamics, the velocities of motion over the vertical and over the chord could be assumed to obey the same ratio as the mechanical moments, and by the second Aristotelian kinematic proposition, Galileo could then conclude that the times of motion were equal.[26]

The provisional character of the sketch on folio 151 recto as well as its close relation to considerations on 131 recto suggest that Galileo immediately expanded an idea that had occurred to him in the course of his work on 131 recto into a complete argument. This was then quickly noted in abridged form for use as a blueprint from which a clean, publishable version of the ex mechanicis proof could be elaborated.[27] Compelling as this picture may be, it would remain speculative if

---

[26] In their own interpretations of folio 151 recto, Wisan and Galluzzi have likewise assumed that the argumentative steps, which remained implicit, must be the same as those of the published ex mechanicis proof, i.e., the application of the basic principle of inclined plane dynamics and of the second Aristotelian kinematic proposition. Galluzzi states "[t]uttavia, la proporzionalità implicita, ma evidentissima, *momento-velocità*, indica che lo scienziato pisano è tornato a lavorare sull'ipotesi dinamica utilizzata nel *De motu*. (Galluzzi 1979, 267). Wisan remarked "[t]he conclusion follows from the converse of the mediaeval assumption, which we have seen in the *Liber Karistonis* and in BRADWARDINE, that in equal time intervals, speeds are proportional to distances traversed. Speed is not distinguished from *momentum* (Wisan 1974, 164)."

[27] Galluzzi (1979, 266) has observed that the proof on folio 151 recto may "conservarci la dimostration [della quale Galileo dichiarava a Guidobalo nel novembre del 1602] esattamente come Galileo la concepì al tempo della lettera a Guidobaldo." He explicitly rejects making a decision on whether the entry was drafted before or after Galileo's conceptual shift but clearly leans toward the second alternative when he states "[n]on esiste elemento preciso che ci consenta di stabilire se Galileo stia pensando, in questi theoremi, a moti accelerati o uniformi; anche se la ricomparsa del teorema del *De motu*, precedentemente abbandonato, potrebbe far supporre che Galileo abbia

it weren't for the fact that a relation can be established between Galileo's work on the two folios, showing beyond doubt that when writing on 151 recto Galileo must have had folio 131 recto on his desk.

Despite the sparse information it conveys, it can indeed be shown that the content of the small table on 151 recto is directly related to 131 recto. It bears the results of Galileo's determination by means of the double distance rule of the times of motion along the paths dgc and dlc depicted in the central motion diagram on 131 recto.[28] As the double distance rule necessarily presupposes that the initial motion was accelerated, this alone would suffice to demonstrate that Galileo's work on 131 recto and 151 recto must postdate his conceptual shift toward the assumption that motion on inclined planes was naturally accelerated. As argued in Chap. 5, Galileo's interest in the time of motion along the paths dgc and dlc, which he had calculated and noted in the table on 151 recto, is to be situated within the general context of his approach to exhausting motion along an arc by motion along polygonal paths.

After Galileo had sketched this first proof idea for the ex mechanicis proof of the law of chords on folio 151 recto, all that remained was to elaborate this sketch into a well-ordered draft of a proof. Galileo drafted such a proof on folio 160 recto, from where it was eventually copied to folio 172 recto.[29]

Folios 160 recto and 172 recto The draft of the ex mechanicis proof which Galileo jotted down on folio 160 recto compares to the version which he later published in the *Discorsi* and thus a detailed discussion of the argument is not required. Folio

---

voluto verificare la validità nell'ipotesi discese accelerate. (Galluzzi 1979, 267)" Wisan (1974, 163), in her discussion of the proof on folio 151 recto, carefully circumnavigates the question as to whether Galileo was already working under the assumption that motion on inclined planes was naturally accelerated. Yet, only some pages later, she states that a number of propositions and proofs, among them the ex mechanicis proof, "may precede the times-squared law. (Wisan 1974, 171)" In her periodization of the content of the *Notes on Motion* presented in the form of a list, she finally assigns the ex mechanicis proof to the period antedating Galileo's conceptual shift. Cf. Wisan (1974, 163, 171 and 296).

[28] 151 recto, C01.

[29] The first sketch of the argument on 151 recto, the draft on 160 recto and naturally its copy on 172 recto, all regard the motion situation where motions along the chords under consideration end at the lowest point of the circle. An earlier proof of the law of chords contained on folio 172 recto, in contrast, considered the situation where the motions along the chord all started from the highest point of circle. Both cases are of course eventually equivalent yet regarding motions starting from the same point seems to have been somewhat more natural for Galileo than regarding motion ending at the same point. The focus put on motion along chords ending at the same point in the construction of his ex mechanicis proof argument was apparently inherited from his considerations regarding the relation between swinging and rolling that had prompted the proof idea for the ex mechanicis proof.

8.1 The "Ex Mechanicis" Proof in the Sources   225

151 and folio 172 bear the same watermark, a small star. Folio 160 bears a large star as a watermark.[30] This could suggest that folios 151 and 172 both stem from the same single batch of paper and, if this indeed is the case, that not much time had passed between taking down the first sketch on 151 recto to the final copying of the draft of the proof to 172 recto.

The unusual distribution of the entries on folio 160 recto into two columns, where by means of an insertion mark, Galileo indicated that the text of the left column was supposed to be appended to that of the right column, betrays the impression that the content of the folio was worked out in two separate successive steps. The upper half of folio 160 is cut off with most of the central diagram on the page.[31] Yet as the diagram was copied together with the proof to folio 172 recto, we can infer how it must have looked.[32] With reference to the proof diagram on 172 recto (the diagram is nearly identical to the one later published in the *Discorsi* and given here as Fig. 8.1), in order to be able to apply the law of the inclined plane directly, Galileo introduced a point b on the chord ca such that the inclined plane ba was equal in length to the chord da. This allowed him to infer by the law of the inclined plane that the moment on this plane ba had to the moment on chord da the same ratio as the vertical heights be had to the vertical height df. Since ba and ca by construction are parts of the same inclined plane, the moment on ba is the same as that on the chord ca. Thus Galileo could conclude that the mechanical moments on the chords ca and da were in the ratio of be to df.[33]

Compared to his first quick sketch on 151 recto, Galileo had thus introduced a modification into his argument. Whereas on folio 151 recto, just as on folio 131 recto, Galileo had determined the ratio of the mechanical moments on a chord and on a vertical by making reference to the construction of the bent lever proof, to determine this ratio in the entry on folio 160 recto Galileo took recourse to the law of the inclined plane. He justified the application of the law of the inclined plane by the phrase "constat ex elementis mec[h]anicis," familiar from the published version of the proof. The argumentative steps necessitated by this new way of inferring the ratio of the mechanical moments take up almost all of the right column. Conceptually, however, it does not alter the proof. Galileo's entry on 160 recto thus apparently does not have an explorative character. Rather, it seems to have been motivated by the aspiration to devise a version of the argument apt for inclusion in the deductive

---

[30] As detailed in Chap. 10, it is somewhat unclear whether the watermark of folio 151 is in fact a small star or, alternatively, a large star. These different marks may in fact merely be different manifestations of what is, in principle, the same watermark, suggesting that all folios bearing these marks are affiliated.

[31] The average height of those folios sharing the large star watermark with 160 is 281.5 mm; what remains of folio 160 measures exactly 140 mm in height; hence the folio was apparently cut exactly in half.

[32] 172 recto, D02A.

[33] The fact that the mechanical moment of a body on an inclined plane depends merely on the inclination of the plane is reflected in Galileo's use of expressions such as "momentum ponderis super plano secundum lineam *abc* elevato ...".

structure of an envisioned treatise. The new version of the proof produced no longer required the entire bent lever proof but only its result, namely, the law of the inclined plane needed to be invoked as a premise "ex ... mechanicis." As will be argued in Chap. 10, at about the same time that he devised the ex mechanicis proof, Galileo added a statement of the law of the inclined plane to his working notes.[34]

The entry in the right column ends as follows:

> quare, spacia quae pertransibit idem pondus temporibus aequalibus super inclinationibus ca, da, erunt inter se ut lineae be, df. At ut be ad df, ita demon[s]tratur se habere ac ad da: ergo idem mobile temporibus aequalibus pertransibit lineas ca, da.

The proof seems to be completed. Yet the new way of determining the ratio of mechanical moments had yielded the ratio of the mechanical moments in terms of the heights be and df, and not in terms of the lengths of the chords as required by the argument. It, hence, formally remained to be shown that be and df were in the same ratio as the length of the chords ac and da. Galileo merely claimed this to be provable (ita demon[s]tratur). The left column, in which it is demonstrated that the distances be and df indeed have the same ratio as the chords ac and da, was added to the folio only afterward. That the argument had originally been left incomplete indeed suggests that the prime incentive for drafting the entry on folio 160 recto had been to introduce the law of the inclined plane as a premise in the argument.[35]

Once the proof had been completed by supplementing the geometrical argument in the left column, it was quickly allotted its place in Galileo's attempt to turn his early results into the nucleus of a publishable science. As is further discussed in Chap. 10, Galileo copied the draft from folio 160 recto to folio 172 recto one

---

[34] Statements of the law of the inclined plane have been preserved on the cut sheets 179a verso for an inclined plane and its vertical height and on 179b verso for two inclined planes of equal height. In his argument on folio 160 recto, Galileo employed the law of the inclined plane as formulated on 179a verso and inferring the ratio of mechanical moments for the more general case of two inclined planes of equal height takes up most of the entry on 160 recto. Handwriting and layout suggest that the content of the second pasted sheet, 179b may indeed have been added at a later date and that, at the same time the first sentence on 179a verso containing the general statement was added. Cf. also Chap. 10.

[35] Guisti has pointed out that whereas in his argument on 151 recto Galileo inferred that the motions under considerations were completed in equal times, in the argument on 160 recto, he presupposed the times of motion to be equal, inferring the ratio of the distances that would be covered to find that this corresponded to the ratio between the chords. He claimed that "There is no reason to compare spaces traversed in equal times, unless one knows that the mechanical method can be only applied to the comparison of motions taking place in equal times" and takes this as a "sign that Galileo has become aware of the fact that the mechanical method has a limited validity...(Giusti 2004, 125)" However, as will be argued in Chap. 10, Galileo's insight into the restricted applicability of his mechanical method almost certainly postdated his entries on 160 recto.

of the star group folios on which he had started to collect and systematize the results achieved as an outcome of his considerations regarding the relation between swinging and rolling.[36]

Provisional summary The detailed reconstruction of the individual steps by which Galileo established the ex mechanicis proof provided above can be summarized as follows: The construction of the proof took its point of departure from elaborations documented on folio 131 recto. By chance, Galileo's work on this folio resulted in a diagram that stimulated the central idea for the argument of the ex mechanicis proof. On the one hand, the diagram comprised paths of motion along appropriate chords. On the other hand, the construction resembled that of the bent lever proof, which directly allowed to infer the ratio of the mechanical moments on these chords. Galileo quickly sketched the idea, thus engendered on folio 151 recto, a folio which, as has been shown, he had already used before to take notes in conjunction with his work on 131 recto. This sketch was elaborated into a full, self-contained proof on folio 160 recto in what appears to be two consecutive steps and subsequently copied to 172 recto, one of the star group folios. Allegedly, at the same time, an entry stating the law of the inclined plane which Galileo had recruited to as a premise in the ex mechanicis proof was added to that group of folios.

8.2 The Role of the Ex Mechanicis Proof for the New Science of Motion

As has become apparent in the preceding section the sequence of steps by which Galileo established the ex mechanicis proof can be minutely reconstructed.[37] What remains to be done in the present chapter is to anchor this developmental sequence

---

[36]That the entry from 160 recto was copied to 172 recto and not vice versa is entirely clear as in the former the phrase "extensam dg usque ad circumferentiam," which does not occur in the latter, is written out in the former and then deleted. Otherwise only two spelling errors have been corrected, the punctuation was changed and "demonstratur" substituted for "probatur" when copying. The virtual identity of the two entries indicates that writing down the new entry was not engendered by the intention to introduce a change in the argument but rather served the insertion of the proof in its correct intended position in the collection of material being prepared at this time. Cf. Chap. 10.

[37]The leaps in the sequence of individual steps that, as reconstructed here, led Galileo to establish the ex mechanicis proof, are so small that it is hardly conceivable that he wrote more down in this respect than has been preserved. In other words, the ex mechanicis proof seems to provide one of the rare cases where Galileo's considerations have been preserved in the manuscript without gaps. It is, thus, entirely clear that Galileo is not recapitulating or exploring an existing argument here but that he is developing it.

in the overall picture and to assess the consequences thereby entailed for a more comprehensive understanding of the emergence of Galileo's new science of motion.

To indicate how Galileo's construction of the ex mechanicis proof, as reconstructed above, fits seamlessly with the overall interpretation established in this book some of the results that will be presented only in the second part of the book need to be anticipated here in a nutshell. As it turns out both the start and endpoint of the construction of the ex mechanicis proof can be directly linked to Galileo's earliest attempt to collect and systematize the results of his work swinging and rolling discussed in Chap. 10.

As has been argued, the argument of the ex mechanicis proof had first been conceived when Galileo, in the context of his investigation into swinging and rolling, determined the ratio of the mechanical moments on chords inscribed in a suitable manner in the same circle. From this ratio of mechanical moments, an application of Galileo's basic principle of inclined plane dynamics and of a kinematic proposition led directly to the final argument. In this manner, a statement about the spacio-temporal behavior of naturally accelerated motions was derived based on an assumption about the causes or active forces giving rise to these motions. Galileo, for the first time, had constructed a dynamical argument regarding natural acceleration.

How did Galileo come to include the concept of mechanical moment, with its origin in statics, into his considerations regarding the pheno-kinematics of naturally accelerated motion? How could he base his argument on the very dynamical assumption he had resorted to in *De Motu Antiquiora*, when he, however, still believed that motion along inclined planes was characteristically uniform? As will be demonstrated in Chap. 10, Galileo did not simply extend his earlier argumentative scheme to the new type of motion in an unreflected manner. Rather the understanding that the basic principle of inclined plane dynamics would apply in the case of naturally accelerated motion had been elicited by Galileo's analysis of his new results regarding the pheno-kinematics of naturally accelerated motions.

As part of his work on the star group folios Galileo had engaged in a consideration which suggested that in naturally accelerated motion along inclined planes of different inclination but equal height, the velocities would be in inverse proportion as the lengths of the inclined planes. Galileo recognized that this entailed that velocities be in proportion to the mechanical moments. He had, thus, recovered his basic principle of inclined plane dynamics, now valid, however, for naturally accelerated motion. This implied that the direction of argument could be reversed and that inferences regarding the pheno-kinematics of naturally accelerated motion could be based on this principle. Galileo programmatically formulated his intention to proceed in such a manner in an entry which commences with the phrase "[s]i hoc sumatur, reliqua demonstrari possent."[38]

---

[38] 172 verso, T1C. Cf. Chap. 10.

8.2 The Role of the Ex Mechanicis Proof for the New Science of Motion

Galileo compiled the content of the star group folios contemporaneously to his work on the problem of swinging and rolling, the results of which he indeed collected on this group of folios. It thus becomes apparent that with the construction of the ex mechanicis proof, Galileo had answered the task he had so programmatically set for himself in the si hoc sumatur note. Once it is thus acknowledged that Galileo must consciously have been seeking a way to apply the basic principle of inclined plane dynamics in an argument regarding naturally accelerated motion, it is no longer surprising that his construction on 131 recto must have struck him as providing a good starting point for doing so.

This is strongly confirmed by the final step in the sequence reconstructed above. As we have seen, once Galileo had completed a draft of the ex mechanicis proof on folio 160 recto, he copied and inserted this on folio 172 recto. Thus the newly found proof was inserted right in front of the considerations on the verso page of the folio which had shown that the basic principle of inclined plane dynamics could be assumed to hold in the case of naturally accelerated motion. As will be detailed in Chap. 10, these considerations must already have been noted when Galileo copied the ex mechanicis proof and squeezed it into the remaining space on the recto side of folio 172 whose verso side was already covered with content.

Moreover, the reason why Galileo inserted the newly found ex mechanicis proof into his collection on the star group folios and why he inserted it in exactly the place he did is apparent. After all, with the ex mechanicis proof Galileo had achieved part of the goal he had set for himself, namely, to "reliqua demonstrari" based on the dynamical assumption. As mentioned, folio 172 recto, where the ex mechanicis proof was copied to, already contained a proof of the law of chords which depended on both the length time proportionality and the law of fall as premises. The proof is entirely correct. Yet once the ex mechanicis proof had been copied to the folio, it was canceled out by Galileo by striking it through to indicate that it was to be replaced by the ex mechanicis proof in the deductive structure of perceived future publication. This indeed represented a step toward a solid foundation of the new science. As will be argued in Chap. 11, Galileo at that time had already found proofs for the length time proportionality, the law of chords and the law of fall, among them the proof of the law of chords contained on 172 recto, where each of these proofs was based on the respective other two propositions as premises. To escape the logical circle thus entailed alternatives proof for at least two of these propositions had to be constructed. It thus becomes entirely understandable why Galileo, once the ex mechanicis proof had been found, marked it as replacing the existing proof of the law of chords, perfectly valid but not apt for the foundation of a new science. This, as has been claimed, is what Galileo indicated by striking the older proof through. Yet Galileo's application of the traditional scheme of dynamical argumentation to his new insights concerning naturally accelerated motion was soon to reveal serious conflicts as will be demonstrated in the second half of this book.

That the ex mechanicis proof replaced an older proof of the same proposition, in particular, also shows that Galileo was aware of the law of chords and had even proven it before he came up with the ex mechanicis proof. In contrast to what is claimed by the standard interpretation, insight into the law of chords must thus have been achieved independently of the dynamical argument Galileo later devised for it. There would be no reason to further deliberate about the standard interpretation if it weren't among the few ideas regarding Galileo's early considerations on motion to have found rather wide acceptance and to have become an integral and largely unchallenged part of a more comprehensive understanding of the genesis of Galileo's new science.

According to the standard interpretation, Galileo derived the law of chords from premises of his *De Motu Antiquiora* theory while indeed still being convinced that the motion of bodies falling along inclined planes was in principle uniform. It is, moreover, commonly maintained that the law of chords was among the insights from which Galileo commenced his considerations regarding naturally accelerated motion on inclined planes. According to the standard interpretation, it is, thus, implied that Galileo must have had reasons to believe that his statement originally derived for uniform motion would equally be valid for naturally accelerated motion along chords, yet this is usually not addressed.[39]

In fact, the standard interpretation is usually neither exposed in detail nor substantiated with suitable evidence by those authors who adhere to it.[40] The tenets,

---

[39] Laird (2001, 267), for instance, states: "The conclusions of Galileo's new science of motion may have been purely kinematic and modelled after the statics of Archimedes, but they are ultimately founded upon ...the principle of the balance ...". As the ex mechanicis proof is indeed the only one of Galileo's arguments which directly relates the principle of balance to the kinematics of naturally accelerated motion, this is a perfect example of how repeatedly, the ex mechanicis proof has been assigned a crucial role in Galileo's new science with rather broad-brushstroke claims. Such claims, as a rule, are not the result of a careful analysis of the *Notes on Motion* but rather appear to be based on an understanding drawn mainly from Galileo's correspondence.

[40] Numerous studies have touched upon the questions related to the ex mechanicis proof and its genesis and only those that have examined the proof and its relation to the general framework of the new science of naturally accelerated motion in some detail are briefly surveyed here. Wisan refers to the law of chords as the oldest of the propositions that Galileo published in the *Discorsi* and claims that the ex mechanicis proof was the first proof Galileo had found. In her discussion of the proof, she hesitates to decide whether it was conceived before or after Galileo's conceptual shift but at a later point of her analysis assumes the former. She argues somewhat circularly when she claims the least time propositions must have been drafted early, because they exclusively employ the law of chords as a premise, just to conclude some pages later that the law of chords must be the earliest proposition exactly because it is used as a premise in the proofs of these propositions. According to Wisan the ex mechanicis proof must stem from the "quite early" Paduan period, and she expresses little doubt that it is, in fact, datable to 1602, based on the letter to Guidobaldo del Monte. Cf. Wisan (1974), in particular pages 162–163. Galluzzi (1979, 266) likewise leaves the question of whether the ex mechanicis proof was conceived before or after Galileo's paradigm shift open but leans toward the second alternative. Drake (1978, 67) held that the ex mechanicis proof was originally anchored in the framework of the *De Motu Antiquiora*: "Given Galileo's two erroneous conclusions in *De motu*, that acceleration may be neglected and that their 'speeds' of descent (regarded as constant) along two different incline planes of equal height are inverse to the lengths of the planes, any Euclidean geometer could easily reach Galileo's theorem [the law of chords] by inspection of the diagram." Renn follows the standard interpretation, when he states, for instance, that the law of chords is "a theorem on motion along inclined planes which directly follows from his theory in *De Motu*..." Some pages later he reiterates that "Galileo had obtained the first presupposition, The Isochronism of Chords [law of chords], as a direct consequence

8.2 The Role of the Ex Mechanicis Proof for the New Science of Motion

upon which the standard interpretation is founded and which have contributed to its general acceptance, each of them problematic in itself, can be summarized as follows:

- The argument of the ex mechanicis proof itself is indifferent to whether motion is considered uniform or naturally accelerated.[41] Yet the kinematic proposition employed as a premise in the argument was believed to presuppose uniform motion. Thus, it is surmised that the argument originally emerged in the context of the *De Motu Antiquiora* theory.
- The argument of the ex mechanicis proof is thought to represent the line of reasoning which first led Galileo to his insight into the law of chords.
- Based on a letter to Paolo Sarpi from the end of 1604, it is speculated that Galileo first "discovered" the law of fall around that time and that this marks the onset of his investigation of natural acceleration. As the ex mechanicis proof was referred to by Galileo in his 1602 letter, the assumption is nourished that the proof and indeed Galileo's entire research program alluded to in the letter must have predated the conceptual shift toward the assumption of natural acceleration.

---

of the theory of motion along inclined planes in *De Motu* ..." Naylor (2003, 156–159), when addressing the "remarkable developments" revealed by Galileo's letter to Guidobaldo del Monte in 1602, expresses his belief that the ex mechanicis proof must hence have succeeded Galileo's conceptual shift toward the assumption of natural acceleration but states that there is a "complete lack of direct evidence" for these developments in the sources. (Humphreys 1967, 234), without any consultation of the manuscript sources, speculates about a way in which the law of fall may have been inferred as an "out-growth of his [Galileo's] juvenile speculations on dynamics" and states with regard to the law of chords that it is "plausible to assume that the proof he had was the dynamical one given in *Two New Sciences* and based on the work done around 1598 for the book *On Mechanics*." Hooper (1998), who in his Ph.D. thesis has quite carefully examined parts of the *Notes on Motion*, essentially follows Humphreys' account. Hooper (1992) has included a discussion of the manuscript sources in his analysis of the conceptual underpinnings of the ex mechanicis proof, but remains conspicuously silent about whether Galileo, when he first came up with his ex mechanicis proof, still regarded motion on inclined planes as essentially uniform or already as naturally accelerated. Souffrin (2001) finally, despite having convincingly argued that the assumption that led people to believe that the law of chords was conceptually anchored in the *De Motu Antiquiora* theory in the first place was erroneous, sticks to the assumption that the ex mechanicis proof predated Galileo's conceptual shift toward assuming natural acceleration.

[41] Put anachronistically, for two uniformly accelerated motions from rest completed in the same time the average velocities are in the same ratio as the accelerations. Hence distances traversed in naturally accelerated motions will likewise be traversed in the same time by two uniform motions, provided their velocities are in the same ratio as the constant accelerations of the accelerated motions.

That the kinematic proposition employed in the ex mechanicis proof presupposes uniform motion has been refuted convincingly by Souffrin who, as mentioned above, has compellingly demonstrated that the proposition was considered by Galileo and his contemporaries to be valid and applicable to any motion. It was indeed part of the "operative definition" of a universally valid concept of a kinematic velocity.[42]

The second tenet entails a difficulty which is typically overlooked. While an argument can be made along the line of reasoning of the ex mechanicis proof to determine distances on different planes traversed by the same movable body from rest in equal time, it is not very likely that such an argument would result in the specification of these distances as chords inscribed in the same circle sharing either its vertex or foot-point, i.e., in the law of chords. In general, even though it is not entirely impossible starting from the *De Motu Antiquiora* premises to arrive at the law of chords, it is also far from being trivial, as, for instance, Drake suggested when he stated that "any Euclidean geometer could easily reach Galileo's theorem [the law of chords]..."[43]

Finally, the assumption that the conceptual shift toward assuming naturally accelerated motion took place in 1604, which rests primarily on a particular understanding of Galileo's remarks in a letter sent to Paolo Sarpi in October 1604, has been incontrovertibly refuted here.[44] The research program Galileo had alluded to in the letter of 1602 to Guidobaldo del Monte, as has amply been demonstrated, has left abundant traces in the *Notes on Motion* which unmistakably manifest that at that time Galileo already perceived of motions of heavy bodies along inclined

---

[42] Cf. Chap. 2.

[43] Drake (1978, 67).

[44] That the standard dating of the discovery of the law of fall to the year 1604 rests primarily on Galileo's remarks in the letter sent to Paolo Sarpi in October 1604, and not on a careful analysis of other sources such as the *Notes on Motion* has been argued by Damerow et al. (2001). Resorting to the work of Drake as an example, they state: "But even after Drake had extensively studied Galileo's working papers, and after repeatedly changing his views on Galileo's discoveries, he rather accommodated the dating of Galileo's manuscripts to the standard dating than the other way around (Damerow et al. 2001, 304–305)." Analyzing the material related to Galileo's insight into the parabolic trajectory of projectiles the authors state that "according to common historiographic criteria, Galileo must be credited with having made this discovery already as early as 1592" and that the law of fall "was merely a trivial consequence of this discovery (Damerow et al. 2001, 300)." The authors question the notion of discovery and call instead for a careful reconstruction of the "network of interdependent activities which only as a whole make an individual step understandable as a meaningful 'discovery'." Their stipulation has greatly influenced the current work.

planed as naturally accelerated.[45] An appropriate reinterpretation of the letter to Paolo Sarpi will be provided in Chap. 11.

Crucially, one of the central claims of the standard interpretation, namely, that Galileo first discovered the law of chords by the very argument preserved in the ex mechanicis proof, must be rejected. Thus a new question regarding the origin of Galileo's insight into the law of chords, which can no longer be sought in the line of argument of the ex mechanicis proof, has opened up and a plausible, albeit speculative, answer is provided in the succeeding section.

8.3 The Origin of the Law of Chords as a Heuristic Assumption

According to the interpretation advanced, not only was the law of chords known to Galileo before he conceived of the ex mechanicis proof argument but it stood at the very beginning and played a crucial role in his investigations into the relation between swinging and rolling. In fact, if it hadn't been for the belief that fall along chords appropriately inscribed into the same circle was isochronous, the relation between the two types of motion Galileo perceived could not have been discerned by him in the first place. The question of how Galileo first came to realize the law of chords and how he came to be convinced of its soundness, then gains urgency with regard to an understanding of the genesis of the new science, even more so in view of the fact that the answer traditionally given to this question has now been ruled out.

---

[45] Assuming the onset of Galileo's investigations into naturally accelerated motion to be 1604 has forced a number of authors to adopt rather unorthodox assumptions in their exegesis of the *Notes on Motion*. They acknowledge that some of the considerations documented in the manuscript conform well to Galileo's announcements in the letter of 1602 and indeed date these considerations to this time. Based on the understanding that Galileo began to conceive of natural acceleration only in 1604, they are thus forced to assume that such considerations must be anchored in the understanding that motion along inclined planes is essentially uniform. Confronted with indications of the contrary, the additional assumption is usually invoked that years later Galileo must have returned to his older considerations and reassessed them from the perspective of his new conceptualization, adding new entries to older ones. A good example of this is Wisan. As has been done here, she as well has located the point of departure for Galileo's construction of the ex mechanicis proof in the central diagram on folio 131 recto. She too dates this work to approximately 1602 but assumes that at this time Galileo still assumed motion along inclined planes to be characteristically uniform. At the same time she of course realized that the diagram on the page in the form of the points v and o, which mark mean proportionals, embodies an application of the law of fall. Thus in order to salvage her interpretation she simply but somewhat artificially claimed that these points were added much later in an alleged second attempt at the same problem after 1604: "It may be conjectured then that the central part of this fragment exhibits a very early stage in the investigation of the brachistochrone and that the first sentence (together with the smaller circle and the points O and V) is a later addition stemming from a new attack that makes use of the corollary on mean proportionals [i.e., the law of fall] (Wisan 1974, 177)".

As it turns out, this question cannot be answered conclusively based on the evidence preserved in the *Notes on Motion*. As is the case for the law of fall, Galileo's insight into the law of chords also seems to predate the earliest entries in the manuscript. Spelling out a way in which Galileo could potentially have conceived of the law of chords therefore remains speculative. It is a worthy exercise; however, as it demonstrates that starting out from a position comparable to the one Galileo would have found himself in around the turn of the century, it would have been possible to rather directly chance upon the law of chords.

Interspersed in the Third Day of the *Discorsi* is a scholium devoted to the discussion of what will be referred to as *isochrone curves*. Isochrone curves are defined by the points reached by a given type of motion from a given starting position in the same time.[46] The isochrone curve of uniform motion trivially is a circle around the starting point of motion. As Galileo expressed it:

> If, from any point ... straight lines be drawn extending indefinitely in all directions, and if we imagine a point to move along each of these lines with constant speed, all starting from the fixed point at the same instant and moving with equal speeds, then it is clear that all of these moving points will lie upon the circumference of a circle which grows larger and larger ... (EN VIII, 224, trans. Galilei et al. 1954.)

Next Galileo discusses the isochrone curve of naturally accelerated motion from rest:

> But imagine a ... point ... [from] which are drawn lines inclined at every angle and extending indefinitely; imagine also that heavy particles descend along these lines, each with a naturally accelerated motion and each with a speed appropriate to the inclination of its line. If these moving particles are always visible, what will be the locus of their positions at any instant? Now the answer to this question surprises me, for I am led by the preceding theorems to believe that these particles will always lie upon the circumference of a single circle, ever increasing in size as the particles recede farther and farther from the point at which their motion began. (EN VIII, 224, trans. Galilei et al. 1954.)

Stating, as Galileo does, that the isochrone curve of naturally accelerated motion is the circle with motion starting from rest at its zenith is of course equivalent to the law of chords formulated for motion along chords starting from the highest point of the circle. The scholium could thus all too easily be dismissed as being concerned with a curious but otherwise irrelevant observation. Yet manuscript evidence shows that Galileo systematically pursued the question regarding the isochrone curves of different types of motion in the early phase of his investigation into natural acceleration.

Indeed, two constructions can be identified on folios of the *Notes on Motion* that show that Galileo also tried to find the isochrone curve of naturally accelerated motion starting with an initial velocity, i.e., of a motion which was preceded by fall through a certain distance. Constructions on 147 verso and on 155 recto show arrays

---

[46]The points reached from a given starting position in the same time of course form a surface. However, due to the rotational symmetry of Galileo's problems it suffices to study the problem in two dimensions, i.e., to seek isochrone curves.

8.3 The Origin of the Law of Chords as a Heuristic Assumption

of inclined planes with increasing inclination, all emanating from the same point above which a vertical indicates the distance initially fallen through. As discussed in Chap. 11, the construction on folio 147 verso must have been composed at about the same time that also the content of the star group folios was written down. The construction was left incomplete, yet it is repeated on 155 recto where the points reached on the inclined planes after initial fall in the same time as the time consumed during this initial fall have indeed been constructed and marked on the inclined planes. This indicated to Galileo that in this case, the isochrone curve had a complicated shape and was thus not specifiable with reference to an elementary curve in the Euclidean tradition.

Above all, the considerations preserved on 155 recto show that Galileo was already interested in the question of isochrone curves in the earliest phase of his investigations into naturally accelerated motion. In the *Discorsi*, Galileo credited the fact that the isochrone curve of naturally accelerated motion from rest is the circle to the "preceding theorems," i.e., in particular, to the law of chords. The question regarding the shape of this curve can of course, however, be asked before being aware of the answer provided by the law of chords. How would it be answered then?

If Galileo indeed posed this question, he would, as briefly argued in the following, almost inevitably have arrived at the law of chords.[47] Suppose an array of inclined planes is drawn, starting from a common point of origin, whose inclination varies from 0 to 90 degrees measured against the horizontal and the question is asked which points on those various inclined planes are reached in equal time by a body moving along them in naturally accelerated motion. Possible solutions are determined, to a great extent, by boundary conditions resulting from plausible assumptions about the motions under scrutiny. A body placed on the horizontal should not move at all, and hence the space covered in a given time should be zero on the horizontal. Furthermore, a body should move further in the same time on a more steeply inclined plane, i.e., the distance covered should increase monotonously with increasing inclination to reach a maximum on the vertical.

Only two continuous, smooth curves from the traditional canon of curves apply at all as candidates for a geometrical construction embodying these constraints; the Archimedean spiral around the common starting point with a horizontal tangent at this point, and a circle where the starting point of motion lies at the vertex of the circle (compare Fig. 8.6). In the first case, the distance covered in a given time would depend linearly on the angle of inclination according to the definition of the spiral. In the second case, it would depend on the sine of that angle. In contrast to the spiral, the circle could be constructed by means of a compass and a ruler, and it was

---

[47]Once Galileo had become convinced that the spacio-temporal behavior of the motions he had treated as uniform earlier was correctly specified by the law of fall, what would have been missing for a more comprehensive treatment of naturally accelerated would be an argument allowing to relate naturally accelerated motions along inclined planes of different inclination. Such an argument is of course exactly what an investigation of the isochrone curve of naturally accelerated motion can potentially provide.

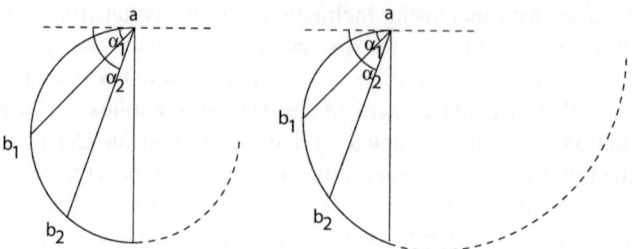

Fig. 8.6 Realizing the requirement of a monotonous relation between the angle α and the length of the inclined plane ab which must hold for the iso-temporal surface by a circle (left) and a spiral (right)

traditionally viewed as the more elementary curve. It would thus not be surprising if Galileo, in such a scenario, had preferred the circle as a heuristic solution to the problem.[48] It is thus entirely perceivable that Galileo first came to accept the law of chords as a heuristic assumption by an argument along this line.

If Galileo first inferred the law of chords along the line of argument sketched or whether he based it on a similar plausibility argument, it would, in any case, initially have represented a rather weak heuristic assumption. It seems hard to reconcile this with Galileo's confident use of the law of chords in his research on swinging and rolling, in particular, in the pendulum plane experiment which, as has been demonstrated, embraced the law of chords not only in the theoretical evaluation of the results but already in its design. The assumption that Galileo must have probed the law of chords empirically is thus almost inevitably thrusted upon us.

As already mentioned in Chap. 4, the experimental setup of the pendulum plane experiment could, with minor modifications, indeed be used to empirically test the law of chords. Mersenne, after Galileo had first communicated the law of chords to a wider audience in the *Dialogue*, in fact conducted an experiment to test the law.[49] Whereas it thus seems exceedingly plausible that Galileo too probed the law of chords experimentally, it seems very unlikely that insight into the law of chords was gained as the result of mere exploitative experimentation. Unaware of the law

---

[48] In the seventeenth century a circle was recognized as a geometrical curve, whereas the spiral was seen as belonging to the class of the mechanical curves. Geometrical curves could be determined from one motion and produced with the conventional tools, compass and ruler. The mechanical curves were defined by two or more motions and hence required more complex instruments to produce them. Descartes (2001) vehemently attacked the distinction. Cf. Büttner et al. (2003, 5–6).

[49] The experiment is described in Mersenne (1963, 111) and analyzed in detail in Settle (1966) and Naylor (2003). Naylor (1974, 121) claims that "it is almost certain that from 1602 onward Galileo tested this statement [the law of chords]." Using an inclined plane he had built to recreate Galileo's projection experiments, he also conducted an experimental test of the law of chords. His results compare to the ones given by Mersenne, and "it seems credible to allow that Galileo could have obtained observations similar to these (Naylor 1974, 122)." As the difference between rolling and free falling shows in the results, these observations, however, do not unreservedly support the law of chords, at least as long as a free falling motion is compared to motion along an inclined plane.

of chords, Galileo may have experimentally determined the starting positions on planes of varying inclinations such that the motions observed were completed in equal time. Yet to infer the law of chords from such experimentally determined positions would have meant recognizing that they all lie on the same circle, which given the inevitable experimental errors, does not seem likely.

As stated in the beginning, the argument by which Galileo originally gained insight into the law of chords seems to have left no trace in the *Notes on Motion*. Likewise, an experiment to confirm the law of chords, if one was conducted at all, is not documented. Thus, ultimately, we may never be able to tell how Galileo arrived at the law of chords as a heuristic assumption, let alone to specify the moment when exactly this happened. As argued in Chap. 3, Galileo quite obviously mentioned the law of chords for the first time to Guidobaldo del Monte in his letter of 1602. As this letter had been preceded by at least one other letter from Galileo to Guidobaldo written shortly before, this may suggest that when the antecedent letter was written Galileo was not yet aware of the law of chords or else that he may already have been aware of it but felt that his assumption was not yet sufficiently supported to be communicated to Guidobaldo del Monte.

Galileo's research on swinging and rolling essentially started out from merely two assumptions about naturally accelerated motion, the law of fall and the law of chords. Only in the course of his investigation were these two initial assumptions expanded into a host of insights which later became the subject matter of a new science of motion. It was only when the law of fall, which Galileo had accepted as the law governing the spacio-temporal behavior of falling motion likely before 1600, was supplemented by the law of chords, that the challenging problem could be realized which eventually assisted the birth of a new science. Given the important role the law of chords played in the emergence of the new science of motion we certainly are better off with a plausible conjecture as to how Galileo first came to accept the proposition than with merely declaring that by 1602 he was aware of it.

8.4 Summarizing Conclusion

In the present chapter, the genesis of the ex mechanicis proof of the law of chords has been reconstructed. To this end, the argument of the proof as published by Galileo in the *Discorsi* has first been analyzed. This has allowed the identification of a number of entries in the *Notes on Motion* as pertaining to his attempt to construct the ex mechanicis proof. Almost all of the entries have already been identified as related to the ex mechanicis proof in the existing literature. Yet their reanalysis has revealed previously unnoticed relations among them allowing for a reconstruction of the developmental sequence, from the first decisive idea to Galileo's drafting of a polished publishable version of the proof.

It has been revealed that the proof was constructed when Galileo first started to collect and systematize the results achieved as a consequence of his investigation into the relation of swinging and rolling on the star group folios. Galileo's analysis of the results achieved so far had suggested that his principle of inclined dynamics, previously applied in the context of the *De Motu Antiquiora*, would also be applicable in the case of naturally accelerated motion. On the other hand, Galileo's attempts to construct a proof of the law of the broken chord had resulted in a construction which, due to its similarity to the geometrical arrangement of Galileo's older bent lever proof, did lend itself to an argument by means of which the ratio of mechanical moments on inclined planes inscribed as chords in a circle could be inferred. This resulted in the proof idea for the ex mechanicis proof. From a first, immediate sketch of the idea, a full-fledged version of the ex mechanicis proof was elaborated. The proof thus established was copied to one of the star group folios to replace an already existing proof of the law of chords.

As the ex mechanicis proof could be considered a proof based on fundamental principles, in contrast to Galileo's earlier proof of the law of chords, a first step toward providing the insights on naturally accelerated motion with a foundation and thus a first step toward a new science had been ventured. It may have been while under the impression of this initial success that Galileo not only mentioned the proof in his letter to Guidobaldo del Monte but also, somewhat exaggeratedly, announced that he had proven everything thus far without transgressing the terms of mechanics.[50] Galileo's success was rather short-lived. As will be argued in the second part of this book, a dynamical foundation, which looked so promising for the time being, was indeed soon to become a challenging problem in Galileo's attempt to find viable pathways through the conceptual labyrinth toward the foundation of the new science of motion.

As it is no longer possible to maintain that it was an argument along the line of reasoning preserved in the ex mechanicis proof by which Galileo initially gained insight into the law of chords, an alternative conjecture has been presented. It may have been the question regarding the isochrone curve of naturally accelerated motion that first prompted Galileo to accept the law of chords as a heuristic assumption, one which Galileo successively very likely probed and satisfactorily confirmed in an experiment.

---

[50] Alternatively, it has been claimed that Galileo's emphasis on having proven everything thus far without transgressing the terms of mechanics was in fact a reaction to Guidobaldo del Monte's position that mechanical principles "did not only comprise the theory as it is exposed in the ancient texts but also a strict correspondence between theory and practical experience," an attitude which "led him to be skeptical with regard to studies involving motion." Hence Galileo needed to convince Guidobaldo that "in spite of the novelty of the subject for traditional mechanics, he is still adhering to the principles of this mechanics which had been the starting point of their exchange (Damerow et al. 2001, 358)."

References

Büttner, J. (2009). Wie auf Erden, so im Himmel: Zwei Welten – eine Physik. *Sterne und Weltraum,* 4: 52–62.

Büttner, J., Renn, J., Damerow, P., & Schemmel, M. (2003). The challenging images of artillery: Practical knowledge at the roots of the scientific revolution. In W. Lefèvre, J. Renn, & U. Schoepflin (Eds.), *The power of images in early modern science* (pp. 3–28). Basel: Birkhäuser.

Damerow, P., Renn, J., & Rieger, S. (2001). Hunting the white elephant: When and how did Galileo discover the law of fall? In J. Renn (Ed.), *Galileo in context* (pp. 29–150). Cambridge: Cambridge University Press.

Damerow, P., Freudenthal, G., McLaughlin, P., & Renn, J. (2004). *Exploring the limits of preclassical mechanics.* New York: Springer.

Descartes, R. (2001). *Discourse on method, optics, geometry and meteorology* (Reprint ed.). Indianapolis: Hackett Publishing Company Inc.

Drake, S. (1978). *Galileo at work: His scientific biography.* Chicago: University of Chicago Press.

Galilei, G., Crew, H., & Salvio, A.D. (1954). *Dialogues concerning two new sciences.* New York: Dover Publications.

Galluzzi, P. (1979). *Momento.* Rome: Ateneo e Bizzarri.

Giusti, E. (2004). A master and his pupils: Theories of motion in the Galilean school. In C.R. Palmerino & J.M.M.H. Thijssen (Eds.), *The reception of the Galilean science of motion in seventeenth-century Europe* (Boston Studies in the Philosophy of Science, Vol. 239, pp. 119–135). Dordrecht: Kluwer.

Hooper, W.E. (1992). *Galileo and the problems of motion.* Dissertation, Indiana University.

Hooper, W. (1998). Inertial problems in Galileo's preinertial framework. In P. Machamer (Ed.), *The Cambridge companion to Galileo.* Cambridge/New York: Cambridge University Press.

Humphreys, W. (1967). Galileo, falling bodies and inclined planes: an attempt at reconstructing Galileo's discovery of the law of squares. *British Journal for the History of Science, 3*(11), 225–244.

Laird, W.R. (2001). Renaissance mechanics and the new science of motion. In J. Montesinos & C. Solís (Eds.), *Largo campo di filosofare: Eurosymposium Galileo 2001* (pp. 255–267). La Orotava: Fundación Canaria Orotava de Historia de la Ciencia.

Mersenne, M. (1963). *Harmonie universelle, contenant la théorie et la pratique de la musique.* Paris: Centre national de la recherche scientifique.

Naylor, R.H. (1974). Galileo and the problem of free fall. *The British Journal for the History of Science, 7*(2), 105–134.

Naylor, R.H. (2003). Galileo, Copernicanism and the origins of the new science of motion. *The British Journal for the History of Science, 36*(2), 151–181.

Settle, T. (1966). *Galilean science: Essays in the mechanics and dynamics of the Discorsi.* Ph.D. thesis, Cornell University.

Souffrin, P. (1990). Galilée et la tradition cinématique pré-classique. La proportionnalité momentum-velocitas revisitée. *Cahier du Séminaire d'Epistémologie et d'Histoire des Sciences, 22,* 89–104.

Souffrin, P. (2001). Proceedings of the XXth international congress of history of science: (Liège, 20–26 July 1997) Bd. 08: Medieval and classical traditions and the Renaissance of physico-mathematical sciences in the 16th century. In P.D. Napolitani & P. Souffrin (Eds.), *De diversis artibus* (Vol. 50, p. 149). Turnhout: Brepols.

Souffrin, P., & Gautero, J.L. (1992). Note sur la démonstration 'mécanique' du théorème de l'isochronisme des cordes du cercle dans les 'Discorsi' de Galilée. *Revue d'histoire des sciences et de leurs applications, 45,* 269–280.

Torricelli, E. (1919). De motu gravium naturaliter descendentium, et projectorum. In G. Loria & G. Vassura (Eds.), *Opera di Evangelista Torricelli* (Vol. 2). Faenza: Montanari.

Wisan, W.L. (1974). The new science of motion: A study of Galileo's De motu locali. *Archive for History of Exact Sciences, 13,* 103–306.

Chapter 9
Whatever Happened to Swinging and Rolling: Faint Echoes and a Late Insight

Whatever happened to Galileo's research on swinging and rolling? As will be argued, it went on to bear a new science of motion. When Galileo published this new science, the traces of its origin in his research agenda on swinging and rolling were almost completely obliterated. However, before discussing how Galileo started to transform his new insights into a new science in the second half of this book, some more immediate, albeit late consequences of his original research on swinging and rolling, which yielded those insights in first place, will be discussed in this last chapter of the first part of this book.

The theme of the pendulum provides an undertone to the entire *Discorsi*. The interpretation advanced in the preceding chapters has, moreover, put us in the position to recognize that many of the propositions drafted around 1602 in the context of Galileo's research on swinging and rolling were later included in the *Discorsi*, some of them verbatim. Yet these propositions exclusively regard the pheno-kinematics of naturally accelerated motion on inclined planes. They thus do not provide their reader with a direct clue to as to the research program they originated from and which Galileo summarized in his letter to Guidobaldo del Monte in 1602.

In his examination of the relation between swinging and rolling, Galileo had encountered ever bigger problems causing him at some point, as shown in Chap. 6, to discontinue this work entirely. When he finally finished off his *Discorsi* some 35 years later, he had apparently not been able to advance his research program beyond the point where he had abandoned it in or shortly after 1602. The *Discorsi*, in particular, mention the law of the pendulum and the isochronism of the pendulum. These are presented, however, as empirical facts. Galileo makes no attempt to provide a proof for either of them or to relate them to the motion of fall along inclined planes in the way, as we now know, he had attempted to do much earlier. Moreover, both laws are introduced on the First Day of the discussion, i.e., not in direct connection with the new science of motion presented on the Third Day.

As discussed, the idea that pendulum motion would be kinematically equivalent to fall along a corresponding arc was central to Galileo's research agenda. The latter motion, he believed, could be approximated by motion along polygonal paths made up of a series of inclined planes. The motion along such paths could in turn be studied, harking back to his assumptions about naturally accelerated motion. This fundamental idea of Galileo's early research resonates in the *Discorsi* in but one argument, the so-called brachistochrone argument.[1]

In Sect. 9.1, it will be demonstrated that the brachistochrone argument, far from being a mainspring for Galileo's considerations on naturally accelerated motion as has previously been claimed, must be considered to be another spin-off from his broken chord approach. It was likely a rather late attempt to make his considerations regarding fall along an arc bear fruit directly for his new science. With regard to the brachistochrone argument, it will be demonstrated that Galileo had reason to surmise that his argument was flawed. He published it nevertheless, presumably in a leap of faith, convinced that the path traversed in least time between two points was indeed the arc as he had argued. If this had been the case, it should eventually have been possible to set the argument straight and construct an unblemished proof. Yet, as we know, the brachistochrone is not an arc of a circle, and from a modern perspective, it is recognized that Galileo's argument rests on a false assumption.

The inclusion of the brachistochrone argument in the *Discorsi* was, however, not the absolute final episode in Galileo's engagement with the problem of the relation between swinging and rolling. In the second half of this chapter,

---

[1] At one point in the discussion in the First Day of the *Discorsi* it is first asked whether pendulums do in fact swing isochronously. From this, Salviati immediately jumps to recounting that Galileo had demonstrated that motion along appropriate chords subtending arcs of the same circle was isochronous, i.e., the law of chords. After this, he goes on to state that the isochronism of the pendulum is confirmed by experience ("mostra...l'esperienza"), motion along chord and motion along the arc spanned are directly compared, and it is emphasized that, counterintuitively, motion along the arc, despite being made over the longer distance, is completed in shorter time ("E quanto al primo dubbio, che è, se veramente e puntualissimamente l'istesso pendolo fa tutte le sue vibrazioni, massime, mediocri e minime, sotto tempi precisamente eguali, io mi rimetto a quello che intesi già dal nostro Accademico; il quale dimostra bene, che 'l mobile che descendesse per le corde suttese a qualsivoglia arco, le passerebbe necessariamente tutte in tempi eguali, tanto la suttesa sotto cent' ottanta gradi (cioè tutto il diametro), quanto le suttese di cento, di sessanta, di dieci, di due, di mezzo e di quattro minuti, intendendo che tutte vadano a terminar nell'infimo punto, toccante il piano orizontale. Circa poi i descendenti per gli archi delle medesime corde elevati sopra l'orizonte, e che non siano maggiori d'una quarta, cioè di novanta gradi, mostra parimente l'esperienza, passarsi tutti in tempi eguali, ma però più brevi de i tempi de' passaggi per le corde; effetto che in tanto ha del maraviglioso, in quanto nella prima apprensione par che dovrebbe seguire il contrario: imperò che, sendo comuni i termini del principio e del fine del moto, ed essendo la linea retta la brevissima che tra i medesimi termini si comprende, par ragionevole che il moto fatto per lei s' avesse a spedire nel più breve tempo; il che poi non è, ma il tempo brevissimo, ed in consequenza il moto velocissimo, è quello che si fa per l'arco del quale essa linea retta è corda. (EN VIII, 19–20)"). Besides the brachistochrone argument, nowhere in the *Discorsi* does Galileo's early research agenda on swinging and rolling resonate so clearly as in this passage.

starting with Sect. 9.2, it will be shown how, stimulated by Giovanni Battista Baliani's *De motu naturali gravium solidorum* published in 1638, only a few months before the *Discorsi*, Galileo briefly resumed his work on swinging and rolling. In his book, Baliani had derived the laws of naturally accelerated motion based on assumptions about the properties of pendulum motion.[2] He had thus seemingly accomplished what Galileo had failed to achieve despite his impassioned struggle. Galileo must have studied the book carefully as he indeed formulated a critique of Baliani's approach in a manuscript that hitherto has remained virtually unstudied.

Analyzing the document, it will be shown that Galileo's critique was based on the very understanding he himself had gained in his own investigations into the relation between swinging and rolling. It will, moreover, be shown that as a result of his reflection upon Baliani's approach, Galileo realized how he could advance the problem he had put aside more than 35 years previously. In fact, inspection of Baliani's approach demonstrated that the question regarding the isochronism of the pendulum could be decoupled from that regarding the law of the pendulum. Thus, Galileo finally managed to come up with a proof of the law of the pendulum based on his laws of naturally accelerated motion and particular assumptions concerning the relation between swinging and rolling. This proof was not only included in the draft of a reply to Baliani but was also added into the margin of Galileo's hand copy of the *Discorsi* and thus earmarked for inclusion in a second revised edition of the book.[3]

9.1 The Brachistochrone Argument

In the *Discorsi*, in a scholium immediately succeeding the law of the broken chord, Galileo argued that "the path of quickest descent from one point to another is not the shortest path, namely, a straight line, but the arc of a circle."[4] Galileo had already made this claim in the *Dialogue*, there, however, without providing an argument for it.[5]

Galileo's brachistochrone argument has already been recognized in the literature as being problematic, and it has been claimed that it contains a false conclusion, false even with respect to Galileo's own framework. Based on a reanalysis of the argument, it will be addressed whether Galileo was in fact aware of its problematic character. It will thus be shown that in all likelihood, Galileo had himself already recognized the argument as being deficient. In addition, this analysis will refute the

---

[2] See Baliani (1638).
[3] We briefly touched upon this episode in Büttner et al. (2004, 111–112).
[4] EN VIII, 263, trans. Galilei et al. (1954).
[5] Cf. EN VII, 476.

strong claim, as has been made in particular by W. Wisan, that Galileo's search for the brachistochrone argument substantially influenced the early development of the new science of naturally accelerated motion.[6]

In the *Discorsi*, the somewhat sketchy argument reads:

> From the preceding, it is possible to infer that the path of quickest descent [lationem omnium velocissimam] from one point to another is not the shortest path, namely, a straight line, but the arc of a circle.
>
> In the quadrant $BAEC$, having the side BC vertical, divide the arc AC into any number of equal parts, AD, DE, EF, FG, GC, and from C draw straight lines to the points A, D, E, F, G, draw also the straight lines AD, DE, EF, FG, GC. Evidently descent along the path ADC is quicker than along AC alone or along DC from rest at D. But a body, starting from rest at A, will traverse DC more quickly than the path ADC, while, if it starts from rest at A, it will traverse the path DEC in a shorter time than DC alone. Hence descent along the three chords, $ADEC$, will take less time than along the two chords ADC. Similarly, following descent along ADE, the time required to traverse EFC is less than that needed for EC alone. Therefore descent is more rapid along the four chords $ADEFC$ than along the three $ADEC$. And finally a body, after descent along $ADEF$, will traverse the two chords, FGC, more quickly than FC alone. Therefore, along the five chords, $ADEFGC$, descent will be more rapid than along the four, $ADEFC$. Consequently, the nearer the inscribed polygon approaches a circle, the shorter is the time required for descent from A to C. (EN VIII, 263, trans. Galilei et al. (1954).)

Using numbers, as indicated in Fig. 9.1 in brackets, to identify the points on the arc instead of the letters Galileo used, the argument can be formalized more easily.[7]

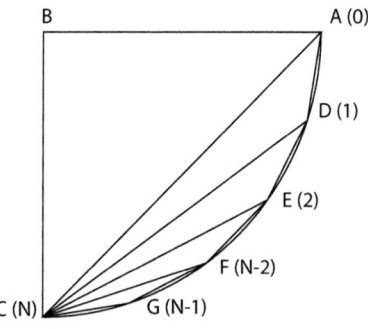

Fig. 9.1 Diagram associated with the Scholium succeeding Theorem XXII, Proposition XXXVI of the Second Day of the Third Book of the *Discorsi*. The lettering in the square brackets has been added to aid the discussion in the present section (cf. EN VIII, 263)

---

[6]In her study of the *Notes on Motion*, Wisan (1974) somewhat reservedly claimed: "Possibly then it was a traditional problem, at least among some mechanicians of the time, but one receiving little attention in written works. In fact it is possible that it was the brachistochrone which originally turned Galileo's attention to motion on inclined planes." Yet in a later paper she lost her reservations and firmly and explicitly stated: "To solve the brachistochrone, however, Galileo had to create a new science, one based on new and fundamental propositions (Wisan 1984, 270)." More recently Palmieri has followed Wisan's lead in assuming that the search for brachistochrone played a central role in Galileo's considerations documented in the *Notes on Motion*: "Galileo sought to prove his conjecture [that the arc of the circle is the brachistochrone], as the many remnants of related theorems and problems surviving in the folios of Manuscript 72 suggest. (Palmieri 2009, 47)".

[7]The analysis given leans on an analysis of the argument in Ariotti (1971/1972).

9.1 The Brachistochrone Argument

For an arc divided by $N+1$, junction points into N conjugate chords of equal length, where the highest point is indexed to 0, the argument can be summarized as follows:
By law of the broken chord:

$$t(01N) < t(0N). \tag{9.1}$$

Probably ("verisimile est")[8]:

$$t(12N|0) < t(1N|0) \tag{9.2}$$

and hence:

$$t(012N) < t(01N). \tag{9.3}$$

By above:

$$t(012N) < t(01N) < t(0N). \tag{9.4}$$

Steps (9.2) and (9.4) can be applied in an iterative fashion:
Again similar to step (9.1) above ("[v]erum similiter"):

$$t(n\ n+1\ N|0) < t(n\ N|0) \tag{9.5}$$

and hence:

$$t(0 \ldots n\ n+1\ N) < t(0 \ldots n\ N|0). \tag{9.6}$$

By above:

$$t(0 \ldots N) < t(0 \ldots n-1\ N) < t(0 \ldots n-2\ N) < t(0N). \tag{9.7}$$

The argument does not depend on the actual number of chords N, and hence this number can be increased above all limits. Galileo thus concludes that "the nearer the inscribed polygon approaches a circle, the shorter the time required for descent."

As has been known since the end of the seventeenth century, the sought-after path of least time is a cycloid and not the arc of a circle as Galileo argued.[9] Indeed,

---

[8] For the use of verisimile (or veri simile) in the sense of probable or plausible, at least in classical Latin, cf. Fuhrer (1993).

[9] In 1696 John Bernoulli challenged the mathematicians of his day to solve the brachistochrone problem. He himself gave two solutions. His brother James sent a solution as well as Leibniz, and another solution was submitted anonymously by Newton. Cf. Goldstine (1980).

aside from its sketchiness, Galileo's argument does exhibit two major problems. The first concerns the justifiability of Steps 9.1 and 9.2 in the above analysis of the proof. In Galileo's own words the conclusions drawn in these steps are justified because:

> manifestum est, lationem per duas ADC citius absolvi quam per unam AC ... sed per duas DEC ex quiete in A verisimile est, citius absolvi descensum quam per solam CD. (EN VIII, 263–264)

That the broken chord ADC is traversed in less time than the chord AC is indeed exactly what was expressed in the law of the broken chord, which had been advanced as a proposition and proven right before the scholium containing the brachistochrone argument. Thus Galileo's use of the expression "manifestum" to characterize this conclusion is justified. Then, however, Galileo seems to imply that it is likewise by virtue of the law of the broken chord ("verisimile est") that motion along the broken chord DEC is made in a shorter time than that along the single chord DC, both motions being made from rest at A. This last conclusion, however, is not justifiable by the law of the broken chord as the proposition requires that the two motions under consideration, that along the chord and that along the corresponding broken chord, have to start from rest. Thus for the law of the broken chord to be applicable to the motions along DC and DEC, those motions would have to have been made from rest at D, which, however, they were not.

The brachistochrone argument thus rests on the assumption that if motion along any chord ending at the nadir of a circle were replaced by motion along a broken chord spanning the same arc, then motion along the broken chord would always be completed in less time than on the single chord, regardless of whether the motions proceeded from rest or had been preceded by another motion. In other words, the brachistochrone argument rested on a generalized law of the broken chord as a premise which, however, as is easily recognized from a modern perspective, does not hold. Galileo's work on the broken chord approach had indeed provided him with strong indications that such a generalized version of the law of the broken chord could not hold and whether Galileo had recognized this will discussed below.

Yet there is another problem with Galileo's argument. Even if a generalized version of the law of the broken chord were to hold, Galileo would not have proved what he claimed to, namely, that the path of absolute quickest descent is the arc of a circle. What he had in fact proved, given the validity of the premises, was that motion along a polygonal path of infinitely many sides, i.e., fall along the arc, would take less time than motion along any polygonal paths that met the constraints of his construction, i.e., polygonal paths whose junction points lie on the arc.[10]

---

[10] That Galileo's argument proves only a weaker statement than the one he actually makes was already realized by Dijksterhuis who, however, did not mention the second, more serious problem with Galileo's argument. Cf. Dijksterhuis and Maier-Leibnitz (2002, 385)

9.1 The Brachistochrone Argument

Thus merely a necessary but not a sufficient condition for the general claim made is proved in the argument. Based on Galileo's argument, it can indeed not be excluded that yet other paths exist, not considered by the argument, which connect the same start and end but which are traversed in even less time. Notably, Galileo was careful enough to restrict his final statement to what he had actually proved, namely, that of the paths considered by the argument the one which most closely approached the arc, i.e., the arc itself, was traversed in least time, rather than restating what he claimed in the prefatory statement to the scholium that the arc was the path of least time.[11] This may indicate that he was aware that his argument was not strong enough to support the general claim that the arc was the brachistochrone and this may also provide the reason why he formulated the brachistochrone argument merely as a scholium and not as a proposition.

But even if Galileo had in fact realized that his argument was not quite strong enough to support the claim he had made, he would have had reason not to be too troubled by this recognition. As argued in Chap. 7, Galileo had considered the question in which geometrical configuration a bent plane would be traversed in least time. Even though he was not able to solve this problem, his consideration had (falsely) suggested that it was precisely the broken chord, whose junction point lay on the upright quarter arc cutting the arc into two equal halves, that was traversed in least time. From a modern perspective, this is not true, yet the correct solution lies so close to the solution Galileo assumed that the trial calculations he had made indeed seemed to support his assumption. Given that if considering a path of two conjugate planes, the geometrical configuration of the path traversed in least time was indeed such that the junction point lay on the arc, the restriction that was imposed on what was proven as opposed to what had been claimed by only looking at polygonal paths whose junction points actually lay on the arc cannot have weighed too gravely on Galileo.

No draft, not even a sketch of an argument related to the brachistochrone can be identified in the *Notes on Motion*. Structurally, however, the brachistochrone argument is of course related to Galileo's considerations in the context of the broken chord approach, and the argument is suggested in particular by Galileo's calculations of the times of motion along polygonal paths composed of a high number of inclined planes, documented on folio 166 and related folios that have been discussed in detail in Chap. 5. The results of these calculations, collected by

---

[11] Galileo's argument suggests that under slight variation of the path, the time of motion along the arc is indeed minimal. It can be argued, based on the understanding that nature acts in a simple manner, that it would not be very plausible to assume that this was a local minimum and that for a path substantially different from the arc the time of motion is minimized absolutely.

Galileo on page 183 recto (see Table 13.1), indeed indicated that time of motion along a polygonal path inscribed into an arc monotonously decreased with the increasing number of nodes considered.

Galileo had only carried out his calculations up to the eight-chord approximation of the quarter arc. Yet the data thus produced could indeed be interpreted as suggesting that the tendency observable in the calculated data should persist and that the time would keep getting smaller with an ever-increasing number of chords being considered, to finally converge to the time of motion needed to fall along the arc.

Moreover, the results of Galileo's calculations were compliant with the assumption of a generalized law of the broken chord as he applied it in the brachistochrone argument. With regard to the calculated times, motion along each and every broken chord ending at the equilibrium position was indeed completed in less time than motion along the single chord spanning the same arc, even when motion was not made from rest but had been preceded by another motion of fall. Incidentally, minor deviations due to the rounding of the numbers Galileo gave as the results as compared to those recalculated by modern means blurred a tendency in the data which may at least have raised doubts. The time of motion along the chord gc from rest at a calculated by Galileo was clearly longer than the time of motion along the broken chord $g8c$ likewise from rest at a. Yet modern recalculation shows that both motions are made in virtually the same time.[12] Still a trend in the data was observable that may have cast doubt on the validity of a generalized law of the broken chord. A clear violation of the assumption of a generally valid law of the broken chord would only have cropped up when calculating the times of motion for the 16-chord approximation of the arc.[13]

Notably it is not only in the *Discorsi* that Galileo argued for the arc of the circle as being the path of descent along which least time is consumed in natural motion from one point to another. A letter concerning the regulation of the river Bisenzio

---

[12]With respect to the central diagram on 166 recto, Galileo calculated for motion along gc from rest at a a time of motion of 19,896, corresponding almost precisely to the recalculated value of 19,895. For motion along $g8c$, likewise from rest at a, Galileo calculated a smaller time of motion of 19,821 which should be 19,894 according to my recalculation. Compare Table 13.1.

[13]There is no indication that Galileo ever calculated the time of motion composed of 16 conjugate chords and inscribed into a quarter arc. Yet, he at least replaced the lowest chord of his eight-chord approximation of the arc on 166 recto, the chord $8c$, by the broken chord $89c$. The times of motion along the chords 89 and $9c$, both motions after fall from rest in a are not listed in the table of results on 183 recto, yet the time of motion along $9c$ is given as 9821 in the diagram on 166 recto. The number is written next to the number 19,605 representing the time Galileo had calculated for motion along $8c$ made from rest in a.

9.1 The Brachistochrone Argument

also contains an argument to this effect.[14] The argument in the letter is less polished and possibly even more sketchy than the one Galileo provided in the *Discorsi*.

There is, however, also a decisive difference between the way Galileo argued for the Brachistochrone in the letter and the way he argued in the *Discorsi*. In the Bisenzio letter, Galileo essentially presupposed that whenever a chord was replaced by a broken chord, motion along the latter path was completed in less time. This renders the argument somewhat more straightforward at the expense of introducing an even stronger premise. With regard to the law of the broken chord, the argument of the letter stipulated not only that the condition that motion should start from rest could be dropped but also the condition that the motion had to end at the nadir of the circle. That this could not be the case should, in principle, have been clear to Galileo based on his prior calculations.

The data in Galileo's table of results on 183 recto in fact shows that the time of motion along the chord fe made from rest at a was 26,721 time units, but that if motion was made along the broken chord $f3c$ instead, also from rest at a, the time elapsed would have been 26,732 time units. With regard to motion between the points f and e, it thus turned out to be the single chord and not the respective broken chord $f3c$ on which motion was completed in less time, providing a counterexample to the premise Galileo had invoked in the argument he gave in the Bisenzio letter.

It is tempting to speculate that Galileo changed the brachistochrone argument of the Bisenzio letter indeed as a consequence of reflecting upon his prior results and insights. Yet even in the revised argument he gave in the *Discorsi* Galileo could not

---

[14]EN. VI, 627–647. The letter was sent to Raffaello Staccoli in January 1631. For a discussion of the broader context in which this letter was sent, see Westfall (1989). Galileo's brachistochrone argument in the letter starts on p.643. His diagram shows a quarter arc spanned by a chord EC, a broken chord EFC, and a polygonal line with all junction points equally distributed on the arc EGFNC: '...che, posta l'istessa pendenza tra due luoghi tra i quali si abbia a far passare un mobile, affermo, la più spedita strada e quella che in più breve tempo si passa non esser la retta, ben che brevissima sopra tutte, ma esservene delle curve, ed anco delle composte di più linee rette, le quali con maggior velocità ed in più breve tempo si passano. E per dichiarazione di quanto dico, segniamo un piano orizontale secondo la linea AB, sopra 'l quale intendasi elevata una parte di cerchio non maggiore di un quadrante, e sia CFED, sì che la parte del diametro DC, che termina nel toccamento C, sia perpendicolare, o vogliamo dire a squadra, sopra la orizontale AB; e nella circonferenza CFE prendasi qualsivoglia punto F: dico adesso, che posto che E fusse il luogo sublime di dove si avesse a partire un mobile, e che C fusse il termine basso al quale avesse a pervenire, la strada più spedita e che in più breve tempo si passasse non sarebbe per la linea o vogliàn dire per il canale brevissimo EC, ma preso qualsivoglia punto nella circonferenza F, segnando i 2 canali diritti EFC, in più breve tempo si passeranno questi che il solo EC; e se di nuovo ne gli archi EF, FC si noteranno in qualsivoglia modo 2 altri punti G, N, e si porranno 4 canali diritti EGFNC, questi ancora si passeranno in tempo più breve che li 2 EFC; e continuando di descrivere dentro alla medesima porzion di cerchio un condotto composto di più e più canali retti, sempre il passaggio per essi sarà più veloce, e finalmente velocissimo sopra tutti sarebbe quando il canale fusse curvo secondo la circonferenza del cerchio EGFNC. Ecco dunque trovati canali che hanno la medesima pendenza (essendo compresi tra i medesimi termini E, C), e che son di differenti lunghezze, ne i quali i tempi de' passaggi sono (al contrario di quello che comunemente si stimerebbe) sempre più brevi ne i più lunghi che ne i più corti, e finalmente lunghissimo è il tempo nel brevissimo, e brevissimo nel canale lunghissimo."

avoid, as we have seen, invoking a premise that was stronger than what he could and had actually proven. Moreover, the premise invoked in the Bisenzio letter and that invoked in the *Discorsi* lead, via otherwise very similar arguments, to the same conclusion. It is thus suggested that the seemingly weaker premise of the *Discorsi* argument directly implies the seemingly stronger one of the Bisenzio argument whose validity, in turn, was refuted by Galileo's own data. Showing that this is indeed the case is in fact rather straightforward. Yet there is no direct evidence to indicate that Galileo realized this and that he could thus have potentially noticed that his calculation directly contradicted not only the argument in the Bisenzo letter but also that in the *Discorsi*.

With regard to the brachistochrone argument, it thus emerges as a very plausible scenario that Galileo grasped that the arc should be the brachistochrone based on indications provided by his considerations in the context of the broken chord approach. As appealing as the statement was that the path traversed in least time was the arc, Galileo was in all likelihood well aware that the argument he gave for it required a premise stronger than what he had been able to prove.[15] This is suggested by the fact that he did not issue the argument as a proposition but as part of a scholium in which he chose his wording carefully. If Galileo indeed saw that the argument was problematic, one might expect him to have searched for ways to improve upon it. Yet no consideration that can be interpreted unambiguously as having served this end can be identified in the *Notes on Motion*.[16] We are thus allowed to speculate that the brachistochrone argument was not conceived while Galileo was engaged in his research agenda on swinging and rolling but later, likely even when he was already composing the *Discorsi*, at a time when he would not have found himself in a position to be able to deeply immerse himself in the problems raised by the argument. Yet as he was apparently convinced that its conclusion was true, he did not shy away from including the problematic argument in the *Discorsi*.

---

[15]Erlichson (1998) has come to the same conclusion. "We are inclined to hypothesize that Galileo knew full well that he did not have a complete proof, and that he also knew that it would be quite difficult to prove his 'yet it seems that …' (Erlichson 1998, 347)." He shows that the argument that Galileo provided in his proof of the law of the broken chord cannot simply be extended to the case where motion along the broken chord does not start from rest and challenges his readers to "try to prove the unproven assumption of Galileo … preferably by methods available to Galileo (ibid.)." As we have seen, however, the assumption does not hold and hence the challenge can be closed.

[16]A number of considerations in the *Notes on Motion* could potentially be related to the brachistochrone argument. In a construction on folio 190 recto, Galileo exploited an approach developed on 130 verso where he had sought the bent plane traversed in least time connecting two points whose vertical and horizontal distances were the same. On 190 recto Galileo applied the same construction to two points whose vertical distance was bigger than their horizontal distance. Apparently he was inquiring if, in this case, the junction point defining the bent plane traversed in least time still lay on the arc connecting start and endpoint of motion, as he (falsely) believed to be the case in the situation under scrutiny on 130 verso. Should this have been found to be the case the insight could have potentially been used to flesh out the brachistochrone argument. Yet it seems very unlikely that this is what motivated the construction on 190 recto and related constructions on 191 verso.

9.1 The Brachistochrone Argument

The irony of the brachistochrone argument In her comprehensive study of Galileo's *Notes on Motion*, Wisan stated:

> [the brachistochrone was] a traditional problem, at least among the mechanicians of the time, but one receiving little mention in written works. In fact it is possible that it was the brachistochrone which originally turned Galileo's attention to motion along inclined planes.[17]

In her subsequent discussion, she dismissed her initial caution and resolutely referred to Galileo's attempts to construct a proof of the law of the broken chord preserved in the *Notes on Motion* as "attempts at the brachistochrone." In essence, she took for granted that from the start Galileo assumed the arc to be the path of quickest descent, that he wanted to prove this statement, and that the law of the broken chord was constructed as a first step toward such a proof. Speaking somewhat pointedly, Wisan assumed that the challenging problem that engendered Galileo's new science of naturally accelerated motion was the search for the brachistochrone and not, as claimed here, Galileo's search for a relation between swinging and rolling.

Yet, as has been demonstrated in the previous chapters, it was the latter problem that stood at the beginning of Galileo's investigations into natural acceleration, which engendered Galileo's investigation into motions along paths composed of conjugate chords, which led him to state and construct a proof for the law of the broken chord, and which finally gave rise to the brachistochrone argument as a late spin-off. In a leap of faith, Galileo included the argument in the *Discorsi* despite its problematic character and could thus, at least in one argument, exploit the idea underlying the broken chord approach, which despite tremendous efforts, had not led him to his desired goal.

In hindsight, there is a certain irony to Galileo's brachistochrone argument. What Galileo had aimed at, as explicitly stated by him in 1602, was not to demonstrate that the arc was the path of least time but that it was the isochrone curve, also referred to as the tautochrone curve, i.e., that the curve upon which the time of motion was independent of the starting point chosen. As it turns out, under the condition of constant force, tautochrone and brachistochrone coincide. Had Galileo assumed the identity of tautochrone and brachistochrone and had he had an explanation for this identity, his attempt to prove that the arc was brachistochrone would in fact have amounted to showing that pendulum motion was isochronous. Yet the identity of the tautochrone and brachistochrone is by no means evident, and there is, it seems, no intuitive way of understanding it. It was indeed completely surprising to the men who first noticed it. Bernoulli remarked:

---

[17]Wisan (1974, 176).

Antequam niam non possum quin iterum admirationem meam prodam, animo revolvens inexpectatam illam identitatem isochronae Hugenianae nostraeque Brachystochronae[18];

There can be no doubt that Galileo did not have the slightest idea of this surprising correspondence. Should he even have remotely speculated about it we can assume that he would not have remained silent about such a marvelous property of motion of heavy bodies, even if it were a mere conjecture.

9.2 A Reply to Baliani Drafted ca. 1638: An Unexpected Twist of Swinging and Rolling

When composing the *Discorsi*, Galileo was able to include but one argument retaining the basic idea of his original research program, the idea of an exhaustion of motion along the arc by motion along a series of inclined planes.[19] As discussed above, Galileo was in all likelihood even aware of the problematic character of his brachistochrone argument but decided to put it in print nevertheless. Compared to the immense effort which, as we have seen, he had invested in his research agenda, this output was rather scant. Galileo had in particular not been able to carry his investigations to the point where the properties of pendulum motion could be derived as a consequence of the properties of naturally accelerated motion on inclined planes or vice versa, as was his declared goal.

Yet there was an unexpected twist. In 1638, only a few months before the publication of the *Discorsi*, Giovanni Battista Baliani published his *De motu naturali gravium solidorum*, in which he put forth a theory of naturally accelerated motion.[20] In many respects what Baliani laid out in this book strikingly resembles Galileo's considerations on the same subject.[21] The propositions Baliani had put forward in his book are, indeed, almost a subset of the propositions contained in the Second Book of the Third Day of the *Discorsi*, a similarity which can undoubtedly be adduced to the intellectual contact between the two men.[22]

---

[18] The Acta Eruditorum of 1697 contain Johann Bernoulli's solution to the brachistochrone problem on pp.206–211. The quote is on p.210. Quoted from www.nlb-hannover.de/Leibniz/Leibnizarchiv/Veroeffentlichungen/III7A.pdf. Accessed 28 Nov 2016.

[19] Yet based on studying the *Discorsi* perspicaciously, Settle (1966, 99) surmised that "the exposition of the Third Day is such that it seems that Galileo was looking for a demonstrative link between linear and circular natural motions."

[20] By the time Giovanni Battista Baliani's book came out Galileo's manuscript had been sent off to Leiden and he was expecting the return of the printed copies. For a detailed account of how the *Discorsi* came to be printed, see Raphael (2012).

[21] For Giovanni Battista Baliani's work on the problems of motion, see above all Moscovici (1967).

[22] On 20 February 1627, Baliani wrote to Benedetto Castelli: "... Io altre volte feci un trattato de' moti dei solidi, e della loro maggiore o minore velocità ne' piani più o meno declinanti: volli poi far quello de' liquidi, e lasciai l'opera imperfetta, perchè mi si accrebbero le difficolà. La causa principale è la seguente. Facendo il trattato de' solidi che ho detto, avvenne che, senza cercarla,

9.2 The Letter to Baliani

Baliani's contribution to the history of science has hence often been discounted as a mere imitation of Galileo's achievements. In consequence, a decisive difference in the way both men attempted to found their sciences of naturally accelerated motion went almost completely unnoticed.[23] In contrast to Galileo, Baliani based his own theory entirely on assumptions about pendulum motion, among them the isochronism of the pendulum and the law of the pendulum. Starting from these assumptions, Baliani constructed his theory and in particular derived the law of fall as well as a proposition comparable to Galileo's law of chords.[24]

It is thus recognized that Baliani apparently achieved what Galileo was not able to, namely, exploiting the perceived relation between swinging and rolling to make inferences about either type of motion from assumptions about the respective other. Instead of demonstrating the isochronism of the pendulum, Galileo's proclaimed goal, Baliani had done the reverse and demonstrated the laws governing naturally accelerated motion along inclined planes from assumptions about pendulum motion, among them the isochronism of the pendulum. Hence, if Baliani's argument could be inverted, it would, if indirectly, provide the solution to the problem Galileo had tried so desperately to solve in 1602.

This did not escape Galileo's attention. A hitherto virtually neglected draft of a reply to Giovanni Battista Baliani exists in which Galileo formulated a critique of Baliani's approach. I will argue in the following that Galileo's critique can be understood as owing to the insights he had gained in his own research on the same subject. Reflecting upon Baliani's approach apparently triggered an insight which led Galileo to rethink his own old problem once more. In his reply, Galileo proposed an alternative to an argument by Baliani. As notes written in the margin in his own

---

mi riuscì, a parer mio, ben dimostrata una proposizione per una via molto stravagante, la quale già il Sig. Galileo m'avea detta per vera senza però addurmene la dimostrazione; ed è, che i corpi di moto naturale vanno aumentando le velocità loro con la proporzione di 1, 3, 5, 7, ec., e così in infinito: me ne addusse però una ragione probabile, che solo in questa proporzione più o meno spazi servano sempre l'istessa proporzione. Non mi dichiaro maggiormente, perchè so che parlo con chi intende. Però io l' ho dimostrata con principi molto diversi; ...(EN XIII, 348–349; letter 1806)" This shows first that Baliani had been working on his treatise for quite a while before 1638; second, that Galileo had communicated results to him; and third, that Baliani had a clear awareness that his approach differed fundamentally from that of Galileo with respect to the principles on which it was based. Capecchi (2014, 177), in contrast, assumes that "Baliani's claims ... according to which all heavy bodies fall with the same temporal law and the periods of the pendulums are proportional to the square roots of their lengths, were independent of the results obtained by Galileo."

[23] With regard to *De motu naturali gravium solidorum*, Richard S. Westfall has, for instance, remarked that "[t]he level of discussion in Baliani does not begin to approach Galileo's, so that issues of plagiary have inevitably arisen." See http://galileo.rice.edu/Catalog/NewFiles/baliani.html. Accessed 4 Sep 2016.

[24] Pendulum motion was clearly among the topics that Galileo and Baliani discussed. A letter (EN XIV, 342–344; letter 2258) from 23 April 1632, sent by Baliani to Galileo, for instance, testifies that the two men debated the use of a pendulum for measuring seconds.

copy of the *Discorsi* show, the argument, which amounts to a proof for the law of the pendulum based on the laws of naturally accelerated motion, was earmarked for publication in a revised edition of the *Discorsi*. It was probably due only to Galileo's death that the novel argument, which brought to completion at least in part what Galileo had set out to attain more than 35 years earlier, was never published.

Baliani's proof of the law of fall from assumptions about pendulum motion
Baliani's book opens with a list of definitions concerning pendulums and their motions. The definitions particularly relevant for the ensuing discussion are his definition of a "vibration" as the motion of a pendulum from one turning point to the other, of "integras," full vibrations, i.e., pendulum swings through a full 180-degree arc, and of a "portio" as a portion of a vibration referring to the arc of a smaller sector.[25] Finally, Baliani defines the "portionem priorem" of the vibration as the "minimal portion which a full vibration has at its beginning." Baliani did not further specify his concept of a "minimal portion," and this was in fact to become one of Galileo's major points of critique.

The list of definitions is followed by a list of four "Suppositiones." The suppositions or principles Baliani lists are the independence of the period from the weight of the bob, the isochronism of the pendulum and the law of the pendulum, and, last but not least, a version of the law of the inclined plane. Hence the properties of pendulum motion which Galileo deemed in need of demonstration were simply advanced by Baliani as principles, which he described in the preface of the book as known from experience.

The list of suppositions is succeeded by six "Petitiones, seu Postulata" of which the postulates one, three, five, and six are particularly relevant for understanding how Baliani derived the laws of naturally accelerated motion on inclined planes from the properties of pendulum motion he had presupposed as evident principles in his suppositions.

The first postulate essentially states a scale invariance property of pendulum motion, namely, that the times of motion along similar portions of the vibrations of pendulums of unequal length are in the same proportion as the periods of the "full vibrations." "Similar portions of vibrations" had been specified as the arcs of two pendulums of unequal length suspended at the same point, which are enclosed by the same two lines radiating from the point of suspension. Thus with respect to Fig. 9.2, the arcs BC and EF are "similar portions." Then, according to the first postulate, the times of motion along these two arcs are in the same ratio as the periods of the two pendulums.

Baliani's third postulate states that minimal portions of a vibration are to be conceived as if they were straight lines. Again, Baliani does not precisely define

---

[25] These and all following quotes from *De motu naturali gravium solidorum* are made according to Baliani et al. (1998); translations, where given, are mine.

Fig. 9.2 Figure associated with Baliani's proof of the law of fall (see Baliani 1638, 14)

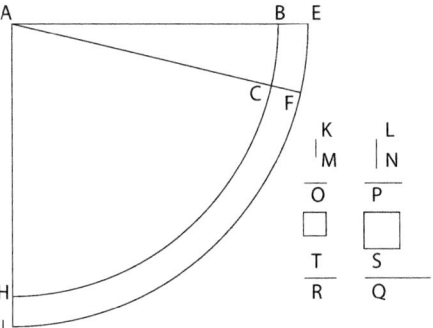

exactly what a "minimal portion" of the arc is and how exactly the straight line—which if motion along it is considered, can be treated as interchangeable to motion along the minimal portion of an arc—should be constructed. In an annotation to his third postulate, Baliani admits that no part of an arc, however small it may be, can formally be a straight line but that just as the difference between the arc and the straight line vanishes for extremely small distances, it vanishes in terms of sensory perception for whatever physical property is being regarded. He supported his argument by drawing the analogy to the science of mechanics, where lines which the theoretical framework stipulates should meet at the center of the universe were considered to be parallel for all practical purposes.[26]

The fifth postulate states that bodies falling freely along the vertical are carried with the same velocity and "in tali proportioni" as if they were pendulum bobs moving over the first portion of their vibration. The precise meaning of this postulate is clarified only by inspection of the use Baliani makes of it in his proof of the law of fall discussed below. Here it suffices to say that basically the first minimal parts of pendulum motion can, according to Baliani, not only be considered as if they were made on appropriate straight lines but also as if the fact that the bobs were attached to a point of suspension by means of a string had otherwise no effect at all. In other words, the motion of a pendulum along the first minimal part of a full vibration can be considered as being kinematically equivalent to the motion of a freely falling body over the appropriate vertical distance. As is easily surmised, this postulate will play a crucial role.

The sixth and last postulate repeats the fifth, except for the difference that Baliani here considered minimal portions of the arcs positioned anywhere between the turning point and the equilibrium position of the two pendulums, and the corresponding motions on inclined planes tangent to these minimal portions of the arc.

---

[26] For the role played by the observation that lines from weights at different positions at a balance or lever to the center of the world cannot be parallel in discussions of mechanics in Galileo's day, see Damerow and Renn (2012).

Thus the prerequisites of Baliani's proof of the law of fall, the third proposition, were laid out. The proposition itself reads[27]:

PROPOSITIO TERTIA.
Lineae descensus gravium, dum naturali motu perpendiculariter feruntur, sunt in duplicata ratione diuturnitatum. Sint LN, KM linea descensus gravium L, K, & sint P O ipsorum diuturnitates. Dico LN, KM esse in duplicata ratione ipsarum P, O. Sint pendula AH, AI, dependentia a puncto A, & eleventur ad libellam ipsius A usque ad E, B, quae in elevatione producant arcus HB, IE, & sint talis longitudinis, ut ducta ACF, secet arcus BC, & EF, portionis minimae, aequales quo ad sensum lineis LN, KM, & sit S, quadratum diuturnitatis P, & T quadratum O, & Q, R, diuturnitates vibrationum BC, & EF. Quoniam diuturnitates Q, R sunt aequales diuturnitatibus P, O; S, T, sunt etiam quadrata ipsarum Q, R, & quia vibrationes integrae pendulorum AH, AI sunt ut quadratum T ad quadratum S, portiones BC, EF, sunt pariter inter se ut quadratum T ad quadratum S, sed BC, & EF sunt aequales lineis KM, LN, ergo etiam K M, LN sunt ut quadrata S, T, & proinde in duplicata ratione P, O, temporum seu diuturnitatum earumdem. Quod, etc.

Baliani's proof can be paraphrased as follows: If one considers the first "minimal portions" BC and EF, which are cutoff of the full vibrations of two pendulums by the radii AE and AF, the motions over these minimal portions can be replaced by motion along the straight lines KM and LN according to his third and fifth postulate. Then, however, Baliani makes a crucial additional assumption, which he had not secured in his list of prerequisites—the lengths of the lines KM and LN had to be equal to the length of the "minimal portions" BC and EF of the arcs they assimilated. With this additional assumption, the proof of the law of fall becomes straightforward. Since according to the first postulate the times of motion along the "minimal portions" were as the periods of the pendulums, they had to be in the ratio of the roots of the radii according to the presupposed law of the pendulum, and hence also of the roots of the arc lengths BC and EF. Since according to the fifth postulate motion over these minimal arcs could be treated as interchangeable to the motion over the straight line which assimilated these arcs, the times of motion over the distances KM and LN were the roots of these distances, and hence the law of fall had been proved.

That a quadratic relation held between the distances covered and the times elapsed during motion, not only for free fall along the vertical but also for motion along inclined planes, was subsequently proved by Baliani in Proposition VII. The argument was largely equivalent to that of Proposition III. Instead of considering the first part of the motion of a full vibration, Baliani considered an arbitrary minimal portion of the vibration of the pendulum which, by means of the sixth postulate, could be linked to the motion along an inclined plane tangent to this minimal portion of the arc.

---

[27] Baliani (1638, 14).

9.2 The Letter to Baliani

A relation between the naturally accelerated motion on the vertical and along an inclined plane of arbitrary inclination was finally established by Baliani in Proposition XIV by means of an argument which is reminiscent of Galileo's ex mechanicis proof of the law of chords. It is important to note that the isochronism of the pendulum that Baliani had secured in his suppositions is not explicitly required by any of these proofs, a fact, which, as will be seen, Galileo apparently realized and was able to make constructive use of.

Galileo's critique of Baliani's proof: a late proof of the law of the pendulum
A letter dating from the 17 December 1638 shows that Giovanni Battista Baliani had sent his book to Galileo in Arcetri. He had asked Galileo to read it and to let him know of his opinion.[28] Given that Baliani's book appeared so shortly before the *Discorsi*, we can be quite certain that Galileo read the book with great interest and, as one can imagine, some discontent.[29] Upon cursory inspection, Baliani's proof of the law of fall from assumptions about pendulum motion must undoubtedly have provoked Galileo's interest and initially have looked quite appealing. Yet since Galileo himself had failed so badly to prove the reverse, he certainly must have inspected the proof with great care.

In the letters Galileo sent to Baliani after December 1638, the latter's book is alluded to only sparsely. Yet, Galileo apparently complied with Baliani's plea for a direct comment as the draft of a reply in which a critique of Baliani's approach is formulated can indeed be found in the collection of Galilean manuscripts.[30] The

---

[28] In 1615, Baliani traveled to Florence where he visited Galileo and also met with Benedetto Castelli. Based on a proposal by Galileo, Giovanni Battista Baliani became a member of the Accademia Lincei in the following year. Correspondence between Baliani and Galileo continued sporadically for many years. The letter sent in December 1638 reads: "Havendo io risoluto di mandar fuori un'operetta del moto naturale de' corpi gravi mi parrebbe far mancamento se non la mandassi subito a V. S., pregandola che a tanti favori fattimi voglia aggionger questo di legerla e corregerla e dirmenc il suo parere. Son sicuro che, se non per altro, la stimerà almeno degna di comparirle dinanti per conoscer la fattura di autore che, ancorchè da lontano, si ingegna di seguir le sue pedate; et io in tanto starò con desiderio di veder uscir in luce le opere di V. S., in cui spero di vedere ridotto a perfettione ciò che io ho abbozzato così alla grossa. E pregandola conservarmi nella sua buona gratia, le baccio per fine le mani, e priego dal Signor ogni vero contento (EN XVII, 413–414; letter 3824)." On 7 January 1639, Galileo wrote back indicating that he had received the book and read it, respectively, had it read to him due to his worsening eyesight, as he noted in a letter to Renieri on 28 March 1639. Cf EN XVIII, letter 3829 and 3858.

[29] The echo the publication of Baliani's book provoked in Galileo's circle in the years 1638 to 1639 was characterized by "dramatic tones of bewilderment and sometimes outrage. (Istituto della Enciclopedia italiana (Roma) 1963)" Galileo himself does not appear to have been too bothered. He kept up his correspondence with Baliani, retaining a polite tone. Occasionally, he interspersed remarks in the conversation to the effect that the *Discorsi* was submitted to the printer much earlier than Baliani's book and that what was demonstrated therein goes beyond what Baliani had demonstrated. Cf. EN XVIII, letters 3829, 3897 and 3912.

[30] The draft is preserved in the Biblioteca Nazionale Centrale in Florence as part of Ms. Gal. 74, folios 35v -38v. Favaro did not include the document in the EN as he did not consider Galileo to be the author. Cf. EN VII, 36–37. It was first published by Caverni, who attributed the authorship to Galileo. See Caverni (1972, IV, 313–314). The document is likewise discussed in Moscovici

document is in Viviani's hand thus raising the question of its origin and authorship. The clearest indication in this regard is given by the first sentence: "Sopra i principii del Signor Baliani. disteso da me ad mentim G." If the draft was not dictated by Galileo to Viviani, then it seems Viviani at least authored it as the result of a prior exchange with Galileo. Compared to Galileo's rather concise style, at times the text appears somewhat wordy and unnecessarily complicated. Hence the latter of the two alternatives seems more likely. There is, however, little reason to doubt that in principle the intellectual authorship lies with Galileo. The considerations laid out in the document will hence be addressed as Galileo's own in the following. A full transcription as well as a translation of the document is given in the Appendix (see Chap. 14).

In the first paragraph, Galileo uttered a principle methodological critique of Baliani's approach. Why is it, Galileo asks, that Baliani, intending to establish a theory of naturally accelerated motion on inclined planes, started from an assumption about a motion that is not made over straight lines of an assignable inclination? Why instead did he start from assumptions about pendulum motion? According to Galileo, this problem is aggravated by the fact that Baliani's approach forced him to accept the isochronism of the pendulum as well as the law of the pendulum as indubitable principles, which in Galileo's opinion was very controversial.

Yet for the sake of argument, Galileo concedes with Baliani's general approach and proceeded with some more direct and constructive criticism. First he made a suggestion as to how Baliani's proof of the law of fall could be improved, and then he went on to criticize Baliani's procedure in more detail. Since Galileo's improved version of Baliani's proof avoided precisely those aspects of Baliani's approach he criticized as being problematic in the second part of the document, I will first discuss Galileo's points of critique and only afterward analyze his proposal for how to modify Baliani's proof.

Galileo's objection to Baliani's approach was directed against the latter's claim that "the motions through the minimal parts of the arcs are as if they were made through lines assumed to be straight." Galileo articulated three distinct points of critique. First, he noted that Baliani had not indicated what constituted a minimal arc. Secondly, even if this could have been specified, it would not have been at all clear, according to Galileo, how the straight line upon which motion were made as if over the minimal part of the arc, should be constructed.

Of the infinite ways to construct such a line, he explicitly mentions two, a chord and a tangent to the arc at its endpoint. For a tangent to the arc at its endpoint, Galileo provides a brief argument why this could not be the construction sought. For the sake of his argument, Galileo considered the minimal part of the arc which connected directly to the equilibrium position of the pendulum, in contrast to Baliani, who

---

(1967) who, without providing further evidence, assumes that the text was dictated by Galileo to Viviani. Capecchi (2014, 180) even states that the document reproduces "a letter of Galileo to Vincenzo Renieri (1606–1647) which is lost but was reported by Vincenzo Viviani." He too, however, provides no direct evidence for his claim.

9.2 The Letter to Baliani

in his proof of the law of fall had considered the first minimal arc of what he had called a full vibration, i.e., the minimal part adjacent to the 90-degree turning point of pendulum motion.

From Baliani's perspective, Galileo's choice of example would, however, have been justified as he himself had argued in a similar fashion in his own proof of Proposition VII. In this case, Galileo argued that the straight motion to be compared to motion on the arc could trivially not be along the horizontal tangent since there the body would remain at rest. This point of critique is somewhat unsound. Even if Giovanni Battista Baliani had indeed not explicitly specified how to construct the straight line upon which motion was comparative to motion on the arc, his arguments insinuated that it had to be tangent to the beginning and not the end of the arc.

Galileo's third point of critique, finally, is the most fundamental. If up to this point Galileo had committed himself, to some extent, to Baliani's idea of replacing motion on a minimal arc by motion on a straight line and had merely pointed to the problems such an approach entailed, he finally argued that Baliani's idea was altogether unsound. Indeed, according to Galileo, for every arc, however small, "there are so many different inclinations as there are tangents and these are as much as the points, that is infinitely [many]." For this reason "one cannot claim motion through the arc to be similar, let alone the same, as through an equal [medesima] inclination of it."

This frontal attack against Baliani's central idea that motion on a minimal arc could be replaced by motion on an appropriately chosen straight line should by now come as no surprise to us. As we have seen, Galileo initially had also hypothesized the time of a quarter swing along the arc to be the same as that of falling along a particular straight line, namely, the chord spanning the arc. His investigation based on his revised idea that motion along the arc could be exhausted by motion along a series of inclined planes inscribed as polygonal lines into the arc indicated that the difficulties he had encountered in relating motion along an arc to that along a straight path did not depend on the size of the arc and would not thus vanish if only small enough arcs were considered in the way Giovanni Battista Baliani had done. The law of the broken chord, for instance, states motion on a broken chord to always be completed in less time than motion on the chord, no matter how small the arc considered.

The proposal Galileo made for improving Baliani's argument is likewise recognized as owing to the understanding gained in his own earlier research. Galileo proposed the following proof (see Fig. 9.3 for the associated diagram):

> The line AB is supposed to represent the thread hanging from above, and supposing the highest point A to be fixed, then the mobile carried to B describes the arc of the quadrant BC; similarly, taking ab as a smaller pendulum, the arc of the quadrant bc would be the one which the mobile placed in b would describe, and of these quadrants let the supporting arcs be BC. bc. and let the horizontal tangents be BD. and bd. and the perpendiculars CD. cd. Now since the two declines are all in all similar, one can very reasonably take and assume as a known principle that the proportions of the motions that happen to be made over the straight lines AB. BC. and through the arc CB. would be the same as in the smaller figure through the homologous lines ab. bc. and through the arc cb. Therefore by permutation the

Fig. 9.3 Diagram from the draft of the reply to Baliani

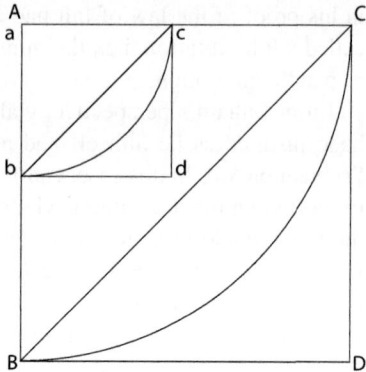

motion through the arc cb. has to the motion through the arc CB. the same proportion which the motion through the perpendicular ab. has to the motion through the perpendicular AB. And therefore if you assume that the times through the arcs are in subduplicate proportion to the lengths of the strings, then it is already manifest that with the same reason one can assume that the times through the perpendiculars ab. AB. are in subduplicate proportion to the same lengths ab. AB. (Gal. 74, folio 36–37; my translation)

Like Baliani, Galileo considered two pendulums of different length hung at the same point of suspension. Both pendulums swing from rest through the same angle but otherwise no particular restriction applies with regard to the angle considered. Instead of motion over Baliani's notoriously ill-defined straight lines taking the place of minimal portions of an arc, Galileo considered motion over well-defined straight lines, namely, the chords spanning the arcs of the pendulums' swings. At this point, Galileo introduced the decisive idea of his proof. He simply postulated that the problem had a scale invariance property. He posed as a "known and reasonable principle" the assumption that, since the motion situations were "all in all similar," the "proportions of the motions" should be the same for the small and the larger figure. Whatever the ratios between the times of motion through arc, chord and vertical radius for the small pendulum, the same ratios should also be characteristic for the larger pendulum. In the following, this will be referred to as the *size independence assumption*.

Then by permutating the ratios, the time of motion along the smaller chord has to the time of motion along the longer chord the same ratio as the period of the smaller pendulum to the period of the larger pendulum. Since, moreover, both chords have the same inclination, the law of fall can be applied, according to which the ratio of times of motion along the two chords equals the ratio of the roots of their lengths. Exploiting the size independence assumption a second time, Galileo concluded that the ratio of the times of motion along the vertical radii of the pendulums was the same as that of the times of motion along the chords. Thus the periods of the pendulums must be in the same ratio as the times to fall through the respective pendulum lengths, that is, in the ratio of roots of the distances covered.

9.2 The Letter to Baliani

In effect, what Galileo offered to Baliani was thus not an alternative to, or an improved version of the latter's proof of the law of fall. Indeed, rather than proving the law of fall, Galileo's argument presupposed the law and derived the law of the pendulum. Yet in Galileo's argument, the role of antecedent and consequent can be exchanged without further complications, resulting in the proof Baliani had been aiming for. In this sense, the sketched argument fulfilled its purpose of indicating to Baliani how he could have alternatively argued his case.

That in his argument Galileo derived the law of the pendulum from the law of fall and not vice versa is not surprising. For Galileo, as he made clear in the letter to Guidobaldo del Monte in 1602, it had always been the predominant aim to exploit the perceived relation between swinging and rolling, to explain the properties of pendulum motion from assumptions about the properties of naturally accelerated motion. Thus the proof sketched in reply to Baliani indeed accomplished in part what Galileo had sought to achieve some 35 years earlier, and this did not remain without consequence.

Two marginal notes in Galileo's private copy of the *Discorsi* indicate that Galileo planned to include the modified proof in a revised edition of the *Discorsi*.[31] A note at the bottom of page 92, which was supposed to be inserted in Galileo's discussion of the law of the pendulum, reproduces the argument communicated to Baliani in abbreviated form (cf. Fig. 9.4).[32] The note reads:

> Adunque di due pendoli diseguali, il tempo per l'arco del'uno al tempo per l'arco dell'altro sta come il tempo pel [sic] seno dell'uno al tempo pel seno d'un arco simile all'altro, i quali seni formano un sol piano inclinato, per i quali i mobili naturalmente discendenti scorrono in tempi che hanno subdupla proporzione di esse seni, perché questi son proporzionali a'lor raggi, che sono le lunghezze de' pendoli. (Galilei and Giusti 1990, 108)

Apparently examination of Baliani's approach had opened an understanding to Galileo that he had not reached before. At first sight, this is somewhat puzzling as

---

[31] Bound as Gal. 79, an annotated copy of the *Discorsi* is preserved in the Biblioteca Nazionale Centrale di Firenze. A leaf bound with the book refers to it as a copy with corrections notes an addenda by Viviani. Damerow et al. (2001, 314) refer to it as Galileo's own copy, and this assumption is shared here. Gal. 79 is one of the texts Guisti used as the basis for his edition (cf. Galilei and Giusti 1990) of the *Discorsi*. Therein the handwritten annotations of Gal. 79 are given as footnotes.

[32] On a separated leaf (Gal. 79 folio 58) inserted into the copy of the *Discorsi*, the proof idea cursorily added in the lower margin is expanded upon. A full transcription of this page is contained in Galilei and Giusti (1990, 105–106). The marginal note directly underneath the thumbnail figure and which is to be appended after "troverò la lunghezza della corda" in the main text as indicated by an insertion mark reads: "perché facendo come il quadrato del piccol numero delle vibrazioni del lungo pendulao al quadrato del gran numero dell vibrazioni del corto, così la lunghezza nota di questo ad un'altra, essa sarà l'ignota lunghezza del lungo (Galilei and Giusti 1990, 108)".

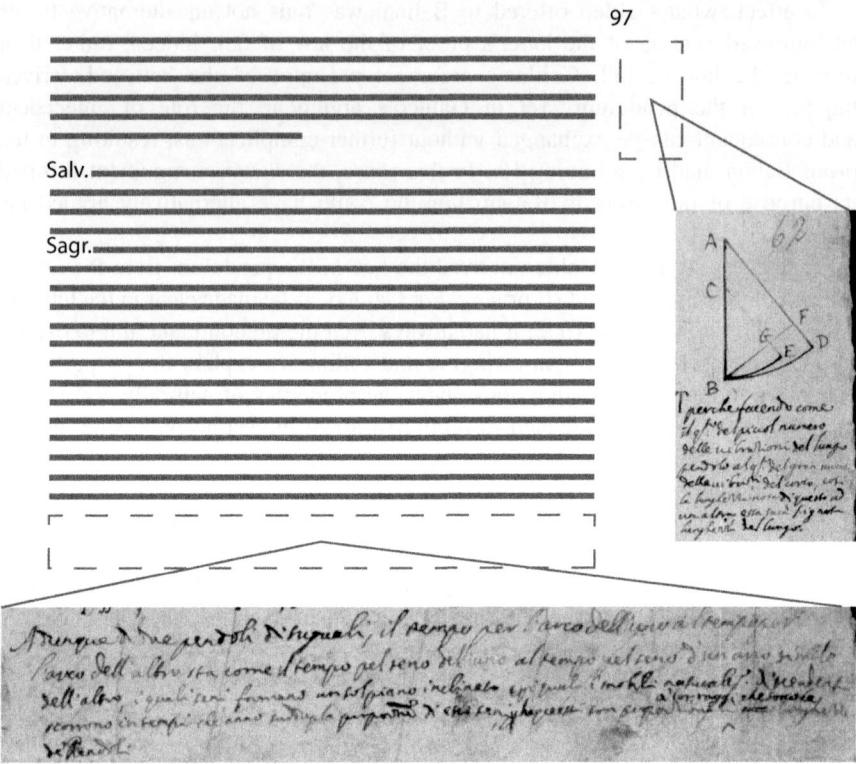

Fig. 9.4 Folio 62 recto of Ms.Gal.79 containing Galileo's own copy of the *Discorsi*; the page number is 92. The two marginal notes relevant for the present discussion are placed next to a schematic representation of the page. The motion situation of two pendulums of different lengths with the same point of suspension is depicted in the right margin close to the top of the page. The second note discussed here is written in the lower margin. The page bears some additional short notes concerned mainly with minor linguistic changes. (Courtesy of Ministero dei Beni e le Attività Culturali – Biblioteca Nazionale Centrale di Firenze. Reproduction or duplication by any means prohibited)

it appears that the argument proposed to Baliani and earmarked for inclusion in a second edition of the *Discorsi* is a rather trivial consequence of assumptions that had been underlying Galileo's research on swinging and rolling all along. In particular, the size independence hypothesis as the crucial element of the alternative proof of the law of the pendulum had been an integral part of Galileo's considerations regarding swinging and rolling from the start.

What was achieved by the new proof for the law of the pendulum and how?
Everything that Galileo stated in the reply to Baliani could indeed, in principle, have already been stated by him in late 1602. The question of why the proof Galileo proposed to his longtime correspondent emerged only in response to the latter's

approach thus has to be raised. First of all, as already indicated above, the critique of Baliani's approach is all too understandable based on the reconstruction that has been given of Galileo's own approach toward the problem of the relation of swinging and rolling. After all, Galileo, just as Baliani, had initially supposed that motion along an arc would be completed in the same time as motion along an appropriately chosen plane. In contrast to Giovanni Battista Baliani, however, Galileo had not restricted his assumption to what the latter referred to as minimal portions of the arc.

After the single chord hypothesis had been rejected experimentally, Galileo devised his modified conceptualization of the relation between swinging and rolling. Central to his new conceptualization was the idea that motion along the arc was equivalent to motion along a polygonal path with infinitely many sides approaching the arc. Against the backdrop of this conceptualization, Galileo's point of critique is clear. Any arc, however small, can still be exhausted by a polygonal line of infinitely many sides.

From a modern perspective, informed by infinitesimal calculus, it may be tempting to reconcile Baliani's and Galileo's approaches. For infinitesimally small distances, the difference between motion along the arc and motion along the tangent vanishes, and this is indeed the basis of the equations of motion of classical mechanics. It could thus be claimed that what Baliani had in mind were infinitesimals and that the same is implicit in Galileo's conceptualization. Such a claim would, however, be whiggish, as a concept of the infinitesimal was simply not available by the mid-seventeenth century. Exhausting an arc by a polygon of infinitely many sides foreshadows infinitesimal calculus but can be formulated coherently based on the mathematical framework of the time as it was indeed rooted in Greek mathematics.[33] Yet it is one thing to claim and prove that an arc and a polygon of infinitely many sides fall together and that the same holds for motion along these paths, but it is quite another thing to speculate about the mathematical and physical properties of motion along such an infinitesimal side.

From Galileo's perspective, the problem would have been solved if he had found a way to argue for the time of motion along a polygonal path of infinitely many sides. This, however, in no way entailed an understanding of a side smaller than any measurable quantity but not zero with regard to which the difference between motion on the arc and along a straight line vanishes as Baliani had in essence claimed. From the perspective of their day, Galileo and Giovanni Battista Baliani's approaches are indeed rather irreconcilable. Whereas Baliani's "minimal portion" could not be stringently specified, Galileo's conceptualization was meaningfully anchored in the extant mathematical framework, and Galileo's critique is absolutely justified.

---

[33]Capecchi (2014, 187) similarly remarks that Baliani's approach "evidences an understanding of infinitesimal analysis at least on an intuitive level. It is quite acceptable for a modern, but it was certainly not such for Baliani's contemporaries. Although mathematics was evolving in the direction of the infinitesimal calculus, this assumption of Baliani still seems very daring and interesting."

Galileo's conceptualization may have been favorable in that it could more stringently be expressed. Yet this carried along with it the disadvantage that the problem thus created could not be solved. As we have seen, Galileo did not come even remotely close to being able to devise an argument that would have allowed him to infer the time of motion for the limiting case of motion along a path of infinitely many sides. This, finally, then leads us toward the answer of what exactly it must have been that Galileo recognized when he reflected upon Baliani's approach and what it was that allowed him to devise his own proof of the law of the pendulum from assumptions about motion along inclined planes.

Galileo's proof crucially hinged on the size independence principle, i.e., the assumption that for an arc of a given angle, the ratio of the time of motion along the chord spanning this arc to the time of motion along the arc was independent of the size of the system, i.e., the radius of the arc. This assumption, as has been argued, had always been embraced by Galileo's own conceptualization of the relation of swinging and rolling. As all his propositions or laws of naturally accelerated motion are size independent, the ratio of the time of motion along a chord to the time of motion along any polygonal path must likewise be size independent. Thus, even if the time of motion for the limiting case of motion along a polygonal path of infinitely many sides cannot concretely be determined, it is nevertheless clear that its relation to the time of motion along a corresponding single chord spanning the same arc cannot depend absolutely on the size of the system considered.

What Galileo crucially recognized upon studying Giovanni Battista Baliani's proof can hence hardly have been size independence per se. What he seems to have realized instead is that size independence, which for him thus far had been a consequence of his conceptualization of the relation of swinging and rolling, could instead be formulated as a principle based upon which as a premise proving the law of the pendulum from law of fall or vice versa was but a trivial exercise. The size independence principle compares to Galileo's earlier single chord hypothesis, in that here as there an assumption about the relation between motion along the chord and along the arc spanned is being expressed. Yet the too strong assumption that these motions are always completed in equal time is reduced to the moderate claim that the ratio of these times is independent of size. As regards the justification of the size independence principle, Galileo merely stated that "molto ragionevolmente si puo' prendere e come principio noto supporre," that it holds. For an external reader such as Giovanni Battista Baliani, this must have been a rather bold claim. From Galileo's perspective, it was of course still justified by his own conceptualization which, however, did not lend itself to being communicated as long as he had not been able to construct the argument by exhaustion for the time of motion along the arc he was seeking.

Notably, the size independence principle, while being strong enough to act as a premise in Galileo's new argument, was weak enough an assumption to be indifferent to the question whether motion along the arc was isochronous or not. For Galileo, as we have seen, the isochrone of motion along the arc had been the focal point of his investigation. This entailed the ratio of the times of motion along chord and along arc spanned having to be independent of the angle of the arc.

From Galileo's perspective, size and angle independence were inextricably coupled, together forming the basis of his research agenda on swinging and rolling. Angle independence had been presupposed in the single chord hypothesis but needed to be recovered theoretically under Galileo's new conceptualization, which he had tried in vain to achieve. This certainly obscured the recognition of size independence as an independent principle and thus contributed to the fact that Galileo did not come up with the argument given in the reply to Baliani earlier. What Baliani's approach demonstrated was that both assumptions were completely independent and could be decoupled. Indeed, despite posing the isochronism of the pendulum as a supposition, Baliani had not used the assumption as a premise in his proofs regarding the laws governing the naturally accelerated motion along inclined planes. Galileo's approach mimicked Baliani's by arguing for the structural similarity between the law of fall and the law of the pendulum based on the assumption of size independence alone such that the argument provided was unaffected by the question of whether motion on the arc was isochronous or not.

With the proof of the law of the pendulum based on the law of fall he thus arrived at, Galileo had, from a modern perspective, finally achieved all that was achievable with regard to his aim of explaining or inferring the properties of pendulum motion from those governing naturally accelerated motion along inclined planes. Indeed, an explanation, as Galileo was seeking, for the perceived similarity between the law of chords and isochronism of the pendulum cannot be given. Not because of a lack of theoretical understanding on Galileo's side but because this similarity is a delusion that cannot be explained. Pendulum motion is simply not isochronous.

Did Galileo with his proof of the law of the pendulum thus after all renounce the isochronism of the pendulum and with it the very foundation of his research agenda on swinging and rolling? Apparently not. No marginal note can be identified in his private copy of the *Discorsi* to indicate that he had ceased to regard the isochronism of the pendulum as true and all passages of the text in which allusion is made to the isochronism of the pendulum remained unaltered. From Galileo's perspective, the proof of the law of the pendulum drafted in the reply to Baliani and noted in a nutshell in the margin of his copy of the *Discorsi* was not the ultimate verdict it may appear to be from a modern perspective. It was but a partial step toward the aim he had set for himself almost 40 years earlier with his research agenda on swinging and rolling. The crucial question regarding the isochronism of the pendulum remained unresolved.

9.3 Summarizing Conclusion

As has been disclosed in the previous chapters, Galileo's investigation into the relation of swinging and rolling came to a halt after an intense but presumably rather short period of work in view of insurmountable problems that had piled up. The *Notes on Motion* provide no evidence that Galileo ever returned to his original question. Part of the reason for this is certainly to be sought in the fact that his work on swinging and rolling had opened up a new prospect, that of establishing a new science of motion. How Galileo engaged in transforming his new insights into a new science is discussed in the second part of the book, which begins after this chapter.

Yet Galileo had not given up on the agenda that had occasioned his new science in the first place. He returned to it in the *Discorsi*, albeit in but one argument. Moreover, when the *Discorsi* had already been submitted to Elsevier, Giovanni Battista Baliani's *De motu naturali gravium solidorum* appeared in print. Prompted by his reading of the book, Galileo returned to his old research agenda one last time. It is Galileo's late engagements with his early research on swinging and rolling that have been discussed in the present chapter.

In the first part of the chapter, it has been asserted that the argument Galileo gave in a scholium regarding the brachistochrone, i.e., the curve along which motion is completed in least time, had been inspired directly by his consideration in the context of the broken chord approach. Galileo's argument was problematic in two ways. It proves a necessary but not a sufficient condition for his claim that the arc is the brachistochrone. His argument is furthermore based on a premise for which Galileo had not provided a proof and which is indeed false from a modern perspective. Galileo, as has been argued, was likely aware of both these problems. Whereas he had reason to assume that the first problem would not jeopardize his conclusion, his earlier considerations had provided him with rather clear indications that the major premise employed in the argument was untenable. In the *Discorsi*, the argument was changed with respect to an earlier version that Galileo had communicated in a letter, and this change indeed seems to have been due to him acknowledging that his earlier considerations were not compliant with the assumption made. Yet the reformulation of the argument, as has been shown, did not solve the problem.

It seems that Galileo was so vehemently convinced that the brachistochrone was the arc, that he issued the argument in the *Discorsi* despite having recognized its flaws. Yet as we know, Galileo's belief was off the mark. Late in the century, it was shown that the brachistochrone was not the arc of a circle but a cycloid. Huygens had earlier shown that the isochrone curve was likewise a cycloid, and, thus, much to the astonishment of Jean Bernoulli brachistochrone and tautochrone were seen to coincide. In hindsight, there is a certain irony to Galileo's contemplation of the matter. What Galileo had set out to demonstrate was that the arc was the isochrone curve, what he ended up with was the conviction of a proof, albeit flawed, that the arc was the brachistochrone. As it later turned out, solving the brachistochrone problem coincided with finding the isochrone curve. Yet this curve was not the arc of circle as Galileo believed, but a cycloid.

In a second, revised edition of the *Discorsi*, besides the brachistochrone argument, a second argument directly related to Galileo's research agenda on swinging and rolling would likely have been included had this not been prevented by Galileo's death. In the second half of the chapter, the draft of a reply has been discussed by which Galileo responded to Giovanni Battista Baliani's request for comment on his *De motu naturali gravium solidorum*. In this book, Baliani derived the laws governing naturally accelerated motion from assumptions about pendulum motion, most notably the law of the pendulum and the isochronism of the pendulum.

Galileo criticized Giovanni Battista Baliani's approach, and this critique, as has been shown, was based on the very understanding he himself had gained in his own investigations into the relation between swinging and rolling. Not only did Galileo issue critique, he also made a proposal to Baliani as to how one of his arguments could be improved. In doing so, Galileo devised a proof of the law of the pendulum from the law of fall. This proof is not only included in the draft of the reply but a nutshell version of it was added in the margin of Galileo's hand copy of the *Discorsi* and was apparently destined to be included a second, revised edition of the book.

Baliani's specific approach acted as a stimulus, allowing Galileo to decouple the question of the isochronism of the pendulum from that of the law of the pendulum, which in his own approach on swinging and rolling had been thoroughly entangled. This decisively cleared the way for a new argument resting on an assumption Galileo had long shared. With regard to his aim of providing a demonstration for the properties of pendulum motion, i.e., the law of the pendulum and the isochronism of the pendulum, Galileo had thus achieved the extent of what, from a modern perspective, was achievable. Pendulum motion simply is not isochronous; the isochronism of the pendulum cannot be demonstrated from Galileo's assumptions about natural acceleration along inclined planes, which are correct from a modern perspective, by a valid argument.

Galileo was unable to perceive of things this way. He could not give up his belief in the isochronism of the pendulum, however problematic the assumption may have by then appeared. From his perspective, the proof of the law of the pendulum devised in reaction to Giovanni Battista Baliani must have been a late addendum to, but not the ultimate verdict of, his research program on swinging and rolling. Compared to the ambition which had led Galileo to embark on his program, he had thus achieved but little after 40 years. Yet the insights his ultimately futile program on swinging and rolling had brought forth allowed Galileo to achieve something much bigger, something that would guarantee him his place in the hall of fame of the history of science. Based on these insights, he lays the foundation for a new science of motion, as will be revealed in the second part of this book.

References

Ariotti, P.E. (1971/1972) Aspects of the conception and development of the pendulum in the 17th century. *Archive for History of Exact Sciences, 8*, 329–410.

Baliani, G.B. (1638). *De motu naturali gravium solidorum Joannis Baptistae Baliani*. Genuae: Farroni.

Baliani, G.B., & Baroncelli, G. (1998). *De motu naturali gravium solidorum et liquidorum*. (Biblioteca della scienza italiana, Vol. 19). Firenze: Giunti.

Büttner, J., Damerow, P., & Renn, J. (2004). Galileo's unpublished treatises. In C.R. Palmerino & J.M.M.H. Thijssen (Eds.), *The reception of the Galilean science of motion in seventeenth-century Europe* (Boston Studies in the Philosophy of Science, Vol. 239, pp. 99–117). Dordrecht: Kluwer.

Capecchi, D. (2014). *The problem of the motion of bodies*. New York: Springer.

Caverni, R. (1972). *Storia del metodo sperimentale in Italia*. New York: Johnson Reprint.

Damerow, P., & Renn, J. (2012). *The equilibrium controversy: Guidobaldo del Monte's critical notes on the mechanics of Jordanus and Benedetti and their historical and conceptual backgrounds*, Vol. 2, Ed. Open Access). Berlin: Max Planck research library for the history and development of knowledge sources.

Damerow, P., Renn, J., & Rieger, S. (2001). Hunting the white elephant: When and how did Galileo discover the law of fall? In J. Renn (Ed.), *Galileo in context* (pp. 29–150). Cambridge: Cambridge University Press.

Dijksterhuis, E.J., & Maier-Leibnitz, H. (2002). *Die Mechanisierung des Weltbildes* (2nd ed.). Berlin: Springer

Erlichson, H. (1998). Galileo's work on swiftest descent from a circle and how he almost proved the circle itself was the minimum time path. *The American Mathematical Monthly, 105*(4), 338–347.

Fuhrer, T. (1993). Der Begriff *veri simile* bei Cicero und Augustin. *Museum Helveticum, 50*(2), 107–125.

Galilei, G., & Giusti, E. (Eds.) (1990). *Discorsi e dimostrazioni matematiche intorno a due nuove scienze attinenti alla mecanica ed i movimenti locali* (1990th ed.). Torino: Einaudi.

Galilei, G., Crew, H., & Salvio, A.D. (1954). Dialogues concerning two new sciences, New York: Dover Publications.

Goldstine, H.H. (1980). *A history of the calculus of variations from the 17th through the 19th century*. New York: Springer.

Istituto della Enciclopedia Italiana (Roma). (1963). *Dizionario biografico degli Italiani*, (baccabaratta, Vol. 5). Istituto della Enciclopedia Italiana, Rome

Moscovici, S. (1967). *L'Expérience du mouvement: Jean-Baptiste Baliani, disciple et critique de Galilée, Collection Histoire de la pensée* (Vol 16). Paris: Hermann.

Palmieri P (2009). A phenomenology of Galileo's experiments with pendulums. *The British Journal for the History of Science, 42*(4), 479–513.

Raphael, R.J. (2012). Printing Galileo's *Discorsi*: A collaborative affair. *Annals of Science, 69*(4):483–513.

Settle, T. (1966). Galilean science: Essays in the mechanics and dynamics of the Discorsi. Ph.D. thesis, Cornell University.

Westfall, R.S. (1989). Floods along the Bisenzio: Science and technology in the age of Galileo. *Technology and Culture, 30*(4), 879–907.

Wisan, W.L. (1974). The new science of motion: A study of Galileo's De motu locali. *Archive for History of Exact Sciences, 13*, 103–306.

Wisan, W.L. (1984). Galileo and the process of scientific creation. *ISIS, 75*(2), 269–286.

Part II
Novel Insights and Old Concepts: From Exploration to Formalization

Chapter 10
Toward a New Science: Gathering Results and the Rise and Demise of a Dynamical Foundation

Beginning with this chapter, the second part of this book is devoted to showing how around 1602, and in any case before 1604, Galileo started to collect and systematize the results he had obtained while pursuing his challenging research program on swinging and rolling. On the one hand, he certainly aimed to consolidate what had been achieved. At the same time, as will be argued, the work was clearly conducted under the prospect of a potential future publication. Changes made and novelties introduced in this stage of work indeed owe almost exclusively to the fact that Galileo was no longer merely exploring but was aiming at molding his results into a form in which they could properly be communicated in print in the future. This entailed casting the insights achieved into the shape of a mathematized demonstrative science meeting the established standards of his time. Thus, in particular, Galileo was required to establish a firm basis to found his new results upon, i.e., to organize them into a deductive structure grounded in fundamental assumptions that would find unreserved acceptance. This necessitated linking the new results concerning the pheno-kinematics of naturally accelerated motion to the existing conceptual frameworks of mechanics. This brought about new problems and in turn occasioned further conceptual development.

The reconstruction of this stage of work is facilitated by the fact that Galileo recorded most of his considerations on a set of folio pages readily distinguishable by several distinct features. Due to the occurrence of the same watermark on all of them, a star, this group of folios will be referred to as the star group folios. Almost every entry on these folios introduces some kind of novelty, be it the first appearance of a proposition with a proof, the introduction of a new proof method, or the noting of a new insight. A detailed discussion of all entries on the star group folios is provided in this chapter.

Not only do the folios of this group share the same watermark, but the way Galileo wrote on them shows a striking resemblance on a formal level. Galileo drew a margin which he for the most part respected when writing. He often opened paragraphs with majuscules, and many of the propositions he noted follow the schema of a classical geometrical proof, that is, Galileo added enunciation to the actual proof or apodeixis and a summary in conclusion. Besides the fact that in his exposition of the content Galileo thus mimicked a treatise, a number of additional aspects suggest that we must perceive of this stage of work as the first step aimed at the formation and eventual publication of a new science. Galileo, for instance, marked conditions entering proofs he wrote down by phrases such as "ut infra demonstratur." More than just a marker that the respective condition was already demonstrated, such phrases show that the logical dependencies among the individual propositions were already conceived of from the perspective of an exposition in a book in which one proposition succeeds the next. On a collection of loose sheets, such as it is represented by the majority of folios of the *Notes on Motion*, such an attribution is rather meaningless. It is this linear order superimposed on the logical structure by putting it down as a deductive science in the context of which attributions such as "infra" receive their meaning.

Almost every single entry on the star group folios can be directly related to results achieved by Galileo in the context of his research agenda on swinging and rolling. Quite possibly he was still working on some of the problems this agenda had bestowed him with while he was contemporaneously collecting and systematizing his results on the star group folios. Thus, the year 1602 provides an approximate terminus post quem for the composition of the material on these folios.

Galileo, however, did not merely collect his new results. He started to analyze them with respect to concepts anchored in the traditional conceptual frameworks. By way of such analysis, Galileo sought to provide his new insights with a foundation and to thus transform them into a new science proper. The attempt to establish a foundation documented by the star group folios together with folio 147, which will be included in the analysis in the succeeding chapter, must predate Galileo's fresh approach at a foundation evidenced by a letter to Paolo Sarpi sent in 1604 and by material related to it in the manuscript. This provides a terminus ante quem for the composition of the star group folios.

As detailed in the present chapter, as a result of his analysis, Galileo concluded that the basic principle of inclined plane dynamics he had applied to motion along inclined planes in the *De Motu Antiquiora* would still be applicable even if this motion was conceived of as being naturally accelerated. More concretely, he concluded by means of the so-called tarditas length proportionality argument that the velocities of naturally accelerated motion along two inclined planes of different inclination were in proportion to the mechanical moments on these planes. Based on this insight, Galileo devised the ex mechanicis proof of the law of chords which, once completed, he copied to the star group folios.

Yet further analysis yielded a result that furnished the ex mechanicis proof, and thus, ultimately, the attempt to found the new science of naturally accelerated motion on dynamical arguments, with a problematic character. Reanalysis of the famous mirandum paradox argument in the present chapter exposes that as a result of the conflict revealed by the argument, Galileo had to renounce his assumption of a general applicability of the basic principle of inclined plane dynamics. This deprived the ex mechanicis proof of its explanatory character, and in lack of an alternative, Galileo, for the time being, had to interrupt his search for a foundation of the new science, and correspondingly the collection of material on the star group folios broke off after the mirandum paradox.

10.1 The Star Group Folios: A First Step Toward a New Science

Manuscript evidence Five folios of the *Notes on Motion*, folios 151, 163, 164, 172, and 179b, bear the watermark of a small star.[1] None of these folios is a double folio; hence, identification of a countermark, if it existed, has not been possible. Furthermore, neither the size nor the paper quality of these folios is distinctive enough to reliably identify folios without a watermark as belonging to this group. The sole exception is a cut piece of paper, folio 179a, which, based on its formal resemblance as well as its relation to the cut and pasted piece 179b bearing the watermark small star, will be discussed with the folios of the small star group. Indeed, both pieces likely stem from just one folio that had been cut.

Three of these folios, folios 163, 164, and 172, furthermore display striking resemblances with respect to the layout of their content. On all three folios, Galileo drew a margin by means of an uninked line, which he respected for the most part when writing on these folios.[2] Rather exceptionally, if compared to other folios of the Paduan period, the recto and the verso sides of all of these three folios are almost completely covered with content. As a further peculiarity, the beginnings

---

[1] The watermark shows a small star with mountains. Drake (1979) interpreted what I take to be a star to be a cartwheel and further differentiated the watermarks of the pages under consideration into: "Mountains below cartwheel 30 mm; overall height 53 mm," folio 151; "Mountains below cross in circle 12 mm; overall height 45 mm," folios 163 and 164; "Mountains below circle 10 mm; overall height 38 mm" for folios 172; and "Mountains 18 × 18mm" for folio 179b. A second watermark also shows as a star with mountains. This is, however, considerably larger and is hence referred to here as the large star watermark. Drake refers to it as "Mountains below cartwheel 30 mm; overall height 53 mm." According to him this watermark is present on folios 149, 151, 154, and 160. According to a working group from the Max Planck Institute for the History of Science that inspected the manuscript in 1996, folio 151 in contrast bears the small star watermark. When examining the manuscript, I missed the opportunity to check the watermark of this folio and defer to the judgment of the working group. In any case the folios bearing either of the two watermarks may indeed stem from the same batch of paper as is also suggested by an analysis of the content. Thus, for instance, an early draft of the ex mechanicis proof is contained on folio 151 (small star) that was elaborated into a more complete version on folio 160 (large star), a version which was copied, in turn, to folio 172 recto (small star).

[2] The margins have the same width on the verso and the recto sides of these folios. If the individual folios are compared among each other, one notices a variation in the widths of the margins.

of the individual sections on these pages are generally accentuated by the use of a majuscule for the first letter of the first word of each section, where the first word is often also set off slightly to the left with respect to the marginal line. This particular style of opening individual sections of text, which is otherwise not very common in the *Notes on Motion*, thus provides a further criterion to discriminate these three folios, which are here referred to as the star group folios.

With respect to their content, for the most part the star group, folios represent a collection of propositions, nearly all of which are equipped with a proof. The entries generally show little or no sign of corrections or revisions. This suggests that they were either copied to the star group folios or at least drafted there based on elaborate preliminary versions.

The way in which Galileo wrote on the two cut pieces 179a and 179b, at some point pasted by him onto folio 179 verso, closely resembles the way in which he wrote on the other folios of the group. In particular, the propositions on these two snippets open with majuscules, and in writing them down, a margin demarcated by an uninked line was respected by Galileo. Furthermore, these two snippets contain drafts of the law of the inclined plane and of its generalized version. The inclusion of the principle had become obligatory since it served as the major premise of the dynamical argument of the ex mechanicis proof which Galileo constructed as a result of his work on the star group folios and included into the collection of material on these folios.

The relations among the folios of the small star group and their relations to folios documenting Galileo's work on the broken chord approach are schematically depicted in Fig. 10.1. In arranging the star group folios from left to right, the figure furthermore embraces a hypothesis concerning their order of composition, which has been established based on the following assumptions: First, Galileo wrote on these pages sequentially, i.e., he started writing on a folio page and kept adding information. Usually he turned the page when he had reason to assume the remaining space on the page would not suffice for the next entry. Secondly, as an exception, Galileo continued an exceptionally long entry begun on 172 verso on folio 163 recto.[3] Thirdly, Galileo, for reasons to be investigated, discontinued his work on 164 verso, leaving half of the page empty, indicating that this was the last page of the star group folios that he worked on. Taken together, these three assumptions imply the following order: Galileo started writing on folio 172, carried

---

[3]On folio 163 recto Galileo wrote smaller and smaller until, toward the end of the page, he even disrespected the margin. A possible explanation, which runs against the order of composition I reconstruct, could be that Galileo was forced to do so because folio 163 verso was already filled when he started to write on the recto side. Alternatively, it can be assumed that Galileo squeezed the content onto the page because he wanted to avoid having the proposition wrap onto the next page as he generally seemed to have done on the star group folios. As will be detailed below, Galileo in fact twice added content on space left blank on one side of a folio of the star group folios when the respective obverse side of the folio was already covered with information.

10.1 The Star Group Folios: A First Step Toward a New Science

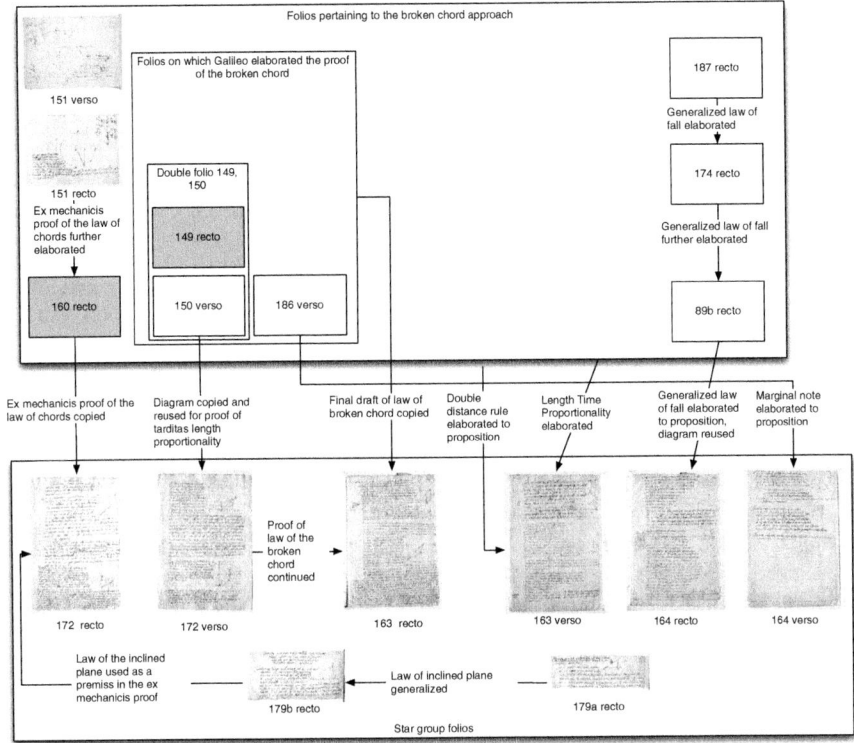

Fig. 10.1 Schematic representation of the folios bearing the small star watermark and their relations to folios whose content is related to Galileo's broken chord approach. Folios bearing the small star watermark are rendered as thumbnail images, those bearing the large star as gray rectangles. The star group folios are assembled in the box at the bottom row

on with his work on folio 163 and after this, on folio 164. In each case he started to write on the recto and continued to write on the verso side of the respective folio.[4]

Content

Folio 172 recto The folio page opens with a proof of the law of chords (first two paragraphs).[5] Galileo's line of reasoning is equivalent to that of the third of the three proofs that he gave for the law of chords in the *Discorsi*. The proof is based on the

---

[4] As argued in the succeeding chapter, it is also possible that the content of 172 recto was written down only after Galileo had completed work on 163 recto.

[5] T1A and T1B.

length time proportionality and the law of fall as premises.[6] Galileo makes no use of integrated time representation. Thus, times of motion are not represented in the diagram, and only proportions between ratios of like magnitudes are considered in the argument.

Added to the proof of the law of chords, more or less as a corollary (the paragraph opens with the words "Ex his colligitur"), is a statement that was published by Galileo as Proposition VII of the Second Book of the Third Day of the *Discorsi*. In the published work, the proposition is stated as follows:

> If the heights of two inclined planes are to each other in the same ratio as the squares of their lengths, bodies starting from rest will traverse these planes in equal times. (EN VIII, 226, trans. Galilei et al. 1954, 194)

This merely provides an alternative geometrical specification for distances on two inclined planes such that these distances are traversed in equal times from rest.[7] The statement, thus, does not add anything substantially new, and its formulation as a corollary appears appropriate. It is indeed somewhat striking that, in contrast to the manuscript, Galileo incorporated the statement into an independent proposition in the *Discorsi*.

With regard to Galileo's aim of transforming his new insights into the nucleus of a new science, the last two paragraphs on the page are the ones which are most revealing.[8] As indicated by the first sentence, Galileo provides a proof for the law of chords, which he presents as an alternative to the proof of the same proposition he had already noted at the top of the page:

> Aliter ostendemus mobile temporibus aequalibus pertransire *ca da* ...

The ex mechanicis proof which follows has been discussed in detail in Chap. 8. It was copied to page 172 recto from a prior draft on folio 160 recto. The proof idea on which this draft was based had in turn initially been noted on folio 151, which shares the star watermark of the star group folios.[9] When copying the proof, Galileo no longer respected the margin he had drawn and observed when writing the first paragraphs on the page. Apparently the verso side of the folio was already filled

---

[6]The third proof of the law of chords in the *Discorsi* opens with the phrase "Aliter idem magis expedite demonstrabitur sic (EN. VIII, 222)." Whereas the manuscript version of the proof is formulated for motions over two arbitrary chords starting from the apex of a circle, in the *Discorsi* version one of the motions is made over an appropriate chord, while the other proceeds over the vertical diameter of the circle. In the course of his proof on 172 recto, Galileo exploits a geometrical proposition which he regularly used and supposed his readers to be aware of, namely, that in right triangles the height over the hypotenuse divides the latter in such a way that the cathetus is the mean proportional between the hypotenuse and the corresponding part of the divided hypotenuse. With respect to the diagram associated with the proof, it thus holds that $ad/ab \sim ab/ac$, i.e., $ab = mp(ad|ac)$.

[7]172 recto, T1C.

[8]172 recto, T2A – T2D.

[9]Cf. Chap. 8.

10.1 The Star Group Folios: A First Step Toward a New Science

with content, and Galileo was thus forced to exploit all remaining space on the recto side to fit in the material he wanted to copy.

As will be detailed below, after drafting the first proof of the law of chords on 172 recto, Galileo had initially turned the page and proceeded by elaborating a direct consequence. This resulted in the very idea on which the construction of the ex mechanicis proof was based. Once the proof had been fully elaborated, its copy was then squeezed in between the existing proof of the law of chords on 172 recto and the considerations noted on top of the verso side of this folio by which it had been engendered. That Galileo squeezed the ex mechanicis proof into the remaining space instead of copying it to a fresh page reflects the fact that in the deductive structure of a publication, this proof from basic assumptions was to replace the proof of the law of chords already contained on the page and indeed the latter was crossed out by Galileo.

A new paragraph was finally added to the copied proof. Here Galileo indicated how to turn the law of chords into a problem, i.e., into a construction task where given two inclined planes and a distance on one of these planes, a distance on the second plane had to be constructed in such a manner that the times of descent were equal.[10]

Folio 172 verso Folio 172 verso opens with two paragraphs that sketch a proposition referred to here as the *tarditas length proportionality*.[11] Directly attached is a third paragraph in which Galileo formulated an immediate consequence.[12] This paragraph will be referred to here as the "si hoc sumatur" note. The tarditas length proportionality and the si hoc sumatur note have received considerable attention and have been discussed controversially.[13] Analyzing these entries, not as isolated fragments but in the context of Galileo's overall agenda documented by the star group folios, will advance a new consistent interpretation.

The tarditas length proportionality opens with the following statement (for the diagram associated with the proposition see Fig. 10.2):

> Sit planum orizzontis secundum lineam *abc*, ad quam sint duo plana inclinata secundum lineas *db*, *da*. Dico, idem mobile tardius moveri per *da* quam per *db* secundum rationem longitudinis *da* ad longitudinem *db*.

---

[10] 172 recto, T2D. From a physical point of view, the transformation of the theorem into a problem adds nothing. In Galileo's time, deductive treatises modeled on the paradigm of Euclid's *Elements* commonly presented propositions in the form of problems. The allusion to reformulating the law of chords into a problem can thus be taken as another indication that when drafting the star group folios Galileo was working under the prospect of publication.

[11] 172 verso, T1A – T1B.

[12] 172 verso, T1C.

[13] Interpretations of the tarditas length proportionality argument have, among others, been given by Wisan (1974, 222–223), Drake (1978, 80–81), Galluzzi (1979, 288–291), and Souffrin (1990, 96–100).

Fig. 10.2 Diagram associated with the tarditas length proportionality argument on folio 172 verso

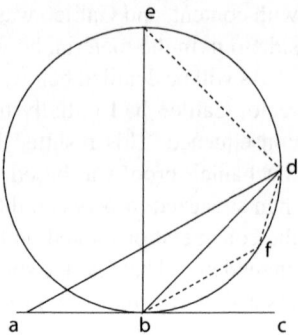

Hence it is to be demonstrated that the slownesses of motions along inclined planes of equal height (da and db) are in the same ratio as the length of the planes traversed. A first noteworthy detail is Galileo's use of the adverb "tardius," and thus the introduction of the concept "tarditas" or slowness into his considerations, a concept otherwise virtually absent from the *Notes on Motion*.[14] From the way Galileo used "tarditas" in the successive argument, it becomes incontestably clear that in this context, it is understood simply as the inverse of velocity. More precisely, Galileo held the ratio of the slownesses of two motions to be inversely proportional to the ratio of their respective velocities. Galileo's claim in the tarditas length proportionality can be represented as:

$$Tard(da)/Tard(db) \sim da/db$$

This first step of Galileo's argument is the one which is most problematic:

Patet, tarditatem per fb esse consimilem tarditati per da...[15]

The slowness of motion along fb is said to be similar ("consimilem") to the slowness of motion along da of equal inclination, but different length and Galileo will treat both motions as interchangeable with respect to their slowness. I transcribe this as:

---

[14] We encounter "tarditas" on three other folios: on folio 34 recto in a copy of the entry from 172 verso, as the adjective "tardius" on folio 147 verso in the so-called hollow paradox argument, and finally once more on another of the star group folios, in an entry on folio 164 verso. In *De Motu Antiquiora* "tarditas" was used more frequently. In the increasingly quantitative considerations concerning naturally accelerated motion, however, the concept, which had previously denoted the contrary of velocity, was reduced to being simply the inverse of the latter and thus became obsolete.

[15] Galileo's use of "per" as opposed to the "super" he used in conjunction with mechanical moments indicates that he perceived of slowness (and thus also velocity) as the property of a concrete motion, i.e., a motion made over a concrete distance completed in concrete time and not as a general property specific to naturally accelerated motion along an inclined plane of a given inclination as he had done in *De Motu Antiquiora*, at least by implication.

10.1 The Star Group Folios: A First Step Toward a New Science

$$Tard(fb) \simeq Tard(da)$$

From a modern perspective, if the slownesses (and thus also the velocities) of the motions along parallel planes such as fb and da are entirely similar, it seems to be inescapably implied that they are equal. This in turn would imply that tarditas (and thus velocity) depends solely on inclination which seems to be irreconcilable with the assumption that this motion is naturally accelerated. Must we therefore assume that Galileo, just as in the *De Motu Antiquiora*, is here considering motion along inclined planes as uniform?[16] Certainly not, as the remaining part of the argument, but also the general context in which the tarditas length proportionality is set, make patently clear. Galileo was without doubt considering the case of naturally accelerated motion.

What exactly then did it mean for Galileo to state that the slownesses (and velocities) of two naturally accelerated motions were entirely similar and why did this understanding not immediately imply the contradiction, at least not for him, which from a modern perspective seems to result so self-evidently? As argued in Chap. 2, for Galileo *the* velocity of *a* motion did not matter and was indeed not well defined, nor was it easily definable based on the underlying mathematical framework. What he considered and what he could express consistently based on his mathematical framework were ratios of velocities. These ratios could be in proportion to ratios of other magnitudes such as, for instance, spaces traversed in equal time.

Galileo's statement concerning the similarity of velocities has to be interpreted with regard to ratios of velocities considered as a shorthand way of expressing that the ratio of the velocities of motions made over planes of different inclinations should not be affected if one of the motions was replaced by another motion made over a different inclined plane, however, one of identical inclination. That this is what is insinuated is indeed indicated by Galileo's use of "similar" rather than "equal" to characterize the relation between the velocities along fb and da.

Understood in this way, Galileo's surmise of a similarity of the velocities of motions on inclined planes of the same inclination does not contradict the assumption that motion on inclined planes is accelerated, at least not directly.[17] Indeed, the assumption that accelerated motions on planes of different inclinations

---

[16]Drake in fact stated: "Another such sheet, f. 172, is of special interest because it was started shortly before Galileo became concerned with acceleration as such; ... Galileo first wrote this entire proposition [the tarditas length proportionality] in terms of overall speeds, ignoring acceleration ...(Drake 1978, 80–81)."

[17]Similar to the example Galileo had given in his letter to Guidobaldo del Monte in 1602, a geometrical figure, for instance, a triangle, can be perceived as growing while always remaining geometrically similar. In this case the ratio of two sides of such a figure remains constant despite the fact that both magnitudes from which it is formed are changing.

are characterized by a constant ratio of their velocities is backed by an intuitive dynamical understanding. The steeper the plane, the greater the cause of motion. It is thus entirely plausible to assume not only that motion on a steeper plane will always be faster than motion on a less inclined plane but that indeed the velocities of motion on planes of different inclination obey a fixed relation.

Next, applying the law of chords, Galileo inferred that the times of motion along the chords db and fb had to be equal, i.e., that:

$$t(db) = t(fb).$$

From the equality of the times, Galileo immediately concluded:

patet, velocitates per db ad velocitates per fb esse ut db ad fb

This last step demands scrutiny, in particular, because of the unexpected use of the plural in the expression "velocitates per" an inclined plane. One would expect Galileo to use the singular, just as he had done to denote the slowness ("tarditatem per fb esse consimilem tarditati per da"), in which case Galileo's conclusion would indeed directly follow from an application of the second Aristotelian kinematic proposition to the motions along the chords db and fb completed in equal time:

$$t(db) = t(fb) \Leftrightarrow v(db)/v(fb) \sim db/fb.$$

However, Galileo's use of the plural can be explained. A number of passages in the *Notes on Motion* but also in the *Discorsi* attest that Galileo occasionally used the expression "velocitates per" as a short way of referring to all degrees of velocity of a motion over a given distance and that this totality of the degrees of velocity was held by him to be interchangeable with the kinematic velocity.[18]

Galileo's use of the plural testifies to his understanding of the relation between the concept of velocity entailed by the doctrine of the configurations of motion and the kinematic concept of velocity as it was implicitly defined by the kinematic propositions. At least for the case of naturally accelerated motion, Galileo held that the quality, whose change was expressed by the change of the degrees of velocity over an extension, corresponded to the overall or kinematic velocity. As will be seen, Galileo initially believed that the area defined by the changing degrees in an Oresme-type diagram and traditionally referred to as the quantity of the quality was proportional to the kinematic velocity. Yet this understanding was modified in the further course of conceptual development. As for the present purpose, it is not important how exactly Galileo conceptualized the relation between all degrees, i.e., the totality of the degrees of velocity and the overall kinematic velocity. I use the term *aggregate velocity* to refer to the totality of the degrees understood

---

[18]Galileo had apparently first written "velocitas per" and then corrected it for the plural to "velocitates per".

10.1 The Star Group Folios: A First Step Toward a New Science

to be proportional to the velocity and transcribe this as $A_{v_g}(ab)/A_{v_g}(cd) \sim v(ab)/v(cd)$.[19]

Indeed, when Galileo formulated his conclusion, he dropped the use of the plural and referred to what had previously been referred to as "velocitates per db" as "velocitas per db" without altering the meaning, demonstrating that for him both expressions essentially denoted the same concept.
Thus, indeed:

$$A_{v_g}(db)/A_{v_g}(fb) = db/fb$$

The next step of the argument represents a dead end. It expresses a consequence of the foregoing which is then, however, not used further in the proof:

> ita ut semper iisdem temporibus duo mobilia, ex punctis d, f venientia, linearum db, fb partes integris lineis db, fb proportione respondentes peregerint.

From the wording it remains equivocal whether "partes integris" refers to two arbitrary distances on the chords fb and db whose lengths obey the same ratio as the chords or to distances taken from the respective starting points d and f of the motions on the planes. For the case of naturally accelerated motion, only the latter holds and can be transcribed as:

$$\text{for } W \subset db, Y \subset fb, dW/fY \sim db/fb \Rightarrow t(dW) = t(fY)$$

Galileo finally completed his argument by geometrical reasoning and by eventually exchanging the ratio of the velocities for the inverse ratio of the slownesses. By geometry:

$$bd/bf \sim ad/db$$

from which it follows by $A(v_g) \sim V$ and $V \sim 1/Tard$ that:

$$V(db)/V(ab) \sim ad/db \sim Tard(ab)/Tard(db).$$

Thus, Galileo could state in conclusion:

> Ergo ut ad ad db, ita velocitas per db ad velocitatem per da, et, ex opposito, tarditas per da ad tarditatem per db.

The argument can be summarized as follows: Velocity and tarditas are considered as inversely proportional; velocity is furthermore used interchangeably with

---

[19] Applied to the case of uniform motion, the assumption that the aggregate velocity is proportional to the kinematic velocity would immediately run into the contradictory statement that the kinematic velocity of uniform motion is growing. For the case of uniform motion, as will be seen, Galileo instead assumed the kinematic velocities to be in the same ratio as the constant degrees of velocities characterizing the respective motions. My use of the term aggregate velocity is inspired by Settle (1966, 164), who refers to the "aggregate of parallels" instead.

aggregate velocity, i.e., the totality of the degrees of velocity of motion. The ratio of the velocities is held to depend solely on the inclinations of the respective planes. Thus, it is neither altered by parallel transposition nor by extension of the distance covered by motion on a plane of given inclination. By means of the law of chords, it is concluded that accelerated motions on two inclined planes are completed in equal times and, by the second kinematic proposition, that the ratio of the velocities on these inclined planes is the inverse of the lengths of these planes. By appropriate parallel translation and an extension of one of the motions considered, it is thus proven that, for naturally accelerated motion on inclined planes of equal height, the slownesses are in the same ratio as the lengths of the planes.

Opposed to most of the material assembled on the star group folios, the tarditas length proportionality appears to be strikingly unrelated to Galileo's agenda on swinging and rolling, at least at first sight. An additional construction line in the proof diagram, not used in the argument itself, however, directly links the tarditas length proportionality to Galileo's consideration in the context of the broken chord approach. In fact, in the proof diagram (Fig. 10.2), the points d and f are connected by a dotted line, such that the arc spanned by the chord db is also spanned by the broken chord dfb revealing that the diagram had originated in a different context from which the line df remained as a relict.[20]

As it turns out, a diagram from which the diagram on folio 172 verso derives can be identified in the manuscript. It is contained on double page 149 recto – 150 verso. As shown in Chap. 5, the diagram was indeed part of considerations situated in the context of the broken chord approach. Moreover, Galileo's considerations on the double folio already comprised the idea of a parallel transposition of one of the motions under consideration, and it is exactly this idea that reoccurred and was to become central in the tarditas length proportionality argument. There can, hence, be little doubt that the argument was borne out of Galileo's considerations documented on 149 recto – 150 verso and thus more generally, like almost the entire content of the star group folios, out of Galileo's research concerning the relation between swinging and rolling.

Appended to the tarditas length proportionality is the si hoc sumatur note, for which Galileo did not start a new paragraph.[21] Instead he connected the entry to the tarditas length proportionality by means of a horizontal line, indicating that the note expressed a direct consequence. Rather emphatically, Galileo opened by asserting that if what followed is assumed, it would be possible to demonstrate the rest.[22] The full entry reads:

> Si hoc sumatur, reliqua demonstrari possent. Ponatur igitur, augeri vel imminui motus velocitatem secundum proportionem qua augentur vel minuuntur gravitatis momenta; et

---

[20] 172 verso, D01A.

[21] 172 verso, T1C.

[22] The close relationship between the "si hoc sumatur note" and the tarditas length proportionality, which Galileo even accentuated graphically, has often been overlooked. Some interpretations advanced even imply a major conceptual break between the two entries.

10.1 The Star Group Folios: A First Step Toward a New Science

cum constet, eiusdem mobilis momenta gravitatis super plano *db* ad momenta super plano *da* esse ut longitudo *da* ad longitudinem *db*, idcirco velocitatem per *db* ad velocitatem per *da* esse ut *ad* ad *db*.[23]

Exactly which assumption did Galileo appraise as so valuable for proving the rest? The answer is not entirely straightforward.[24] Again, the interpretation of the passage is impeded by the use of a plural, this time in the expression moments of heaviness ("momenta gravitatis") on one inclined plane.[25]

An inclined plane of given inclination is characterized by a moment of heaviness (see Chap. 2). Galileo's use of the plural moments has thus resulted in the claim that with the si hoc sumatur note, he introduced a crucial conceptual modification. According to this interpretation, in the note Galileo was expressing his new assumption that the moment of heaviness increases on one and the same inclined plane in such a way that the growing degrees of velocity of naturally accelerated motion on the plane are in proportion to the growing moments of heaviness. This interpretation appears to be inherently plausible as it ascribes a conception to Galileo that would naturally extend the *De Motu Antiquiora* framework to naturally accelerated motion. Where in the *De Motu Antiquiora* a constant ratio of mechanical moments was held to be responsible for the constant velocities on inclined planes, it would seem natural, once it is accepted that motion is naturally accelerated and that such motion is characterized by increasing degrees of velocity, to likewise assume increasing moments to be the cause of such motion.[26]

---

[23]"If we assume this, we can demonstrate the rest. It be therefore assumed that the velocity of motion augments or diminishes according to the ratio in which the moments of heaviness augment or diminish; and since it is certain (constet) that the moments of heaviness of the same movable on the plane *db* are to the moments over the plane *da* as the length *da* to the length *db*, therefore the velocity on *db* is to the one on *da* as *ad* to *db*." Revised translation based on Wisan (1974, 222).

[24]For Souffrin: "Ce passage a semblé particulièrement difficile à interpréter..." His interpretation corresponds by and large to the one presented here. See Souffrin (1990).

[25]When considering the static problem of forces on inclined planes, Galileo usually used a singular to denote the force. In *De Motu Antiquiora*, he used expressions such as "momentum ponderis" or simply "momentum," and in his formulation of the law of the inclined plane on folio pages 179a recto and 179b recto, he used "momentum gravitatis."

[26]Wisan (1974, 223) adheres to a "growing moments" interpretation and states: "use of the plural form indicates that the 'moment of gravity' is something that increases along a given plane." Her interpretation of the si hoc sumatur note strikes me as fairly imaginative. Galluzzi also advocated a growing moments interpretation. He wrote: "Ciò significa che il *momento della gravità* aumenta constantemente dall'inizio del moto," and more strongly on the successive page, "La supposizione è chiarissima e di grande interesse. L'uso del plurale *gravitatis momenta* per un moto lungo uno stesso piano lascia intendere, senza possibilità di equivoco, che Galileo ha ristrutturato il *momento* 'meccanico' constante in termini di accelerazione (Galluzzi 1979, 289–290)." Galluzzi speculates that when writing the note, Galileo assumed the degrees velocities grew in proportion to the distance traversed and thus must have concluded that the moments of heaviness likewise grow in proportion to the distance traversed from the start of motion. Yet in the note Galileo clearly indicated that a conclusion was made from a statement about the moments of heaviness to the behavior of motion and not the other way around as implied by Galluzzi's interpretation.

However, upon closer scrutiny the conception ascribed to Galileo by the *growing moments interpretation* is revealed untenable. First of all it is not stated by Galileo in the note that the moments of heaviness grow with the distance traversed nor that they grow with the times elapsed during naturally accelerated motion, let alone that this growth is in proportion to the growth of the degrees of velocity. It is merely stated that the ratio of the moments changes (grows or diminishes) as the (ratio of) velocities of motion change(s). With regard to the change of what other magnitude the ratios examined are considered to change, this is not explicated. As will be argued, it is change with regard to inclination and not with regard to times elapsed or spaces traversed that Galileo had in mind. At any rate, the assertion that Galileo believed that on one and the same inclined plane the moments of heaviness increase as the degrees of velocity is a conjecture of the growing moments interpretation that is not directly supported by his note.

Indeed, nowhere in the *Notes on Motion*, nor in fact in Galileo's entire oeuvre, does the alleged concept of moments of heaviness growing in proportion to the degrees of velocity reappear.[27] Thus, in particular, no attempt to demonstrate the rest ("reliqua demonstrari"), based on this alleged conception, can be identified, in striking contrast to the emphatic and programmatic declaration which Galileo opened the note with.

Moreover, if, as claimed by the growing moments interpretation, it was here that Galileo indeed first introduced the concept of growing moments of heaviness, one would expect him to specify how this growth takes place. Instead, Galileo began the third sentence of the note by stating that it was already certain ("constet") how the alleged growing moments behaved, namely, that the moments of the same mobile on db were to the moments on da as the distances da to db. This statement can hardly be reconciled with the growing moments interpretation as it would be rather meaningless for Galileo to specify the growth of a magnitude by stating that it was in proportion to a fixed ratio of distances.

An alternative interpretation, which I follow and supplement, has already been proposed in the literature. According to this interpretation, with the note Galileo intended to express his insight that the ratio of mechanical moments on two inclined planes as given by mechanics, was in proportion to the ratio of the velocities of the naturally accelerated motions along these planes and that this implied that his basic

---

[27]Galileo regarded the moment of percussion or impact of a body to depend on its velocity and, hence, as growing in the case of natural acceleration. When Galileo wrote the si hoc sumatur note he did not draw a connection, however, between the moment of impact, held to be proportional to the degree of velocity, and the mechanical moment on an inclined plane depending on the inclination. Indeed Galileo drew a clear distinction between "momento interno e naturale" and the "momento estrinsico o violento" in the Sixth Day on the force of percussion, added to the *Discorsi*.

principle of inclined plane dynamics would be valid for, and applicable to, the case of naturally accelerated motion.[28]

By the tarditas length proportionality argument, Galileo had in fact exactly reproduced the velocity theorem of the *De Motu Antiquiora*, now, however, valid for the case of naturally accelerated motion. In the *De Motu Antiquiora*, the velocity theorem had been adopted as a direct consequence of Galileo's basic principle of dynamics and had thus allowed the perception of the kinematics of the ensuing motions as a direct result of the forces on the inclined planes. In the *Notes on Motion*, starting from the kinematics of naturally accelerated motion, the tarditas length proportionality argument had disclosed that the velocities of motion along inclined planes of equal height were in indirect proportion to the length of the planes and thus also in direct proportion to the moments of heaviness on these planes, suggesting that the basic principle of dynamics held in this case. Galileo's "si hoc sumatur, reliqua demonstrari possent" then refers to the idea of reversing the movement of thought. Instead of perceiving the tarditas length proportionality as implying the validity of the basic principle of inclined plane dynamics, the latter could be perceived as the premise based upon which the former could be proved. As, moreover, the tarditas length proportionality itself had been derived as a consequence of the law of chords, Galileo could be rather confident that the law of chords should likewise be provable based on the basic principle of inclined plane dynamics as a premise and, thus, ultimately, based on assumptions rooted in mechanics.[29]

As we have already seen, this is in fact exactly what Galileo did. As shown in Chap. 8, based on the basic principle of inclined plane dynamics as a premise, Galileo devised the ex mechanicis proof of the law of chords. As his "si hoc sumatur" indicates, his note on 172 verso must mark his very first realization that the dynamical reasoning of his older *De Motu Antiquiora* analysis would in principle also apply in the case of naturally accelerated motion. This at last allows a contextualization of the sequence of considerations which culminated in a draft of the ex mechanicis proof, as has already been indicated in Chap. 8 and will be repeated in some more detail here. The following remarkably complete and consistent picture results:

On 172 recto, as part of his attempt to collect, systematize, and analyze the results of his research on swinging and rolling, Galileo noted a proof of the law of chords, which was one of the fundamental propositions forming the basis for his considerations in the context of the research program. He turned the

---

[28] Without further scrutiny Giusti asserts that in the si hoc sumatur note, Galileo stated the basic principle of dynamics, which according to him "clearly lies behind a number of his [Galileo's] proofs. Moreover, we find it in Toricelli's *De motu*, with explicit reference to Galileo . . . The same statement can be found in Baliani (Giusti 2004, 121–122)."

[29] Souffrin (1990) shares this understanding and phrased it thus: "Le début du passage exprime la conviction de Galilée qu'il doit être possible de donner à ce principle un rôle fondateur dans la science du mouvement."

page and, setting out from the law of chords, engaged in the tarditas length proportionality consideration. The argument indicated that the basic principle of inclined plane dynamics would apply to the case of naturally accelerated motion and that it should hence be possible to derive at least some of his assumptions concerning the pheno-kinematics of naturally accelerated motion based on this principle.

Prior considerations in the context of his research on the relation between swinging and rolling and, indeed, noted on a folio sharing the star watermark, lend themselves to the formation of an argument based on the application of the principle. Beginning from there, Galileo constructed the ex mechanicis proof. The final draft of the ex mechanicis proof of the law of chords was copied to folio 172 recto, squeezing it in between the tarditas length proportionality argument from which the construction of the proof had taken its outset and the already existing proof of the law of chords, which it was supposed to replace in the deductive structure of a new science.

The basic principle of inclined plane dynamics presupposed the law of the inclined plane, which determined the ratio of mechanical moments on inclined planes, and it was the latter which, as argued, Galileo was referring to when he stated that it was already certain ("constet") how the ratio of mechanical moments depended on the inclinations of the planes considered. This made it mandatory to insert this principle into the deductive structure of a new science Galileo was aiming for. Indeed, evidently, in the same period of work, Galileo wrote down a statement of the principle in a restricted and more general version. Only later was/were the folio/folios on which Galileo had written these entries cut and those cut pieces (179a and 179b) pasted to folio 179.

The interpretation implies that Galileo's use of the plural "momenta gravitatis" in the entry was simply the result of a grammatical imprecision. In the first sentence of the passage, the plural is unproblematic as there Galileo was referring to two different moments each on a different inclined plane. His use of the plural in the second sentence can be explained, assuming that he originally intended to refer to the ratio of the moments and thus concurrently to two moments each on one of the two planes, just as he had done in the first sentence but that he inconsistently repeated the specification of "momenta" for the second plane. In other words, it is implied that Galileo in his sentence conflated the expressions "the ratio of the moments over db and da" and "the ratio of the moment on db to the moment on da," both of which correctly render what, according to the interpretation given, Galileo wanted to express and it is the conflated sentence thus brought about that has given rise to the growing moments interpretation.

On the remaining space on folio 172 verso, Galileo began to draft the version of the law of the broken chord and its proof, including all of its lemmata, as it was later published in the *Discorsi*. Folio 172 verso contains drafts of lemmata one and two, which Galileo continued writing on folio 163 recto.

10.1 The Star Group Folios: A First Step Toward a New Science

Folio 163 recto The draft of the law of the broken chord, begun on folio 172 verso, continues on folio 163 recto where Galileo noted the third lemma associated with the law of the broken chord.[30] This is followed by the proposition together with the proof.[31]

Folio 163 verso The verso side of folio 163 comprises an entry in which Galileo established what will be referred to as the *sesquialterum version* of the double distance rule, a proof of the length time proportionality, as well as an entry in which the double distance rule is generalized. The entries are written on the page in the listed order. Corresponding to the close relation of their content, a line in the left margin connects the first with the last entry.

The double distance rule has a pivotal status as it links naturally accelerated and uniform motion. It states that if a body, after an initial accelerated motion of fall through a certain distance, is deflected into the horizontal, where it continues its motion uniformly, the distance traversed in a time equal to that of the initial motion of fall amounts to twice the distance initially fallen.[32] The sesquialterum version of the proposition which Galileo noted on 163 verso states for the same motion situation, i.e., deflection of a falling motion into a horizontal uniform motion, that the time of the initial fall together with that of the successive uniform motion through a distance equal to that of the initial distance fallen through is equal to one and a half (sesquialterum) times that of the initial motion of fall alone. The double distance rule trivially follows from the sesquialterum version and vice versa.[33]

Galileo's double distance rule is closely related to the medieval Merton or mean degree rule, and the exact nature of this relation has been intensely debated by historians of science.[34] Central to this debate has been the question whether Galileo had knowledge of the Merton rule or whether he formulated his double distance rule more or less independently. Galileo's entries on folio 163 recto, as will be seen, strongly suggest that he was familiar with the Merton rule. The opposite would indeed be surprising in view of the fact that Galileo was certainly familiar with the conceptual tradition of which the Merton rule was part.[35]

---

[30] Entry T1A and T1B.

[31] Entry T2A and T2B. For Galileo's construction of the proof, including the associated lemmata, see Chap. 5.

[32] The double distance rule was not explicitly formulated by Galileo as a proposition in the *Discorsi*. It is merely disclosed in a scholium appended to Proposition XXIII of the Second Book of the Third Day of the *Discorsi*: [h]ence we can infer that, if, after descent along the inclined plane ... the motion is continued along a horizontal line ..., the distance traversed by a body, during a time equal to the time of fall through AC, will be exactly twice the distance AC. (EN VIII, 242, trans. Galilei et al. (1954, 214))."

[33] Galileo may initially have had a slight preference for the way this essentially same physical fact was expressed in the sesquialterum version of the double distance rule. This is implied by his application of the sesquialterum variant on folio 151 recto. See the discussion in Chap. 8.

[34] For a survey of the discussions regarding the relation between Merton rule and Galileo's double distance rule, see, for instance, Damerow et al. (2004, 175–179).

[35] Based on the study of his early notebooks Wallace has examined in a number of studies (see in particular Wallace 1984, 1981), the influence of the teachings of the professors at the Jesuite Collegio Romano on the young Galileo. The early notebooks contain "the mention of Heytesbury

However, as the following analysis will show, the double distance rule is more than just Galileo's own rephrasing of a long established insight. Indeed, by embedding the statement in his considerations concerning accelerated motion, Galileo far exceeded the conceptual scope of the traditional Merton rule. The Merton rule represents merely a specific application of a general, abstract theory of processes of change to the special case of local motion which, moreover, remained without any further consequences for the theory itself. Galileo's double distance rule, in contrast, was a proposition applicable and in fact applied by him to actual as well as theoretically perceived motions.[36] The way Galileo phrased the sesquialterum version of the double distance rule in the manuscript indeed noticeably mirrors this difference, indicating that Galileo was well aware of it.

Correspondingly, the entry in which Galileo argued for the sesquialterum version comprises two readily distinguishable parts. In the first part of the entry, Galileo in fact expressed the Merton rule and gave an abridged argument. In the second part of the entry, the Merton rule is transformed into the double distance rule by giving the abstract elements of the former an actual physical interpretation in terms of motions as treated by Galileo in the context of his emerging new science.

The first part of the entry reads (for the associated diagram see Fig. 10.3):

> Factus sit motus ex *a* in *b* naturaliter acceleratus: dico, quod si velocitas in omnibus punctis *ab* fuisset eadem ac reperitur in puncto *b*, duplo citius fuisset peractum spacium *ab*; quia velocitates omnes in singulis punctis *ab* lineae, ad totidem velocitates quarum unaquaeque esset aequalis velocitati *bc*, eam habent rationem quam [tri]angulus *abc* ad rectangulum *abcd*.[37]

---

and the 'Calculator', the discussion of degrees of qualities in which Galileo uses the expression *uniformiter difformis* and explicit references to the *Doctores Parisienses* (Wallace 1981, 192)." Yet Wallace (1981) claimed that in his early work on the new science of naturally accelerated motion Galileo did not resort to the doctrine of the configurations of motion. According to my interpretation, this claim cannot be upheld. With regard to the question of how Galileo concretely became aware of the double distance rule, Renn stated: "However, although, ... the early modern tradition of the configuration of qualities did not include its use as a calculational tool, it cannot be doubted that some of its basic concepts were part of common knowledge, and that hence no particular medieval text has to be assumed to be the privileged source of Galileo's familiarity with this tradition (Damerow et al. 2004, 166)."

[36]The double distance rule played a crucial role in particular in Galileo's experiments concerning a theory of projectile motion, as documented, for instance, on folio 116 verso. For an interpretation of this experiment which, as indicated, may indeed have been conducted at the time when Galileo was working on his challenging agenda on swinging and rolling, see Hahn (2002).

[37]Entry T1. "Let the motion from *a* to *b* be made in natural acceleration: I say, if the velocity in all points *ab* were the same as that found in the point *b*, the space *ab* would be traversed twice as fast; because all velocities in the single points of the line *ab* have the same ratio to just as many velocities each of which is equal to the velocity be as the triangle *abe* has to the rectangle *abed* (trans. Damerow et al. (2004, 176))."

Fig. 10.3 Proof diagram associated with the proof of the sesquialterum variant of the double distance rule on folio 163 verso

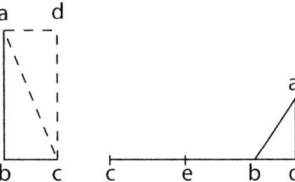

Galileo's reasoning is extremely abridged and falls short of a proof. What is to be proved is that naturally accelerated motion from a to b proceeds twice as fast as a uniform motion over the same distance, whose degree of velocity in every point is the same as the final degree of velocity of the accelerated motion represented by the line bc. The core of Galileo's argument consists in merely one line of reasoning in which it is said that all degrees of velocity of the accelerated motion have the same ratio to all degrees of velocity of the uniform motion as the area of the triangle abc has to the area of the rectangle $abcd$ and thus has the ratio of one to two.

A number of the assumptions required to formally complete the reasoning remain implicit. It is thus, in particular, not explicitly stated according to which rule the degrees of velocity are supposed to grow in naturally accelerated motion. No reason is given why the ratio of all degrees of velocity of one motion to all degrees of velocity of the other motion is in proportion to the ratio of the areas identified, and finally, it is not made explicit why, given this is the case, the uniform motion can be said to proceed with double the *velocity* of, or else twice as fast as, the accelerated motion.

All the assumptions which remained implicit in Galileo's brief argument are embraced in the conceptual tradition of which the Merton rule was a part, i.e., the doctrine of the configurations of motion. The cursory nature of Galileo's entry thus lends additional support to the hypothesis that Galileo was here resorting to the Merton rule as part of this conceptual tradition, as familiarity with this tradition allowed to straightforwardly amend the implicit assumptions: that the degrees grew in proportion with the extension considered which, as the diagram indicates, is here conceived of as the distance fallen. Thus, in the entry, the degrees of velocity are, at least implicitly, assumed to grow according to the Sarpi letter principle.[38] The areas of the figures defined by the growing degrees of velocity over the extension represent the quantity of the quality which, as before, is addressed as the aggregate velocity, equivalent to the kinematic velocity. Galileo's proposition then directly follows from a kinematical argument.

---

[38] Galileo's erroneous principle of naturally accelerated motion according to which the degrees of velocity grow in proportion to the distance covered in free fall was first expressed in a letter to Paolo Sarpi written on 16 October 1604. It is referred to here as the Sarpi letter principle.

The second part of the entry is marked as a consequence of the foregoing:

> Sequitur ex hoc, quod si ad orizontem *cd* fuerit planum *ba* elevatum, sitque *bc* dupla ad *ba*, mobile ex *a* in *b*, et successive ex *b* in *c*, temporibus aequalibus esse pervencturum: nam postquam est in *b*, per reliquam *bc* uniformi velocitate et eadem movetur, qua in ipsomet termino *b* post casum *ab*. Patet rursus, totum tempus per *abe* ad tempus per *ab* esse sesquialterum.[39]

Whereas in the first, the Merton rule, part of the entry, the motions were characterized in abstract terms, essentially by reference to the arrangement of their degrees of velocity, in the second part of the entry, these characterizations have been replaced by description of the motions in terms of motion situations occurring, at least ideally, in physical reality: a motion along an inclined plane and a successive motion on a horizontal plane. In consequence, the abstract descriptions of motions in the first, Merton rule, part of the entry are identified with real or potentially real motions: the rolling of a ball down an inclined plane is identified as a naturally accelerated motion and the successive horizontal motion after deflection as a uniform motion. Thereby, an assumption about the deflection of naturally accelerated motion into uniform motion is entailed, namely, that the degree of velocity is preserved, as otherwise the conclusions of the first part would not be applicable.

Galileo's interpretation of the Merton rule as the double distance rule is by no means natural or self-evident. That the motion of a falling body is indeed characteristically accelerated in such a way, moreover, that the degree of velocity grows uniformly may have been stated before Galileo. However, it was Galileo who turned this into a central claim based upon which a new science of motion was to be erected.[40] That "natural," force-free horizontal motion proceeds uniformly, and is not decelerated, was a theoretical consequence Galileo had first elaborated in the *De Motu Antiquiora*.[41]

That the degree of velocity finally can be assumed to be preserved upon the deflection of motion does not hold true on sharp corners such as the one Galileo depicted at point *b*, due to the loss of kinetic energy the deflected body suffers by impact. In his experiments concerning projectile motion, Galileo had used smoothed corners for deflecting the motion of bodies falling along inclined planes. His

---

[39]Entry T1. "From this it follows, that if there were a plane *ba* inclined to the horizontal line *cd*, and *bc* being double *ba*, then the moving body would come from *a* to *b* and successively from *b* to *c* in equal times: for, after it was in *b*, it will be moved along the remaining *bc* with uniform velocity and with the same with which [it is moved] in this very terminal point *b* after fall through *ab*. Furthermore, it is obvious that the whole time through *abe* is 1 1/2 the time [sesquialterum] for *ab* (trans. Damerow et al. (2004, 176))."

[40]The first to have related the Merton rule as an abstract statement founded in the doctrine of the configurations of motion to an actual motion, namely, that of free fall, was apparently Domingo De Soto. Cf. Barbour (2001, 374) and Wallace (1990).

[41]For Galileo's assumption that a vanishingly small force is needed to set a body in motion on a horizontal, see Chap. 2. On how this assumption may have led to what has come to be called Galileo's "circular inertia," see Wolff (1994). For Galileo and rotational motion see Büttner (2008).

10.1 The Star Group Folios: A First Step Toward a New Science

assumption of the preservation of degree of velocity on sharp corners thus emerges as a theoretical abstraction.[42]

After the identification of the abstract with a real motion situation, the proposition follows trivially by application of the first Aristotelian kinematic proposition: as established in the first part of the entry, the uniform horizontal motion proceeds twice as fast as the initial motion of fall, which, since both motions are made over equal distances, is equivalent to stating that it is completed in half the time of the first motion. As a consequence, the overall time of both motions successively had to be one and a half times that of the initial motion of fall alone.

At the bottom of folio 163 verso, Galileo stated an immediate consequence of the sesquialterum version double distance rule[43]:

> Si post casum per aliquod planum inclinatum sequatur motus per planum orizontis, erit tempus casus per planum inclinatum ad tempus motus per quamlibet lineam orizontis ut dupla longitudo plani inclinati ad lineam acceptam orizontis. Sit linea orizontis cb, planum inclinatum ab, et post casum per ab sequatur motus per orizontem, in quo sumatur quaelibet linea bd. Dico, tempus casus per ab ad tempus motus per bd esse ut dupla ab ad bd. Sumpta enim bc ipsius ab dupla, constat ex praedemonstratis, tempus casus per ab aequari tempori motus per bc: sed tempus motus per bc ad tempus motus per bd est ut linea cb ad lineam bd: ergo tempus motus per ab ad tempus motus per bd est ut [du]pla ab linea ad lineam bd.

Essentially, Galileo stated that given the motion situation of the double distance rule, i.e., the deflection of an accelerated motion into a uniform horizontal motion, it holds for an arbitrary distance traversed in the horizontal that the time of motion to cover this distance is to the time of motion of the preceding fall in the same ratio as the distance covered in the horizontal is to twice the distance initially fallen. This generalization followed as an immediate consequence of the sesquialterum version of the double distance rule, by application of a kinematic proposition to the uniform motion along the horizontal. According to the Archimedean kinematic proposition, the times elapsed during uniform motions proceeding with one and the same velocity are as the distances covered. Hence, given the time to traverse a given distance, it is always possible to infer the time needed to cover any other given distance, and with this Galileo's statement followed without further complications.

The generalization of the double distance rule, which allowed motion over an arbitrary horizontal distance to be considered, could be the outcome of unconstrained theoretical speculations. Yet it was more likely driven by Galileo's grappling with a real challenging problem that he had encountered in the context of the emerging new science. Indeed, the generalized variant of the double distance rule Galileo formulated on this folio was incontrovertibly needed in the context of his considerations concerning projectile motion and was, in particular, underlying

---

[42] The use of a smoothed corner in the projection experiments is unambiguously hinted at by a diagram on folio 175 verso.

[43] Paragraph T3C.

his experimental investigation in this regard which, as indicated in Chap. 4, Galileo had likely already commenced by the end of 1602.

As it turns out, yet another problem can be identified in the *Notes on Motion* whose solution required an inference to the time of uniform motion over an arbitrary distance from knowledge of the distance and time of a preceding motion of fall that had generated this uniform motion. Galileo's attempt to prove the empirical adequacy of his so-called cosmogonical hypothesis required precisely the type of considerations on which the generalization of the double distance rule was also based.

Galileo ventured two attempts to verify the empirical adequacy of his cosmogonical hypothesis. As I have argued elsewhere, his first attempt must fall roughly into the same period of work as the composition of the star group folios, as indicated by the employment of the Sarpi letter principle in this attempt. The generalized double distance rule noted on folio 163 recto was in fact very likely a spin-off of Galileo's work on his cosmogony.[44]

In the entry on page 163 verso, which is inserted between the two entries just discussed, Galileo provided a proof of the length time proportionality. Just as he would do later in the *Discorsi*, Galileo first proved the proposition for the case where one of the two motions was made along an inclined plane and the other through a vertical distance equal to the height of the inclined plane. In a corollary, the proposition was afterward generalized to motion along two arbitrary inclined planes of equal height. The first part of the entry reads (the diagram associated with this entry is reproduced as Fig. 10.4)[45]:

> Tempus casus per planum inclinatum ad tempus casus per lineam suae altitudinis est ut eiusdem plani longitudo ad longitudinem suae altitudinis.
>
> Sit planum inclinatum *ba* ad lineam orizontis *ac*, sitque linea altitudinis perpendicularis *bc*. Dico, tempus casus quo mobile movetur per *ba* ad tempus in quo cadit per *bc* esse ut *ba* ad *bc*. Erigatur perpendicularis *ad* orizontem ex *a*, quae sit *ad*, cui occurrat in *d* perpendicularis ad *ab* ducta ex *b*, quae sit *bd*, et circa [tri]angulum *abd* circulus describatur: et quia *da*, *bc* ambae sunt ad orizontem perpendiculares, constat, tempus casus per *da* ad

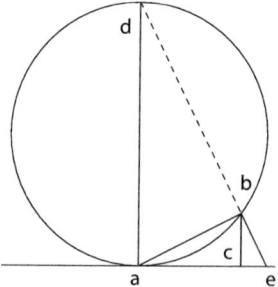

Fig. 10.4 Proof diagram associated with the proof of the length time proportionality on folio 163 verso

---

[44]Cf. Büttner (2001).
[45]Entries T3A and T3B.

10.1 The Star Group Folios: A First Step Toward a New Science

tempus casus per *bc* esse ut media inter *da* et *bc* ad ipsam *bc*. Tempus autem casus per *da* aequatur tempori casus per *ba* : media vero inter *da* et *bc* est ipsa *ba*: ergo patet propositum.

The general strategy of the proof is the following: First, exploiting the law of chords, the vertical distance *da* traversed in the same time as the inclined plane *ba* is constructed. By an application of the law of fall to the motions over this vertical distance and over the vertical height *bc* of the plane, Galileo arrived at the desired result, namely, that the times of motion over an inclined plane and over a vertical distance equal to the height of this plane are in the inverse ratio of the distances covered. The proof can be represented as follows:

Statement of length time proportionality:

$$t(ba)/t(bc) \sim ba/bc$$

Construction of point *d* such that by the law of chords:

$$t(ba) = t(da)$$

By the law of fall:

$$t(da)/t(bc) \sim mp(da|bc)/bc$$

By geometry:

$$mp(da|bc) = ba$$

By the theory of proportions and above:

$$t(ba)/t(bc) \sim ba/bc$$

The generalization of the proposition to planes of equal height and arbitrary inclination proceeded via two successive applications of the length time proportionality restricted to the case where one of the distances traversed is a vertical. Thus, two proportions between ratios of times and distances were yielded from which the desired result followed as a proportion ex aequali:

By restricted length time proportionality:

$$t(ba)/t(bc) \sim ba/bc \text{ and } t(bc)/t(be) \sim bc/be$$

Proportion ex aequali:

$$t(ba)/t(be) \sim ba/be$$

As argued in Chap. 7, Sect. 7.2, by all accounts Galileo first became aware of the length time proportionality as a consequence of his attempts to prove the law of the broken chord. The occurrence of a proof of the length time proportionality on the star group folios is thus in good agreement with the overarching hypothesis according to which this group of folios essentially represents Galileo's attempt to collect and systematize the results he had attained in the context of his work on the broken chord approach and thus more generally in the context on his research agenda on swinging and rolling.

However, the proof of the length time proportionality contained on page 163 verso differs somewhat from the partial line of the argument, part of the proof of the law of the broken chord, and which taken by itself, represents a proof of the length time proportionality. It would thus seem that the proof of the length time proportionality drafted on the star group folios was composed from scratch, based on Galileo's principal insight into the proposition. As already indicated in Chap. 7, a proof that is structurally much closer to the partial line of reasoning of the proof of the law of the broken chord, which itself represents a proof of the length time proportionality, is contained on folio 147, whose relation to Galileo's work on the star group folios will be discussed in detail in the succeeding chapter.

Folio 164 recto Folio 164 recto contains the draft of a geometrical lemma as well as of two propositions concerning accelerated motion, both of which deal with motions that do not start from rest. With regard to their physical content, both propositions correspond to each other closely, and only closer analysis reveals why Galileo noted them as separate entries. As part of the first entry, Galileo stated (for the associated diagram, see Fig. 10.5):

> Patet insuper, tempora casuum [from *a*]per *gb*, *f c*, *ed* esse ut lineas *gb*, *f c*, *ed*

and in the second, he stated (for the associated diagram see Fig. 10.6):

> Dico, tempus per *bc* post casum *ab* ad tempus per *bd* post eumdem casum *ab* esse ut linea *bc* ad *bd*.

As already indicated in Chap. 7, both statements essentially correspond to the generalized length time proportionality issued by Galileo as Proposition X of the

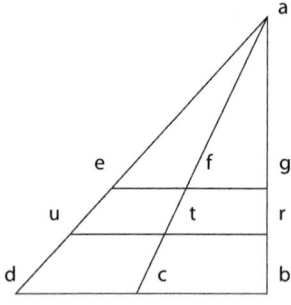

Fig. 10.5 Proof diagram (D01A) associated with the generalized law of fall on folio 164 recto

Fig. 10.6 Proof diagram associated with the generalized length time proportionality on folio 164 recto

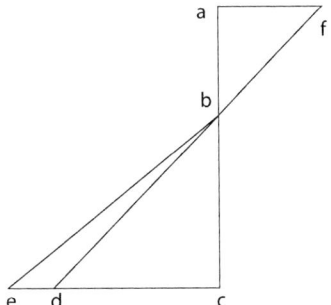

Second Book in the *Discorsi*. The proposition asserts that the proportion between spaces traversed and time elapsed for motion along planes of equal height, which the length time proportionality asserts for motions starting from rest, holds, even if the motions considered do not start from rest but are preceded by fall through the same distance.

The generalized length time proportionality and, in particular, the first entry on 164 recto were drafted, as argued in Chap. 7, as a direct consequence of Galileo's considerations documented on 174 recto, from where, in particular, the diagram associated to the first entry was taken over. It has already been indicated in this chapter why Galileo drafted the two closely related entries on 164 recto, and this will be elaborated upon in some more detail in the following.[46]

Indeed, in the first entry, Galileo combined the generalized law of fall, which likewise had emerged as a result of his considerations leading up to those documented on 174 recto, with the generalized length time proportionality, in such a manner that a somewhat confusing consequence of his new technique of integrated time representation could be elaborated upon further.

Galileo's first entry on 164 recto commences as follows [47]:

> Sint ad orizontem *db* quotcumque lineae ex eadem altitudine *a* demissae *ab, ac, ad*, et sumpto quolibet puncto *g*, per ipsum orizonti parallela sit *gfe*, sitque media inter *ba, ag* ipsa *ar*, et per *r* altera parallela *rtv*: constat, lineas *at, av* esse medias inter *ca, af* et *da, ae*. Dico, quod si assumatur *ab* esse tempus quo mobile cadit ex *a* in *b*, tempus *rb* esse illud quo conficitur *gb*, *tc* vero esse tempus ipsius *cf*, et *vd* ipsius *ed*. Id autem constat: nam, cum *ar* sit media inter *ba, ag*, sitque *ba* tempus casus totius *ab*, tempus *ar* erit tempus

---

[46]Wisan (1974, 191) holds that both propositions pertain to the generalized length time proportionality.

[47]Entry T1. A sketchy elaboration of this proposition, associated with a diagram identical to that on folio 164 recto together with an additional numerical calculation of the times of motion, can be identified on folio 174 recto. This, in all likelihood, served as the blueprint for drafting the proposition on folio 164 recto. Hooper (1992, 364) claims that, "The first proposition on the recto side [of folio 164] is a prototype for Theorem Three of Book Two of *On Local Motion*," i.e., of the length time proportionality. His assessment must be wrong as the length time proportionality treats motion from rest, whereas the proposition on folio 164 recto treats motions that have been preceded by fall through a certain distance.

casus per *ag*; ergo reliquum temporis *rb* erit tempus casus per *gb* post *ag*; et idem dicetur de aliis temporibus *tc*, *vd* et lineis *fc*, *ed*.

As explicitly stated in the first sentence, it is the purpose of this passage to demonstrate that if *ab* is taken to represent the time along *ab*, then *rb* represents the time along *gb*, *tc* that of motion along *fc*, and finally *vd* that along *ed*, all motions being conceived as starting from rest in point *a*. Thus, as far as the physical content is concerned, what is proved is indeed first and foremost the generalized law of fall. For the argument, Galileo employed his new technique of integrated time representation paired with distance time coordination. The proof itself consists of a straightforward, slightly abridged line of reasoning, which can be explicated in as follows:

If by distance time coordination:

$$t(ab) = ab$$

then statement of the generalized law of fall:

$$t(gb|a) = rb, t(fc|a) = tc \text{ and } t(ed|a) = vd.$$

By the law of fall and above:

$$t(ag) = ar.$$

From above:

$$t(gb|a) = rb.$$

By above and length time proportionality:

$$t(ac) = ac, \ t(ad) = ad.$$

By the same line of argument as above:

$$t(fc|a) = tc, \ t(ed|a) = vd.$$

Thus, what Galileo stated in the determinatio had been proved. It is only at this point of the argument that the sentence in which the generalized law of fall is stated, and that was already quoted above, follows. Despite not being marked as the aim of the proof, the latter proposition is thus proved as part of the argument by the triple application of what had already been proved up to that point:

Statement of the generalized law of fall by above and geometry:

$$t(gb|a)/t(fc|a)/t(ed|a) \sim gb/fc/ed$$

10.1 The Star Group Folios: A First Step Toward a New Science

The entry carries on:

non tamen a magnitudinibus ipsarum linearum *gb*, *fc*, *ed* esse determinandas eorumdem temporum quantitates, si temporis mensura ponatur *ab*, in quo tempore conficiatur linea *ab*, sed desumendas esse a lineis *rb*, *tc*, *vd*.

It is from this last sentence that the reason why Galileo included a statement of the generalized length time proportionality in this entry in the first place can be gleaned. Galileo issued a warning to himself and also, it would seem, to the envisaged reader of a future publication, concerning a potential pitfall which can be seen as arising from the application of integrated time representation in his argument.

With regard to his last conclusion, Galileo warned himself not to draw from the proportionality between the times and the distances of motion along inclined planes of equal height traversed after fall through a certain distance, the suggested but erroneous conclusion to conceive of the lengths of these inclined planes as representing the times of motion along them. In fact, with the initial choice of distance time coordination, i.e., in the case under scrutiny with the decision to represent the time along *ab* by the line *ab* itself, a measure had implicitly been determined. With regard to this measure, the times of motion along the lower parts of the inclined planes did not numerically coincide with the distances covered by these motions, as may seem suggested, but with the corresponding mean proportionals.

The use of integrated time representation indeed entailed that one and the same line in a diagram (and consequently also in the verbalized geometrical reasoning on the basis of such diagrams) could in some cases represent a distance and concurrently a time, while in others this would not be the case. The expressions used by Galileo on folio 164 recto to identify times elapsed and distances covered indeed facilitated the type of erroneous conclusion Galileo had alluded to in his warning. When in his first entry on the page Galileo talked about the time *rb*, for instance, he was indubitably referring to the time represented by the line *rb* and not the time needed to traverse the distance *rb*. Such a terminology was perfectly unambiguous in the context of his old technique in which times of motion were represented by an external time line. With the introduction of the new technique, however, expressions such as "tempus *rb*" had become ambiguous, as they could now mean the time represented by the line *rb* or, alternatively, the time needed to traverse the distance *rb*. Against such equivocalness Galileo's warning not to confuse the two alternatives comes into view as absolutely appropriate.

Not only would Galileo never repeat such a warning but, as argued in Chap. 7, he even adapted his terminology so as to avoid confusing the time represented by a line and the time needed to traverse the distance represented by this line. The entry on folio 164 recto must hence represent one of the earliest instances of Galileo's use of integrated time representation in a proof, in perfect agreement with a dating of the work on the star group folios to 1602, contemporaneous to, or shortly after, Galileo's work on the problem of swinging and rolling.[48]

The second proposition noted on folio 164 recto reads:

> Sit *ac* perpendicularis ad orizontem *cde*, ponaturque inclinata *bd*, fiatque motus ex *a* per *abc* et per *abd*. Dico, tempus per *bc* post casum *ab* ad tempus per *bd* post eumdem casum *ab* esse ut linea *bc* ad *bd*. Ducatur *af* parallela *dc* et protrahatur *db* ad *f*; erit iam tempus casus per *fbd* ad tempus casus per *abc* ut *fd* linea ad lineam *ac*: est autem tempus casus per *fb* ad tempus casus per *ab* ut linea *fb* ad lineam *ab*: ergo tempus casus reliquae *bc* post *ab* ad tempus casus reliquae *bd* post *fb* erit ut reliqua *bc* ad reliquam *bd*. Sed tempus casus per *bd* post *fb* est idem cum tempore per *bd* post *ab*, cum *af* orizonti aequidistans sit: ergo patet propositum.
> Colligitur autem ex hoc, quod tempora casuum per *bc* et *bd*, sive fiat principium motus et termino *b*, sive praecedat motus, ex eadem tamen altitudine, eandem inter se servant rationem, nempe eam quae est lineae *bc* ad *bd*.

Here Galileo indeed stated the generalized length time proportionality and provided a proof which can be transcribed as follows:
Statement of the generalized length time proportionality:

$$t(bc|a)/t(bd|a) \sim bc/bd.$$

By length time proportionality:

$$t(fbd)/t(abc) \sim fd/ac.$$

By length time proportionality:

$$t(fb)/t(ab) \sim fb/ab.$$

By compounding and division of ratios:

$$(t(fbd) - t(fb))/(t(abc) - t(ab)) \sim (fd - fb)/(ac - ab).$$

By above:

$$t(bd|f)/t(bc|a) \sim bd/cb.$$

By path invariance:

$$t(bd|f) = t(bd|a).$$

---

[48] Application of the new technique of integrated time representation may explain why this particular version of the generalized law of fall was included in the star group folios collection in the first place. An alternative proof of the proposition, which in all likelihood antedates the drafting of the star group folios and which is formulated purely in the mathematical language of the theory of proportions, i.e., in particular, without recourse to integrated time representation and distance time coordination, is contained on folio 126 recto.

10.1 The Star Group Folios: A First Step Toward a New Science

By above:

$$t(bc|a)/t(bd|a) \sim bc/bd.$$

The step in which ratios are compounded and divided in the above formalization remained implicit in Galileo's reasoning and has been added to formally complete the argument. As we have seen in the argument he had noted on the top of the page, Galileo had drawn a similar inference. However, this argument had been based on integrated time representation and distance time coordination such that the corresponding step in that case had merely required the subtraction of two pairs of distances, one pair representing times elapsed and the other representing distances traversed. The proof of the generalized length time proportionality, in contrast, was formulated without recourse to integrated time representation, and thus this step is to be replaced by the equivalent operation for proportions, namely, a combination of compounding, inverting, and dividing ratios.[49] Both operations lead to the same result, yet the subtraction of distances representing time strikes as being more straightforward.[50] This may have provided the reason why Galileo reformulated the proof of the generalized length time proportionality once again and offered this reformulated version in the *Discorsi* which, like the majority of proofs of the Second Book of the Third Day, was then based on the combination of integrated time representation and distance time coordination.[51]

Placed between the two entries just discussed is a geometrical lemma in which Galileo stated that if a chord in a circle, comprising either the lowest or the highest point of that circle, is projected onto any line parallel to the vertical diameter by two lines orthogonal to the chord, this projected distance is equal in length to the diameter itself. The lemma results as a rather trivial consequence of Thales' theorem and Proposition 34 of Book One of Euclid's *Elements*. An unlettered version of the proof diagram accompanying the lemma is contained on 168 recto, a folio page that otherwise contains elaborations of the least time propositions, which had emerged

---

[49]Wisan (1974, 194) already remarked that the entry seems to imply "one may simply subtract FB and AB, respectively, from FD and AC it is clear ...that this is not intended ." However, her interpretation of the entry is clearly erroneous as she claims that Galileo uses the integrated time representation: "on the bottom of folio 164r, the new technique ...is used ..."

[50]An improved version of a proof of the generalized length time proportionality in the hand of Mario Guiducci is contained on folio 38 recto. It may have been copied there from a now lost original by Galileo, or else Mario Guiducci, possibly under Galileo's supervision, redrafted the more straightforward version on the basis of the entry on folio 164 recto.

[51]The generalized length time proportionality was published as Proposition X of the Second Book of the Third Day of the *Discorsi*, followed by the generalized law of fall which Galileo issued as Proposition XI. The generalized law of fall is applied in the proof of Proposition X, and this is the only instance in the *Discorsi* where the order of propositions does not follow the deductive structure. The proof Galileo gave for the generalized law of fall in the *Discorsi* differed from the proof of the proposition noted on 164 recto, however, and was based on integrated time representation and distance time coordination, making the twist in the deductive structure introduced by the order of propositions much less manifest.

as a spin-off result of the broken chord approach.[52] From there, it seems to have been copied to page 164 recto. A marginal note to the left of the entry, likely added at a later stage, presents a slightly more efficient alternative proof.

As discussed in Chap. 5, the geometrical lemma noted on 164 recto had been employed by Galileo in his attempt to construct an alternative proof for the law of the broken chord that he could, however, not complete. Since the lemma is used in no other context in the manuscript, it is suggested that its insertion in the collection of material on the star group folios owes to Galileo's intention to complete the unfinished proof. This, as I have argued, was indeed indicated, as it would have allowed him to move ahead with his broken chord approach.

Folio 164 verso Folio 164 verso opens with a proposition where we recognize a precursor of a proposition that Galileo later published as Proposition VIII of the Second Book of the Third Day of the *Discorsi* (the diagram associated with the entry is reproduced as Fig. 10.7):

> Si in semicirculo [, cuius diameter sit ad perpendiculum, ducatur linea?] quae cum perpendiculo non habeat terminum communem, motus per illam cicius [citius] absolvitur quam per diametrum perpendicularem.[53]
>
> Si enim *bb* fuerit perpendiculus, ducta quaelibet linea *ca* in semicirculo non terminetur ad *b*: patet quod, si connectatur linea *cb*, erit *ca* ipsa *cb* brevior et minus inclinata; ex quo patet propositum.[54]
>
> Si in circulo, cuius diameter sit ad perpendiculum, ducatur linea quae a diametro secetur, motus per ipsam tardius absolvetur quam per diametrum perpendicularem.
>
> In praecedenti enim figura sit linea quaelibet; et quia ipsa erit longior quam *cb* et magis inclinata, propositum fit manifestum.

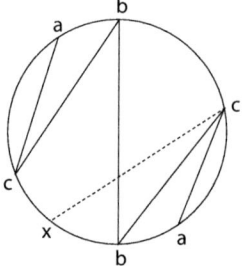

Fig. 10.7 Reconstruction of the diagram on top of folio 164 verso

---

[52]Cf. Chap. 5, Sect. 7.3.

[53]The first paragraph and the drawing were damaged when a strip of paper was cut off of the upper margin of the folio. The content of the recto side of the folio was not affected. This may imply that the folio was cut before Galileo wrote on its recto side, in disagreement with the order of composition reconstructed.

[54]The inclined plane *ca* is said to be less inclined than the plane *cb*, i.e., Galileo here considers the angle of inclination against the vertical instead of against the horizontal. This seems to be the only case in the *Notes on Motion* where Galileo used this convention, which in the published version of the proposition is changed to Galileo's standard way of specifying inclinations.

Galileo claims that descent along an inclined plane inscribed in a circle is made faster or slower than along the vertical diameter itself if the plane intersects the vertical diameter of the circle or does not intersect it, respectively. The expression "motion made faster" here translates, just as in other cases where fast (citius) is used adverbially, unambiguously into "motion made in shorter time," and the latter expression was in fact substituted for the former in the *Discorsi* version of the proposition.[55] Another difference with regard to the version of the proposition that was published in the *Discorsi* is that the two cases alluded to are dealt with separately; otherwise the proof essentially equals that given in the *Discorsi*.

In this entry, Galileo generalized a note he had put in the margin of folio 186 verso.[56] It was there that he had noted his insight that motion on an inclined plane inscribed in a circle which does not intersect the vertical diameter of this circle is completed in less time than motion on the vertical diameter itself. This insight had resulted as a consequence of Galileo's elaboration of the law of the broken chord contained on the same page. Thus, formally the same restrictions applied for the statement noted on 186 verso that applied for the law of the broken chord. In particular, the statement could only be considered valid if the inclined plane in question could be inscribed into an appropriate arc of 90 degrees or less.

Accordingly, in his entry on 164 verso, Galileo generalized his initial note in two respects. Firstly, he dropped the restriction to arcs smaller than 90 degrees, and secondly he included the case of motion along inclined planes intersecting the vertical diameter. These generalizations required a new line of argument. The proof Galileo came up with and that he noted on 164 verso reflects the simplicity and the generality of the proposition.

Galileo introduced two chords which, according to the law of chords, would be traversed in the same time as the vertical diameter itself (chords *bc*). Motion along these chords subsequently served him as a reference point for comparing times of motion.[57] Next, the inclined plane not cutting the vertical diameter was said to be shorter and steeper than the respective chord, and from this, without any intermediate step, Galileo concluded that it would be traversed in shorter time. The argument is cursory and remained so in the version of the proposition published in

---

[55] The statement of Proposition VIII of the Second Book of the Third Day of the *Discorsi* reads: "The times of descent along all inclined planes which intersect one and the same vertical circle, either at its highest or lowest point, are equal to the time of fall along the vertical diameter; for those planes which fall short of this diameter the times are shorter; for planes which cut this diameter, the times are longer (EN VII, p 262, trans. Galilei (1914))."

[56] Cf. Chap. 5 Sect. 5.1.

[57] In the entry on 164 verso, Galileo's application of the law of chords as a premise of the argument remains implicit. In the proof of the respective proposition in the *Discorsi*, explicit reference is made to the law of chords as a premise of the argument.

the *Discorsi*. Galileo could easily have given an unabridged argument, for instance, by applying the length time proportionality and the law of fall.[58]

Yet he does not do so and it was in fact likely another line of reasoning that reassured Galileo that motion on the shorter, more steeply inclined plane was completed in less time than that on the respective chord. It could be assumed that motion along the more steeply inclined plane under investigation proceeded with a higher velocity than motion on the chord with smaller inclination. With this supposition—which seems to be irrefutably implied by thinking about the dynamics of motion—Galileo's statement indeed results trivially[59]: Let a third motion be made on the more steeply inclined plane over a distance equal to the length of the chord. According to the kinematic propositions, this third motion would be completed in less time than the motion along the chord. Moreover, motion over the original, shorter distance on the same plane would be completed in even less time and the statement follows.[60]

There is no direct evidence that the line of argument reconstructed above correctly renders the logic which Galileo had in mind with his entry on 164 verso. Yet the very next entry on the page, which, following the literature, will be referred to as the mirandum paradox, is devoted to exactly the question of whether accelerated motion on a vertical can indeed be said to proceed with a higher velocity than motion on an inclined plane as implied by dynamics. This renders it more than likely that the argument reconstructed above, indeed, correctly represents the reasoning which Galileo had in mind but which remained implicit in his cursory entry at the top of the page.[61]

Before proceeding to the discussion of the mirandum paradox, however, it should be noted that the result established in the entry at the top of 164 verso

---

[58] If, with respect to Fig. 10.7, the inclined plane *ca* is prolonged until it has the same vertical height as the chord *cb*, then, by the length time proportionality, motion on this prolonged plane is completed in less time than motion on the chord. Furthermore, since the initial plane *ca* is shorter than the prolonged plane, it is traversed in naturally accelerated motion in less time than the latter by the law of fall.

[59] What has to be presupposed for the argument reconstructed in the text is a very basic and minimal dynamical assumption, namely, that if compared to an inclined plane of smaller inclination the cause of motion and, in consequence, the velocity of motion is greater on a plane of greater inclination. This assumption is weaker than Galileo's basic principle of inclined plane dynamics. As we have seen in his treatment of motion on inclined planes in the *De Motu Antiquiora* (EN I, pp. 293–302), Galileo had declared such a monotonous relation between the inclination and the velocity of motion along inclined planes (the weaker statement) as obvious, where the real problem was that of determining their exact mathematical relation (the basic principle of inclined plane dynamics). The difference between the weaker and the stronger assumption was thus amply clear to him.

[60] Explicating the logic of Galileo's conclusion makes the argument appear somewhat complicated and lengthy, whereas it is intuitively immediately clear that motion which proceeds with higher velocity over a shorter distance is completed in shorter time.

[61] Hooper (1992, 367) has already assumed that it was a re-reading of the first proposition on folio 164 verso which inspired Galileo's considerations, documented by the mirandum paradox.

10.1 The Star Group Folios: A First Step Toward a New Science

is established in greater generality by Galileo's least time propositions. This may imply that the proposition on 164 verso was drafted before Galileo engaged in his considerations concerning motion along broken planes discussed, from which the least time propositions originated. Yet even in the *Discorsi* where the least time propositions were issued as propositions XXX, XXXI, and XXXII, Galileo retained the proposition originally sketched on 164 verso and published it as Proposition VIII. This may be seen as supporting my interpretation according to which the reasoning underlying Proposition VIII was, from Galileo's perspective, essentially dynamical, whereas in the proofs he gave for the least time propositions, his argument was purely kinematical. Galileo may thus have transformed the entry from 164 verso into an independent proposition because from his perspective it rested on a different premise and thus on a different kind of argumentation. In short, *what* was proved in the note and thus, subsequently, also in Proposition VIII may have been redundant but not *how* it was proved.

The entry that follows contains the mirandum paradox. It reads:

> Mirandum. Numquid motus per perpendiculum *ad* velocior sit quam per inclinationem *ab*? Videtur esse; nam aequalia spacia citius conficiuntur per *ad* quam per *ab*. Attamen videtur etiam non esse; nam, ducta orizontali *bc*, tempus per *ab* ad tempus per *ac* est ut *ab* ad *ac*: ergo eadem momenta velocitatis per *ab* et per *ac*. Est enim una eademque velocitas illa quae, temporibus in[a]equalibus, spacia transit inaequalia, eandem quam tempora rationem habentia.[62]

The interpretations this entry has received cover a wide spectrum, ranging from Galileo's acknowledgement of a contradiction in his conceptual framework, to his having found a solution for one of the most pressing problems of his emerging new science of naturally accelerated motion. In a nutshell, these varying interpretations have found their repercussion in how various authors have translated the word "mirandum" that opens the entry. The translations vary from "remarkable" to "it has to be seen."

I commence the analysis of the mirandum paradox with those aspects of the entry that are uncontested. The entry makes a statement about the velocities of motion on an inclined plane and on a vertical. First, Galileo established that motion on the vertical *ad* is faster (velocior) than motion on the inclined plane *ab* and then he showed that motion on the vertical along the distance *ac* proceeded with the same velocity as motion along the inclined plane over the distance *ab*.

As regards Galileo's statement that motion over the vertical was faster than that along the inclined plane, he merely remarked that if equal distances traversed [from rest] on the inclined plane and on the vertical were considered, motion on the vertical would be made in less time than on the inclined plane and hence, according to the

---

[62]Entry T2. "It has to be seen [*Mirandum*], whether the motion along the perpendicular *ad* is not perhaps faster than that along the inclined plane *ab*? It seems so; in fact equal spaces are traversed more quickly along *ad* than along *ab*; still it seems not so; in fact, drawing the horizontal *bc*, the time along *ab* is to the time along *ac* as *ab* is to *ac*; then the moments of velocity along *ab* and along *ac* are the same; in fact, that velocity is one and the same which in unequal times traverses unequal spaces which are in the same proportion as the times (trans. Damerow et al. 2004, 201)."

kinematical understanding, also with greater velocity. That less time was needed to traverse the same distance on the vertical than on the inclined plane was not reasoned. It follows, however, from the very line of reasoning, which according to the interpretation given also formed the basis of Galileo's dynamical reasoning in the proposition drafted above.

Galileo's wording allows for two different readings. Either his statement "motus per ... *ad* velocior sit quam per inclinationem *ab* ... nam aequalia spacia citius conficiuntur per *ad* quam per *ab*" refers to the two concrete motions over the concrete distances *ad* and *ac*. Given his justification, this would imply that these two distances were equal, which indeed they approximately are in the diagram. However, "motus per perpendiculum *ad*" and "[motus] per inclinationem *ab*" could also refer to motions along the vertical and the inclined plane in general, i.e., regardless of the concrete distance covered. This is suggested by the fact that Galileo marked his conclusion as deriving from the consideration of (infinitely many) different motions over different equal distances on ("per") *ad* and *ab*. This would entail that conclusion being valid for motions along the vertical and the inclined plane in general. On the other hand, as we have seen, Galileo usually used an expression such as "planum secundum lineam" to unspecifically refer to an inclined plane or more precisely to an inclined plane of given inclination without specifying its specific length. As will be seen, the question of which of these readings is correct ultimately has little bearing on the overall interpretation of the mirandum paradox, as Galileo's argument and, in particular, the conflict it revealed, hinges on his assumption that what held with regard to the velocities of two concrete motions on the inclined and the vertical planes should characterize the relations of the velocities of motions along the inclined and the vertical plane in general. At which step of the argument Galileo made the transition from the concrete to the general case is thus of somewhat secondary importance for the overall interpretation.

Galileo's second statement, namely, that "eadem momenta velocitatis per *ab* et per *ac*," is based on the third Archimedean kinematic proposition and the length time proportionality.[63] If the height of the inclined plane is marked on the vertical (point *c* in Fig. 10.8), then it holds according to the length time proportionality that the times

Fig. 10.8 Diagram associated with the mirandum paradox

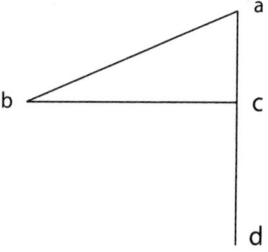

---

[63]The statement that the velocities of motions over planes of different inclination but equal height are equal is in direct conflict with the statement of the tarditas length proportionality Galileo had

of motion through the vertical height ac and through the inclined pane ab are as the distances covered. Consequently, by virtue of the third Archimedean kinematic proposition, the velocities of those motions have to be equal. As before on the star group folios, the expression "momenta velocitatis" is used by Galileo to denote the aggregate velocity understood to be equivalent to the kinematic velocity.[64] Indeed, in the last sentence, "momenta velocitatis" is readily exchanged by the "velocitas" again whereby the same meaning is retained.

Thus, up to this point, Galileo had argued on the one hand that motion along the vertical ad was faster than along the inclined plane ab and on the other hand that motion along the inclined plane ab was made with the same velocity as that on the vertical ac. These two statements refer to the relation of motion along the inclined plane ab to different motions along the vertical and from a modern perspective do not seem to imply a conflict. Yet for Galileo they did as his expression "[a]ttamen videtur etiam non esse" indicates.

Why the two statements derived as part of the mirandum paradox were incongruous for Galileo is rather obvious. As we have seen, Galileo plausibly assumed that motion along a steeper inclined plane was faster than that along an inclined plane of lesser inclination. In the tarditas length proportionality, Galileo had even employed a somewhat strong variant of this assumption when he assumed that just as the causes of motion along two inclined planes of different inclination were in a fixed ratio, so should the velocities of the resulting motions be in a fixed relation. As a result of his tarditas length proportionality argument, he had concluded that his basic principle of inclined plane dynamics should apply in the case of naturally accelerated motion and he constructed his ex mechanicis proof based upon this assumption.

It was thus, ultimately, his dynamical understanding which implied that whatever the relation of the velocities along ab and ad, the same should also apply to the relation of the velocities along ab and ac. It was this assumption which, as the mirandum paradox revealed, turned out to not be compliant with his insights about the pheno-kinematics of naturally accelerated motion. As will be argued, this revelation had far-reaching consequences as it pointed to a crucial problem in the conceptual underpinnings, forcing Galileo to interrupt his work on the foundations of the new science.

---

just drafted on folio 172 verso. The realization of this conflict may have provided another motive leading Galileo to consider the problem treated in the mirandum paradox.

[64] The assessment that Galileo here—just as in the cases of the tarditas length proportionality and the sesquialterum version of the double distance rule—referred to the totality of the degrees of velocity, or aggregate velocity held to be equivalent to the overall velocity, is shared by all authors who have discussed the mirandum paradox except for Hooper. According to Hooper, Galileo stated that the moments of velocity in b and c are equal, i.e., the velocity proportional to vertical height principle. As a result of his discussion of the mirandum paradox, Hooper (1992, 368–373) concludes that the argument may represent a "point of initiation for the 1604 principle."

Indeed, as has already been argued in the literature, the collection of material on the star group folios originally broke off with the mirandum paradox. It is likely that the final entry was added to the page only later.[65] It reads[66]:

> Momenta velocitatum cadentis ex subblimi [sublimi] sunt inter se ut radices distantiarum peractarum, nempe in subduplicata ratione illarum.

This statement represents an immediate consequence of Galileo's correct principle of fall, according to which the degrees of velocity increase in proportion to the time elapsed. Since according to the law of fall the distance traversed increases with the square of the time, this immediately also entails that the degrees or moments of velocity increase as the roots of the spaces covered, just as stated.

Here then, "[m]omenta velocitatum" clearly does not refer to the aggregate velocity but to the individual moments of velocity reached at different times (or after different spaces traversed).[67] The entry is thus clearly based on a thorough analysis of naturally accelerated motion based on the understanding of velocity as a changing quality as entailed by the doctrine of the configurations of motion. As will be argued in the succeeding chapter, Galileo first engaged in such an analysis only around the end of 1604. As is well known, even then he initially assumed that the moments of velocity increased with the distance fallen, i.e., according to the Sarpi letter principle. It took him some time to reject this assumption in favor of the correct principle of acceleration intrinsic to the last entry on 164 verso. The entry thus must have been drafted substantially later than 1602, i.e., much after the composition of the content of the star group folios and with it the drafting of the mirandum paradox.

The likely reason why Galileo added this entry after the mirandum paradox, i.e., where his considerations had originally broken off, is strongly suggested. By the time the entry was added, the problem imposed by the mirandum paradox on the attempt to create a foundation for the new science had been resolved. Galileo now acknowledged that the intensive velocity increased with the square root of the distances fallen and that this entailed the aggregate velocities and thus, according to his understanding, the kinematic velocities of motion along inclined planes of different inclination could not obey a fixed ratio. The statements derived as part of the mirandum paradox had thus changed their character. Instead of being conceived as conflicting they could now be envisaged as harmonizing with the new underlying conceptualization. The brief entry on 164 verso can thus be understood as a reminder

---

[65] In my perception, the ink and hand of the entry are similar to the two entries above. Hooper, however, remarked: "...I think I do see differences in the hand and ink when I compare 164–5 and 164–6 that are sufficient to suggest that the two were written at widely separated times (Hooper 1992, 61)."

[66] Entry T3.

[67] That this entry formulates a direct consequence of the correct principle of fall was also realized, for instance, by Damerow et al. (2004, 191) or Hooper (1992, 372).

that the problem created by the mirandum paradox had been resolved at least to the extent that it concerned the problem of finding a foundation for the new science. In a nutshell, the entry also indicated how it had been resolved.[68]

What had not vanished with Galileo's new approach was the problem which the mirandum paradox had created with regard to the attempt to provide a dynamical argument for naturally accelerated motion. However, with Galileo's new approach to the problem of the foundation, this problem had lost its pressing character. Why the conflict revealed by the mirandum paradox initially represented such a crucial problem and why its resolution undermined the attempt to found the new science on dynamical arguments will be discussed in the following section.

10.2 The Rise and Demise of a Dynamical Foundation

In the mirandum paradox, Galileo concluded that, with respect to Fig. 10.8, motion along the inclined plane *ab* and the vertical *ac* proceed with the same velocity. This conclusion was based on the length time proportionality and the Archimedean kinematic proposition. On the other hand, the application of one of the Aristotelian kinematic propositions had yielded that motion along the vertical *ad* was faster than that along the inclined plane *ab*. This collided with Galileo's assumption that the velocities of motion along the inclined planes of different inclinations should obey a fixed ratio which in turn would be in proportion to that of the mechanical moments on these planes.

At least one of the assumptions which had entered the argument and furnished the result with its problematic character thus had to be discarded, but which one? After his investigations into the relation between swinging and rolling, Galileo had all reason to be convinced that the length time proportionality, which had entered his argument, represented a true statement concerning the pheno-kinematics of naturally accelerated motion and consequently he did not scrutinize the proposition as a result of the mirandum paradox. It has been held in the literature that for Galileo, the mirandum paradox cast doubt on his kinematic understanding of velocity, in particular, on the applicability of the Archimedean kinematic proposition to naturally accelerated motion.[69]

---

[68] According to Drake's opinion, the mirandum paradox and the last entry on folio 164 verso were made contemporaneously in 1607. For him, these entries do not testify to Galileo having discovered a conflict but rather show "that Galileo had reached complete clarity on puzzles of accelerated motion that had long plagued him (Drake 1978, 124–125)."

[69] "But since one of the two conflicting statements depends on Galileo's generalized definition of equal velocities, while the other depends on the more traditional Aristotelian concept of velocity, the paradox casts doubts on the applicability to accelerated motion of the statement that equal velocities are characterized by equal proportions between distance and times (Damerow et al. 2004, 202)." What is referred to as "Galileo's generalized definition of equal velocities" in the quotation, I refer to as the Archimedean kinematic proposition. According to Renn, the proposition

The way in which Galileo resolved the mirandum paradox shows, however, that he did not give up the Archimedean kinematic proposition and the consequences for the understanding of kinematic velocity it entailed. Rather he accepted the implied conclusion, namely, that the velocity of a naturally accelerated motion along an inclined plane changes with and thus depends on the concrete part of motion considered as a true and given fact.[70]

From a modern perspective, the application of the kinematic propositions to the case of naturally accelerated motion which gave rise to this conclusion is indeed unproblematic. The kinematic concept of velocity entailed by the kinematic propositions is consistent. It compares to the modern average velocity. What holds true for Galileo's kinematical velocity holds true cum grano salis for the modern average velocity but certainly not vice versa.[71] All three kinematic propositions were in fact still being applied to naturally accelerated motion in Galileo's later, published works. In particular, their use as premises in proofs in the *Discorsi* clearly indicates that Galileo did not reject the propositions and accepted the kinematical understanding of velocity they entailed.

In the *Dialogue*, Galileo returned to the problem raised by the mirandum paradox explicitly. Galileo's spokesman Salviati addressed Sagredo[72]:

---

was originally applicable to uniform motion only. However, Galileo apparently took for granted the application of the proposition to the case of naturally accelerated motion and did not scrutinize such an application in his published works.

[70]Renn has voiced the opinion that the fact that all of the kinematic propositions are still applied to accelerated motion in the *Discorsi* is because despite Galileo's alleged realization of the problematic character of applying, in particular, the Archimedean kinematic proposition, he could not avoid doing so in his attempts to erect a foundation for the new science due to a lack of alternatives. Cf. Damerow et al. (2004, 273–274). Drake, in the conclusion of his discussion of the mirandum paradox, stated, "The inherent difficulty to treat speeds in natural acceleration as truly mathematically continuously changing is hard for us to appreciate. Fundamentally the difficulty arose from the concept of cause which had to be abandoned for continuous change of speed (Drake 1978, 125–126)." Drake's conclusion, despite grave differences in the underlying interpretations, corresponds precisely to the one presented here.

[71]The use of mean or average velocity to characterize Galileo's concept of velocity has been criticized as the source of serious misunderstandings. Cf. in particular Giusti (1981). Guisti's claim that a concept of "mean or average velocity" cannot be expressed in Galileo's framework is certainly correct not least because it presupposes the modern concept of velocity, which Galileo did not share. If it is claimed here that Galileo's kinematic understanding of velocity is comparable to the average velocity of modern physics, this is understood to apply modulo the restrictions imposed by the framework in which Galileo expressed and applied that concept. Thus, statements about the "average velocity" that can be made in the framework of classical mechanics are not always meaningful for Galileo. Statements which hold true with regard to Galileo's kinematical understanding of velocity, on the other hand, will, if translated into the modern concept of average velocity, hold true.

[72]In the early 1630s, Galileo refereed a dispute between the cousins Niccoloò and Andrea Arrighetti which had unfolded about questions concerning the straightening of the Bisenzio river. He was asked by the Grand Duke to judge the plans of two rivaling engineers. His expertise was sent to a court official, Rafaello Staccoli, on 16 January 1631. For details see Maffioli (2008). In his expertise (EN V, 627–647) Galileo applied results of his new science of naturally

10.2 The Rise and Demise of a Dynamical Foundation

Then it would seem to you still more false if I should say categorically that the speeds of the bodies falling by the perpendicular and by the incline are equal, Yet this proposition is quite true, just as it is also true that the body moves more swiftly along the perpendicular than along the incline. (EN VII, 48, translation quoted according to Damerow et al. (2004, 256))

Salviati stresses that both statements that furnished the mirandum paradox with its seemingly paradoxical character are true. In the discussion which ensues from this initial intervention by Salviati, two definitions of velocity are introduced which are basically equivalent to the Aristotelian and the Archimedean kinematic propositions Galileo had used in the mirandum paradox. It is then shown that the seemingly paradoxical statement from which the discussion set out in fact follows. No doubt is cast on either of the two definitions of velocity, neither by Salviati nor by his dialog partners, and Salviati concludes:

...Now since we can conceive distances and velocities along the incline and along the perpendicular such that the proportion between the distances will now be greater and now less than the proportion of the times, we may very reasonably admit that there are also spaces along which the times of motion bear the same proportions as the distances. (EN VII, 50, translation quoted according to Damerow et al. 2004, 257)[73]

Thus, Salviati resolves the paradox by pointing out that parts of the motion over the vertical and over the inclined plane can be identified such that the first motion is either slower, faster, or just as fast as the second motion. In other words, the assumptions entering the argument are accepted as true. In consequence, it has to be accepted as a true and necessary conclusion that the velocity and, hence, the ratio of velocities depend on which part of the naturally accelerated motion is considered. It thus no longer makes sense to speak of the relation of the velocity along the vertical to that along an inclined plane in general. Sagredo, who was initially doubtful, conceded:

---

accelerated motion to the hydrodynamical problem. Among others he repeated the mirandum paradox argument. Giving a complete numerical example of the spatiotemporal behavior of the accelerated motions along two inclined planes of equal heights which he had elaborated on folio 100 verso, Galileo concluded, just as in the mirandum paradox and later in the *Dialogue*, that if the total motions over the vertical and the inclined plane of equal height were considered, the velocities were in fact equal. Yet the very same example likewise shows that "in the higher part of the long canal [the longer inclined plane] (which, in this example, is only its fourth part) the motion is slower, but in the remaining 3/4 is likewise faster. (EN V, 627–647)" Depending on which partial motions along the inclined planes are being compared, motion along a certain distance on one plane can be faster, as fast or even slower than motion along a certain distance on the other plane. In essence, in his Bisenzio expertise, Galileo thus resolved the mirandum paradox in the same fashion as in the *Dialogue*.

[73] In the quoted passage, Galileo speaks about distances and velocities in the plural. This indicates that he is perfectly aware that besides an inclined plane and a vertical of equal height, innumerable pairs of distances traversed, one on the inclined plane and the other in the vertical, can be identified such that the times elapsed in the motions over these distances are as the distances covered and hence the velocities of both motions are equal according to the Archimedean kinematic proposition.

I am already freed from my main doubt, and perceive that something which appeared to me a contradiction is not only possible but necessary. (EN VII 50, translation quoted according to Damerow et al. (2004, 257))[74]

Hence, like so often in the *Dialogue* and later also in the *Discorsi*, in the course of the discussion, Sagredo goes through a learning process, which mirrors the process Galileo himself had undergone.[75] Galileo, as can be argued, drew the consequences which Sagredo accepts as true more or less as an immediate reaction to his considerations documented by the mirandum paradox. But whereas Sagredo was merely enlightened by the discussion, for Galileo, accepting that in naturally accelerated motion the kinematic velocity of naturally accelerated motion was ever changing had a far reaching ramification in 1602. In essence, it undermined the attempt to found his new science on dynamical considerations.

As has been shown, by the time the mirandum paradox was drafted, Galileo had already come up with the ex mechanicis proof of the law of chords. It rested on the basic principle of inclined plane dynamics whose applicability to naturally accelerated motion had just been suggested to him by his work preserved on the star group folios. The inference made in the proof from the underlying forces to the pheno-kinematical nature of naturally accelerated motion represented a first successful step in Galileo's journey to providing his new insights with a foundation and thus turning them into the core of a new science, as will be made clear in the succeeding chapter.

However, the mirandum paradox jeopardized the assumption that the kinematic velocities of accelerated motion are in a fixed relation and thus possibly also the basic principle of inclined plane dynamics. As the ex mechanicis proof rested on

---

[74] Renn interprets this particular passage as follows: "At the end of this explanation, his spokesman Salviati comes to the conclusion that his concept of velocity does not allow for a consistent comparison of the speeds of these two motions, because such a comparison does not yield univocal results: ... In this way, Galileo has identified the application of his kinematic rules to accelerated motion as the source of the contradiction in the Mirandum Paradox ... This analysis does not, however, offer a solution to the Mirandum Paradox that would allow for an understanding of accelerated motion in terms of a more advanced concept of velocity that does not led to contradictions or ambiguities. See (Damerow et al. 2004, 257). The presented conclusions are, however, not marked by Galileo as inconsistent. As will be demonstrated, an "advanced concept of velocity" was in fact needed, according to my interpretation not, however, because the kinematical concept was inconsistent but because it didn't allow Galileo to advance with the program of establishing a foundation for the new science which had suffered a serious setback with the mirandum paradox.

[75] Commenting upon Souffrin's analysis of Galileo's concept of velocity in Souffrin (1990), Giusti remarked: "Pierre Souffrin has recently addressed this issue, introducing the term 'holistic velocity' to express this general concept. According to his interpretation, velocities are measured by the spaces traversed *in equal times*. ... With this proposal Souffrin has managed to clarify many of Galileo's most obscure passages and has opened the way for further research. Other texts, however, such as the well-known *Mirandum* passage, can hardly be reconciled with Souffrin's interpretation... (Giusti 2004, 121)." Against this it has been argued here that it is exactly Galileo's concept of velocity that Souffrin has reconstructed which Galileo applied and which led to the conflicting result.

10.2 The Rise and Demise of a Dynamical Foundation

the basic principle as its crucial and fundamental premise, this had an important bearing. The proof was not simply rendered false. In fact, an assumption weaker than Galileo's basic principle of inclined dynamics sufficed as a premise for the proof. If, however, the proof was conceived of as resting on the weaker assumption, it was deprived of its explanatory value and could thus, in particular, no longer be considered a fundamental proof.

For the argument of the ex mechanicis proof, it merely needs to hold that the velocities of naturally accelerated motions along inclined planes appropriately inscribed into the same circle as chords and thus traversed in the same time are in the same ratio as the mechanical moments on these planes, as the dynamical principle in the argument is only applied to motions along such chords. In light of the fact that the basic principle of mechanics could no longer be maintained as generally true, in principle, nothing spoke against this weaker assumption.[76] Yet the truth of the weaker assumption could no longer be considered as following from the truth of the general principle of inclined plane dynamics embracing the weaker assumption. Thus, an alternative explanation as to why the weaker assumption should hold if the more general assumption didn't would have to be given. Such an explanation, however, so obvious from a modern perspective, was not at hand. Thus the ex mechanicis proof of the law of chords and, more generally the dynamics of naturally accelerated motion, had been transformed from an explanans to an explanandum.

Why this, indeed, weighed so gravely on Galileo that he broke off his work on the star group folios leaving roughly half of folio page 164 verso blank, will only become fully apparent in the next chapter, where it will be demonstrated that Galileo's resort to dynamical arguments was not incidental but part of a systematic attempt to provide his new insights with a foundation. What the problem revealed by the mirandum paradox thus, ultimately, undermined, at least for the time being, was the prospect of being able to create a new science.

The rise and rather immediate demise of Galileo's attempt to provide a dynamical foundation for his new science of naturally accelerated motion can be summarized as follows: Galileo's consideration in the context of his research agenda on swinging a rolling had been pheno-kinematical in nature and so were the insights it had engendered. On the star group folios, Galileo began to analyze these new insights harking back to the concepts embraced by the existing frameworks traditionally applied to describe motion and its causation. He started to employ, in particular, the concept of velocity into his considerations. Even though an understanding of velocity as a changing quality as entailed by the doctrine of the configurations of motion started to figure in these considerations, he mainly resorted to what has been referred to as the kinematic concept of velocity as implicitly defined by

---

[76]This provides the likely reason why Galileo felt entitled to include the ex mechanicis proof in the *Discorsi*, where he issued it as the third proof for Theorem VI, i.e., the law of chords. As Settle (1966, 240) expressed it: "I take it, then, that the causal use of the mechanical principle in the subsidiary demonstration of Theorem VI of the Third Day is a vestige of this early period of confidence."

the kinematic propositions. The kinematic understanding of velocity recognizes velocity as an external characteristic of motion, neglecting internal characteristics such as acceleration.

His analysis initially suggested that the ratio of the kinematic velocities of naturally accelerated motion along inclined planes was in proportion to the ratio of the mechanical moments on these planes and that thus, assuming the basic principle of inclined plane dynamics, his insights about naturally accelerated motion could eventually be founded on dynamical considerations. This allowed him to formulate the ex mechanicis proof, in which a central statement concerning the pheno-kinematics of naturally accelerated motion was proven based on dynamics. Yet further analysis of his new results did not lead to the expected seamless integration of the new class of phenomena under the extant conceptualizations but yielded unanticipated results which eventually proved to be irreconcilable with the dynamical assumptions that had been employed.

The mirandum paradox indicated that the velocities along two inclined planes could not be held to be in a fixed ratio, but the velocity of motion over a distance on one plane was faster, equal, or slower than that of motion along a certain distance on the other plane depending on the distances considered. This conflicted with the basic intuition that the bigger force on the steeper plane should correspond to a higher velocity on the steeper plane. All the more, it conflicted with the stronger dynamical assumption which found its expression in the basic principle of inclined plane dynamics, namely, that the ratio of velocities could be assumed to be in proportion to the ratio of the mechanical moments.

When Galileo worked on the star group folios, his insights concerning the pheno-kinematics of naturally motion, obtained as a result of his research on swinging and rolling, had already achieved the status of well justified rather incontestable facts before and outside the attempt to further analyze them. Thus, the conflict made apparent through the mirandum paradox argument could not easily be dismissed. In particular, Galileo could not simply reject the assumption about the pheno-kinematics of naturally motion which had entered the argument. Rather he had to accept that a traditional dynamical understanding, which for Galileo found expression in the basic principle of inclined plane dynamics, was not compliant with the kinematics of naturally accelerated motion as he had worked out.[77] As will be discussed in the next chapter, the quandary was resolved when around 1604 Galileo systematically attempted to found his theory on an understanding of velocity as it was entailed by the doctrine of the configurations of motion which allowed him to sideline the question of the dynamics of naturally accelerated motion entirely.[78]

---

[77] In fact, Galileo's entire search for a foundation for his new science of naturally accelerated motion, from the very beginnings to the time of the publication of the *Discorsi*, can be regarded as dominated to a large extent by his attempt to accommodate assumptions about velocity from all three domains to the theoretical description of naturally accelerated motion in such a way that they would allow for a coherent foundation of the theory. Cf. Ravetz (1972).

[78] In 1602, Galileo had resorted to the increase of degrees of velocity in the double distance rule. Yet, as has been argued, this mimicked traditional application, and the proposition remained

10.2 The Rise and Demise of a Dynamical Foundation

The blow the mirandum paradox had delivered to the attempt at dynamical reasoning for naturally accelerated motion would remain final, at least as far as Galileo is concerned. By applying the old conceptualizations to the new insights achieved in the context research on swinging and rolling, Galileo had probed the traditional frameworks, and thus the limits of preclassical mechanics had become apparent. It was not for Galileo to ultimately resolve the conundrum the mirandum paradox had led him into and to transgress the limits and accept that it was not motion but change of motion which is to be explained by the action of forces.

As is plain to see from a modern perspective, the constant force on an inclined plane of a given inclination indeed finds its repercussion in the constancy of a kinematical magnitude, characterizing the naturally accelerated motion along this plane. The force is however not proportional to velocity as Galileo held but to the rate by which velocity changes, i.e., to acceleration. It is only when distances traversed by naturally accelerated motions from rest in equal time are concerned that the accelerations are in the same ratio as the velocities and thus also as the forces. Thus, the law of chords indeed picks out the one situation where Galileo's, from a modern perspective, wrong assumption is, again from a modern perspective, accidentally correct. Galileo recognized the problem with the ex mechanicis proof but did not renounce it. Rather he published it as one of three alternative proofs given for the law of chords in the *Discorsi*. He did thus not present the proof as part of the foundation of the new science and concealed the problem he had encountered with regard to a dynamical reasoning about natural acceleration.[79] For Galileo to solve the problem and thus to transgress the limits of preclassical mechanics would have meant conceptualizing acceleration as the rate of change of velocity. This was beyond what he accomplished, and indeed it took almost a century until a decisive step toward the solution of the problem was made when Newton stated with reasonable clarity that force was proportional to the rate of change of velocity.[80] What Galileo contributed was his new science of motion whose conclusions rather directly bespoke the limit of traditional dynamics and thus bore the seed of its transgression.

---

isolated in a certain sense, i.e., Galileo did not, and in fact could not, exploit it any further in his attempt to erect a foundation for his new theory. In consequence, his use of the concept of velocity conveyed by the doctrine of the configurations of motion, at least for the time being, had hardly any further bearing.

[79] Shea (1983, 54) shares this position: "Galileo never abandoned, however, his initial conviction that gravity is the essential cause of motion. Unable to account for the role of gravity, he skilfully skirts around the subject in the Discourses ...The historical Galileo, however, only gave up the investigation of the cause of acceleration because he was unable to identify it."

[80] "Mutationem motus proportionalem esse vi motrici impressæ..." Motus is here to be understood as momentum. Cf. Harman et al. (2002, 353).

10.3 Summarizing Conclusion

The *Notes on Motion* comprise four folios which share the same watermark, a small star. Three of these folios display a striking formal similarity with respect to the way in which Galileo wrote on them. These so-called star group folios have been analyzed in detail in this chapter. On these folios Galileo orderly and rather systematically collected results which had emerged from his research agenda on swinging and rolling. Thus, his work served the consolidation of what had been achieved. It has been argued that the driving force behind this consolidation was not so much the prospect of a continued exploration of the challenging problem of the relation of swinging and rolling but that of turning his results into a new science to be published in print. To this end, in some of the entries of the star group folios, Galileo started to analyze his new insights taking recourse to the concepts encompassed by the extant theoretical frameworks.[81]

The most extensive entry on the group of folios, virtually identical to the version later published in the *Discorsi*, is Galileo's proof of the law of the broken chord together with its three lemmata. The proof had been a central and one of the most profound results of Galileo early research program. The folios of the group, furthermore, contain proofs of two elementary propositions, the law of chords and of the length time proportionality. The latter proposition had emerged as an important consequence of Galileo's considerations in the context of the broken chord approach. Once established, the proposition was immediately used as a premise for many of Galileo's successive arguments concerning naturally accelerated motion.

The star group folios, furthermore, contain two propositions concerning motion not starting from rest but equipped with an initial velocity, the generalized law of fall and the generalized length time proportionality, which likewise had resulted from his research agenda on swinging and rolling. With an entry that Galileo later elaborated into Proposition VIII of the Second Book of the Third Day of the *Discorsi*, the star group folios contains, in addition, an entry concerned with paths traversed in naturally accelerated motion in less or least time, a type of problem which likewise had originated from his early research program. In his proof of the generalized law of fall, we find Galileo grappling with the intricacies of his newly established technique of integrated time representation.

Besides the material mentioned, the star group folios contain a number of entries of noticeably different in character. These entries generally attest an exploration or analysis of Galileo's new findings based on concepts rooted in the traditional frameworks. In particular, the concept of velocity, which is virtually absent from Galileo's considerations regarding swinging and rolling, starts to figure. Among

---

[81] Naylor (1990, 700) refers to the "search for a physical principle capable of explaining the mathematical characteristics" of experimentally established phenomena as Galileo's "method of analysis and synthesis."

the entries in which Galileo engaged with the concept of velocity are the tarditas length proportionality, the ex mechanicis proof of the law of chords, and the double distance rule.

In all three considerations, Galileo applied one of the kinematic propositions to make inferences concerning the relation between distances covered and times elapsed. The kinematic propositions imply what has been referred to as a kinematical understanding of velocity. In his proof of the double distance rule, Galileo availed himself of an understanding of velocity as a changing quality as it was entailed by the framework of the doctrine of the configurations of motion. This shows that by that time he was well familiar with the tradition and it has indeed been argued that the double distance rule has to be regarded as Galileo's concrete physical interpretation of the traditional Merton rule. In some entries on the star group folios, moreover, Galileo treated kinematical velocity and the totality of these degrees of velocity, the aggregate velocity, as interchangeable. Yet the proof of the double distance rule is the only argument on the star group folios in which an understanding of velocity as a changing quality is made constructive use of and here Galileo basically reiterated an established result. In fact, as will be shown in the next chapter, Galileo only started to systematically analyze naturally accelerated motion based on an understanding of velocity as a changing quality in 1604.

The analysis of his new insights about the pheno-kinematics of naturally accelerated motion on the star group folios opened the prospect of being able to provide a dynamical argument regarding the kinematics of naturally accelerated motion. Galileo's considerations, as documented by the tarditas length proportionality argument, suggested that his basic principle of inclined plane dynamics would be applicable to the case of naturally accelerated motion. This was emphatically and programmatically expressed by Galileo in the si hoc sumatur note. The assumption that the velocities of accelerated motion on inclined planes are in the same ratio as the mechanical moments on these planes indeed became the essential asset for his construction of the ex mechanicis proof. Once a decent version of the proof had been drafted, it was immediately copied to the star group folios where it was to replace an older proof of the law of chords already contained on the page, since, as a fundamental proof from a basic principle, it would have replaced the latter in the deductive structure of a new science.

Galileo's progress regarding the foundation of the new science was momentary. With the mirandum paradox, the star group folios contain an entry documenting a consideration by Galileo which led him to recognize a serious problem in the conceptual underpinnings of the ex mechanicis proof. Galileo realized his kinematical understanding of velocity as applied to naturally accelerated motion yielded conclusions that were not compliant with his dynamical assumptions. Thus, the ex mechanicis proof had lost its basis and with that its explanatory value. Why this was particularly devastating for Galileo in 1602 will be revealed in the succeeding chapter where it will be demonstrated that Galileo's construction of a dynamical argument was not incidental. Rather it will be shown that in 1602, Galileo had begun to work systematically on the creation of the foundation for his new insights to turn them into a new science. This attempt had run into an impasse with the mirandum paradox, forcing Galileo to interrupt this work for the time being.

References

Barbour, J. (2001). *The discovery of dynamics: A study from a Machian point of view of the discovery and the structure of dynamical theories.* Oxford: Oxford University Press.

Büttner, J. (2001). Galileo's cosmogony. In J. Montesinos & C. Solís (Eds.), *Largo campo di filosofare: Eurosymposium Galileo 2001* (pp. 391–401). La Orotava: Fundación Canaria Orotava de Historia de la Ciencia.

Büttner, J. (2008). Big wheel keep on turning. *Galilaeana, 5*, 33–62.

Damerow, P., Freudenthal, G., McLaughlin, P., & Renn, J. (2004). *Exploring the limits of preclassical mechanics.* New York: Springer.

Drake, S. (1978). *Galileo at work: His scientific biography.* Chicago: University of Chicago Press.

Drake, S. (1979). *Galileo's notes on motion arranged in probable order of composition and presented in reduced facsimile* (Annali dell'Istituto e Museo di Storia della Scienza Suppl. Fasc. 2, Monografia n. 3). Florence: Istituto e Museo di Storia della Scienza.

Galilei, G. (1914). *Dialogues concerning two new sciences.* New York: Macmillan.

Galilei, G., Crew, H., & Salvio, A.D. (1954). *Dialogues concerning two new sciences.* New York: Dover Publications.

Galluzzi, P. (1979). *Momento.* Rome: Ateneo e Bizzarri.

Giusti, E. (1981). Aspetti matematici della cinematica galileiana. *Bolletino di storia delle scienze matematiche, 1*(2), 3–42.

Giusti, E. (2004). A master and his pupils: Theories of motion in the Galilean school. In C.R. Palmerino, & J.M.M.H. Thijssen (Eds.), *The reception of the Galilean science of motion in seventeenth-century Europe* (Boston Studies in the Philosophy of Science, Vol. 239, pp. 119–135). Dordrecht: Kluwer.

Hahn, A.J. (2002). The pendulum swings again: A mathematical reassessment of Galileo's experiments with inclined planes. *Archive for History of Exact Sciences, 56*, 339–361.

Harman, P.M., Shapiro, A.E., & Whiteside, D.T. (2002). *The investigation of difficult things: Essays on Newton and the history of the exact sciences in honour of D.T. Whiteside.* Cambridge: Cambridge University Press.

Hooper, W.E. (1992). *Galileo and the problems of motion.* Dissertation, Indiana University.

Maffioli, C.S. (2008). Galileo, Guiducci and the engineer Bartolotti on the Bisenzio river. *Galilaeana, 5*, 179–206.

Naylor, R.H. (1990). Galileo's method of analysis and synthesis. *ISIS, 81*(4), 695–707

Ravetz, J.R. (1972). Galileo and the mathematization of speed. In G. Canguilhem (Ed.), *La mathématisation des doctrines informes*, Hermann, Paris, (pp. 11–32)

Settle, T. (1966). *Galilean science: Essays in the mechanics and dynamics of the Discorsi.* PhD thesis, Cornell University.

Shea, W.R. (1983). The Galilean geometrization of motion: Some historical considerations. In: Shea W.R. (eds) *Nature Mathematized.* (pp. 51–60). Dordrecht: Springer Netherlands.

Souffrin, P. (1990). Galilée et la tradition cinématique pré-classique. La proportionnalité momentum-velocitas revisitée. *Cahier du Séminaire d'Epistémologie et d'Histoire des Sciences, 22*, 89–104.

Wallace, W.A. (1981). *Prelude to Galileo: Essays on medieval and sixteenth-century sources of Galileo's thought.* Dordrecht: Reidel.

Wallace, W.A. (1984). *Galileo and his sources: The heritage of the Collegio Romano in Galileo's science.* Princeton: Princeton University Press.

Wallace, W. (1990). Duhem and Koyré on Domingo de Soto. *Synthese, 83*(2), 239–260.

Wisan, W.L. (1974). The new science of motion: A study of Galileo's De motu locali. *Archive for History of Exact Sciences, 13*, 103–306.

Wolff, M. (1994). "Neutrale Bewegung" beim jungen Galilei. In L. Schäfer & E. Ströker (Eds.), *Naturauffassungen in Philosophie, Wissenschaft, Technik* (Renaissance und frühe Neuzeit, Vol. 2). Freiburg: Alber.

Chapter 11
Toward a New Science: Axiomatization and a New Foundation

The previous chapter analyzed the content of the star group folios. On these folios Galileo collected and structured his new insights, and it has been argued that his primary motivations for this work lay in his intention to render the results achieved until that point into a form that would allow their publication as a new science in the future. As has been shown, the content of these folios was drafted around 1602, contemporaneously to Galileo's investigations into the relation of swinging and rolling, or shortly after these investigations had come to a halt. We have seen how the analysis of the new insights, adopting concepts anchored in the extant theoretical frameworks, allowed Galileo to construct the ex mechanicis proof of the law of chords. Thus, one of his fundamental propositions on accelerated motion was provided with a demonstration based on dynamical considerations. Yet further analysis showed that the dynamical principle he had employed was not consistent with his new insights concerning the pheno-kinematics of naturally accelerated motion. A foundation of the new science on dynamics, which had briefly appeared possible, thus, moved beyond reach. In the absence of an alternative approach, for the time being, Galileo was forced to discontinue his attempt at providing a foundation for his new results.

In this chapter, one additional folio, folio 147, is included into the analysis. It will be demonstrated that the content of this folio must have been drafted by Galileo contemporaneous to his work on the star group folios. The analysis corroborates the interpretation given and enhances the picture conveyed in the previous chapters. In particular, Galileo's systematic attempts to turn his new insights about the pheno-kinematics of naturally accelerated motion into a new science, by providing them with a deductive structure rooted in fundamental principles or assumptions, will emerge more clearly than before.

In the current chapter, it will be shown, in particular, that even before he had been able to demonstrate any of the propositions that had formed the basis of his research program on swinging and rolling from fundamental principles, Galileo systematically explored the question of which of his statements could serve as a minimal yet strong enough set from which all remaining propositions could be derived. This will be referred to as his search for an *axiomatic foundation*. Galileo thus recognized that the law of fall, the law of chords, and the length time proportionality represented a coherent yet overdetermined foundation, as he had been able to derive any of these three propositions based on the remaining two as premises.

This, in particular, implied that in order to transform his new results into a mathematized demonstrative science, two of these three propositions needed to be furnished with demonstrations from basic principles firmly rooted in generally accepted conceptual traditions.[1] Galileo's search for such demonstrations is referred to as his search for an *analytic foundation*. As demonstrated in the previous chapter, this search bore out the ex mechanicis proof and, thus, more generally, the perspective of founding the pheno-kinematics of naturally accelerated motion on dynamics.

As will be illustrated, most of the entries on folio 147 are indeed concerned with the exploration of the application of dynamical arguments to naturally accelerated motion on inclined planes. Thus, in an entry that will be referred to as the hollow paradox, Galileo tried to secure his assumption that the kinematics of accelerated motion on inclined planes did not depend on the absolute weight of the moving body. This assumption, ultimately due, from a modern perspective, to the identity of inertial and gravitational mass, was implied by his conceptualization of swinging and rolling and his observation that the period of swing of a pendulum did not depend on the weight of the bob.

Moreover, folio 147 contains a set of considerations that can be interpreted as an attempt by Galileo to resurrect the dynamical argument of the ex mechanicis proof after it had already become clear to him via the mirandum paradox that the dynamical principle originally employed in the argument was not compliant with his insights about the pheno-kinematics of naturally accelerated motion. Yet these considerations obviously did not provide Galileo with a clue as to how to reformulate the underlying dynamics. In addition, his search for an axiomatic

---

[1] According to the Aristotelian conception, a demonstrative science has to be based on principles "true, primary, immediately better known than, prior to and grounds of the conclusion" (*Posterior Analytics*, I.2 71b20-22). For a discussion of the Aristotelian notion of basic principles, see McKirahan (1992, 21–50). In the contemporary philosophical debates, the thought movement from an effect to the existence of cause of this effect followed by an inference from this very cause to the effect was referred to as the *regressus* method. It was considered to be the foundation of scientific reasoning, yet usually not discussed in relation to physico-mathematical sciences such as Galileo's new science of motion. For the *regressus* method in general, see, for instance, the corresponding entry in the Routledge Companion to Sixteenth Century Philosophy (Palmieri 2017). For the question to what extend Galileo can be said to have followed this method, see in particular Wallace (1992) and Jardine (1976).

foundation had shown that it would be necessary for at least a second proposition to be equipped with a proof from basic principles in order to root the deductive tree. In the absence of an alternative approach, Galileo was forced to discontinue his search for a foundation.

In the last section of this chapter, it will be briefly recounted how Galileo, who by then had come up with an entirely different approach to the analytic foundation, picked up work again in 1604. His return to the problem of the foundation is evidenced by a letter written to Paolo Sarpi toward the end of the year, as well as by material in the *Notes on Motion* that can be related to the program outlined in the letter. As is demonstrated, Galileo's fresh attempt at a foundation exactly respected the requirements which his search for an axiomatic foundation had brought to light approximately two years earlier. Galileo, in particular, managed to avoid arguing from dynamical assumptions. Moreover, besides providing a proof for the law of fall, Galileo also argued for the length time proportionality on the basis of his new approach. Based on his prior considerations, he could, thus, be confident that he had found a way to root the deductive tree and to demonstrate his new insights concerning pheno-kinematics of naturally accelerated motion. In fact, Galileo insinuated in the letter a number of times that he was finally able to demonstrate what he had essentially already previously established.

As is well known, the principle of acceleration that Galileo communicated to Paolo Sarpi in 1604, and on which he had then based his demonstrations, was later recognized as false. In consequence, the fundamental proofs devised in this period of work had to be rejected as well or at least to be revised. Thus, the problem of the foundation was open once again. How it was finally solved by Galileo is not discussed in this book.

11.1 1602: Immersed in Foundational Problems

Manuscript evidence Folio 147 bears a crown watermark and can hence securely be attributed to Galileo's Paduan period. Both sides of the folio are almost completely covered with content. The recto side contains a short memo, the elaboration of a proof of the law of fall, as well as an entry that will be referred to as the hollow paradox. In this entry, Galileo applied an argument to motion along inclined planes that had its origin in the study of motion in media. The first entry Galileo made on the verso side of the folio was a diagram related to his search for iso-temporal surfaces. The remaining space on the page is filled with considerations related to motion along inclined planes completed in equal time. This set of considerations will be referred to as the isochronism of planes or, for short, *isochronism considerations*. The diagram Galileo had initially drawn on the page was reused for these considerations.

Folio 147 recto The content of 147 recto can be broken down into three clearly distinguishable blocks. First of all, the page contains a small motion diagram concerning the descent of bodies on planes of equal length. The diagram is drawn upside down with respect to a modern orientation of the page.[2] It is accompanied by an abbreviated notation of a proportion that holds between times elapsed and distances covered with respect to the motion situation under consideration.[3] Judging from the layout, this was most likely the first entry on the page. All successive entries were written with the page turned 180 degrees.

The most extensive entry on the page, which was likely written second, documents Galileo's construction of a proof of the law of fall. It comprises a central motion diagram as a part of which a numerical example is coded, giving lengths of distances covered and the times elapsed during motion.[4] Galileo, moreover, sketched a small table as well as two distance-time ratio diagrams as part of his attempt to construct the proof.[5] Directly underneath the diagram, he wrote a short statement concerning a ratio of times and distances, which marks the key insight for the construction of the proof then eventually elaborated toward the bottom of the page.[6] The last entry that Galileo added was the hollow paradox. This was written on the remaining space at the top of the page and floats around the proof diagram already contained on it.[7] Underneath, the word "Paralogismus" is written.[8] It may pertain to the hollow paradox but could also refer to a fault in the numerical example that Galileo devised as part of his attempt to prove the law of fall.

Proposition V memo On the top of the page, Galileo drew a small diagram of two inclined planes of equal length, *ab* and *cb*, whose heights are given by *ae* and *cd*, respectively.[9] Above the diagram he noted the proportion $t(ab)/t(ae) \sim bc/ae$, which directly results from an application of the length time proportionality and the substitution of *ab* by the equal distance *cb*.[10] Galileo left his considerations unfinished. This short entry was elaborated to a full proposition on folio 189 verso and was published as Proposition V in the *Discorsi*.

---

[2] 147 recto, D01A.

[3] 147 recto, T2.

[4] 147 recto, D02A.

[5] 147 recto, T3, D03A and D03B.

[6] 147 recto, T4 and T5.

[7] 147 recto, T1A and T1B.

[8] 147 recto, T1C.

[9] 147 recto, D01A.

[10] 147 recto, T2.

11.1 1602: Immersed in Foundational Problems

The hollow paradox In *De Motu Antiquiora*, Galileo was concerned with a theory of motion in media. Extended passages of Galileo's later works are still concerned with this topic, partly restating older *De Motu Antiquiora* results, partly introducing new considerations, and partly revising or contradicting Galileo's older unpublished results.[11] The hollow paradox is one of only two passages in the *Notes on Motion* directly referring to motion in media.[12]

In the first paragraph Galileo outlined his agenda:

> Contempletur quod, quemadmodum gravia omnia super orizonte quiescunt, licet maxima vel minima, ita in lineis inclinatis eadem volicate [velocitate] moventur, quemadmodum et in ipso quoque perpendiculo; quod bonum erit demonstrare, dicendo quod, si gravius velocius, sequeretur quod gravius tardius, iunctis gravibus inaequalibus, etc.[13]

Thus, what Galileo set out to demonstrate was that bodies of different absolute weight move with the same velocity on one and the same inclined plane.[14] It is not explicitly stated that these motions are considered to be naturally accelerated but that Galileo assumed this to be the case is unambiguously clear, particularly in view of the remaining content of the page.

In particular the phrase "quod bonum erit demonstrare" here indicates that Galileo was not primarily exploring consequences of earlier considerations. Rather his aim was to justify his results. The necessity to do as can be argued had arisen from the intention to communicate his results as a deductive demonstrative science. Galileo signaled that such demonstration could be achieved along the line of an argument he had made in the *De Motu Antiquiora* to refute those who believed

---

[11] Part of Galileo's consideration concerning motion in media of *De Motu Antiquiora* was published in the *Discorso intorno alle cose che stanno in sull'acqua* in 1612. It was there that Galileo first introduced his assumption that all bodies would show the same kinematical behavior in a vacuum.

[12] The second entry in the *Notes on Motion*, which relates directly to motion in media, is contained on folio 66 recto. It is written in the hand of Mario Guiducci. In the entry, the change of the natural heaviness of a body on an inclined plane depending on the inclination is compared to the change of the heaviness of a body depending, according to the principle of buoyancy, on the density of the medium in which it is immersed. It is insinuated that the kinematics of falling motion should be comparable in both cases. After the consideration documented by the mirandum paradox, Galileo would have perceived the argument as problematic which may speak in favor of Mario Guiducci being the author. Indeed, no original draft in Galileo's hand is extant from which the entry could have been copied.

[13] 147 recto T1A. "It should be considered, that, just as all heavy [bodies] rest on the horizontal, be they very big or very small, so they are moved on inclined lines with the same velocity, just as along the perpendicular itself; it will be beneficial to prove that by saying that if a heavier [body] [be] faster it would follow that a heavier [body] [be] slower, having combined differently heavy [bodies], etc. (trans. Damerow et al. 2004, 203)."

[14] The wording indicates that the hollow paradox was written before the mirandum paradox. After the considerations documented by the latter, it would not have made sense for Galileo to speak of the equal velocity of two naturally accelerated motions without further specification, as he does in the hollow paradox.

Fig. 11.1 Diagram (D01B) on folio 147 recto associated with the hollow paradox

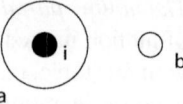

that the velocity of a body falling through a medium depended on the absolute weight of this body. As the *De Motu Antiquiora* argument did, in fact, not depend on the particular conditions of fall, i.e., neither on concrete assumptions about the surrounding medium nor on the assumption that descent proceeded vertically, transferring it to the case of motion on the inclined plane was indeed a natural step.

In the paragraph that follows, reference to the inclined plane as the context of the argument is indeed dropped entirely[15]:

> Movebuntur autem eadem celeritate non solum gravia inaequalia et homogen[e]a, sed etiam eterogenea, ut lignum et plumbum. Cum enim antea ostensum fuerit, magna et parva homogenea aequaliter moveri, dicas: sit *b* sp[ha]era lignea et a plumbea, adeo magna ut, cum in medio habeat cavitatem pro *b*, sit tamen gravior quam sphaera solida lignea ipsi *a* [sic] *a* aequalis, ita ut per adversarium velocius mov[e]atur quam *b*: ergo si in cavitate *i* ponatur *b*, tardius movebitur *a* quam cum erat levior: quod est absurdum.[16]

With respect to the associated diagram reproduced here as Fig. 11.1, where *b* is a sphere made of wood and *a* a hollow sphere made of lead such that the wooden sphere would fit exactly into the hollow part *i* of the lead sphere *a*, Galileo's argument can be paraphrased as follows: let us imagine a homogeneous wooden sphere, equal in size to the lead sphere *a*. Let the lead sphere, despite being hollow, be heavier than this wooden sphere. An adversary who claims that the velocity of fall depends solely on the heaviness of the moving body would, thus, have to conclude that the hollow lead sphere moves faster. The same would, however, also hold true according to Galileo's own assumption, according to which the velocity depends on

---

[15] In the hollow paradox, Galileo uses the expressions "celeritas" and "velocitas" synonymously. Morphologies of the noun *celeritas* or the adjective *celer* occur 16 times on 7 folio pages in Galileo's *Notes on Motion*, with one of the occurrences being a duplication by copying. Most of the occurrences congregate on one text contained on folio 91 verso, which contains early considerations pertaining to projectile motion. Just as *celeritas* is used by Galileo synonymously with *velocitas*, so *momentum celeritatis* is, albeit rarely, used synonymously with *momentum velocitatis*. The disappearance of the former expression from Galileo's vocabulary seems to owe to his attempt to unify terminology.

[16] "Moreover, not only homogeneous and unequal heavy bodies would move at the same speed, but also heterogeneous ones such as wood and lead. Since as it was shown before that large and small homogeneous bodies move equally, you argue: Let *b* be a wooden sphere and *a* be one of lead so big that, although it has a hollow for *b* in the middle, it is nevertheless heavier than a solid wood sphere equal [in volume] to *a*, so that for the adversary it should move faster than *b*; therefore if *b* were to be put into the hollow *i*, *a* would move slower than when it was lighter; which is absurd. (trans. Damerow et al. 2004, 204)."

the specific weight.[17] Both spheres have the same volume and hence the ratio of their specific weights simply corresponds to the ratio of their weights. The smaller wooden sphere b, in turn, would either move as fast as the larger wooden sphere (Galileo's position, because of their equal specific weight) or slower (the position of his adversaries, because of smaller absolute weight).

If the wooden sphere b is now placed within the hollow of the lead sphere a, the weight and the specific weight of this composite body are increased as compared to the hollow lead sphere a. To complete the argument, an additional assumption is invoked which remained implicit but is familiar. In the *Discorsi*, Galileo expressed this assumption as follows:

> If then we take two bodies whose natural speeds are different, it is clear that on uniting the two, the more rapid one will be partly retarded by the slower, and the slower will be somewhat hastened by the swifter. (Galilei et al. 1954, 63)

In other words the velocity of a composite body should lie between that of its parts. Thus, the velocity of the composite of a lead and small wooden sphere should lie between that of the hollow lead sphere a and that of the slower wooden sphere b. In consequence, the velocity of the composite body must, in any case, be slower than that of the lead sphere a alone. On the other hand, however, both the overall weight and the weight per volume are greater in the compound body. Hence, the adversaries (but also Galileo himself) were forced to conclude that the compound body falls faster than the hollow sphere a and a contradiction had been derived from the assumptions.

Accordingly, one of the underlying assumptions must be refuted. If it is maintained that the velocity of a compound body must lie between that of its individual parts, it necessarily follows that bodies of different weights cannot fall with different velocities. The argument does not hinge on the concrete circumstances of fall and should hold, thus, regardless of whether motion proceeds along the vertical or along an inclined plane. This is what Galileo intended to demonstrate.

From a modern perspective, it is rather obvious that Galileo's argument holds true for bodies of different specific weights as well and, therefore, also implies an independence of the kinematics of fall on the specific weight. Thus, the hollow paradox argument contradicted, at least implicitly, Galileo's own opinion that the velocity of fall depended on the difference in weight per unit volume between the medium and falling body. It has indeed been argued that the word "paralogismus," false conclusion, written underneath the hollow paradox, indicates that Galileo had discovered a mistake in his argument.[18]

In forming the composite body by placing the wooden sphere in the hollow of the lead sphere a in the thought experiment, Galileo had expelled the air contained in the hollow part of sphere a. This volume of air, however, was part of the hollow sphere

---

[17] For the concept of specific weight in Galileo, see Van Dyck (2006) (Weighing falling bodies. Galileo's thought experiment in the development of his dynamical thinking, unpublished. Available at http://www.sarton.ugent.be/index.php).

[18] Cf. Damerow et al. (2004, 202–204).

a and, together with it, formed a composite body. For his argument to be valid, Galileo would have, thus, needed to include the volume of air in his considerations, for instance, by considering a composite body formed by attaching it to the surface of the lead sphere, instead of putting the small wooden sphere inside the hollow. Had this been done, the argument would have remained inconclusive as can easily be shown.[19] There is, however, no evidence that Galileo had realized this particular flaw in his argument. The word "paralogismus" written beneath the hollow paradox, as will be seen, most likely refers instead to a mistake made by Galileo as part of his attempt to construct a proof of the law of fall on the same page as is implied by how and where the word is written.

As will be argued below, the content of folio 147 and, thus, the hollow paradox was drafted at about the same time as that of the star group folios. A likely reason why Galileo thought it fit to prove that fall along inclined planes did not depend on absolute weight thus becomes apparent. As Galileo emphasized in his letter to Guidobaldo del Monte in 1602, he was aware, and had indeed likely observed, that the period of a pendulum did not depend on the absolute weight of the pendulum bob. Based on the way Galileo conceptualized the relation between swinging and rolling, this immediately implied that fall along inclined planes likewise had to be independent of the weight of the moving body. The hollow paradox, thus, comes into view as an attempt to secure a fundamental premise of his research program on swinging and rolling. This was particularly indicated once Galileo had begun to immerse himself in considerations regarding the dynamics of falling motion. The hollow paradox is a dynamical argument, after all, in that it excludes or at least aims at excluding that change of absolute weight has an effect on the kinematics of motion on an inclined plane.[20]

The application of his old argumentative schema in a new context may have been occasioned by a concrete event. On 24 March 1604, Costanzo da Cascio wrote a letter to Galileo in Padua:

---

[19] If the air included in the hollow sphere is taken into consideration when forming the composite body, the argument breaks down because then the example can be constructed such that the specific weight of this composite body lies between that of the wooden sphere *b* and the hollow lead sphere *a*. Cf. Damerow et al. (2004, 204, footnote 126).

[20] In the *Discorsi* the statement that the velocity of fall is independent of weight is directly adjoined to a discussion of the independence of pendulum frequency from the weight of the pendulum bob:

> ...and if Simplicio is satisfied to understand and admit that the gravity inherent [interna gravità] in various falling bodies has nothing to do with the difference of speed observed among them, and that all bodies, in so far as their speeds depend upon it, would move with the same velocity, pray tell us, Salviati, how you explain the appreciable and evident inequality of motion; (EN VIII, 131, trans. Galilei et al. 1954, 87.)

This accords well with the conjecture that Galileo had originally conceived the independence of the kinematics of motion along inclined planes on the weight of the falling body, which he argued for in the hollow paradox, as a consequence of the independence of the period of the pendulum on the weight of the bob together with his conceptualization of the relation between swinging and rolling.

11.1 1602: Immersed in Foundational Problems

Dipoi, quando fui costà in Padova, mi ricordo che li domandai come si poteva dimostrare che dui corpi d'una medesima specie et figura, equali o vero inequali, per il medesimo mezzo havessero la medesima velocità di moto; et lei mi assegnò dui ragioni, per le quali si conduceva l'aversario a dui inconvenienti. Hora, per essere già tanto tempo che fu questo, me le sono scordate, e perchè me ne fa bisogno a un certo mio proposito, la prega si degni di novo accennarmele; (EN X, 108–109, letter 98)

On the occasion of a prior visit by da Cascio to Padua, Galileo had communicated an argument to him by means of which he had refuted that the velocities of bodies falling through a medium depended on their absolute weights. As the argument had escaped da Cascio's memory, he asked Galileo for it again, and it may well have been his return to the old argument, thus occasioned, that led Galileo to apply it in the context of his emerging new science of naturally accelerated motion. Notably, da Cascio uses the exact same expression to refer to the opponent, "l'aversario," which Galileo himself uses in Latin in the hollow paradox. Should the drafting of the hollow paradox indeed have been inspired by da Cascio's question, this would date it to sometime around the spring of 1604, a dating that is, as will be seen, absolutely reconcilable with the interpretation advanced in the present chapter.

The law of fall The remaining space on the page is filled with Galileo's elaboration of a proof of the law of fall. As has already been argued in Chap. 7, the proof reverses a line of argument which Galileo had first established as part of his attempt to construct a proof for the law of the broken chord. Galileo started by drawing the central motion diagram which, as illustrated in Fig. 7.8, is simply a copy of the proof diagram of the law of the broken chord reduced by some elements and turned upside down. He furnished the motion diagram with a numerical example, i.e., he added concrete, exemplary values for the lengths of the relevant distances and the times needed to traverse these distances in naturally accelerated motion (Fig. 11.2).

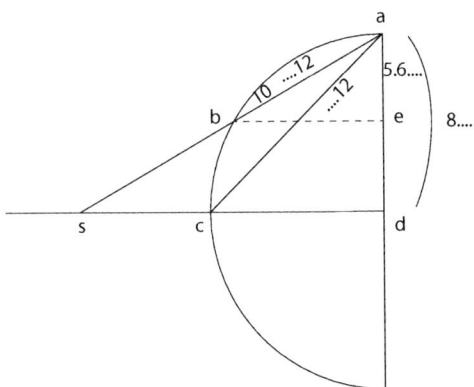

Fig. 11.2 Schematic representation of the diagram associated to Galileo's attempt to prove the law of fall on 147 recto

Galileo made use of integrated time representation and represented spatial as well as temporal aspects of the motions considered in the same diagram. The different numbers representing distances and times of motion are written next to the respective lines in the motion diagram. In order to distinguish whether a number is representing a time or a distance, Galileo preceded the numbers representing times by a dotted line, a convention repeated in one of the distance-time ratio diagrams drawn underneath.

The numerical example was constructed as follows:

- Chord ac is assumed to measure 12 units and, exploiting distance time coordination, the time unit is so chosen that the time to traverse this chord measures 12 units as well. The number 12 is written next to chord ac with a dotted line indicating that it represents a time.
- Chord ab is assumed to measure 10 units and its vertical height ae, to measure 5 units. The numbers are written next to the respective distances.
- By the law of chords, the time along chord ab is inferred to measure 12 units. The number is written next to chord ab with a dotted line indicating that it represents a time.
- By length time proportionality, the time along ae is inferred to measure 6 units ($t(ab)/t(ae) \sim 10/5$). The number 6 is written next to ae and is again marked with a dotted line.
- Vertical ad is taken to measure 8 units. By length time proportionality, the time along ad is inferred to measure 8 units as well ($t(ac)/t(ad) \sim 12/8$). The number 8 representing a time as well as a distance of motion is written next to the vertical height ad.

Superficially, Galileo's example appears to be sound. Yet the ratio of the distances ae and ad (5/8) is by no means the square of the ratio of the times of motion along these distances ($6^2/8^2 = 9/16$), as required by the law of fall. Galileo had made a rather trivial mistake. According to his construction, the planes ac and ab had to be chords in the same circle, as only thus could it be inferred by the law of chords that the times of motion along these planes are equal. The planes defined by the numerical values for heights and lengths Galileo had chosen for calculational convenience simply did not satisfy this condition. Hence, his application of the law of chords to infer that the times of fall along ab and ac are equal was not justified and Galileo could not conclude that these planes are traversed in equal time. As his further proceeding shows, Galileo recognized his mistake and it is thus possible and indeed plausible that the word "paralogismus" refers to this mistake and not to a flaw in the reasoning of the hollow paradox.

Next Galileo wrote out a small table and a short entry:

te[mpus] : *ae ac ad*
                ab
li[nea]   *ae ab*
temp[us] da ad temp[us] ac ut lin[ea] da ad lin[ea] ac.

The table and entry together define but do not solve the problem. In the first row, Galileo symbolically noted the times through ae, through ac equal to the time through ab by the law of chords, and finally the time through ad. What was needed

11.1 1602: Immersed in Foundational Problems

to complete the proof was to establish a relation between the time in the first and the third column. In the second row, Galileo listed the lines which represented or measured the times of motion listed in the first row. By distance time coordination the time ea was simply taken to be measured and, thus, represented by the line ea. With this choice it was implied by the length time proportionality that the line ab provided a measure of the time along ab and what remained to be done was to fill the last cell with a line that represented the time of motion along ad.

However, as the fault in Galileo's numerical example demonstrated, this third cell could not easily be filled in. He knew that by the length time proportionality, the times through ac and ad had to be in the same ratio as the distances covered and noted this result in the short textual entry below. However, since the time along ac was not represented by the line ac, ad could not be taken as representing the time of motion along the distance ad. From a formal point of view, the last cell would have to contain the mean proportional between the lines ae and ad. This mean proportional was, however, not part of Galileo's construction at all.[21]

Yet, probably by means of the left of the two diagrams underneath the central diagram, Galileo arrived at the insight which allowed him to finally complete the proof. He noted:

Concludendum: tempus da ad tempus ea esse ut linea ca ad ab.

Galileo's statement is correct, as shown by the following line of argument, which must not necessarily correspond to the way that he arrived at the insight:

By length time proportionality:

$$t(ae)/t(ab) \sim ae/ab \text{ and } t(ad)/t(ac) \sim ad/ac$$

With $t(ab) = t(ac)$ compounding both ratios yields:

$$t(ad)/t(ae) \sim ab/ae * ad/ac$$

By geometry, where D is the diameter of the circle:

$$ab/ae \sim D/ab \text{ and } ad/ac \sim ca/D$$

---

[21] Galileo's construction of a proof of the law of fall on folio 147 bears striking resemblance to his elaboration on folio 189 verso of a proposition that was later to be published as Proposition V in the *Discorsi*. In both cases, for instance, Galileo employs distance-time ratio diagrams and uses different types of lines to distinguish between lines representing times of motion and such lines representing distances covered by motion, lending additional plausibility to the assumption that work on the two folios was conducted roughly at the same time.

By above:

$$t(da)/t(ae) \sim ca/ab$$

Based on this insight completing the proof was a rather trivial endeavor. Galileo noted:

> Postea quam ostensum fuerit, tempora per *ab*, *ac* esse aequalia, demonstrabitur tempus per *ad ad* tempus per *ae* esse ut *da* ad mediam inter *da*, *ae*. Nam tempus per *da* ad tempus per *ac* est ut *da* ad *ac* lineam; tempus autem per *ac*, idest per *ab*, ad tempus *ae* est ut linea *ba* ad *ae*, hoc est ut *sa* ad *ad*: ergo, ex aequali in analogia perturbata, tempus per *ad* ad tempus per *ae* est ut linea *sa* ad lineam *ac*. Cumque *ac*, ex demonstratis, sit media inter *sa*, *ab*, et ut *sa* ad *ab*, ita *da* ad *ae*, ergo tempus per *ad* ad tempus per *ae* est ut *da* ad mediam inter *da*, *ae*: quod erat probandum.

His argument can be paraphrased as follows:
Statement:

$$t(ad)/t(ae) \sim ad/mp(ad|ae)$$

By length time proportionality:

$$t(ad)/t(ac) \sim ad/ac$$

By law of chords ("postea quam ostensum fuerit"):

$$t(ac) = t(ab)$$

By above and length time proportionality:

$$t(ac)/t(ae) \sim ab/ae$$

By geometry:

$$t(ac)/t(ae) \sim as/ad$$

By theory of proportions:

$$t(ad)/t(ae) \sim as/ac$$

"Ex demonstratis":

$$ac = mp(sa|ab)$$

11.1 1602: Immersed in Foundational Problems

By above and geometry:

$$t(ad)/t(ae) \sim ad/mp(ad\backslash ae) \text{ q.e.d.}$$

Two aspects are particularly noteworthy with regard to the argument.[22] Firstly Galileo started by noting that his argument will amount to a demonstration only "after it will have been shown that the times through ab, ac are equal," i.e., he unconventionally uses the indicative future II to introduce a premise of the argument. This implies that a proof for the law of chords, or at least a proof for the proposition, considered fundamental, was lacking when Galileo wrote this entry on folio 147. This entry in particular is thus likely to have been written before Galileo came up with the ex mechanicis proof of the law of chords.[23]

Second, the decisive step from the insight expressed in the concludendum statement to the complete argument consists in replacing the ratio ac/ab with the ratio as/ac. The line ac is indeed the mean proportional between the distances as and ad and this geometrical fact was explicitly marked by Galileo as "ex demonstratis." A lemma demonstrating this is comprised on folio 58 recto. The entry is written in the hand of Mario Guiducci and may be the copy of a no longer extant draft that Galileo was referring to with the expression "ex demonstratis." At least, this is what has been claimed in the literature.

However, an entry in Galileo's own hand is extant on folio 172 verso in which the geometrical statement entering the proof of the law of fall is demonstrated. Here, Galileo noted one of the three geometrical lemmata accompanying the proof for the law of the broken chord. As is argued in some detail in Chap. 7 it was, in all likelihood, indeed, this very lemma noted in 172 verso that Galileo was referring to as "ex demonstratis."[24] Indeed, as is argued in that chapter, the proof of the law of fall Galileo sketched on folio 147 recto essentially reverses a line of argument contained within the proof of the law of the broken chord and, in consequence, both arguments make use of the same geometrical assumption whose validity is demonstrated in the lemma.

---

[22] Based on his discussion of the proof of the law of fall on 147 recto, Renn concludes: "this argument for the law of fall cannot be considered a proof on the basis of the *De Motu* theory, since one of the two premises on which it is based, the Length-Time-Proportionality, actually cannot be justified on the basis of this theory." Yet when discussing the length time proportionality, Renn claims that it "was merely a plausible assumption, which Galileo at first also believed to be a consequence of his [*De Motu Antiquiora*] theory (Damerow et al. 2004, 205)."

[23] Galileo's use of the indicative future II has been overlooked by the authors who have discussed this entry. Renn, for instance, translates "After it has been demonstrated that the times through ab and ac are equals ..." and states "At the beginning of his proof Galileo formulates the Isochronism of Chords as a theorem which he had already shown (Damerow et al. 2004, 205)."

[24] The assessment that the geometrical lemma invoked in the proof on 147 recto is the one on 172 verso is shared by Caverni (1895, 347).

Folio 147 verso Four self-contained entries can be distinguished on folio 147 verso. The order they were drafted in can be reconstructed rather unambiguously from the layout of the page. Two of the entries were written upside down with respect to the modern orientation of the page, and then Galileo turned the page 180 degrees and continued to write. In all likelihood the first entry on the page was the central diagram.[25] This diagram is only partially labeled and there is no textual information associated with it. It pertains to Galileo's search for iso-temporal surfaces. The iso-temporal surfaces diagram was left unfinished, possibly because while working on it Galileo turned his attention to a different problem similar to, and likely inspired by, the one he had originally been working on.

Galileo outlined the new problem in a short entry that opens with the phrase "advertas cur," i.e., he turned his attention to the question why something holds. Therefore, the entry will be referred to here as the *advertas note*.[26] In the succeeding two entries, Galileo elaborated the general agenda outlined in the note. He did so first for the restricted case, in which one of the two motions under consideration proceeds on a vertical and the other on an inclined plane and then for the general case of motion along two inclined planes of arbitrary inclination. For the elaboration of the general case, he drew a new diagram.[27] For this diagram, Galileo reused parts of the construction of the iso-temporal surfaces diagram already contained on the page.

The iso-temporal surfaces diagram The large central diagram on the page, reproduced here as Fig. 11.3, is not accompanied by any text. It was left unfinished by Galileo. The extant information, in particular, in comparison to a diagram on folio 155 recto, allows assessment of the underlying considerations. The original purpose

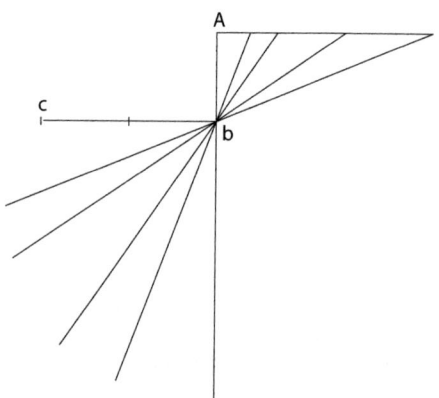

Fig. 11.3 First diagram on folio 147 verso

---

[25] 147 verso, D04A.

[26] 147 verso, T2.

[27] 147 verso, D02A.

of the diagram was to serve the identification of points reached in motion from b in the same time on a set of planes of different inclination, including the horizontal as well as the vertical. The motions under consideration were perceived as proceeded by initial fall through the distance Ab.

Galileo marked off the point c on the horizontal, such that the distance bc amounted to twice the distance Ab and, thus, c would be reached from b in uniform motion in the same time as point b was reached, starting from A in naturally accelerated motion according to his double distance rule. Galileo intended to determine the points on the inclined planes reached after initial fall through Ab in motion from b in the time in which c is reached from b on the horizontal. The considerations seem to have been left unfinished.[28]

Galileo's isochronism of planes considerations This series of considerations on 147 verso has provoked considerable interest and has been furnished with widely varying interpretations. The majority of authors who have discussed the considerations preserved on folio 147 verso have located them before Galileo's conceptual shift toward the assumption of natural acceleration, albeit sometimes with reservations.[29] Clearly, however, Galileo was already working here under the assumption that motion of fall is in principle accelerated. This is not only implied by the fact that the recto side of the folio carries a proof of the law of fall. This proof could, after all, have been added to the page substantially later than the isochronism of planes considerations. First and foremost it is Galileo's application of the double distance rule, evidenced by the original iso-temporal surfaces diagram, which implies with necessity that Galileo was already perceiving of falling motion as naturally accelerated. As the diagram was reused for the isochronism of planes considerations, these must be later and likewise based on the assumption of natural acceleration.

---

[28] When examining the original manuscript, I missed the opportunity to inspect the page for incisions of a compass or uninked lines. Comparison to the related diagram on folio 155 recto suggests that such marks could be present.

[29] Hooper (1992, 345), for instance, lists a number of alternative interpretations and states that "[f]olio 147 may be the record of a historically important moment in Galileo's real conceptual growth, or it may simply be an exercise of something he had already achieved, or it may have been a rehearsal for some public or private demonstration." He, however, strongly suggests that the content of the folio was composed before Galileo's conceptual shift toward the assumption of natural acceleration. Wisan (1974, 174–175), summarizing her detailed analysis of the entries on the page carefully surmised: "In this stage GALILEO may still regard motion as 'naturally' uniform" just to drop her caution and to more strongly conclude just a few sentences later: "[a]s has been remarked there is nothing yet to suggest a line of thought leading to the times-squared theorem, or even to a general law of fall."

The advertas note written above the central diagram reads (for the associated diagram see Fig. 11.4):

> Advertas cur cadentia ex *a* sint semper una in locis sibi respondentibus, ut *o*, *s*, ita ut [angulus] *aos* sit aequalis angulo *bas*.

Galileo called upon himself to direct his attention to the question why, given a certain geometrical constraint determining their mutual positions, the points *o* and *s* on two inclined planes or an inclined plane and a vertical, respectively, are always reached in the same time in naturally accelerated motion starting from rest in *a*. A number of aspects deserve special attention. To start with, Galileo did not state, as he usually did, that it is to be demonstrated that the motions are completed in equal time. This is expressed as a known fact ("sint semper"). Rather, as signaled by the phrase "advertas cur," Galileo aimed at achieving a deeper understanding of why the stated fact held.

A purely geometrical argument would have sufficed to demonstrate what had been stated. Galileo already knew that planes appropriately inscribed as chords in the same circle were traversed in equal time and showing that this was geometrically equivalent to the specification given in the advertas note is rather trivial. The equivalence is particularly obvious for the restricted case represented in the diagram that accompanies the advertas note where one of the inclined planes is a vertical. In this case, the specification of the angles *aos* and *bas* as equal directly entails that *aos* is a right angle. Thus, in turn, according to the Pythagorean theorem, point *o* lies on the circumference of a circle whose diameter is *as*. Demonstrating the essential equivalence of the different geometrical specification for the general case is only slightly more complicated.

Even though Galileo represented the restricted case in the diagram, he did not specify the mutual positions of *o* and *s* by demanding that the angle at *o* had to be a right angle but by specifying that the angles *aos* and *bas* had to be equal. This suggests that upon writing the advertas note, he already had the more general case of two arbitrarily inclined planes in mind but intended to proceed stepwise from the

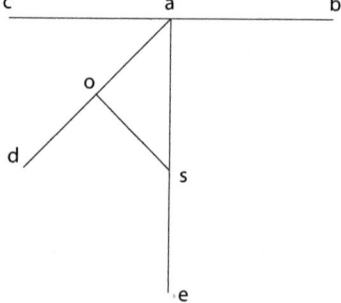

Fig. 11.4 Diagram D03A associated with the advertas note

11.1 1602: Immersed in Foundational Problems   333

Fig. 11.5 First proof
diagram associated to the
isochronism of planes
considerations

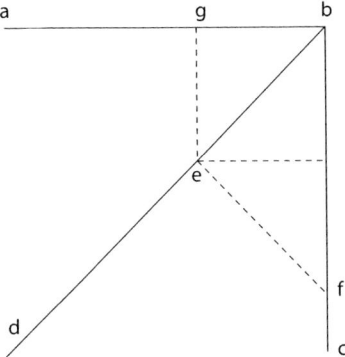

restricted to the more general case. Given that it was Galileo's incentive to arrive at a deeper understanding, such a proceeding was indeed appropriate.[30]

Galileo turned the page and set out to first treat the restricted case (the associated proof diagram is here reproduced as Fig. 11.5)[31]:

> Sit ad orizontem *ab* perpendicularis *bc* et inclinata *bd*, in qua sumatur *be*, et ex *e* ad *bd* perpendicularis agatur *ef*, ipsi *bc* occurrens in *f*. Demonstrandum sit tempus per *be* aequari tempori per *bf*.

Opposed to the advertas note, Galileo now indeed specified the constraining condition by requiring that *ef* be perpendicular to *bd* and thus requiring that *bef* be a right angle.

> Ducatur ex *e* perpendicularis ad *ab*, quae sit *eg*: et quia impetus per *bd* ad impetum per *bc* est ut *eg* ad *be* (ut infra demonstratur); ut autem *eg* ad *be*, ita *be* ad *bf*, ob similitudinem triangulorum *geb*, *bef*; ergo ut *bf* spacium ad spacium *be*, ita impetus per *bf* ad impetum per *be*: ergo eodem tempore fiet motus per *bf* et per *be*.

The essential steps of his argument can be paraphrased as:
Statement ("[d]emonstrandum sit"):

$$t(be) = t(bf)$$

"Ut infar demonstratur" (by the law of the inclined plane):

$$I_{bd}/I_{bc} \sim eg/be^{32}$$

---

[30] In the copy that Arrighetti made of the advertas note on folio 57 verso, a second diagram with identical keying, but depicting the more general case of motion on two inclined planes, was in fact added.

[31] 147 verso, T1A.

[32] As argued in the previous chapter the qualification of "ut demonstratur" by "infra" indicates that Galileo does not merely relate to a logical dependency between statements but is considering these statements as ordered as a consequence of their sequential exposition in a treatise or book. With the expression "ut infar demonstratur," he may thus have been referring to the law of the inclined

By geometry:

$$I_{bf}/I_{be} \sim bf/be$$

By above, the principle of inclined plane dynamics and the second kinematic proposition:

$$t(bf) = t(be)$$

Notably, Galileo uses the expression *impetus* in his argument. An alternative version of essentially the same proof was, however, drafted by Galileo on folio 180 recto at some later point when he replaced the expression "impetus" by "momentum gravitatis" but did not otherwise alter the argument.[33] This suggests that at least with regard to the arguments on 147, Galileo used "impetus" more or less synonymously with "momentum gravitatis." This does not of course mean that both terms denoted the same concept. Indeed, the likely reason why Galileo, when he drafted his entries on 147, saw fit to use impetus rather that momentum will be presented below. Over time, in particular, the concept of impetus developed and it came to denote a concept quite distinct from momentum. Exchanging the no longer fitting expression "impetus" for the expression "momentum gravitatis" likely provided incentive for drafting the alternative version of the proof on 180 recto.

If, indeed, at the time when Galileo wrote his entries on 147, impetus was understood cum grano salis to be interchangeable with momentum, at least with regard to the arguments drafted, the reconstruction of his argument for the restricted case is straightforward. It emerges that structurally the argument is equivalent to that of the ex mechanicis proof. First, it is inferred, exploiting the fact that the ratio of the forces along the planes only depends on the inclinations, that the force along the inclined plane and along the vertical are in the same proportion as the distances of the motion considered.[34] From this, it is immediately concluded that the motions along bf and be are completed in equal time. Just as in the ex mechanicis proof, the assumptions based on which this last step can be justified remain implicit, namely, the principle of inclined plane dynamics, according to which the velocities on bf and be are in the ratio of the mechanical moments (or impeti), and the second

---

plane drafted on a folio bearing the small star watermark and later cut and pasted onto folio 179. This was, however, formulated in terms of mechanical moment and not in terms of impetus.

[33] Hooper, as well as Drake, assumes a reverse relationship between the respective entries on 147 and 180. According to Drake (1978, 63) the results of folio 180 were "recast... in terms of impetus rather than in terms of moment of heaviness" on folio 147. The assumption that the tidily written draft of a proposition should predate its elaboration on a page whose content clearly has provisional character is, however, rather preposterous. For Hooper's position, cf. Hooper (1992, 345).

[34] The considerations on 147 verso start from considering forces along planes of equal length but different heights, whereas in the ex mechanicis proof, Galileo started by considering the moments on inclined planes of equal height and different lengths. The latter corresponds to the formulation of the law of the inclined plane drafted on the sheets pasted to folio 179.

kinematic proposition which implies, given the velocities are in the ratio of the distances traversed, that the times of motion are in fact equal.

Thus, Galileo not only stated essentially the same physical fact as in the law of chords, moreover, he reasoned for this fact in more or less the same way as he had done in the ex mechanicis proof and the difference between them is reduced to the way in which the geometry defining the motion situation was specified.

In his next entry, Galileo generalized the motion situation to the case of motion along planes of different inclination (the associated diagram is reproduced as Fig. 11.6):

> Infra orizontem ab ex eodem puncto c duae rectae aequales utcumque inclinentur cd, ce, et ex terminis d, e ad orizontem perpendiculares agantur da, eb, et lineae cd in puncto d constituatur [angulum] cdf [angulo] bce aequalis. Dico, ut da ad be ita esse dc ad cf.
>
> Ducatur perpendicularis cg: et quia cdf aequatur angulo bce, et rectus g recto b, erit ut dc ad cg, ita ce ad eb: est autem cd ipsi ce aequalis: ergo cg aequatur be. Et cum angulus cdf angulo bce sit aequalis, et [angulus] fcd communis, reliquus ad duos rectos dfc reliquo dca aequabitur, et anguli ad a et g sunt recti; ergo [triangulum] adc [triangul]o cgf est simile: quare ut ad ad dc, ita cg ad cf, et, permutando, ut ad ad cg, hoc est ad be, ita dc ad cf: quod erat probandum.
>
> Cum autem impetus per cd ad impetum per cf sit ut perpendiculus ad ad perpendiculum be, constat, motus per cd et cf eodem tempore absolvi. Itaque distantiae quae in diversis inclinationibus eodem tempore conficiuntur, determinantur per lineam quae (ut facit df) lineis inclinatis occurrit secundum angulos aequales illis quos inclinatae ad orizontem constituunt, permutatim sumptos.[35]

Galileo's argument is somewhat more convoluted than in the restricted case as now it is finally, indeed, the equality of the angles cdf and bce which is constitutive for the underlying geometry. This, however, concerns merely the geometrical part of the argument. With regard to the conceptual underpinnings, Galileo's argumentation is essentially equivalent to the way he had argued in the restricted case and, thus, ultimately, also to the line of reasoning of the ex mechanicis proof.[36]

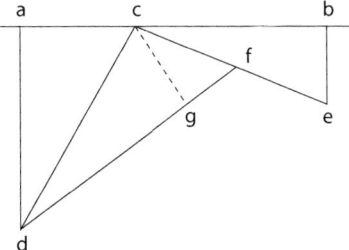

Fig. 11.6 Second proof diagram associated to the isochronism of planes considerations

---

[35] 147 verso, 3A, 3B and 3C.

[36] Wisan has commented upon this seeming duplication as follows: "GALILEO could have given a proof from the law of chords ... and it is somewhat odd that it should be proven as above [i.e., without using the law of chords as a premise]. A note ... suggests the reason why. ... The question is, *why* does this relation hold? Perhaps it is for the answer to this that GALILEO goes back to his dynamical principle (Wisan 1974, 172)."

As already mentioned, Galileo later revised the argument he had advanced for the restricted case. The revised version in which "impetus" was replaced by "momentum gravitatis" was drafted on folio 180 recto. When Galileo had his disciples make copies of selected material from the Paduan period, almost the entire content of folio 147 was copied, including his isochronism considerations from 147 verso.[37] At the same time, however, the revised version of the argument for the restricted case drafted on 180 recto was also copied.[38] Apparently, upon editing the copied material, he realized this and added "haec prima propositio est iam demonstrata, et ideo, ut dupla, demictatur" to the copy made of the original entry from 147 recto onto 57 recto.[39]

Moreover, most likely also when editing the copies made by his disciples, Galileo utilized the verso side of one of the folios used by the disciples, folio 89 a verso, to rework the argument he had provided for the general case on 147 recto. In this reworking, Galileo completely expelled the dynamical assumptions and, thus, all dynamical reasoning which had proven problematic in 1602 from the argument. Instead, he introduced a proposition that corresponds to Proposition VII of the *Discorsi* as a premise for the argument. The proof which thus resulted was later published as the second of the two proofs advanced for Proposition IX of the Second Book of the Third Day of the *Discorsi*.[40]

Had the propositions published in the *Discorsi* been derived in one consolidated effort, the duplication, entailed by disclosing the law of chords and Proposition XI as independent propositions, would be surprising. Yet, as has been shown, the formulation of each of the two propositions had its individual, distinct history within the intellectual pathways that led Galileo to the new science. It thus becomes understandable that in this one case, Galileo ended up publishing two separate propositions that essentially state the same physical fact.

---

[37] The proof of the law of fall from 147 recto was copied by Arrighetti to 50 recto, the isochronism considerations to 57 recto and 57 verso. That almost the entire content of folio 147 was copied shows that Galileo considered it to be of crucial importance. This is not surprising in view of the fact that, as argued here, the folio was part of his earliest attempt to provide his new insights with a foundation and to, thus, erect a new science of motion.

[38] Arrighetti copied the draft from 180 recto to 177 verso.

[39] 57 recto, T3.

[40] An elaboration of the first of the two proofs which Galileo provided for Proposition IX in the *Discorsi* is contained on folio 194 recto. The entry remained incomplete and breaks off in the middle of a sentence. It is written rather thoroughly and the text floats around a blank space intended to hold the proof diagram which was, however, never added. The handwriting differs considerably from most of the entries in the *Notes on Motion* but resembles that of an entry in the lower part of folio 182A recto. According to Favaro, the entry is in Galileo's hand.

Folio 147, provisional summary The content of folio 147, discussed in detail above, can be summarized as follows. With the hollow paradox, the folio contains one of the only two entries preserved in the *Notes on Motion* that combine considerations regarding motion through media, with such, regarding naturally accelerated motion on inclined planes. The aim of the hollow paradox was to demonstrate that the kinematics of motion along an inclined plane did not depend on the weight of the moving body. The argument likewise indicated that kinematics did not depend on the weight per volume of the moving body, thus contradicting his own earlier *De Motu Antiquiora* theory. Yet it is unclear whether Galileo realized this.

The folio, moreover, documents Galileo's attempt to construct a proof of the law of fall. The argument he developed is closely related to and had likely been inspired by the argument by which he had proven the law of the broken chord. The proof Galileo finally gave is based on the length time proportionality and the law of chords, where the latter was introduced as a premise explicitly marked by Galileo as not yet demonstrated. Lastly, Galileo elaborated on the motion situation characterized by two inclined planes traversed from rest in equal times. The arguments he provided follow the argumentative schema characteristic of the ex mechanicis proof. Galileo deploys the term "impetus" instead of "momentum gravitatis" to denote the force on a plane yet both expressions are used interchangeably with regard to the argument. Galileo was apparently searching for a deeper explanation as to *why* under a certain geometrical constraint, dynamics dictated that motion proceeded in equal time. Indeed, in a first entry, Galileo almost programmatically formulated the search for the underlying reason as the goal of his considerations.

11.2 The 1602 Foundational Situation in Context

Folio 147 and the star group folios as pertaining to the same stage of work
The interpretation of Galileo's work on the foundation of a new science of motion likely around the end of the year 1602 but in any case before the end of 1604 to be established in the second half of this chapter is grounded on the assumption that the entries on folio 147 were written roughly contemporaneously to Galileo's work on the star group folios. This assumption is supported by the following arguments.

To begin with, folio 147 bears a crown watermark. It can, thus, securely be attributed to Galileo's Paduan period. The crown watermarks on different folios of the manuscript are rather similar. An attempt to systematically differentiate them with the aim of a more finely grained periodization would presuppose more precise means than mere visual inspection.[41] However, a direct comparison with the crown watermark of the succeeding folio, folio 148, indicates that the watermarks of these two folios are very likely identical, as both crowns depicted share the same conspicuously bent jag.[42] This suggests that both folios could stem from the same batch of paper and, thus, that the entries on them were, in all likelihood, written in a rather limited period around 1602, the date provided, at least indirectly, for the content of folio 148 in Chap. 5.

Secondly, on folio 147 Galileo used a number of technical terms with a particular low frequency in the *Notes on Motion*, some of which, indeed, reoccur only in entries on folios of the star group. Thus, the term "gravia," which Galileo had used frequently in *De Motu Antiquiora*, reoccurs only twice in the entire *Notes on Motion* on 147 as well in a proof of the law of chords drafted on folio 172 recto, one of the star group folios.[43] Likewise, the term "tarditas," used in an entry on folio 147, reoccurs only twice in the *Notes on Motion*, and both entries are comprised on folios of the star group. In the increasingly mathematized new science of naturally accelerated motion, the notion of "tarditas," which from a mathematical point of view was simply the inverse of velocity, became marginalized to the point of complete disappearance. The occurrence of terms with an otherwise low frequency in the *Notes on Motion* on folio 147, as well as on the star group folios, is best explained and thus points to a contemporaneous composition of the content.[44]

Thirdly, the first entry which Galileo wrote on folio 147 verso was a diagram devoted to the search for the iso-temporal surfaces of naturally accelerated motions not starting from rest. A more complete diagram devoted to the elaboration of the same problem has been identified on folio 155 recto, and it was argued in Chap. 5

---

[41] I had the opportunity to examine the watermarks in the *Notes on Motion* cursorily by means of a light-sheet device courteously provided by the Bibliotheca Nazionale Centrale di Firenze. A systematic campaign to analyze the watermarks on the paper used by Galileo is a desideratum.

[42] Drake has identified the watermark of folio 148 as type C18 of his typology. Unfortunately, this type is missing from the table in which he provided detailed descriptions. The watermark of folio 147 is specified by him as "Mountains" 20×17 mm. See Drake (1979, XXXIII). I have measured a width and height of 18 mm for the watermarks on folio 147 and 148. Both folios, moreover, share the exact same height of 305 mm.

[43] Arguments on the relationship between folios based on terms with low frequencies are, as a rule, only meaningful if combined with an hypothesis concerning the reason why a particular term is rarely used. This may, for instance, simply be due to the fact that the term in question pertains to a topic treated only marginally, in which case it is, of course, not suggested that two entries where the term occurs belong to the same period of work. In the case of the Latin "gravia," it can be argued, however, that the low frequency is due to the fact that over time it was replaced by the term "mobilia" to get rid of the dynamical connotations of the former. It is then, indeed, suggested that entries which employ the term "gravia" stem from the same early period of work.

[44] Without a comprehensive linguistic analysis, arguments based on the frequency of words such as the one made in the text need to be taken with a pinch of salt.

how the general type of problem Galileo attacked by means of these diagrams had resulted as a spin-off from his research program on swinging and rolling.[45]

Fourthly, Galileo drew a small diagram at the top of folio 147 recto to remind himself to elaborate a particular proposition and such an elaboration can indeed be identified on folio 189 verso, comprising the experimental record of the pendulum plane experiment which, as the empirical component of Galileo's program on swinging and rolling, must have been conducted in 1602. Fifthly, Galileo's isochronism of planes considerations on the verso side of 147 is based on applying his principle of inclined plane dynamics. They must, thus, postdate Galileo's insight into the applicability of the principle in the case of naturally accelerated motion evidenced by and indeed emphatically expressed in the si hoc sumatur note on folio 172 verso, part of the star group folios.

Evidence for dating the content of folio 147 to roughly the same period of work as the composition of the star group folios is finally provided by the proof of the law of fall contained on 147 recto. In his attempt to construct a proof, Galileo made use of a transitional form of representing times of motion diagrammatically that was occasioned by his adaptation of integrated time representation, which in turn he had developed in the context of his research on swinging and rolling. Such transitional diagrams, in which distances and times were represented in the same diagram but where a graphical convention is used to disambiguate the different types of lines, are rare, and all date to more or less the same period of work. Moreover, the argument Galileo finally formulated was clearly inspired by a line of argument that he had come across as part of his attempt to construct a proof of the law of the broken chord. In consequence, as has been argued, both proofs make use of the same geometrical lemma drafted on one of the star group folios. Last, but not least, Galileo indicated that the law of chords, which he had employed as a premise in the argument, was still in need of demonstration. This implies that either no proof, or at least no sufficiently strong proof, of the law of chords existed for the time being, and, thus, the content of folio 147 must have been composed either before a proof of the law of chords was added to the star group folios or at least before this first proof was replaced by the ex mechanicis proof.

Galileo's first attempt at a foundation Galileo's first attempt to provide a foundation for his new insights and to, thus, turn them into a new science of motion will be discussed in the following. The inclusion of folio 147 in the analysis develops the picture which began to emerge in the previous chapter, based on an analysis of

---

[45]The results of Galileo's considerations regarding the iso-temporal surfaces of different types of motion were presented as part of a longer Scholium appended to Proposition XXIII of the Second Book of the Third Day of the *Discorsi*. Proposition XXIII shows how to construct a plane which, appended to a vertical, is traversed after fall through the vertical in the same time. Cf. Chap. 8.

the star group folios. It will transpire in more clarity that around the end of 1602, but in any case, before 1604, Galileo was deeply and systematically engaged in establishing a foundation for his new insights concerning the pheno-kinematics of naturally accelerated motion. This work represents Galileo's first attempt to mold the results he had achieved by his investigation into the relation of swinging and rolling into a demonstrative science modeled according to the Euclidian exposition of mathematics. With regard to this attempt, two different yet mutually related problems, which Galileo attacked in this period of work, can be recognized. On the one hand, he axiomatized his system of statements concerning accelerated motion on inclined planes, i.e., he rather systematically explored which of these statements could serve as a minimal yet strong enough set from which all remaining propositions could be derived. His exploration of this problem will be referred to as his search for an axiomatic foundation.

At the same time, as has already transpired in the previous chapter, in search of arguments by means of which his fundamental or axiomatic statements could be inferred from basic or fundamental principles, Galileo started to analyze his new results harking back, thereby, to the concepts anchored in the extant conceptual frameworks. Galileo's search for foundational proofs in this sense will be referred to as his search for an analytic foundation, as it was the aim to render the science true and meaningful by virtue of the truth and evidence of the principles resorted to in the foundation.[46]

The insights or rather, as they had indeed been derived, results concerning the pheno-kinematical behavior of accelerated motion Galileo had achieved so far could in no way be considered self-evident. It seems to be a widely shared belief that Galileo must initially have derived at least some of his statements from more basic principles and that these statements were then successively used to derive further propositions. Against this, it has become clear that Galileo progressed more or less in the opposite direction. Starting from a limited number of initially mere heuristic assumptions concerning the pheno-kinematics of naturally motion, Galileo's investigations of the relation between swinging and rolling led him to derive a substantial number of new propositions concerning the spacio-temporal behavior of naturally accelerated motion. These propositions had shown to be consistent in his extensive elaborations. They had, in particular, allowed him to

---

[46] According to the Aristotelian understanding, a science "explains them[facts] by displaying their priority relations (APo. 78a22/28). That is, science explains what is less well known by what is better known and more fundamental, and what is explanatorily anemic by what is explanatorily fruitful (Shields 2016)." A statement that is not proven or demonstrated but considered either self-evident or subject to necessary decision is commonly referred to as an axiom, whereas I refer to this, in accordance with Galileo's own parlance, as a principle or basic principle. When speaking about axioms or more precisely about Galileo's fundamental or axiomatic propositions, I refer to his statements concerning the spacio-temporal behavior of accelerated motion which served as a starting point for inferring all other propositions. The use of axiomatic in the text is thus not fully compliant with its use in contemporary logic.

progress ever further with his consideration without an internal conflict arising. Moreover, his results had received a rather firm empirical basis.[47] Thus, the subject matter of his new science was, in a sense, given to Galileo and rather irrevocably fixed in 1602.[48] It was from there, as will be seen, that he worked his way backward, first to an axiomatic and then to an analytic foundation.

The search for an axiomatic foundation Galileo's considerations on the relation of pendulum swing and naturally accelerated motion along inclined planes had taken their point of departure from the law of fall and the law of chords which had initially been merely well-motivated assumptions that, however, lacked a rigid justification. The law of fall and the law of chords became the premises in some of the earliest proofs that Galileo devised in the course of his investigations on the relation of swinging and rolling and had likely been submitted to an empirical test. Soon, Galileo also became aware of the length time proportionality. As a particularly accessible statement concerning the relation of times elapsed and distances covered in naturally accelerated motion on inclined planes, this proposition quickly became another cornerstone of Galileo's consideration and was frequently used as a premise in proofs but also, for instance, for the calculation of numerical examples regarding spaces traversed and times elapsed in naturally accelerated motions.

In the course of his further work, the law of fall, the law of chords, and the length time proportionality in different combinations became the premises in the proofs of the majority of propositions that Galileo was to derive. In particular, based on these propositions as premises, Galileo established the generalized law of fall and the generalized length time proportionality which made it possible to treat accelerated motion not starting from rest. But what was the relation between these three propositions? Did they have to be independently presupposed or could they possibly be reduced to just one proposition. As will be seen, Galileo reflected systematically upon this question.

Proofs for all three propositions are indeed contained on the star group folios, together with folio 147, where each of the three propositions in question was proved

---

[47] The results of the pendulum plane experiment did not fully comply with Galileo's theoretical expectation but could, as argued, nevertheless be interpreted by him as partially corroborating his exposition of the pheno-kinematics of naturally accelerated motion. Moreover, as discussed in Chap. 6, it is very likely Galileo had performed additional experiments, in particular, an experiment to validate the law of fall as he later described it in the *Discorsi*.

[48] We are thus now in a position to answer in the affirmative the question posed by Settle (1966, 99): "[a]re we to infer that much of the substance of the Third Day was already in hand in 1602?" to which he, based on the limited sources available to him, could then merely state, "there is warrant for entertaining the idea."

based on the remaining two as premises. In these proofs, Galileo took great care, when invoking other propositions as premises, to signal their status as either already demonstrated, i.e., furnished with a proof, or not. The overall picture which thus emerges is rendered schematically in Table 11.1.

In the table, those passages of the respective proofs are quoted in which Galileo introduced either of the other two propositions as premises. As can be seen, Galileo in fact carefully and consistently distinguished between premises already proved and that which he merely stated as fact. The information thus conveyed is consistent with the assumption that all three proofs were drafted in close succession and even allows for reconstruction of the order this must have happened in.

According to this logic, the proof of the law of fall on folio 147 must have been drafted first. Both propositions which enter the proof as premises were qualified as not yet proved by Galileo. Referring to the law of chords as a premise, he even used the otherwise uncommon indicative future II of the verb "demonstrare." Thus, on the one hand, he explicitly signaled that a formal proof for the proposition was not yet present. On the other hand, the use of indicative future II points to his intention to construct such a proof such that, once this had been done, the proposition could indeed be referred to as having been demonstrated.

Table 11.1 Table of the logical dependency among the foundational propositions, including a characterization of the status of the premises entering the proofs. Propositions and their proofs are represented in rows; the propositions entering these proofs as premises are represented in columns. Cells contain the expression by which Galileo referred to the respective premise, as well as an indication of the status of the premise as either stated as fact or else stated as proved

	Law of fall	Law of chords	Length time proportionality
Law of fall		"Postea quam ostensum fuerit, tempora per ab, ac esse aequalia,…" **Stated as fact, not yet proven**	"Nam tempus per da ad tempus per ac est ut da ad ac lineam;" **Stated as fact**
Length time proportionality	"…constat, tempus casus per da ad tempus casus per bc esse ut media inter da et bc ad ipsam bc." **Stated as proven**	"Tempus autem casus per da aequatur tempori casus per ba:" **Stated as fact**	
Law of chords	"…quare ex demonstratis tempus quo mobile ex a cadit in c ad tempus casus ex a in d est ut linea ba ad ad." **Stated as proven**		"Verum similiter ex demonstratis tempus casus ex a in b ad tempus casus ex a in d est ut ba ad ad…" **Stated as proven**

11.2 The Foundational Situation Around 1602

The next proof Galileo drafted must accordingly have been the proof of the length time proportionality contained on folio 163 verso. In this proof the law of fall is referred to by the expression "constat," Galileo's short form for "constat ex demonstratis," and the demonstration referred to must, in all likelihood, be the proof of the law of fall contained on folio 147. The law of chords is still merely stated as a fact. The last of the three proofs Galileo drafted must, therefore, have been that of law of chords contained on folio 172 recto in which both the law of fall and length time proportionality are explicitly referred to as already proved.

That Galileo was consciously exploring the logical dependency among these three fundamental propositions in search of an axiomatic foundation for the new science is strongly implied by the care exercised in a thorough and consistent identification of the status of the premises entering the proofs. It is, moreover, suggested that this happened before any of these propositions had been furnished with a proof from more basic principles, that is, in particular, before Galileo came up with the ex mechanicis proof of the law of chords.

Galileo must have been fully aware of the implications of the logical dependency among these three propositions as it had been revealed by his search for an axiomatic foundation. On the one hand, the law of fall and the law of chords, together with the length time proportionality, formed a coherent, non-conflicting set of statements that could serve as an axiomatic foundation for his new science. Furthermore, any one of the three propositions was redundant in the sense that it could be proved from the remaining two. In the absence of alternative proofs from basic principles for any of them, the mutual logical dependency among these fundamental propositions constituted a logical circle. This amplified the need for an analytical foundation showing, moreover, that a foundation would be achieved if any two of these three propositions were furnished with demonstrations starting from fundamental principles.

The search for an analytic foundation In the previous chapter, we witnessed Galileo engaged in the analysis of his new insights on the pheno-kinematics of naturally accelerated motion. Taking into account the content of folio 147, the picture is completed. It is substantiated that by means of this analysis, Galileo aspired to be able to fashion arguments starting from fundamental principles by means of which at least two of his fundamental propositions could be demonstrated. This, as his search for an axiomatic foundation had demonstrated, was required in order to furnish his new insights with a foundation.

As detailed in the previous chapter, by invoking, in particular, the concept of velocity in this analysis, Galileo realized that his older principle of the dynamics of the inclined plane could be brought to bear in arguments about naturally accelerated motion, i.e., that it could be assumed that the velocities of accelerated motions along inclined planes of different inclination were in the same ratio as the mechanical moments on these planes. Together with the results of his pondering in the context of the broken chord approach, this allowed Galileo to construct the ex mechanicis proof of the law of chords. Thus, at least one of his potential axiomatic propositions had received a proof from basic principles and a first decisive step toward an analytical foundation had been ventured.

Yet, as argued above, Galileo knew that besides the law of chords, at least one more proposition needed to be equipped with a proof from basic principles. Attempting a proof of the law of fall based on the dynamics of the inclined plane was ruled out prima facie, as the law of fall related motions made along one and the same inclined plane (or vertical), that is, under the influence of the same (static) force and how to construct a dynamical argument for this case was completely unclear. Likewise, no attempt to prove the length time proportionality based on dynamics can be identified in the *Notes on Motion*, and achieving a closure for his new science based, ultimately, on dynamical considerations alone would not have appeared possible to Galileo at the time.[49] Yet, with an alternative approach lacking, a deeper investigation of the dynamics of naturally accelerated motion was indicated and indeed, as we have seen, the majority of entries on folio 147 relate to dynamics in one way or another.

With the hollow paradox, Galileo devised an argument intended to show that bodies of different weights show the same kinematical behavior, falling along a plane of the same inclination. The independence of the kinematics of naturally accelerated motion on the absolute weight was, as argued, implied by the independence of the pendulum's period on the weight of the pendulum bob, together with his advanced conceptualization of the relation between swinging and rolling. We recall that Galileo fundamentally assumed that the swinging motion of a pendulum along its arc was equivalent to the falling of a body along this arc supported, however, from below and that this latter motion would in turn be equivalent to the falling along a polygonal path inscribed into this arc with an unlimited number of sides. This inevitably implied that the fall along each of the inclined planes making up the polygonal path likewise had to be independent of the weight of the falling body. Once Galileo had started to avail himself of dynamical arguments, this implication must inevitably have been thrust upon him.

The need to provide a demonstration for the fact that bodies of different absolute weight show the same kinematical behavior on planes of the same inclination will have been felt ever more urgently in view of the fact that in the ex mechanicis proof, Galileo had essentially argued that the change of mechanical moment, i.e., of positional weight, did indeed have a decisive effect on the kinematics of naturally

---

[49]The only page of the *Notes on Motion* in which Galileo uses the expression "momentum" to refer to the force on a inclined plane which has not been discussed here is 173 recto. Based on material criteria, the order in the manuscript, and the diagram on the verso side, it is plausible to assume that the page pertains to the period of work discussed in the present chapter. The notes are too sketchy to reliably reconstruct Galileo's underlying considerations. The diagram in the upper left depicts the motion situation of the law of chords, i.e., of two inclined planes of different inclination traversed in equal times. The diagram below and in the lower right could indicate that Galileo was considering how the dynamical argument could be transferred to the motion situation of the length time proportionality, i.e., to motions over inclined planes of equal height or even to the most general case of motion along inclined planes of different inclination, height, and length.

11.2 The Foundational Situation Around 1602

accelerated motion. To put it more pointedly, Galileo was confronted with the puzzling fact that changing the absolute weight of an object supposedly had no effect on the kinematics of fall, whereas a change in positional heaviness as engendered by placing the moving body on planes of different inclination did. What if a heavier body were placed on the same inclined plane? Wouldn't the positional heaviness increase as well? For the time being, this puzzling situation could be resolved, as the argument provided with the hollow paradox indeed implied that the kinematics of naturally accelerated motion did not depend on the absolute weight. Arguably, what is not clear is whether Galileo realized that his argument likewise implied that the kinematics should not depend on specific weight and that it, thus, also conflicted his earlier considerations on how the effective forces are influenced by buoyancy effects and that, thus, change of specific weight should indeed have an effect on kinematics.

Moreover, with the isochronism considerations, folio 147 comprises extensive reflections upon the question why the positions reached in motion from rest after the same period of time in naturally accelerated motion along inclined planes of different inclination obeyed a particular geometrical relation. As has been argued, the dynamical arguments that Galileo devised are structurally equivalent to the ex mechanicis proof. Existing interpretations have grappled with the reoccurrence of the same line of reasoning in the ex mechanicis proof and the isochronism considerations, to argue, moreover, what is essentially the same physical fact. No really satisfactory explanation, however, has been provided so far. Given that Galileo sojourned in intellectual terra incognita, it is well possible that this duplication eludes further explanation. After all, thoughts sometimes go in a circle, in particular, when exploring new intellectual grounds, and this may well be the case with regard to the isochronism considerations. Yet a different reading is offered here which is inherently more plausible.[50]

As argued in the preceding chapter, when working on the foundation of the new science around 1602, Galileo realized that the principle of inclined plane dynamics was potentially irreconcilable with the kinematical understanding of velocity of naturally accelerated motion along inclined planes. This, as has been argued, did not render the ex mechanicis proof false but deprived it of is explanatory character. The proof could no longer be perceived as being based on the generally valid principle of inclined plane dynamics, but the argument had now to be considered as resting on a restricted dynamical assumption. What would have been needed to reestablish the demonstrative character of the ex mechanicis proof would have been an explanation why, if the velocities of motion along inclined planes could no longer generally be

---

[50]Galileo had all of the isochronism considerations from folio 147 copied. The copying must have been preceded by a rather systematic process in which he selected the material to be copied. The duplication entailed by isochronism considerations with regard to the ex mechanicis proof of the law of chords, if not when writing them down, would most likely have been realized then. That he had the isochronism considerations copied nevertheless suggests that they originally served a somewhat different intellectual goal, which even though eventually not reached, they still represented for Galileo.

assumed to obey the same ratio as the mechanical moments, they obviously did obey this ratio if, and only if, considering motions from rest completed in the same time.

From a modern perspective, the explanation is rather trivial. Forces give rise to change of motion. It is acceleration and not velocity that is proportional to force. Or, to remain closer to the way Galileo could possibly have framed this fact, the accelerations on inclined planes of different inclinations obey the same ratio as the mechanical moments on these planes. In motions with constant acceleration, the velocity grows in proportion to time, with acceleration being the proportionality constant. If such motions are considered starting from rest, it holds that velocities are in the same ratio as the accelerations and in consequence also in the same ratio as the forces if and only if the motions are completed in equal time.

Galileo, however, did not possess a concept of acceleration that would have allowed for a conclusion along this line. Indeed, the mathematical means Galileo availed himself of severely obstructed the development of such a concept. For Galileo, acceleration expressed first and foremost the fact that falling bodies speed up. Even once he started to consider the change of velocity quantitatively, as he later did, his mathematical framework, especially his use of the theory of proportions, made it extremely difficult to analyze what, from a modern perspective, are proportionality constants. Within Galileo's framework, it is rather straightforward to state that the velocity or degree of velocity increases in proportion to time. Formulating a dependence of this proportionality on another parameter, such as the inclination of the plane upon which motion takes place, on the other hand, is hardly possible. Given the difficulties Galileo but also his contemporaries saw themselves confronted with in their attempt to arrive at an understanding of velocity that would allow to coherently describe the relation between spaces traversed and times elapsed in different motions, it becomes understandable that he did not come up with an adequate concept of acceleration which, after all, is nothing but the velocity of the change of velocity in time.

Arguably, in his isochronism considerations, however, Galileo tried with the means available to him to tackle the very problem which the mirandum paradox had bequeathed him with concerning the dynamics of accelerated motion on inclined planes. As has been emphasized, Galileo's incentive in the isochronism of planes considerations on 147 verso was clearly not to ascertain or justify statements but as indicated by the phrase "advertas cur," to gain a deeper understanding of an already established fact. Specifying planes traversed in equal time by the equality of two particular angles, as opposed to specifying that the planes had to be inscribable in the same circle, in a particular manner, must have appeared to be a promising point of departure in this respect. Focusing on a geometrical characteristic such planes shared would, indeed, appear to be a rational strategy in trying to come up with an understanding why, for motion along these and only these planes, the active forces could be assumed to be in proportion to the velocities.

The use of the term impetus in the isochronism considerations finds a natural explanation given that Galileo was indeed exploring the dynamical underpinning. Galileo's concept of impetus underwent significant change from the way it was used in his earliest works to the publication of the *Discorsi* more than 40 years later.[51] What can generally be said is that it was conceived by Galileo, following traditional understanding, as an internal or impressed force, capable of accumulating (or diminishing). Already in *De Motu Antiquiora*, the concept had allowed a bridge between static situations where the forces that hold a body at rest are considered and the investigation of motion in the case where a force holding a body in place is released and the body is thus set in motion. As outlined in Chap. 2, in *De Motu Antiquiora*, Galileo had conceptualized acceleration by assuming that if a body was held in the hand, the hand impressed an impetus equal to the weight of the body. Upon release, this impetus was gradually diminished accounting for the acceleration that prevailed until the impetus is exhausted completely and the body continued to fall with its natural (uniform) velocity.

Thus, if, with regard to his new considerations about naturally accelerated motion along inclined planes, impetus was taken by Galileo to be the magnitude representing the accumulating quantity of accelerated motion on an inclined plane, he could possibly maintain that impetus was proportional to the (changing) kinematic velocity, while at the same time, the ratio of the acquired impetus would correspond to the ratio of the (static) mechanical moments. Yet, a magnitude that always obeys the ratio of the mechanical moments while at the same time being in proportion to the (growing) velocities simply cannot be constructed. Moreover, the concept of impetus as a measure of the quantity of motion changing with the degree of velocity at every given moment, which Galileo was to develop, in particular, in the context of his considerations concerning projectile motion, was, ultimately, not compliant with the use of impetus in the isochronism considerations if the above interpretation is correct.

The above interpretation is plausible, yet, it must remain speculative to some extent. If it is correct, then Galileo had not only realized that there was a limit to the traditional dynamical understanding but had even attempted to explore this limit with the means available, in his isochronism considerations on folio 147 verso.[52] If not the limits, Galileo was certainly exploring the dynamics of naturally accelerated motion along inclined planes and this is what is crucial with respect to the overall interpretation to be established. That his considerations did not provide him with the clue he was hoping to find is illustrated by the fact that, as mentioned, he later reformulated the arguments he had devised, as a result of his considerations on 147 recto in such a way as to free them of any underlying dynamical assumptions. His inability to resolve the conflict which had arisen from the application of the traditional dynamical scheme of argumentation to his new insights concerning the pheno-kinematics of naturally accelerated motion, indeed, had lasting consequences for Galileo.

In the *Discorsi*, the new science of naturally accelerated motion is not founded in dynamics. Galileo included the ex mechanicis proof in the book but did not formally present it as part of the foundation. Rather, he tucked it away as one

---

[51] Cf. Galluzzi (1979).
[52] For the idea of exploring the limits of a conceptual system, see Damerow et al. (2004).

of three alternative proofs adduced for the law of chords. Proposition IX, which Galileo had ultimately formulated as a result of his isochronism considerations, was reformulated so as to omit dynamical assumptions. The way in which the new science was finally presented, thus, testifies to the fact that Galileo had not been able to solve the problem recognized as early as 1602 that had arisen from his attempt to bring the traditional dynamical understanding to bear on his new insights concerning naturally accelerated motion. In an attempt to make the best of the situation, in the *Discorsi*, Galileo expressly stated that he had been able to demonstrate the properties of naturally accelerated motion without regarding their causes. What he concealed from his readers is that he had attempted to do so but had failed and that it was this failure that had pointed to a limit of the traditional dynamical understanding.

It was not for Galileo to transgress the limit, to reject the traditional idea that motion is to be explained by the presence of causes, and to accept, instead, that what is caused by the action of force is change of motion, i.e., acceleration. Yet, his insights concerning the pheno-kinematics of naturally accelerated motion entailed this conclusion and their publication paved the way to a new understanding of dynamics to which many contributed until it was finally formulated, in all clarity, by Newton.

For Galileo, however, rejecting a dynamical approach to the problem of the foundation was necessary in 1602 and had a crucial consequence. For the time being, he was simply not in possession of an alternative approach that would have allowed solving the problem of the foundation and, thus, the road to a new science seemed blocked. As demonstrated in the previous chapter, this, according to all evidence, interrupted his search for an analytical foundation.

In the long run, however, Galileo resolved the problem of the foundation and the new science ended up being published after all. The first and decisive step toward the solution, not published until more than three decades later, was ventured by Galileo in 1604. It ended the compulsory hiatus he had been thrown into by his failed attempt to erect a dynamical foundation 2 years earlier.

By convention I do not consider his work on the foundation in 1604 to be part of the early genesis of Galileo's new science that I have set out to reconstruct with this book. Yet I conclude this chapter with a brief discussion of Galileo's 1604 attempt at a foundation. I do so for two principal reasons. On the one hand, the analysis of Galileo's work in 1604 provides a definite terminus ante quem for the material discussed, in particular, in the second part of this book. On the other hand, Galileo's attempt at a foundation in 1604 was, as will be seen, prefigured by the understanding gained by way of the consideration reconstructed in this and the preceding chapter. The brief discussion of the approach taken by Galileo in 1604, thus, not least, serves to corroborate the interpretation that has been advanced here.

As will be seen, Galileo's new approach to the foundation hinges on his recourse to conceptual means provided by the doctrine of the configurations of motion in the analysis of naturally accelerated motion. As we have seen, Galileo was familiar with the tradition and had already resorted to it in 1602, in particular, in his proof of the double distance rule.⁵³ An, at least temporary breakthrough was achieved in 1604 when Galileo uncovered a way to exploit the understanding of naturally accelerated motion entailed by the tradition and to apply it in the construction of proofs for propositions concerning the pheno-kinematics of naturally accelerated motion. That Galileo believed to have found such a way in 1604 and, thus, saw himself in the position of being able to establish a new science is evidenced by a letter he sent to Paolo Sarpi toward the end of this year.

11.3 Toward a New Foundation in 1604

The letter which Galileo sent to Paolo Sarpi in October 1604 has been in the focus of numerous studies.⁵⁴ It has been claimed to demarcate the starting point of Galileo's investigations concerning naturally accelerated motion.⁵⁵ Yet as has been shown here, Galileo was deeply immersed in the study of naturally accelerated motion definitely already some 2 years earlier. Against the backdrop of the understanding of the emergence of Galileo's new science established here, the letter receives a new interpretation. What Galileo so proudly announced to Sarpi in 1604, as will be argued, was not the "discovery" of the law of fall and with it the commencement of his work on the problem of naturally accelerated motion. Rather he made known to his friend and patron that he had found an escape from the deadlock he had run into approximately 2 years earlier in his attempt to establish a foundation for his new insights concerning naturally accelerated motion. Thus, what Galileo announced in the letter was nothing less than his expression of his, for the time being, justified belief that he was finally in the position to establish a new science.⁵⁶

---

⁵³Galileo's proof of the double distance rule already rested on the assumption that the degrees of velocity increased in proportion to the distance fallen, i.e., the Sarpi letter principle. Yet, for the time being, his argument rested on showing that the kinematic velocity of successive motion was double that of the preceding motion, which seemed reconcilable with both the Sarpi letter principle and the later correct principle. The conflict entailed which led Galileo to abandon the erroneous principle and accept the correct principle was realized only after systematic elaboration. See the discussion in Damerow et al. (2004, 180–188). From a modern perspective, the double distance rule, indeed, implies the law of fall, but for this to become manifest, an appropriate argument has to be constructed. Galileo, however, was never quite able to integrate the double distance rule into his new science, and it is indeed not formally presented as a proposition in the *Discorsi*.

⁵⁴Letter and translation are provided in Chap. 14.

⁵⁵An exception is Renn, who maintained that a deeper investigation of naturally accelerated motion must have antedated the letter. Thus, for instance, he assumes, as is done here, that the proof of the law of fall on 147 was drafted before the letter. Cf. Damerow et al. (2004, 205–208).

⁵⁶The letter sent by Galileo to Paolo Sarpi on the 16th of October 1604 was a direct reply to an antecedent letter by Sarpi, dated 9 October 1604 (EN X, 114, a translation is contained in Drake 1969, 340). The letter by Sarpi indicates that the two men had recently discussed problems of motion in person. Drake assumes that the considerations alluded to in Galileo's letter were inspired by this conversation.

In the letter, Galileo imparted to have found an indubitable principle based on which he was able to construct a proof of the law of fall. He spelled out the principle asserting that in naturally accelerated motion, the degrees of velocity increased in proportion to the distances fallen. The proof alluded to is, however, not part of the letter. Two proofs of the law of fall that employ this Sarpi letter principle have been identified in the *Notes on Motion*. The proofs which are comprised on folios 128 and 85 can rather unambiguously be related to the letter and have been discussed extensively in the literature.[57] In the letter no particular emphasis is put on the proof. Specifically, Galileo did not at all announce having discovered the law of fall, as is often portrayed. What he accentuated instead is to have found (after repeated consideration) a way to construct such a proof based on a "completely indubitable," i.e., fundamental principle. Such a proof, as we have seen, was indeed crucially lacking, in particular, after Galileo's attempt to establish a foundation based on dynamical reasoning had failed.[58]

As a result of his effort to axiomatize his new insights, Galileo was aware that the new science could not be based on the law of fall alone. Indeed, in the letter Galileo stressed as many as three times that besides the law of fall, he had already reached a number of other conclusions concerning naturally accelerated motion and that it was these conclusions as a whole, which his new approach puts him in the position of being able to demonstrate with absolute certainty, based on assuming the Sarpi letter principle. He opened the letter by stating:

> Ripensando circa le cose del moto, nello quali, per dimostrare li accidenti da me osservati, mi mancava principio totalmente indubitabile da poter porlo per assioma, mi son ridotto ad una proposizione la quale ha molto del naturale et dell' evidente; et questa supposta, dimostro poi il resto, cioè gli spazzii passati dal moto naturale esser in proporzione doppia dei tempi, et per conseguenza gli spazii passati in tempi eguali esser come i numeri impari ab unitate, et le altre cose. (EN X, 115, letter 105)

Galileo, thus, indeed, made amply clear that the law of fall ("[the] spaces passed by natural motion are in double proportion to times") was just one among "other things" he could now demonstrate. Moreover, Galileo indicated that he had returned ("ripensando") to the problem of a foundation not solved before ("mi mancava principio") and that the solution was found through active analytical work ("mi son ridotto"). Galileo's declaration in the letter is, thus, seen to correspond precisely to what one would expect, based on the interpretation provided here.

---

[57] I follow the interpretation of folios 128 and 85 established by Renn in Damerow et al. (2004).

[58] As was seen, Galileo had already employed the Sarpi letter principle in 1602. The emphasis in his claim is, thus, not on the principle itself but on the practicability of its use as an axiom ("assioma"), i.e., a basic principle in the foundation of the new science. Wallace (1992, 266) has rightly emphasized that to "serve as a principle for a demonstrative science there would have to be independent evidence of its truth, either as *per se nota* in its own right or as demonstrated on other grounds" and that in "1604 Galileo was optimistic that he could produce such a demonstration."

11.3 Toward a New Foundation in 1604

A little further into the text Galileo declares: "[e]t se accettiamo questo principio, non pur dimostriamo, come ho detto, le altre conclusioni ..."[59] Thus, once more, he emphasized that the new principle had allowed him to demonstrate a number of conclusions previously obtained. This is stressed even once again when he finally remarks, "è vero quanto ho detto et creduto sin qui."[60]

As his search for an axiomatic foundation had made clear, demonstrating "the other conclusions" meant that, besides the law of fall, at least one of his other fundamental propositions, i.e., either the length time proportionality or the law of chords, had to be furnished with a proof from fundamental principles. Such a proof can indeed be identified. Folio 179 recto contains a proof of the length time proportionality in which, just as the proofs of the law of fall on folios 128 and 85, presupposes the Sarpi letter principle. The interpretation of the proof on 179 recto is hampered by the fact that Galileo heavily edited his entry after it had first been drafted. As a result, the interpretations of the proof, which have been advanced, vary widely.

No detailed interpretation of Galileo's argument will be provided here.[61] This remains reserved for a future publication in which I intend to trace the development of the new science from 1602, continuing on toward its publication some 30 years later. What suffices for the present purpose is to acknowledge that the drafting of this proof was part of Galileo's fresh attempt at a foundation in 1604, as has indeed already been argued rather convincingly in the literature. If the entry was originally drafted in 1604, it is implied that the editing of the proof represents a reworking that had become necessary when Galileo realized that the Sarpi letter principle, with respect to which the proof had originally been formulated, was erroneous and needed to be replaced.[62] The underlined passages of the proof on 179 recto are indeed exactly the ones that make statements which are not compliant with the

---

[59] EN X, 115, letter 105. As the remaining part of the sentence shows, Galileo explicitly distinguished the "other conclusions" from conclusions concerning projectile motion and must, thus, here indeed be referring to his conclusions concerning naturally accelerated motion on inclined planes.

[60] EN X, 116, letter 105.

[61] The proof on 179 recto rests on a proposition drafted on folio 138 verso. As part of his argument, Galileo considered a step-by-step increase of the velocity which then remains constant over a minimal distance of fall. An, according to my opinion, essentially correct reconstruction of Galileo's reasoning in the proof on 179 recto has been given by Wisan (1974, 216–219). I, however, disagree with most of the conclusions she draws on the basis of her reconstruction, in particular, her dating of the entry to 1609.

[62] In the original draft, Galileo stated that the degrees grew according to the Sarpi letter principle and this is also represented in the associated diagram. Yet the assumption was not invoked as a premise in the argument which is, in fact, indifferent to the precise way in which the degrees of velocity increase with the distance fallen as long as this increase is related in particular manner for the different motions compared, which facilitated its reworking.

new, correct principle of acceleration and needed to be deleted or altered in the reworking.[63]

The argument in its initial draft form on folio 179 recto, thus, indeed invoked the Sarpi letter principle, rendering it very likely that its drafting was part of the afresh attempt at a foundation Galileo had engaged in at about the time he sent his letter to Paolo Sarpi, in autumn 1604. Besides fitting in with what Galileo announced in the letter, a number of additional arguments can be adduced in support of the assumption that the proof on 179 recto was drafted as part of the new 1604 attempt at a foundation.

Folio 179 bears a crossbow watermark and, thus, the paper on which the entry was written was certainly acquired by Galileo during his Paduan period. Moreover, the entry of folio 179 is strikingly similar to that of folio 128 in a number of ways. The handwriting on both folios is conspicuously broad if compared to the majority of Galileo's other entries in the *Notes on Motion*, almost as if reflecting the certain enthusiasm with which they were written. Moreover, similar to what was observed in the case of the star group folios, in writing on both pages, Galileo respected a margin which he had marked beforehand. The most conspicuous similarity, however, is the use of copious majuscules to open individual paragraphs in which, again, the entries on 178 and 128 resemble those on the star group folios.[64] The reason for this can be assumed to be the same here as there, namely, that upon writing these entries, Galileo had in mind the image of a formal treatise presenting a new science.

Acknowledging that the proof of the length time proportionality on 179 recto was part of Galileo's attempt at a foundation, the situation of the new science in late 1604 can be assessed. Based on his new approach Galileo had been able to construct proofs for the law of fall and the length time proportionality. The former proof rested on the Sarpi letter principle as the crucial premise. In the proof of the length time proportionality, the degrees of velocity were specified as growing according to the Sarpi letter principle, but this assumption was actually not further exploited in the argument, which, instead, crucially rested on the assumption that equal degrees of velocity are reached after fall through the same height, regardless of the inclination of the plane upon which motion takes place. Galileo had availed himself of this assumption already in 1602, as we have already seen.

It, thus, turns out that the decisive difference between the 1602 and the 1604 attempts at creating a foundation is Galileo's move away from building his

---

[63] That the underlined passages were marked for deletion and replacement was first noted by Wisan (1974, 218). The content of folio 179 recto was copied by Arrighetti to folio 88 recto without, however, deleting or otherwise altering the underlined passages. Based on this observation, Wisan (1974, 218) has claimed that if these passages were in fact marked for deletion, then this must have happened after the proof was copied by Niccolò Arrighetti which, however, does not appear to make much sense in view of the fact that the passages in question are underlined in the original entry on 179 recto as well.

[64] Wisan denies a close relation between the law of fall on 128 and the length time proportionality on 179; in fact she dates the former to 1604 and the latter to 1608 or 1609. According to her, the content of 179 reveals "an attempt by GALILEO to rework his foundations using ARCHIMEDEAN methods rather than those of the medievals, which were used to establish the times-squared theorem on folio 128 (Wisan 1974, 216)."

11.3 Toward a New Foundation in 1604

arguments on dynamical assumptions to assuming principles which were formulated with respect to the understanding of velocity as a changing quality, as entailed by the doctrine of the configurations of motion.[65] In 1604, Galileo applied this understanding in an unprecedented manner to argue for concrete results concerning the pheno-kinematics of naturally accelerated motion that he had already achieved beforehand. Proving the law of fall, thus, required comparing motions over different distances, completed in different times. Proving the length time proportionality required finding a way how the information about the dependence of kinematics on the inclination implicit in the statement could be expressed with respect to the doctrine of the configurations of motion so that from that starting point, an argument could be devised to establish the proposition. The arguments that Galileo finally came up with transcended the realm of the traditional application of the method. It was his insights concerning pheno-kinematics of naturally accelerated motions achieved by him considerably earlier and already certain almost beyond doubt that guided him in the search for these arguments.

After he had found and formulated these arguments, Galileo was indeed justified in assuming that he had successfully turned his new insights into a new science, thus rendering them publishable, and it was this belief that he communicated so emphatically to Paolo Sarpi in 1604. The belief was soon to be shattered when Galileo realized that the Sarpi letter principle actually conflicted the law of fall which he had derived from it.[66] A long road leads from this first attempt to create a foundation for his new science based on principles formulated with respect to accelerated motion understood as characterized by a changing velocity represented and treated with the methods of the Oxford Calculators and Oresme, to the solution eventually offered in the *Discorsi*.[67] Reconstructing this road remains, as already announced, reserved for a future publication.

---

[65] That the systematic resort to the doctrine of the configurations of motion in 1604 was stimulated by Galileo's "having reached an impasse in his work on the theory of motion" as a result of which "Galileo may have sought and began seriously to consider some of the writings of the fourteenth century" has congenially been anticipated by Settle (1966, 199).

[66] How Galileo realized that the Sarpi letter principle was not compliant with the law of fall has convincingly been reconstructed by Damerow et al. (2004, 188–197), when this happened exactly is not entirely clear. Direct evidence that Galileo had abandoned the Sarpi letter principle by 1611 is offered by a letter that Daniello Antonini sent to Galileo on 9 April 1611. "Ho pensato alcuna volta a quella sua propositione: *Mobile secundum proportionem distantie, a termino a quo movetur velocitatem acquirens, in instanti movetur* ... (EN XI, letter 512)." Just as he would later argue in the *Discorsi*, Galileo had obviously communicated to Antonini before that if the Sarpi letter principle held motion would be necessarily instantaneous. As Galileo categorically rejected the possibility of instantaneous motion, this implies that before 1611 Galileo renounced the Sarpi letter principle and thus most likely adopted the correct principle of acceleration.

[67] In 1610 in a letter to the Grand Duke's Secretary of State Belisario Vinta (EN X, 348–353, Galileo felt confident enough to announce he had completed three books on local motion in such a way that he could justifiably call it new science established from its first principles ("onde io la posso ragionevolissimamente chiamare scienza nuova et ritrovata da me sin da i suoi primi principii (EN X, 352)." I side with Wisan (1981, 330) who stated, with regard to new science as finally laid out in the *Discorsi*: "In fact, Galileo never completely solved the problem of foundations for the

11.4 Summarizing Conclusion

By including the content of folio 147 into the analysis, it has been the undertaking of the present chapter to complete the picture of Galileo's very first attempt to transform his new insights into a new science, likely in 1602 but definitely before the end of 1604. It has been shown in the previous chapter that on the star group folios, Galileo started to collect, systematize, and analyze the results he had attained through his investigation of the relation of swinging and rolling. It has been argued that this work was undertaken mainly with a view to a future publication in which his new insights would form the subject matter of a new science of motion.

In the present chapter, the content of folio 147 has been discussed in great detail and it has been shown that it must have been drafted roughly contemporaneously to the entries on the star group folios and that, furthermore, this happened under the same general agenda of establishing a new science. This, in particular, required assembling the rather erratically disposed individual insights which had resulted from the investigation of swinging and rolling into a coherent whole, deductively organized.

It has been shown that in this phase of work Galileo systematically explored the logical relation between the law of fall, the law of chords, and the length time proportionality. Those three propositions had been the crucial prerequisites in his previous arguments. Galileo, thus, knew that all insights concerning the pheno-kinematics of naturally accelerated motion he had attained could in principle be deduced from these three propositions. His exploration of their mutual logical relation thus amounts to an axiomatization, i.e., to the search for a minimal yet strong enough set of propositions which, as axioms, could root the deductive tree of the new science. Galileo had been able to construct proofs for all three propositions in such a manner that in each of the proofs the other two propositions were used as premises.

This entailed that the new science could be founded on a combination of any two of these three propositions. Yet none of these propositions was self-evident. None of them could, hence, constitute a starting point for a new science. Alternative proofs for these propositions based on fundamental assumptions had to be furnished so that they could be considered demonstrated according to the understanding of the time. It has been argued that no such demonstration was extant when Galileo first searched for an axiomatic foundation and, in consequence, realized the mutual dependency relation among the three fundamental propositions. The need to construct proofs from basic principles for at least two of his three fundamental propositions thus arose.

It has been argued that Galileo's analysis of the new results, taking recourse to concepts anchored in the extant frameworks of mechanics, owes primarily to his search for such demonstrations. As shown in Chap. 10, by way of such analysis, Galileo realized how the dynamical principle he had applied in his analysis of

---

new science, but his treatise demonstrated the potential inherent in his new mathematical approach to motion."

11.4 Summarizing Conclusion

motion along inclined planes in *De Motu Antiquiora*, then still considered uniform, could be applied to argue for the kinematics of naturally accelerated motion on inclined planes. This insight became the starting point for his construction of the ex mechanicis proof which, once established, was copied to the star group folios and was supposed to serve as a demonstration from basic principles for one of the fundamental propositions of the new science to be established.

Thus, for the time being, a foundation of the new science based on dynamics seemed promising. Most of the entries on folio 147 are, indeed, concerned with the dynamics of naturally accelerated motion in one way or another. With the hollow paradox, Galileo set out to demonstrate that the kinematics of motion along inclined planes was not affected by the absolute weight of the moving body. It has been argued that Galileo was, thus, seeking to secure the basis of his dynamical reasoning. The ex mechanicis proof had been based on the assumption that it was change of positional weight that was responsible for the dependence of the kinematics of motion on the inclination of the inclined plane.[68] At the same time, Galileo's conceptualization of the relation between swinging and rolling, together with the observed independence of the period of a pendulum on the weight of the bob, implied that the kinematics of such motion should not depend on the absolute weight of the body. Providing an argument for the latter was a fitting endeavor in view of the requirements of the envisaged publication.

Alongside the isochronism of planes considerations, folio 147, furthermore, contains a series of considerations related to the dynamics of naturally accelerated motion along inclined planes. The arguments Galileo established in this context are structurally equivalent to the argument of the ex mechanicis proof. Evidence has been provided indicating that Galileo engaged in these considerations after the mirandum paradox had already demonstrated to him that the principle of the inclined plane dynamics he had employed, and which reflected the traditional understanding that velocity as a measure of the effect should be proportional to the cause of motion, could no longer be assumed to be generally true. This deprived the ex mechanicis proof of its explanatory value and the isochronism of planes considerations have been interpreted as an attempt to elaborate the implication of why and how the argument of the ex mechanicis proof could be retained if only a weaker dynamical principle was assumed.

Yet Galileo did not draw the, from a modern perspective, implied consequences as he did not reformulate the dynamical underpinnings of his arguments. This would

---

[68]The ex mechanicis proof explicitly compares two motions made by one and the same falling body. One could claim that, if the same schema of argument were applied to bodies of different absolute weights, this would immediately show that the kinematics of naturally accelerated motion should indeed depend on the absolute weight of the body. Yet there is now evidence that Galileo engaged in such a consideration. Van Dyck (2006, 210) has rightly remarked that essentially the same inconsistency can already be pinpointed in *De Motu Antiquiora* and that it represents "an instance of the dynamical conundrum that threatens the whole of *De motu*." The problem of the dynamical framework of *De Motu Antiquiora* is that "[s]pecific weight appears impotent to cause any effects (Van Dyck 2006, 211)."

have to wait for almost a century. Rather, the problem realized with regard to the application of a dynamical understanding to naturally accelerated motion puts an end to Galileo's search for a foundation in late 1602.

In the last section of this chapter, we have briefly glanced at the further development of Galileo's work on the foundation. As evidenced by a letter to Paolo Sarpi written in October 1604, at that time Galileo had picked up work on the foundation once again. His new approach owed to the understanding gained in 1602, yet no attempt is made to argue dynamically. Instead, the new approach is based on the doctrine of the configurations of motion or, more concretely, on framing naturally accelerated motion as characterized by a changing velocity where the change is expressed by the change of degree over extension. How the velocity was supposed to grow was specified by the Sarpi letter principle, and the growing velocities on planes of different inclination were related by the velocity proportional to vertical height principle. Both principles are elementary yet not without alternative. No further explanation as to why they should hold was provided in 1604.[69]

Based on these principles, Galileo managed to construct proofs for the law of fall and the length time proportionality. By proving two propositions from basic principles, Galileo's 1604 attempt had met the bounding conditions for a foundation which had emerged from his search for an axiomatic foundation some 2 years earlier. At the same time, Galileo had avoided arguing from dynamics. Galileo, hence, for the time being, could be confident that a new science could be erected and this explains the enthusiasm expressed in his letter to Paolo Sarpi.

In the letter, Galileo, indeed, more than once, stressed that a foundation for his exposition of the pheno-kinematics of naturally accelerated motion had been laid. However, his success was short-lived and after he had realized that the Sarpi letter principle was not compliant with the law of fall Galileo had to return to the question of the foundation. Reconstructing how he thus finally ended up with the solution he decided to publish in the *Discorsi* lies beyond the scope of this work.

Galileo's earliest attempt to arrive at a new science, thus, proceeded exactly the other way around than is commonly assumed. It is, indeed, usually held that Galileo must have first deduced his fundamental propositions synthetically, starting from basic assumptions by arguments which, if not identical, must allegedly have been similar to foundational arguments presented in the *Discorsi*.[70] In contrast, Galileo's work on the foundation succeeded the establishment of the subject matter of the

---

[69]In the *Discorsi*, the principle of acceleration according to which the degrees of speed increase in proportion to the time elapsed, which replaced the Sarpi letter principle, is justified by ruling out the Sarpi letter principle as the only sensible alternative. Galileo's refutation of the Sarpi letter principle in the *Discorsi* has recently been interpreted by Norton and Roberts (2012) with a critical response by Palmerino (2012) and Laird (2012). For the velocity proportional to vertical height principle, a plausibility argument is provided in the *Discorsi* which makes recourse to pendulum motion.

[70]The belief that Galileo must have arrived at his results in a way comparable to the way he argued for the results in the *Discorsi* in part rests on insufficiently distinguishing "two types

new science. The latter consisted of propositions concerning the pheno-kinematics that had emerged as a spin-off from his challenging research program on swinging and rolling. These propositions were then deductively organized. His successive analysis of the conceptual underpinning of the fundamental propositions in this deductive structure was motivated by a search for demonstrations of what, for him, were already established facts.[71] This, however, already belongs to the more general implications of the early genesis of Galileo's new science and it is the task of the conclusion, which succeeds this chapter, to lay these out.[72]

References

Caverni, R. (1895). *Storia del metodo sperimentale in Italia* (Vol. IV). New York: G. Civelli.
Clavelin, M. (1983). *Conceptual and technical aspects of the Galilean geometrization of the motion of heavy bodies* (pp. 23–50). Dordrecht: Springer.
Damerow, P., Freudenthal, G., McLaughlin, P., & Renn, J. (2004). *Exploring the limits of preclassical mechanics*. New York: Springer.
Drake, S. (1969). Galileo's 1604 fragment on falling bodies (Galileo gleanings xviii). *The British Journal for the History of Science, 4*(4), 340–358.
Drake, S. (1978). *Galileo at work: His scientific biography*. Chicago: University of Chicago Press,
Drake, S. (1979). Galileo's notes on motion arranged in probable order of composition and presented in reduced facsimile. In *Annali dell'Istituto e Museo di Storia della Scienza Suppl. Fasc. 2, Monografia n. 3*. Istituto e Museo di Storia della Scienza, Florence
Galilei, G., Crew, H., Salvio, & A.D. (1954). *Dialogues concerning two new sciences*. New York: Dover Publications.
Galluzzi, P. (1979). *Momento*. Rome: Ateneo e Bizzarri.
Hooper, W.E. (1992). *Galileo and the problems of motion*. Dissertation, Indiana University.
Hoyningen-Huene, P. (1987). Context of discovery and context of justification. *Studies in History and Philosophy of Science, 18*, 501–515.
Jardine, N. (1976). Galileo's road to truth and the demonstrative regress. *Studies in History and Philosophy of Science Part A, 7*(4), 277–318.

---

of historical processes, namely, the process of discovery and the process of justification of this discovery (Hoyningen-Huene 1987, 504)."

[71] With regard to the *Discorsi*, Clavelin (1983, 36–37) remarked: "the definition of naturally accelerated motion does not occur in any of the theorems or in any of the constructions which will achieve the geometrisation [of naturally accelerated motion]. The conceptual analysis could only remain exterior to the mathematical analysis, incapable of supporting it or of being supported by it ... Whoever makes the effort of reading closely the theorems and propositions will find that the demonstrative apparatus is principally based on two relations of proportionality established independently one from the other [propositions II and III, the law of fall and the length time proportionality]." This state of affairs in the *Discorsi*, so acutely observed by Clavelin, thus finds its explanation in the concrete historical process by which it was brought about.

[72] One cannot help but get the impression that for Galileo the work on the conceptual underpinnings was important primarily in that it was a requirement for establishing a foundation for his new science. Harriot, in contrast, seems to have considered it a theoretical question in its own right and, thus, for instance, diligently investigated the implications of the fact that the Sarpi letter principle and the correct principle of acceleration were irreconcilable. Cf. Schemmel (2008).

Laird, W.R. (2012). Stillman Drake on Salviati's proof. *Centaurus, 54*(2), 177–181.
McKirahan, R.D. (1992). *Principles and proofs*. Princeton: Princeton University Press.
Norton, J.D., & Roberts, B.W. (2012). Galileo's refutation of the speed-distance law of fall rehabilitated. *Centaurus, 54*(2), 148–164.
Palmerino, C.R. (2012). Aggregating speeds and scaling motions: A response to Norton and Roberts. *Centaurus, 54*(2), 165–176.
Palmieri, P. (2017). On scientia and regressus. In H. Lagerlund & B. Hill (Eds.), *Routledge companion to sixteenth century philosophy* (Routledge philosophy companions). New York: Routledge/Taylor & Francis Group.
Schemmel, M. (2008). *The English Galileo: Thomas Harriot's work on motion as an example of preclassical mechanics* (Boston Studies in the Philosophy of Science, Vol. 249). Dordrecht: Springer.
Settle, T. (1966). *Galilean science: Essays in the mechanics and dynamics of the Discorsi*. Ph.D. thesis, Cornell University.
Shields, C. (2016). Aristotle. In E.N. Zalta (Ed.), *The Stanford encyclopedia of philosophy* (Winter 2016 ed.). Stanford: Metaphysics Research Lab, Stanford University.
Van Dyck, M. (2006). *An archaeology of Galileo's science of motion*. Ph.D. thesis, Ghent University.
Wallace, W. (1992). *Galileo's logic of discovery and proof: the background, content, and use of his appropriated treatises on Aristotle's posterior analytics*. Dordrecht: Springer.
Wisan, W.L. (1974). The new science of motion: A study of Galileo's De motu locali. Archive for History of Exact Sciences, 13, 103–306.
Wisan, W.L. (1981). *Galileo and the emergence of a new scientific style* (pp. 311–339). Dordrecht: Springer.

Chapter 12
Conclusion: The Emergence and Early Evolution of Galileo's New Science of Motion

With the institution of a clear-cut account of how Galileo's new science of accelerated motion emerged as the guiding theme, this book has focused on reconstructing and interpreting those parts of Galileo's *Notes on Motion* that document the genesis and early conceptual development of the new science. I am confident that the results presented will prove reliable and hope that they will serve as a basis for the establishment of more comprehensive interpretations of this important manuscript, which despite considerable effort remains insufficiently understood.

The first part of this book identifies the challenging problem which first directed Galileo's attention toward a deeper investigation of the problem of naturally accelerated motion. The research program on the relation of swinging and rolling that arose in response has been reconstructed and dated to 1602. It has been demonstrated how Galileo's work on this paramount challenge starting from but a few isolated assumptions brought about a host of novel insights concerning the pheno-kinematics of naturally accelerated motion along inclined planes. The second part of the book reconstructs how Galileo immediately set about turning his novel insights into a new comprehensive science of motion compliant with the standards of his time, mainly by analyzing and relating them to assumptions grounded in the extant conceptual frameworks which were being employed to describe motion and its causation.

Galileo perceived a challenging similarity between what he held to be true concerning the behavior of pendulum motion on the one hand, and accelerated motion on inclined planes on the other, and this is what first motivated his more extensive investigation of the phenomenon of naturally accelerated motion. Initially, Galileo's understanding was limited to a few assumptions concerning the relations between distances covered and times elapsed for both types of motion. In particular, he assumed that in naturally accelerated motion, the spaces traversed increased with the square of the times elapsed (the law of fall) and that motions along all chords that can be inscribed in the same circle ending at either its apex or nadir are isochronous, i.e., completed in equal times (the law of chords). For the time being, these assumptions stood in isolation and did not bear any further-reaching consequences.

However, when Galileo arrived at the conviction that pendulum motion was isochronous, i.e., that the period of a pendulum did not depend on its amplitude (the isochronism of the pendulum), a palpable similarity between swinging and rolling manifested itself for him. From the recognition of this similarity, the opportunity and challenge arose of conceptualizing the relation between these two types of motion in such a way that it would be possible to demonstrate the assumed properties of pendulum motion from those of naturally accelerated motion along inclined planes and vice versa. Thus, as evidenced by a letter sent to his patron Guidobaldo del Monte in late 1602, Galileo engaged in a research agenda on swinging and rolling.

There are a great number of entries in the *Notes on Motion* documenting Galileo's investigation into the relation of swinging and rolling. This work has been dated to 1602, independently of the letter, and it has thus been shown that it corresponds to the research agenda that Galileo mentioned to Guidobaldo del Monte in that same year. Galileo initially held the particularly simple hypothesis that the time taken by a pendulum to swing through a vertical quarter arc might be identical to the time it takes a body to roll along a chord spanning this same arc. If this single chord hypothesis was proven correct, his assumptions about pendulum motion, the isochronism of the pendulum and the law of the pendulum, would follow as an immediate consequence from his assumptions about naturally accelerated motion, the law of fall and the law of chords, and vice versa. However, in an experiment designed to test it, the pendulum plane experiment, the single chord hypothesis proved empirically untenable.

The single chord hypothesis had not been born out of physical reasoning but had been adopted as the result of an abductive inference based on Galileo's assumptions about swinging and rolling whose alleged mutual relation it was supposed to explain. Its empirical rejection, thus, did not put an end to Galileo's endeavor but resulted in a reconceptualization. Galileo's revised conceptualization of the relation between swinging and rolling was based on the idea that swinging and rolling along the same arc were kinematically equivalent and that rolling along an arc could be perceived as the limiting case of motion along a polygonal path made up of a series of conjugate chords of increasing number inscribed into this arc.

Galileo's investigations of the naturally accelerated motions along such polygonal paths have been termed the broken chord approach. The broken chord approach bestowed Galileo with ever-new problems, many, but by far not all of which, related to the fact that he was searching for an argument that would allow an inference to be made concerning the limiting case that of rolling along the arc. Coping with these problems, Galileo explicated new assumptions, was able to formulate and prove a substantial number of new insights concerning naturally accelerated motion on inclined planes, and devise new methods for solving his pertinent problems.

Some of the results achieved in the context of the broken chord approach strongly suggested that rolling along an arc may not be isochronous but for as long as Galileo had been unable to make an inference about the limiting case of motion on the arc, this conclusion was not mandatory. Indeed, dismissing the isochronism would have deprived his agenda on swinging and rolling of its very basis.

Instead, Galileo returned to his experimental results and reevaluated them in light of his new conceptualization. Once again, the experimental data turned out not to comply with his theoretical expectation. From a modern perspective, two factors can be held responsible for the deviation of the experimental results as compared to the theoretical expectation Galileo had formed on the basis of his revised conceptualization. By the way the experiment had been designed, Galileo had implicitly compared a rolling motion to a sliding one and wrongly assumed that these motions would be kinematically equivalent. Furthermore, he had based the design of the experiment, as well as the theoretical evaluation of its results, on the false assumption of the isochronism of the pendulum.

With the second stage of evaluation of the experimental data proving futile, Galileo's agenda on swinging and rolling had run into a dead end. In view of the convoluted conceptual problems his approach had confronted him with, Galileo discontinued his work but retained most of his notes, likely in the hope of being able to make progress with it in the future.

A solution, however, did not crop up until the publication of the *Discorsi*, in which pendulum motion is an important issue but in which, save for a single argument, Galileo's initial agenda on swinging and rolling left virtually no apparent trace. A scholium, in which Galileo argued for the arc of the circle as the path of quickest descent between two points under the action of gravity, must be considered as his attempt to bear fruit from his early consideration of the relation between swinging and rolling. However, the argument he gave is recognized as flawed, and based on his earlier considerations, Galileo must have been aware of its problematic character.

A late success was imminent when Giovanni Battista Baliani sent Galileo a copy of his *De motu naturali gravium solidorum* in which he had derived the laws of naturally accelerated motion from assumptions about pendulum motion. Based on the insights he had gained in his own investigations on the relation between swinging and rolling, Galileo formulated a critique of Baliani's approach and in consequence devised a proof of the law of the pendulum based on the law of fall as a premise. He thus achieved at least part of what he had set out to do with his challenging research agenda. The proof was put in the margin of Galileo's own hand copy of the *Discorsi* but, despite thus being earmarked for publication, did not make it into the second edition of the book.

However, his work on swinging and rolling had not left Galileo empty handed. It had bequeathed him with a great number of propositions concerning the phenokinematics of naturally accelerated motion on inclined planes. His investigation had engendered a metamorphosis by which a very limited number of networked heuristic assumptions turned into the key insights of a new conceptualization of naturally accelerated motion. Indeed, roughly contemporaneously to his research on swinging and rolling, and in any case before 1604, Galileo started to collect and systematize the results he had achieved thus far. In the second part of this book, it is shown that

Galileo endeavored to turn his results into a publishable science able to meet the standards of the time.

Galileo systematically collected and structured the results achieved. In this period of work, arguments which before had only been sketchily noted were drafted as propositions, meeting the formal standards of exposition, i.e., in such a way as to follow the classical schema of a geometrical proof in the Euclidean tradition. In doing so, despite still writing private notes, Galileo formally mimicked a treatise, for instance, in that he opened paragraphs by majuscules or by drawing a margin on the respective folios which he respected when writing. Besides such formal criteria, a number of additional aspects suggest that he was conceiving of his effort as preparation for a publication.

Galileo, however, did more than merely collect, systematize, and formalize his results. He started to transform them into a mathematized demonstrative science. This attempt manifests itself mainly in two aspects. On the one hand, Galileo inquired into, what has been referred to as, an axiomatization of his propositions on the pheno-kinematics of naturally accelerated motion, i.e., he explored the question of which of his propositions could serve as a minimal sufficient set of statements in which the deductive tree of a demonstrative science could be rooted.

On the other hand, he started to analyze his new insights taking into account concepts other than spaces traversed and times elapsed, concepts as they were embedded in the theoretical frameworks, by means of which motion phenomena were being described and analyzed in his day. His hope was, thus, to be able to construct proofs from basic principles that would find unreserved acceptance.

Invoking, in particular, the notion of velocity, which had played virtually no role in his considerations thus far, Galileo's analysis suggested that his older basic principle of inclined plane dynamics would apply in the case of naturally accelerated motion as well, i.e., that it could be assumed that the velocities of motions on planes of different inclinations were in the same ratio as the mechanical moments on these planes. This dynamical assumption allowed him to construct the so-called ex mechanicis proof of the law of chords. Thus Galileo had arrived at a proof for one of his fundamental or axiomatic statements concerning the spacio-temporal behavior of naturally accelerated motion that pivoted on well-established mechanical results. The proof was later published as one of the three proofs Galileo gave for the law of chords in the *Discorsi*, where it was explicitly classified by him as "demonstratur ex mechanicis."

Galileo's search for an axiomatic foundation had shown that the new science could not be based on the law of chords alone but that another proposition, either the law of fall or the length time proportionality, needed to be equipped with a proof from basic principles. Moreover, the results of further analysis of his new insights revealed an incompatibility of the assumptions applied. The so-called mirandum paradox showed that the velocities of motions along inclined planes of different inclination could not generally be held to be in a fixed relation to each other as Galileo had presupposed in the argument of the ex mechanicis proof. This did not render the proof false, but it did deprive it and, thus, the dynamical approach in

12 The Emergence and Early Evolution of Galileo's New Science of Motion

general, of its explanatory value. Indeed, in the *Discorsi*, the ex mechanicis proof is not formally accredited as part of the foundation of the new science.

The recognition of this complication brought Galileo's search for a foundation to a halt, presumably in late 1602. A letter which Galileo sent to Paolo Sarpi in 1604, rather than marking the discovery of the law of fall, as is commonly held, indicates Galileo's relief at having found a way out of the deadlock that he had run into approximately 2 years earlier when searching for a foundation for his new science. The seeming solution that Galileo had found in 1604 was revealed to be ultimately not tenable when he realized that the principle which he had applied to characterize the change of velocity in naturally accelerated motion was not compliant with the law of fall. The reconstruction provided in this book ends in 1604. By then, core insights concerning the pheno-kinematics of naturally accelerated motion had rather irrevocably been established, and Galileo was in the midst of his search for an analytic foundation for his new science. How this search progressed, until Galileo finally presented the solution in the *Discorsi*, is not discussed.

This picture, which has emerged from the in-depth analysis of a substantial part of Galileo's *Notes on Motion*, has a number of profound implications for an understanding of the emergence, development, and the further fate of Galileo's new science of naturally accelerated motion, which will briefly be sketched.

To begin with, the standard chronology asserted in the literature, namely, that Galileo "discovered" the law of fall late in 1604 and that this essentially marks the onset of his systematic work on naturally accelerated motion is overthrown. This amendment of the perceived view is, however, merely incidental. It is the vital outcome of this study that, for the first time, it presents an actual account of the genesis of Galileo's new science of motion, giving a detailed exposition of the incentives and the mechanisms as well as the individual steps that enabled Galileo to expand some isolated assumptions about naturally accelerated motion along inclined planes into the core of a new science.

In a nutshell the most important ramifications are:

- Galileo's work on the problem of naturally accelerated motion was motivated by a concrete, tangible problem which had arisen in first place because of the close attention Galileo payed to a motion phenomenon that hitherto had attracted very little theoretical attention—pendulum motion. The problem of the relation of swinging and rolling was complex enough to trigger substantial elaboration yet could still be addressed with Galileo's extant means.
- Galileo's perception of the problem was based on the assumption of the isochronism of the pendulum, false from a modern perspective, while being reasonably well in accordance with observable behavior. Thus, the assumption could not easily be debunked as erroneous. The ever-growing problem of recovering this false assumption, theoretically as well as empirically, provided the essential motor for Galileo's early efforts.
- Galileo's work on the problem of swinging and rolling led to the establishment of a set of coherent well-justified statements concerning the pheno-kinematics of naturally accelerated motion on inclined planes. Galileo's attempt to root this

pheno-kinematical description in the concepts of the extant theoretical frameworks had post hoc character. Analysis of the new insights yielded results that collided with existing conceptualizations, thus revealing the limits of preclassical mechanics.

Each of these points in turn has further bearings. In terms of the role of pendulum motion for the genesis of the new science of motion, I have argued elsewhere that the attention Galileo paid to the phenomenon of pendulum motion was not an arbitrary idiosyncrasy that eludes further explanation. Rather it is indicative of a historical development which goes above and beyond the individual level. In the early modern period, the pendulum became one of the *challenging objects* of mechanics. Many of Galileo's contemporaries, mostly independently of one another, started to turn their attention to the problems of pendulum motion. The attention Galileo paid to pendulum motion becomes understandable from this larger perspective, and questions concerning the concrete circumstances and motivation of his engagement with the phenomenon, as interesting as they are, are revealed to be of somewhat secondary importance.

That Galileo's research agenda on swinging and rolling and, thus, ultimately, the birth of his new science of naturally accelerated motion, centrally hinged on Galileo's false assumption that pendulum motion is isochronous, is probably the most surprising twist in the historical development reconstructed here. In 1602, Galileo communicated to Guidobaldo del Monte that it was precisely this admirable ("mirabile") assumption which he was seeking to demonstrate. That scientists go amiss and pursue false assumptions in their investigations is not uncommon and possibly a rule rather than an exception in scientific practice. What is so outstanding in the case of Galileo is that it is only as a result of the wrong assumption that the object of inquiry itself, namely, the similarity between swinging and rolling, is established. Without the assumption of the isochronism of the pendulum, the similarity Galileo fancied, between swinging and rolling, can simply not be perceived. It is an utter illusion.

Had Galileo realized this and accepted that pendulum motion was not isochronous, his investigations would thus have come to a halt and wouldn't have unfolded their full potential. But Galileo stuck to the false assumption, and in both the *Discorsi* and private correspondence, even that postdating the publication of the book, he confidently stated that pendulums swing isochronously. Indeed, Galileo must have had a strong disposition to stick to this assumption despite the indications of its problematic character, as giving up the assumption would have meant retrospectively renouncing the very research agenda which, in the long run, had bestowed him with a new science.

The isochronism of the pendulum is approximately true and thus accords reasonably well with observable behavior. Together with the incipient status of his consideration, confronted with indications that motion on the arc might not be isochronous, this allowed Galileo a forward escape more than once. Instead of accepting the, however, never mandatory implication that motion along the arc is not isochronous based on the indications he had, Galileo sought additional hypothesis and immersed himself in the problems created. The most prominent example of this is his reconceptualization of the relation between swinging and rolling after the single chord hypothesis had been proven empirically untenable. Pointedly speaking it was thus centrally and somewhat paradoxically a wrong assumption which advanced the generation of new and, from a modern perspective, correct insights. Founded on a wrong surmise, Galileo's research agenda on swinging and rolling, thus, became the breeding ground for the new science.

Galileo pushed forward and had found almost all there was to find from a modern perspective. Though he was himself unable to recognize this, still, his investigation into the relation between swinging and rolling eventually came to a standstill. However, as stated above, this did not happen because Galileo rejected the assumptions fundamentally underlying his research program. Rather, he was forced to abandon his research on swinging and rolling because the problems it had confronted him with had become unsurmountable. However, by the time this happened, the seed for the creation of the new science had been sown. A small number of isolated assumptions concerning naturally accelerated motion had by then been expanded into a dense phenomenological description of the kinematics of naturally accelerated motion epitomized by a rather large set of propositions concerning the relation between spaces traversed and times elapsed in such motions. These were deductively organized, and the deductive structure could be rooted in a very limited set of statements from which all other propositions could rigidly be derived, thus allowing the consistent treatment of very complex motions.

Crucially, this still characterizes the way in which Galileo presented his new science in the *Discorsi*. There, once the law of fall and the length time proportionality were established as propositions II and III by arguments starting from fundamental principles as premises, the new science proceeds as a science of the pheno-kinematics of naturally accelerated motion. All of the remaining propositions are essentially inferred from those two fundamental propositions. They exclusively make statements about spaces traversed and times elapsed, and these are basically also the only two physical concepts that figure in the respective proofs. The subject matter of the new science of naturally accelerated motion in the *Discorsi* is, thus, in fact, the pheno-kinematics of this type of motion as it had originally come out of his investigation into the relation between swinging and rolling.

On the surface, from a modern perspective, the 35 propositions and their respective proofs, which Galileo gave succeeding propositions II and III in the Second Book of the Third Day of the *Discorsi*, may appear to be a trivial, somewhat futile exercise. It certainly would not have been so for Galileo. Taken together they delineate the pheno-kinematics of naturally accelerated motion in a novel and unprecedented way. In the *Discorsi* they are presented as synthesized by inferences from basic assumptions invoking concepts rooted in the conceptual tradition of mechanics such as velocity or force. This is, however, neither how his propositions had come into being nor how they had been vindicated. Rather, Galileo's exposition of the pheno-kinnematics of naturally accelerated motion had originated independently of and, indeed, predated any considerations invoking traditional conceptual frameworks in any strong sense. It was internal consistency paired with empirical tests that provided it with an independent justification.

Galileo's attempt to root his new findings in more fundamental principles compliant with extant conceptualizations of motion was inherently post hoc. The synthesis finally presented in the *Discorsi* was conceived from the end in reverse. To turn his new insights into a new science, basic principles had to be found that were generally acceptable and based on which arguments could be devised to demonstrate his, already rather irrevocably fixed, new statements. In a famous passage in the *Discorsi*, Galileo quite explicitly unveiled how he had proceeded:

> but we have decided to consider the phenomena of bodies falling with an acceleration such as actually occurs in nature and to make this definition of accelerated motion exhibit the essential features of observed accelerated motions. And this, at last, after repeated efforts we trust we have succeeded in doing. In this belief we are confirmed mainly by the consideration that experimental results are seen to agree with and exactly correspond with those properties which have been, one after another, demonstrated by us. (EN VIII, 197, transl. Galilei et al. 1954)[1]

Galileo's exposition of the pheno-kinematics of naturally accelerated motion stood outside of extant conceptual traditions and, in particular, was not a logical consequence thereof. It was, thus, not clear beforehand whether his new results could be integrated into and subsumed under the contemporary understanding of motion and mechanics. With the aim of creating a foundation, analysis revealed conflicts and thus showed that this was not the case. Galileo's considerations, documented by the mirandum paradox brought to light, in particular, that the dynamical assumption based on which he had constructed his ex mechanicis proof of the law of chords and on which he effectively hoped to be able to erect his new science, could not be maintained in all generality.

A conflict had thus arisen between Galileo's new understanding of the phenomenon of naturally accelerated motion, which by that time had stabilized to the point where it was no longer called into question, and the traditional dynamical understanding, amounting, ultimately, to the assumption that force is the cause of motion with velocity being a measure of the effect. This showed the limits of the preclassical conceptualization of motion. It would, however, ultimately, not be Galileo who would draw the consequence of understanding from this that gravitational force is proportional to mass, that acceleration is proportional to force per mass, and that it is, thus, in the end, not motion but change of motion that is caused and explained by the action of forces. The transgression of these limits and the reorganization of mechanics were left to Galileo's successors who accepted his pheno-kinematical description of naturally accelerated motion and took it as their starting point.

---

[1] "[T]amen, quandoquidem quadam accelerationis specie gravium descendentium utitur natura, eorundem speculari passiones decrevimus, si eam, quam allaturi sumus de nostro motu accelerato definitionem, cum essentia motus naturaliter accelerati congruere contigerit (EN VIII, 197)."

12 The Emergence and Early Evolution of Galileo's New Science of Motion

Galileo himself concealed the conflict which, as he had realized, resides between the kinematics of naturally accelerated motion and the traditional understanding of dynamics. In his search for a foundation for his new insights, he abandoned the dynamical approach altogether and directed his attention to an alternative way of grounding his new science which he, indeed, did come up with eventually. Galileo did not refrain from publishing the ex mechanicis proof of the law of chords, which had been constructed and called into question 35 years earlier in the *Discorsi*. However, he tucked it away between two alternative proofs of the proposition and specifically did not mention the problem regarding the conceptual underpinning of the proof he had so early realized.

In the *Discorsi*, Galileo famously went so far as to, at least seemingly, reject causal explanation of natural acceleration altogether:

> The present does not seem to be the proper time to investigate the cause of the acceleration of natural motion ... Now, all these fantasies, and others too, ought to be examined; but it is not really worth while. At present it is the purpose of our Author merely to investigate and to demonstrate some of the properties of accelerated motion (whatever the cause of this acceleration may be) ... (EN VIII, 202, transl. Galilei et al. 1954)

This statement has provoked much debate. Galileo did not reject causal explanation outright, and this is indeed not implied by his wording. Clearly neither did Galileo express here a fundamental methodological credo according to which it was necessary for the study of kinematics to come before a causal or dynamical explanation, as claimed, for instance, by McMullin. For Galileo, the study of the kinematics of naturally accelerated motion had come before any attempt at its explanation. This was, however, not due to a methodological conviction but had been the contingent result of the intellectual path he had ventured and whose direction had largely been dictated by the nature of the challenging problem of the relation of swinging and rolling. Galileo had tried to found his new insights on dynamics but had failed. In the *Discorsi* he turned this failure into a strength by stressing that he had been able to erect his new science without recourse to causal explanation.

Galileo's strategy obviously worked. The discussions about the role of dynamics seem to have almost completely eclipsed the realization of how shaky and problematic the alternative foundation of the new science, that Galileo had finally come up with and that he offered in the *Discorsi*, actually is. An account of how the solution emerged and what traces this left in the *Notes on Motion* would by far have exceeded the scope of this book. Suffice it to state, for now, that structurally, the foundation as published in 1638 had already been largely preconfigured in 1604. As we have seen, by then Galileo was already systematically trying to base the new science on an understanding of velocity as a changing quality as implied by the doctrine of the configurations of motion and was, against the backdrop of this understanding, trying to construct fundamental proofs for the law of fall and the length time proportionality.

The fundamental proofs which he gave for these two propositions in the *Discorsi* are based on the principle of acceleration, according to which, the degrees of speed grow in proportion to the time of fall, as well as, the postulate that equal degrees of velocity are reached after fall through the same height. They bear the character of somewhat impromptu drafted fig-leaf solutions. The arguments Galileo gave avoided problems that had arisen before, yet, apart from demonstrating what, for Galileo, were long-established truths, they remained without consequences and, thus, rather meaningless. Descartes was quick to realize problems in the foundation of the new science, and his critique reached Galileo and his disciples via Mersenne. Thus, dynamics was reintroduced into the foundation in the second edition of the *Discorsi* in an esprit de l'escalier. Detailing this remains reserved for a future publication that will deal with the development of Galileo's new science from 1604 onward. For now, the inclined reader will have to content themselves with my reconstruction of its development which ends, thus, in 1604.

Chapter 13
Appendix: Folio Pages

In this appendix the content of those folios upon which the interpretation presented in this book substantially rests is examined in great detail. Folios whose content exhaustively discussed in the text itself such as 131, 147, 160, 163, 164, 172, or 174 are not included to avoid redundancy. The reconstructions and interpretations presented in this appendix are the backbone of the synthesizing interpretations imparted in the individual chapters. At the same time, the chapters have been written in such a manner as to be intelligible without the material included here. The material in this appendix may be consulted to deepen the understanding and may also be a starting point for further study of the manuscript.

To further the readability and to enable the discussion in this appendix to be somewhat independent from an edition of the primary source, the content of the folio pages is reproduced. This is not intended as an edition of any sort. Indeed, I assume that the concerned reader will peruse this book and, in particular, this appendix having direct access to an edition of the manuscript, i.e., currently the electronic representation of Galileo's *Notes on Motion*, produced jointly by the Biblioteca Nazionale Centrale, Florence, the Istituto e Museo di Storia della Scienza, and the Max Planck Institute for the History of Science, Berlin, which has been online since 1996.

In this online representation, blocks of content have been uniquely identified on the folio pages of the manuscript and have been assigned with unique identifiers. I have availed myself of this groundwork and have referenced the content of the manuscript according to the identifiers that have been assigned. Yet the current representation was produced in the incunable stage of putting computer and the Internet to the use of producing and disseminating scholarly editions of historical sources. As such it has its own deficits which will likely lead to its replacement at some point in the future. It cannot be foreseen whether a new edition will retain the same content identification scheme. Secure reference has thus not been the least of my motives to include, at least in rudimentary form, the content of the folios discussed. This should allow the continued identification of the content referenced and analyzed on the folio page itself or a digital facsimile thereof and hence also in any future edition of the *Notes on Motion*.

13.1 Folio 90

Folio 90 has received attention primarily for the content of its recto. On this page Galileo constructed a set of projectile trajectories resulting from horizontal projections. Using this construction he explored the consequences of determining the impetus of projectile motion by scalar addition. This method of adding impetus was later abandoned and replaced with what can be referred to as *pseudo vectorial addition*.[1] To increase the usable space on the folio's recto, Galileo cut a second sheet of paper and pasted the cut part onto the lower margin of folio 90. On the sheet that was cut, Galileo had previously noted considerations pertaining to the second stage of evaluation of the pendulum plane experiment (cf. Chap. 6). These notes were mutilated when the page was cut and pasted. The content that remains on the back of the cut and pasted sheet is now formally part of folio page 90 verso.

Folio 90 verso Galileo's entries on the page are directly related to the content of 189 verso bearing the experimental record of the pendulum plane experiment but also to that of the obverse side, folio page 189 recto. Some of the content which remains on the cut sheet that was pasted onto folio 90 is now covered by overlapping paper from the folio it was pasted onto. The transcription below is based on an inspection of the original manuscript page and hence slightly exceeds what is visible in the digital facsimile:

- T1, results from 189 verso restated:

    [c]u[m] *ab* est 27834. diam[eter]
    q[uo]d erit diameter.[2]

    The line *ab* is the vertical diameter of a circle in which the inclined plane Galileo had used in the pendulum plane experiment could be inscribed as a chord. 27,834 is the length of a chord subtending an angle of 16 degrees in a circle whose diameter measures 100,000.[3]

---

[1] The content of folio 90 recto has been discussed by various authors. The most comprehensive account is Damerow et al. (2004, 224–227). It bears a crowned unicorn as a watermark which it shares with almost all other folios containing consideration that can be attributed to an early phase of Galileo's considerations regarding projectile motion.

[2] 90 verso, T1.

[3] See the interpretation of C02 in the discussion of 189 verso in Sect. 13.20.

13.1 Folio 90

- T2, results from folio 189 restated:

    Tempus quo absolvitur perpendic[ulum]
    280. Quodna[m] e[rg]o erit tempus
    80 2/3.[4]

    $t_{plane} = 280$ is the time Galileo had measured for motion along a long plane in the pendulum plane experiment. $t_{ab} = 80\ 2/3$ is the time to fall vertically through a distance equal to twice the pendulum's length calculated in the units of the experiment on 189 verso.

- T3, state proportion:

    ut cf ad fb. ita tempus ...[5]

    If it is conjectured that this entry expresses in language the very same proportion that is also being exploited in the succeeding calculation on the page, the entry can be completed to read: as cf to fb so is the time of motion through the diameter to the time of motion through the radius (the labels of the points correspond to those of diagram D02B on 189 recto).

    cf is the mean proportional between the radius fb and the diameter $2 * fb$. Thus if it is assumed by distance time coordination that $t_{radius} = fb$, it follows by the law of fall that $t_{diameter} = cf$.

- C02, calculation of time to fall through the radius in the units of the experiment:

$$100 * 80\ 2/3 : 141\ 1/2 = 57.^{6}$$

Scheme of calculation:
$t(rad.)_{unit\ 189r} / t(diam.)_{unit\ 189r} * t(diam.)_{unit\ exp.} = t(rad.)_{unit\ exp.}.$

- C01, calculation (corrupted):

$$27834.^{7}$$

Only this single number (the length of a chord subtending an angle of 16 degrees in a circle of radius 100,000) remains legible of a second calculation comprised on the cut and pasted sheet.

Galileo's considerations documented on 90 verso are part of his attempt to determine, based on his measurements of the time of motion along a long gently

---

[4] 90 verso, T2.
[5] 90 verso, T3.
[6] 90 verso, C02.
[7] 90 verso, C01.

inclined plane (280 time units), the time in the units of the experiment of motion along a path composed of specific conjugate chords (cf. Chap. 6). In the only calculation that is preserved complete, Galileo reckoned that in the time units of his experiment, the time of fall through a distance of 2000 [punti], i.e., the length of the pendulum used in the experiment, should be 57. This result was further exploited on 189 recto.

13.2 Folio 115

Folio 115 bears no watermark. The paper is thin and white, and the dimensions are 208×300 mm. The folio has received attention mainly because of the content of its recto bearing the drafts of two propositions on projectile motion. It is one of only three folios in the manuscript on which Galileo determined the impetus of projectile motion by adding in scalar fashion the impetus of the two motions from which he conceived projectile motion resulted as a superposition.[8] It is primarily the content of the verso that has bearing on the interpretation presented here.

Folio 115 verso Galileo wrote on folio 115 verso in three different directions. The entries pertain to four distinct contexts. The distribution of the content suggests that the considerations pertaining to the pendulum plane experiment which are written with the page upside down with respect to its modern orientation were put on the page first. In these notes Galileo restated and used results from folios 90 verso and 189 verso. Furthermore, the number 235, which Galileo used in his calculations on the page, represents his guess obtained from considerations on 166 recto, of the time of motion along an upright quarter arc of 90 degrees given the time of motion to fall along the radius of this arc was measured by 180 units (cf. Chap. 6).

- T1, essential results from folio 189 verso restated:

    tempus totius diametri est 280, cum eius longitudo fuerit per 48143; ergo tempus diametri, cuius longitudo 4000, erit 80 2/3.[9]

- C02, time of motion along arc calculated (first version):

$$235 * 80\ 2/3 : 360 = 52.^{10}$$

---

[8]The other two instances of a scalar addition are contained on folios 90 and 110. Cf. Damerow et al. (2004, 223–229 and 234). Just like on the obverse of 115, so the obverse of 90 contains notes related to the pendulum plane experiment.

[9]115 verso, T1.

[10]115 verso, C02.

13.2 Folio 115

Based on the assumption that the ratio of the time of motion along an arc to the time of motion through the respective diameter, is independent of the unit chosen, or else of the absolute size of the system considered, Galileo calculates the time of motion along an arc in the units of the experiment from his results on 166 recto pertaining to a base unit of 180.[11]

- C03, previous calculation corrected:

$$235 * 57 : 180 = 74.^{12}$$

On 166 recto Galileo had assumed that the diameter of the circle with respect to which the geometry of the problem was fundamentally defined had a length of 360 units. At the same time, he had assumed that the radius, i.e., half that distance, was traversed in free fall in 180 time units. This was not compliant with the assumption that a diameter of 360 distance units would be traversed in 360 time units on which Galileo had based his first calculation (C02 above). Galileo realized the mistake and corrected it by repeating the calculation. This time, however, instead of relating the time along the arc to the time along the diameter, he related it to time along radius which in units of the experiment had been calculated to 57 time units on 90 verso. The result of the calculation, i.e., the number 74, thus represents Galileo's best guess, expressed in units of the experiment, for the time elapsed during motion along an upright quarter arc whose radius corresponded to the length of the pendulum used in the pendulum plane experiment.[13]

---

[11] Galileo's base unit 180 approach is characterized by the assumption that the radius of the fundamental circle measures 180 length units and is traversed in 180 time units. Cf. the discussion of 166 recto in this appendix.

[12] 115 verso, C03.

[13] That the number 235 represents Galileo's guess of the time or possibly of a lower limit of the time of motion on an arc is supported by the following argument. On folio 189 verso, Galileo had rescaled his numerical result for the time of motion along the path of his best, the eight-chord approximation of the arc, calculated under the assumption that a vertical radius of 100,000 distance units would be fallen through in 100,000 time units, to yield that if the radius was instead assumed to be fallen through in 180 units, then fall along the eight-chord path would be completed in a time of 235 80/100 of these units. Galileo knew that the actual time of motion along the arc should be slightly smaller than the value thus calculated and might thus have rounded the figure to 235. The number 235 moreover reoccurs in a calculation on folio 184 verso, which receives a consistent interpretation if it is assumed that it represented Galileo's best guess for the time of motion on the arc (cf. discussion of 184 verso). Modern analysis shows that the actual value for time of motion along the arc differs only about four in a thousand from the assumed value of 235.

Other content

- Page turned 90 degrees counterclockwise, C01 and D02A:

  137, 91, 46, 183, 99[14]

  $\sqrt{100 * 50} = 70\ [14/100]$.[15]

  The numbers and the calculation of a mean proportional could not be related to any other consideration in the manuscript.

- Page upright, D01A–D01E.

  Study comprising four small diagrams whose purpose could not be reconstructed.[16]

- Page upside down, D01G and variant thereof D01F.

  The lower of the two unlabeled diagrams shows the geometrical constellation characteristic of a proposition which Galileo first elaborated on folio 147 verso and which in the *Discorsi* ended up being published as proposition IX of the Second Book of the Third Day.[17] The upper diagram appears to be a variant.[18]

13.3 Folio 121

Folio 121 bears an encircled sun as a watermark as well as a countermark depicting a crown. The encircled sun watermark is unique in the *Notes on Motion*. The presence of a crown countermark indicates that Galileo obtained the paper during his time in Padua. The folio was cut so as to accommodate the large central diagram on its recto side which was copied with modification from William Gilbert's *De Magnete*. The modifications suggest that Galileo had produced the diagram in reaction to a question that Paolo Sarpi had posed him concerning the construction of this particular diagram in Gilbert's book in a letter sent in September 1602 (cf. the discussion in Chap. 5). The verso side of the folio contains two diagrams of unclear purpose. In addition, some numbers and calculations are scattered over the page some of which can be identified as auxiliary calculations related to Galileo's attempt to determine the times of motion along a series of conjugate chords closely approximating an arc, which is centrally documented on 166 recto.

---

[14] 115 verso, D02A.

[15] 115 verso, C01.

[16] 115 verso, D01A–D01E.

[17] 115 verso, D01G.

[18] 115 verso, D01F.

13.3 Folio 121

Folio 121 recto The large diagram is a modified copy of Gilbert's construction for determining what is today called the magnetic inclination, i.e., the angle of orientation of a magnetic needle with respect to the plane of the horizon as a function of the latitude. The diagram was originally published by Gilbert in chapter 7 of his *De Magnete*. A synopsis of the underlying construction and the way in which Galileo modified Gilbert's original diagram is given in Chap. 5 Sect. 5.3.

Folio 121 verso Folio 121 verso was apparently used as a scratch paper for auxiliary calculations and additional reflection. Some of the content was lost when a part of the folio was cut off. Despite the fact that the meaning of several entries on the page remains unclear, a relation of the content to Galileo's work on folio 166 recto can firmly be established.

- D01B and D01C, two diagrams of unclear purpose.[19]
- C01–C03, calculations of unclear purpose:

$$24 + 24[0] = 264$$

$$27\ 1/3 * 6 = 164$$

$$37\ 1/3 * 6 = 224.^{20}$$

- C04, calculation (base 100,000) of distance $e\theta$ in the geometrical construction on folio 166 recto:

$$100000 - 70711 = 29289.^{21}$$

The calculated distance was further used in calculations on folio 192 recto to determine the time of motion along the broken chord egc in the geometrical construction on folio 166 recto.

- C05 and C13, calculation of distances ao and od (base 180) in the geometrical construction on folio 166 recto:

$$180 - 127 = 53^{22}$$

$$141 * 180 : 200 = 127.^{23}$$

---

[19] 121 verso, D01B and D01C.
[20] 121 verso, C01, C02, and C03. C02 has incorrectly been transcribed in the electronic representation of Galileo's *Notes on Motion*.
[21] 121 verso, C04.
[22] 121 verso, C05.
[23] 121 verso, C13.

The calculated distances were further used in calculations on folio 184 recto.
- C08 and C06, calculation of unclear purpose:

$$70\ 1/2\ 29^{24}$$

$$90 + 21\ 1/2 = 112\ 1/2^{25}$$

$$29 * 112\ 1/2 : 70\ 1/2 = 46.^{26}$$

The number 70 1/2 used in the calculation corresponds to the length of the individual chords of the four-chord approximation of the arc in the base unit 180 approach on folio 166 recto.
- C07 and C09–C12, considerations of unclear purpose.[27] Galileo wrote the natural numbers from 0 to 9 in a column and probed a relationship which involved multiplication and the squaring of these numbers. The rationale behind his procedure could not be reconstructed, but a relationship to the law of fall seems obvious.

13.4 Folio 129

Folio 129 bears no watermark. Paper size (206 × 302 mm) and quality (thin and white) compare to that of other folios of the crown/crossbow group, and the folio can thus, in all likelihood, be attributed to Galileo's Paduan period. The recto side of the folio contains considerations regarding the geometrical situation of a broken chord and a single chord inscribed into the same upright arc; its verso side is blank.

Folio 129 recto Folio 129 recto bears three motion diagrams all of which depict the same type of geometrical situation, in which a broken chord and a chord are inscribed into the same arc.[28] One of the diagrams is merely a sketch that remained unlabeled.[29] The other two diagrams depict basically the same geometrical situation. The smaller diagram has an additional circle *doc*. This circle might be present as an uninked construction in the larger two diagram as well. The smaller diagram is here schematically reproduced with additions as Fig. 13.1. As auxiliary constructions for noting and inferring valid proportions, Galileo noted two distance-

---

[24] 121 verso, C08.
[25] 121 verso, C06.
[26] 121 verso, C06.
[27] 121 verso, C07 and C09–C12.
[28] 129 recto D01A, D02B and D03B.
[29] 129 recto, D01A.

13.4 Folio 129

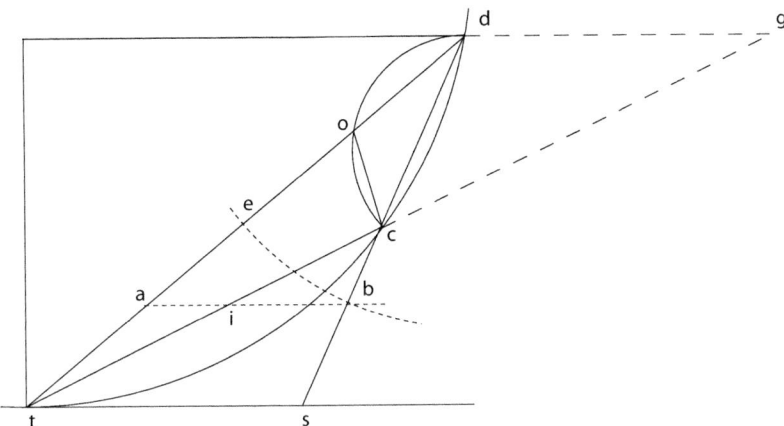

Fig. 13.1 Central drawing on folio 129 recto. Dashed lines with long gaps and point g are not part of Galileo's original diagram but have been added for analysis purposes

time ratio diagrams.[30] From the distribution of the content, it is rather clear that Galileo started to write with the page turned 90 degrees clockwise and only then turned the folio to put down those entries which are written upright with respect to the modern orientation of the page.

No explicit mention is made of motion or times of motion in the entries scattered over the page. However the construction, in particular the presence of the points e and i marking the mean proportionals $mp(do|dt)$ and $mp(gc|gt)$ (point g not marked in Galileo's diagram is defined by the intersection between the lower chord tc and the horizontal through d), uniquely identify the diagram as pertaining to the general context of Galileo's elaboration of a broken chord motion situation. The distances te and ti represent the times to traverse the second part of the single chord and the lower of the two broken chords, respectively (see Fig. 13.1).[31]

Compared to other diagrams, by means of which Galileo analyzed broken chord motion situations in particular in the context of his attempt to construct a proof for the law of the broken chord (cf. Chap. 5, Sect. 5.1), on folio 129 recto, he added a new element to the geometrical construction. He drew a circle around d and thus projected the distance de onto the inclined plane ds. This defined a point b where $|db| = |de|$. Then db is the mean proportional between dc and ds, and point b has

---

[30] 129 recto, D03A.

[31] In D02B Galileo joined the points e and c by a horizontal line. That point e lies at the same height as point c, the junction point of the broken chord, is a mere coincidence of the concrete configuration Galileo had chosen in the diagram. It may well have been the purpose of the rough sketch D01A, by choosing a more extreme position of the junction point, to show that this was in fact not an essential property of the construction.

the same vertical height above ts as point i as Galileo indicated by a dotted line.[32] He noted this as follows:

- T1:

    Sit bd media inter sd, dc, et centro d, intervalo b, secetur de, et per b ipsi st parallela bia.[33]

Galileo next recorded a geometrical property of the construction which, however, does not lead him anywhere:

- T2:

    Quia ts tangit, et tc secat, et ba est parallela ts, erit [triangulum] ati similis [tri]angulo tcd.[34]

Galileo then turns the page and states what he believes to hold and intends to show in the following:

- T3:

    Credo angulum sec bifariam esse sectum per eb.[35]

His proof goes as follows:

- T5:

    Quia est ut sd ad de, ita de ad dc, ergo [triangulum] sde similis est [triangul]o dec, et ut se ad ec, ita sd ad de, et ita est sb ad bc: ergo angulus ces bifariam secatur linea eb.[36]

$de = mp(ds|dc) \Rightarrow \triangle sde \simeq \triangle dec \Rightarrow se : ec \sim sd : de \sim sb : bc$

Thus, according to Euclid, book 6, proposition 3, be indeed bisects the angle sec as was claimed.

Next Galileo shows that the triangles odc and icb are similar:

- T4:

    Angulus tds duabus circumferentiis oc, ct insistit; ergo illae sunt similes, et circumferentia do similis est dct; ergo ut linea do ad oc, ita dt ad tc: et quia [rectangulum] dsc aequatur [quadrato] st, ergo ut ds ad st, ita ts ad sc: ergo [triangul]a dst, tsc similia sunt, quibus et [triangula] odc, icb similia sunt.[37]

---

[32] Compare the discussion of Galileo's work on 174 recto in Chap. 7.

[33] 129 recto, T1.

[34] 129 recto, T2.

[35] 129 recto, T3.

[36] 129 recto, T5.

[37] 129 recto, T4.

Galileo does not explicitly state so, but from the similarity of the triangles odc, icb, tcs, and dst, it follows with Euclid book 1, proposition 32 rather trivially that the line through i and b is a horizontal. Galileo may have been particularly interested in the fact that with his construction, the line through the points i and b is a horizontal because it allows for a qualitative understanding of how the ratio of te and ti, which is a measure for the relative time saving if motion along a broken chord is compared to motion along a single chord spanning the same arc, depends on the position of the junction point of the broken chord. It is indeed striking that Galileo varied the position of the junction point c of the broken chord dct between the first and the second diagram he drew on the page.

13.5 Folio 130

Folio 130 bears the watermark of a small, thick crossbow.[38] The verso side of the folio is dominated by a large, rather complex diagram. This diagram served Galileo's investigation of motions along bent inclined planes under a particular variation of the position of the junction point of the bent inclined plane. In an uninked construction, Galileo first geometrically constructed the times of motion along the various bent planes he had drawn in his diagram. He then turned to the question of theoretically determining the configuration yielding the bent plane which is traversed in least time. The recto side of the folio contains four calculations and a small table of results related to the construction on the verso side of the folio.

130 verso Folio 130 verso large is dominated by a large and rather complex geometrical construction.[39] A long text wraps continuously around the diagram.[40] In addition, the page bears two short notes.[41] Under raking light a number of uninked construction lines become visible on the page (see Fig. 13.4) . Based on these uninked construction lines, the way in which Galileo constructed the points q, g, s, and q' can be reconstructed. The short note close to the left-hand margin of the folio in abbreviated form elucidates the meaning of these points.[42] Galileo is

---

[38] The folios bearing this exact same watermark are 127, 130, 140, 165, 176, and 190. Interestingly, three and possibly even four of these folios are part of a double folio. 127 is connected to 126, 140 to 141, and 190 to 191. Judging from an identical fissure on both pages, 130 and 131 may also have been connected when Galileo wrote on them and then separated only later.

[39] 130 verso, 130 versoD01A.

[40] 130 verso, T1A-E.

[41] 130 verso, T1F and T2.

[42] 130 verso, T2.

inquiring for which position on mx of the junction point of a bent plane connecting points a and c the time to fall from a to c along this the bent plane is the shortest.

- T1A, first geometrical statement noted:

    [rectangulum] iae esse omnium minimum lab, oan, paf etc.[43]

    The lines falling from a to the horizontal cx (ac, ap, ai, al, ao, and ax) are cut by the line mxi. These lines and their respective first part cutoff are understood to define the area of a rectangle, and it is claimed that the area of the rectangle with sides ia and ae (short $\Box iae$) is smaller than the area of all other rectangles and thus in particular than $\Box lab$, $\Box oan$, and $\Box paf$.

- T1A and T1B, proof of first statement:

    cum angulus cax bifariam sectus sit, pendet ex eo, quod angulus aem [tri]ang[ul]i aem est aequalis angulo aix [tri]ang[ul]i aix et, quod consequens, est minor omnium alx, aox, etc., et maior omnium api, aci etc.: probabitur ergo sic, [rectangulum] iae esse minus [rectangul]o lab. Cum enim angulus ame sit aequalis angulo axi, et angulus mae aequalis angulo xai (est enim angulus a bifariam sectus), ergo reliquus mea reliquo xia aequabitur: sed angulus aem maior est angulo abe: ergo [angulus] ail est maior [angulo] eba. Si igitur fiat [angulus] ait [angul]o abe aequalis, erit, ob triangulorum similitudinem, ut ia ad at, ita ba ad ae, et [rectangulum] iae [rectangul]o tab aequale: ergo [rectangulum] iae est minus [rectangul]o lab.
    Similiter ostendetur esse quoque minus [rectangulum] paf. Cum enim [angulus] aef, idest ail, sit maius angulo api, erit reliquus afe minor reliquo aip. Si igitur constituatur aiu [angulus] ipsi afe aequalis, erit [rectangulum] uaf [rectangul]o iae aequale, ex quo patet propositum.[44]

    The proof is relatively elementary, and only the fundamental idea is repeated here. Galileo introduces the points t and u such that angle aiu equals angle afe (angles α in Fig. 13.2) and angle abe equals angle ait (angles β in Fig. 13.2). It can then be shown based on a consideration invoking similar triangles that the areas of the rectangles $\Box uaf$ and $\Box tab$ are equal to the area of the rectangle $\Box iae$, i.e., $ab * at = ae * ai = af * au$. Consequently, since $al > at$ and $ap > au$, respectively, the areas of rectangles $\Box paf$ and $\Box lab$ must also be greater than the area of rectangle $\Box iae$, ie., $ab * al > ae * ai < af * ap$.

- T1C, second geometrical statement noted:

    Demon[s]trabitur etiam, quod rectangula talia quae a lineis ex a ad lineam cx ductis et a linea xm sectis, ea quae fiunt a lineis vicinioribus ipsi aei semper minora sunt illis quae a remotioribus describuntur lineis.[45]

---

[43] 130 verso, T1A.
[44] 130 verso, T1A and T1B.
[45] 130 verso, T1C.

13.5 Folio 130

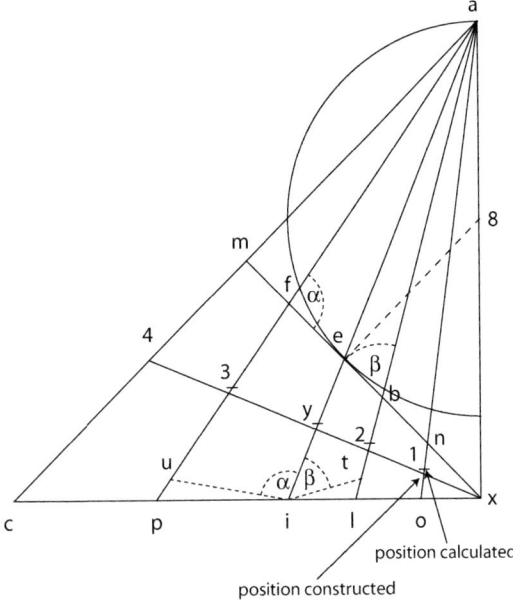

Fig. 13.2 Schematic representation of the central diagram on 130 verso. The diagram has been reduced to the elements necessary for an analysis of the entries T1A-E on the page

Galileo states that the area of the rectangles inspected before increases monotonically with increasing distance of the intersection point from point e. This statement remains without a proof but is evident from the construction employed in the proof above.

- T1D, corollary to first statement noted:

  Constat insuper, quod media inter *iae* est omnium mediarum minima, quae cadunt inter *paf*, *lab*, etc.[46]

  Galileo states that the mean proportional between ae and ai (ay in the diagram) is the smallest of all mean proportionals between lines falling from a to the horizontal $c\chi$ and their first halves cut off by the line $m\chi$. This is a mere corollary to the first statement as the length of the mean proportional between ae and ai is given by the square root of the product $ae*ai$, i.e., the area of the rectangle. Thus, $ab*al > ae*ai < af*ap$ and also, $\sqrt{ab*al} > \sqrt{ae*ai} < \sqrt{af*ap} \Leftrightarrow mp(ab|al) > mp(ae|ai) < mp(af|ap)$.

- T1E, alternative proof for first statement advanced:

  Aliter brevius. Posito angulo $ae8$ aequali [angul]o eam, erit linea $e8$ parallela am; ergo perpendicularis ad $m\chi$, eritque aequalis $8a$: quare, centro 8, intervalo $8e$, circulus tanget $m\chi$ in e: unde patet propositum.[47]

---

[46] 130 verso, T1D.
[47] 130 verso, T1E.

This entry, in which an alternative proof is presented, is characterized by a somewhat different hand and ink and may have been written somewhat later than the rest of the content on the page. A new geometrical element is incorporated into the argument, a circle around point 8 on ax. The radius is chosen so that the circle touches the line $m\chi$ in point e. The entries T1A-D overlap with the circle suggesting that it was constructed only after the text had been written.

With reference to this new construction, Galileo casually concludes that what had been demonstrated before holds. A geometrical argument which can potentially be constructed based on the new geometrical element of the circle to prove that the rectangle □iae has the smallest area does not essentially differ from the argument Galileo had already advanced, and indeed his claim that what had been stated before follows immediately was likely not based exclusively on a geometrical argument.

If Galileo's statement regarding which rectangle has the minimal area is interpreted as pertaining to motions along inclined planes, it becomes immediately clear why the circle around center 8 with radius $8e$ proves the statement. If the lines ax, ao, al, ai, ap, and ac are considered as inclined planes of equal height traversed in accelerated motion, it follows from the length time proportionality that the times to traverse these inclined planes from rest at a have the same ratio to each other as the respective lengths. If by distance time coordination the time to fall over the vertical distance $a\chi$ is assumed to be measured by the length of $a\chi$, it then follows that also the times of motion along the other planes are measured by their respective lengths, i.e., $t(ax) = ax$, $t(ac) = ac$, $t(ap) = ap$, etc. At the same time, the times to fall from rest at a to the intersection points of the inclined planes with the line $m\chi$ are given by an application of the law of fall. Hence, the mean proportionals between the inclined planes and their first parts, i.e., the parts between a and the intersection with xm, are direct measures for the time to fall from point a to the line xm on the various inclined planes, i.e., $t(am) = mp(am|ac)$, $t(af) = mp(af|ap)$, etc. Thus, the statement that the area of the rectangle iae is minimal translates directly into the statement that of all motions along a straight line from a to the line xm, fall along ae requires the least time. Galileo's statement about the monotonic increase of the areas of the rectangles translates into the enunciation that the time to fall along any straight line from a to xm increases monotonously the further away its intersection with xm lies from point e. Galileo indeed transformed this insight gained on 130 verso into an independent proposition about naturally accelerated motion on 127 verso. In the *Discorsi*, the proposition was published as proposition XXXII of the Second Book of the Third Day.[48]

- T2 times of motion along the different bent planes constructed and results noted in abbreviated form:

    [a]s [tempus] per abc
    [a]q [tempus] per aec

---

[48] A second elaboration of essentially the same proposition is contained on 168 recto.

13.5 Folio 130

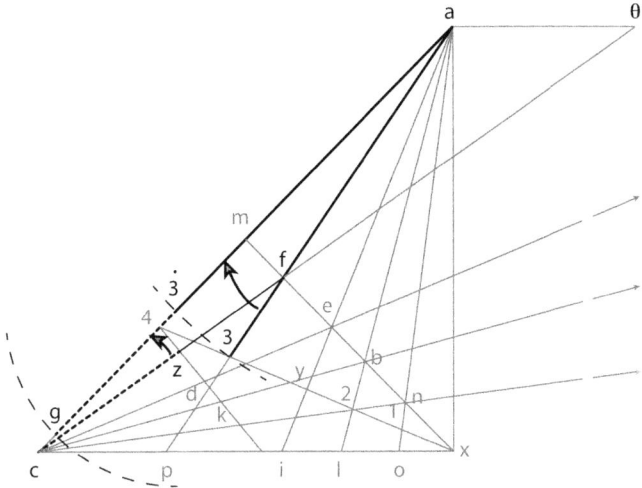

Fig. 13.3 Galileo's construction of the lines as, aq, ag, and aq' representing the times of motion along the various bent planes in the central diagram on 130 verso. The construction is exemplified here with regard to his construction of time of motion along bent panes afc. Bold and dotted bold lines represent the times of motion along the first and the second part of the bent plane, respectively. As indicated these are transferred to and added on ac to yield ag measuring the time of motion along bent plane afc. Dashed lines are present in the construction as uninked construction lines. These arcs were created when Galileo transferred the distances in question to ac by means of a compass

[a]g [tempus] per afc
[a]q' [tempus] per anc.[49]

Galileo noted that the lines $as, aq, ag,$ and aq' represent the times to fall along the bent planes $abc, aec, afc,$ and anc, respectively.[50] Based on an inspection of uninked construction lines (see Fig. 13.3) on the folio, the scheme by which Galileo constructed the points $s, q, g,$ and q' can be identified. It is here explained based on the example of Galileo's construction of the point g such that ag measures the time to fall along the bent plane afc from rest at a given that the time of fall along the inclined plane ac is measured by ac. Galileo's construction scheme is rendered in Fig. 13.3.

First a point 3 is constructed such that $a3$ is the mean proportional between af and ap. If by distance time coordination the time along ac is assumed to be measured by ac, then by length time proportionality, the time of fall through ap is measured by ap. It thus immediately follows by the law of fall that $a3$ measures

---

[49] 130 verso, T2.

[50] Galileo uses the letter q as a label twice. For disambiguation I refer to the second point labeled q as q'.

the time to fall along the inclined plane af from rest at a.[51] Next, as the uninked construction marks show, he transferred by means of a circle the distance $a3$ onto ac thus creating the point labeled $\dot{3}$ in Fig. 13.3. Furthermore, by the same token as above, θc measures the time to fall along θc.[52] With θz being constructed so as to be the mean proportional between θf and θc, it follows by law of fall that θz measures the time to fall along θf.[53] Then, by generalized law of fall, zc is a measure of the time to fall through the remainder fc from rest at θ. By path invariance, finally, this time is the same as the time to fall along fc from rest at a. In a last step, Galileo transferred the distance zc to ac where it was added to $a\dot{3}$ to produce the point g. This was again accomplished by means of a compass as the incisions of the compass at $\dot{3}$ and the uninked arc running through g show.

The distance ag adds the distances $a3$ and zc representing the times of motion over the first and the second part of the bent plane afc, respectively, and hence ag indeed represents the time of motion along afc as Galileo note.

- T1F, open question noted:

    Vide num ya ad $a3$ sit ut dc ad cz.[54]

Galileo does not indicate why he is interested in the question of whether $ya/a3 \sim dc/cz$. However, if this proportion were to hold, it would mean that for motions along the bent planes afc on the one hand and aec on the other hand, the times of motions along the first parts af and ae which are given by ya and $a3$, respectively, would be in the same ratio as the times of motion along the second parts fc and ec which are given by dc and cz, respectively. This in turn would imply that the total times of motion along the bent planes would likewise obey the same ratio.

Thus, what was proven for the times to fall along the first parts of motion from a to line xm would hold also for the times to fall from a to c along the respective bent planes. Thus, in particular, of all the bent planes between ac with the junction point on mx, it would be the bent plane aec which is traversed in least time. An attempt by Galileo to answer whether indeed $ya/a3 \sim dc/cz$ can be identified on folio 187 recto.

---

[51] That the points 4, y, 2, and 1 which mark mean proportionals are all positioned on the same line running from 4 to x initially seems not to have been clear to Galileo, as he first determined the position of the points individually by calculation and only later added the line $4x$.

[52] Galileo added the letter θ just once on the right margin of his diagram, indicating all intersections of the perceived extension of the lower parts of the bent planes cf, ce, cb, and cn to a horizontal running through a.

[53] The points marking the corresponding mean proportionals on the other inclined planes are lettered 4, d, and k. The last point defined by the intersection of $4k$ and cn remained unlettered.

[54] 130 verso, T1F.

13.5 Folio 130

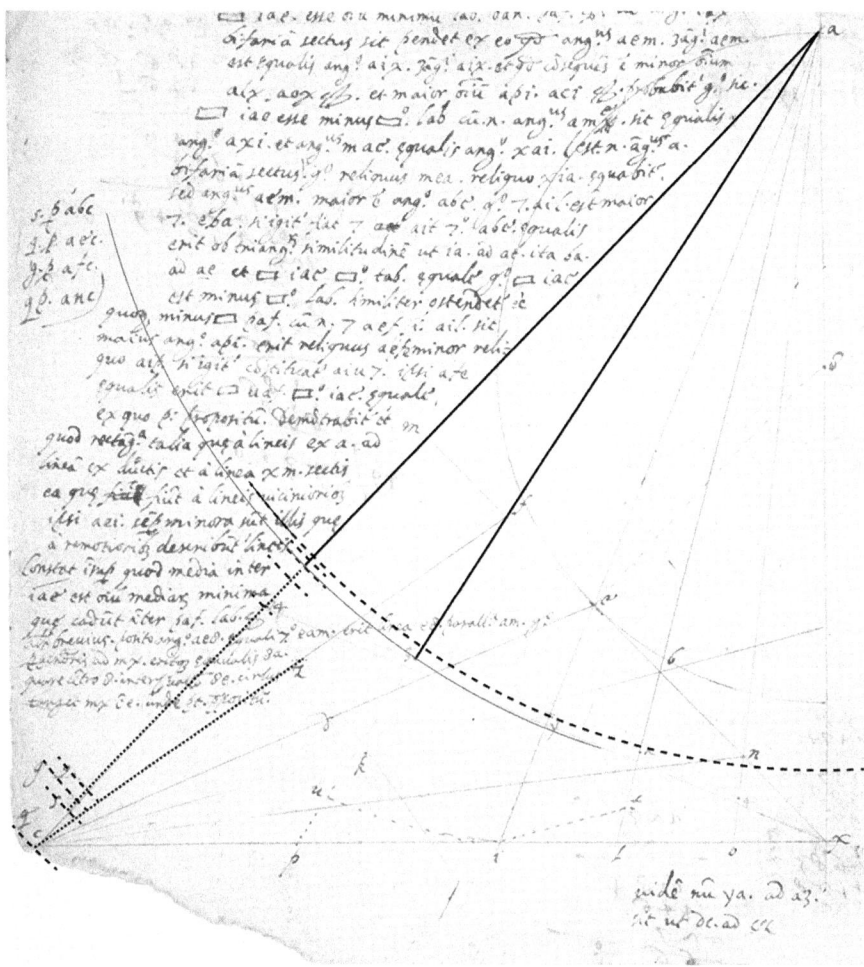

Fig. 13.4 Central drawing on 130 verso. Uninked construction lines that become visible under raking light are marked as bold dashed lines. The scheme of the construction of point *g* as exemplified in Fig. 13.3 has been added using the same graphical conventions for ease of comprehension. (Courtesy of Ministero dei Beni e le Attività Culturali – Biblioteca Nazionale Centrale di Firenze. Reproduction or duplication by any means prohibited)

130 recto

- C01 and T1, distance calculated as mean proportional and result noted:

$$\sqrt{186 * 146\ 1/2} = 165^{55}$$

media inter *ag af* 165.[56]

---

[55] 130 recto, C01.
[56] 130 recto, T1.

- Result expanded into distance time table:

	long.	temp.
ag	186	186
af	1461/2	165
fg	391/2	21.[57]

- C03, calculation of distance as mean proportional:

$$\sqrt{196 * 137} = 163.^{58}$$

- C04, calculation of distance as mean proportional:

$$\sqrt{223 * 128} = 169.^{59}$$

With the three calculations C01, C03, and C04, Galileo seems to have determined the positions of the points 2, y, and 3 in the diagram on the obverse side of the folio. Based on Galileo's unit of the punto, the diagram on 130 verso is drawn to scale with a side length ax of the square on which the construction is based of 180 punti.[60] Yet the labeling of the points cannot correspond to that of the central diagram on the verso side. Possibly the calculations were done for a similar diagram no longer extant on the basis of which the diagram on 130 verso was devised.

- C02 calculation of distance by rule of three:

$$gc\ 131 * [ag]\ 186 : gd\ 49 = 4[97].^{61}$$

13.6 Folio 148

Folio 148 bears the watermark crown and can hence be attributed to Galileo's Paduan period. The verso side of the folio was left blank. All content on the recto side was written with the page turned 90 degrees with respect to its modern

---

[57] This distance time table has not been transcribed in the electronic representation of Galileo's *Notes on Motion*.

[58] 130 recto, C03.

[59] 130 recto, C04.

[60] I have not measured the length of the line ax or any other line in Galileo's diagram from which this length can be inferred directly. Yet from the known width of the sheet of 210 millimeters, I infer a side length of 175 millimeters which divided by 180 gives a value of approximately 0.97 millimeters for Galileo's punto in good accordance with the value of the units that can be inferred from other scaled diagrams. Cf. the discussion of folio 166 in this appendix.

[61] 130 recto, C042.

13.6 Folio 148

orientation in the manuscript. The recto side contains considerations regarding motion along a broken chord, and correspondingly the main diagram on the page comprises the typical constellation of chord dc and broken chord dbc, spanning the same upright arc.[62] Furthermore, the page contains two geometrical constructions revealed to represent timelines, one vertical to the right of the central diagram and one horizontal underneath.[63] The horizontal timeline is geometrically constructed, i.e., it preserves the actual ratios of times of motion that hold for the motion situation at hand in the form of the ratios between the corresponding distances representing times in the timeline. The vertical timeline is not geometrically constructed but is associated with a list of numerical values for the times of motion represented. These values will be referred to as the timeline values. The short textual entry on the page relates the timeline to the main diagram and briefly describes how the distances on the timeline have to be constructed from the corresponding distances of the spacial geometry according to the laws of motion.[64] The motion diagram and timeline are typical for Galileo's considerations concerning the law of the broken chord, and the considerations documented on 148 recto are indeed set in this general context. In addition, the page contains a complete numerical example of the times of motion for the motion situation under consideration. Some calculations are contained on 148 recto. Yet, most of the calculations by which the results listed were produced are contained on folio 153 recto. These calculations will be discussed together here with the content of folio 148. As the content of 153 recto can furthermore be directly related to the content of folio 166 recto, an indirect link between folios 148 and 166 is established allowing the conclusion that Galileo's attempt to construct a proof for the law of the broken chord and his attempts to exhaust motion along the arc by motion along polygonal paths composed of adjacent inclined planes of increasing number, centrally documented on folio 166, were pursued at approximately the same time. The central motion diagram on folio 148 recto was adopted from a diagram on double folio 156, 157 with which it initially even shared the labeling of the points. The point labels were subsequently changed to accommodate the ones used in the diagram associated with the final draft of the proof of law of the broken chord on 163 recto.

148 recto and 153 recto

- D01A, central motion diagram drawn (see Fig. 13.5). Motion along chord dc will be compared to motion along broken chord dbc.

---

[62] 148 recto, D01A.
[63] 148 recto, D01D and D01E.
[64] 148 recto, T1.

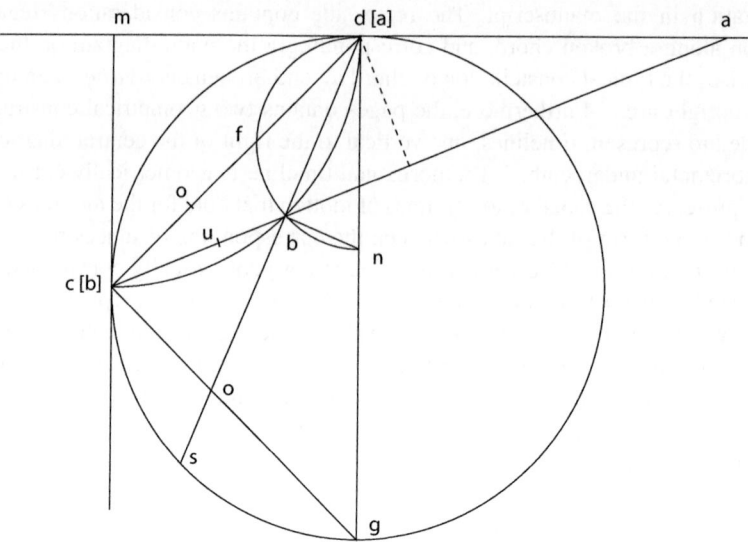

Fig. 13.5 Central motion diagram on page 148 recto. Letters in brackets indicate the original labels of the points that were changed in a revision of the diagram

- D01A and T1, vertical timeline drawn (see Fig. 13.6) and construction principles formulated.[65]

   Sit qp tempus per ac [arbitrary choice of unit] et ut ac ad cd ita fiat pq ad qr erit qr tempus per dc [by length time proportionality] seu per bc [(from rest in b) by law of chords]. Sit ut cd ad do ita rq ad qs erit qs tempus per df [by law of fall] et sr tempus per fc post df [by generalized law of fall]. Fiat rursus ut ca ad au ita tempus pq ad qt erit qt tempus per ab [by law of fall] tp vero tempus per bc post ab [by generalized law of fall].[66]

- C06, numerical values for the relevant distances listed (including mean proportionals which represent times of motion rather than distances traversed):

   sd 111 dg 120, cd 85, db 46 1/2, ca 155 1/3, ab 109, ua 130, bu 21, cu 25 1/2, fd 35 1/2, fc 49 1/2, do 55, fo 19 1/2, oc 30.[67]

---

[65] When Galileo originally drew the timeline, it started at a and ended at b. The point labels a and b were successively changed to q and p (compare Fig. 13.6). Note T1 had been formulated with respect to the original labeling. Thus also in this short text, every a was turned into a q and every b into a p, respectively. A corresponding change cannot be observed in the list of times to the left, which must thus have been produced after the labels were changed in the timeline and in the short entry.

[66] 148 recto, T1.

[67] 148 recto, C06.

13.6 Folio 148

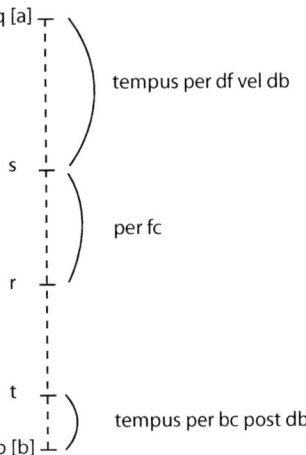

Fig. 13.6 Schematic representation of the vertical timeline on 148 recto. Letters in square brackets indicate the labels Galileo had used in the first version of the diagram and which were subsequently altered

By convention, the diameter dg is assumed to measure 120 units. All other lengths can then be determined, mostly by reading them off from a sine table or else by simple addition and subtraction. The only calculations explicitly carried out and contained on the folio are:

- C01, determination of ca by ratios in similar triangles:

$$85 * 85 : 461/2 = 155\ [1/3].[68]$$

- C02, determination of au as $mp(ca|ab)$:

$$\sqrt{155\ 1/3 * 109} = 130.[69]$$

- C04, determination of fd by ratios in similar triangles:

$$46\ 1/2 * 85 : 111 = 35.[70]$$

- C03, determination of do as $mp(df|dc)$ (C03):

$$\sqrt{85 * 35\ 1/2} = 54\ [9/10].[71]$$

- C07 and C08, all relevant times of motion listed:

---

[68] 148 recto, C01.
[69] 148 recto, C02.
[70] 148 recto, C04.
[71] 148 recto, C03.

tempus	~~qf~~	~~120~~
	qp	219 1/3
	qr	120
	qs	77 2/3
	sr	42 1/3
	qt	183 1/2
	tp	36
tempus per	dc	120
per	dbc	113 2/3.[72]

The time unit in which the times have been calculated is dictated by the choice to have the time to fall through the diameter dg be measured by the length of the distance fallen through, i.e., 120 units. By law of chords, it immediately follows that the time along dc (represented by qr in the timeline) likewise measures 120 units. The calculations necessary are found on 153 recto. They follow Galileo's description of how the times have to be constructed in T1.[73]

Auxiliary calculations on 153 recto

- C02, calculate qp given that $qr = 120$[74]:

$$155\ 1/3 * 120 : 85 = 219\ [1/3].$$[75]

Scheme of calculation: $ca/cd * dg = qp$.
- C03, calculate qs[76]:

$$55 * 120 : 85 = 77.$$[77]

Scheme of calculation: $do/cd * qr = qs$.

---

[72] 148 recto, C07 and C08.

[73] According to the description given in T1, it is the overall length of the timeline qp representing the time along ac which is initially given and with reference to which all other times are determined. Galileo, however, had conventionally decided to have the time along dc represented by qr to be measured by 120. He wrote this down in C07 first and only then determined qp noting the result next. He then crossed out the first entry and rewrote it underneath what had previously been the second line such that the order of the final list thus produced conformed to the order in which the distances had been provided in T1.

[74] Corresponds to: "...et ut ac ad cd ita fiat pq ad qr erit qr tempus per dc seu per bc."

[75] 153 recto, C02.

[76] Corresponds to: "Sit ut cd ad do ita rq ad qs erit qs tempus per df."

[77] 153 recto, C03.

- C04 calculate qt[78]:

$$130 * 219\ 1/3 : 155\ 1/3 = 183\ [1/2].$$ [79]

Scheme of calculation: $va/ca * qp = qt$.[80]

- D01E, construction of conformable timeline. The labels for this new timeline were chosen so as to be compatible with the timelines Galileo had drawn on 185 recto and 186 verso except for the point q which in the two similar timelines is labeled d. In constructing the timeline, the distance sq of the timeline was chosen to be equal in length to the distance dg of the main motion diagram.[81]

  The distance gt, which represented the time of motion over the second broken chord bc from rest at a, was transposed to r. The point t' thus constructed was noted by Galileo in the timeline by means of a dotted line. With this construction the distance qt' in the timeline directly represented the sought time of motion along the broken chord.[82]

- C05, time along dbc calculated and compared to time along dc:

  85
  $55 + 25\ 1/2 = 80\ 1/2$.[83]

  Galileo calculates the sum $do + uc$. If by distance time coordination dc is taken to be a measure of the time along dc, then do represents the time along db, and uc represents the time along the lower chord bc from rest at d. Hence the sum of do and uc represents the time of motion along dbc. The result, 80 1/2, is written next to the number 85, which measures the distance covered but also the time elapsed in motion along the chord dc. This means, in particular, that lines which are part of the spacial geometry of the problem in the first place

---

[78] Corresponds to: "Fiat rursus ut ca ad av ita tempus pq ad qt erit qt tempus per ab."

[79] 153 recto, C04.

[80] Galileo had initially miscalculated $3 * 155\ 1/3$ to 366 instead of 466 and thus had arrived at an overall result of 233. That this could not be right would have been immediately obvious as the overall time pq had amounted to the smaller 219 1/3. The calculation is not transcribed correctly in the electronic representation of Galileo's *Notes on Motion*.

[81] This conforms to the choice based on which also the numerical values for the times of motion listed in C07 and C08 were calculated, namely, to have the time along the chord dc be represented by the length of the vertical diameter dg.

[82] The reason why Galileo may have constructed this second timeline is suggested by the presence of some geometrical elements in the motion diagram on 148 recto which Galileo had adopted from his diagram and the related considerations on double page 156, 157. These comprise the arc $dcsg$, the chords ds and cg, the point o marking the intersection of the latter two chords, as well as the mean proportional between the lines do and ds, which is marked but unlettered. In Fig. 13.5 I have labeled this point r'. As argued in Chap. 5, Galileo's construction was most likely based on the assumption that the ratio of the time of fall along the broken chord to the time to fall along the chord would be given by the ratio of ds to dr'. The conformable timeline, in particular with its construction of the time of motion to fall along the broken chord, could have served to test this assumption and would, if this had been done, have demonstrated it to be unmaintainable.

[83] 148 recto, C05.

are here interpreted by Galileo as also representing time elapsed during motion. This shows that when Galileo drafted the content of folio 148, he had already familiarized himself with the consequences of his new method of integrated time representation (cf. Chap. 7, Sect. 7.1).

13.7 Double Folio 149, 150

Folios 149 and 150 form a double folio. Folio 149 bears the large star watermark.[84] One side of the double folio made up of 149 recto and 150 verso contains but a single diagram. No textual entry or calculations are associated. Hence little can be said except that the diagram belongs to the general context of Galileo's considerations pertaining to motions over chords and broken chords spanning the same arc. Moreover, the diagram apparently served as the blueprint for a diagram on folio 172 verso based on which Galileo elaborated what is here referred to as the tarditas length proportionality (see Chap. 10). On the obverse side of the double folio, made up of 149 verso and 150 recto, Galileo continued work begun on folio 186 verso. The lines of the central diagram have not been inked by Galileo but are scratched in and become visible under raking light. Small holes indicate that the diagram might have been transferred from another page to the double folio by placing the two leaves on top of each other and transferring the positions of the points by means of incisions of the compass. On the right half of the double page, i.e., on 150 recto, Galileo further advanced his construction of a proof of the law of the broken chord. He transformed his statement regarding the inequality of two times of motion into an equivalent statement regarding the inequality of two distances that were part of the spatial geometry of the motion situation. On the left half of the double folio, i.e., on 149 verso, Galileo formulated a geometrical argument which was to become part of his proof of the law of the broken chord.

149 verso and 150 recto

- 150 recto D02A, central motion diagram (see Fig. 13.7) constructed.

   The diagram repeats the construction of 186 verso. The lines of the diagram have been scratched into the paper but have not been inked. Notably the angle subtended by the arc dbc is neither the same as the corresponding angle on 186 verso nor is it a right angle. As compared to the diagram on 186 verso a point k is introduced such that $ck = mp(ac|cb)$. In Galileo's argument the line ck as the mean proportional between ac and ab allows via the law of fall to infer the time

---

[84]See the discussion of the star watermarks in Chap. 10.

13.7 Double Folio 149, 150

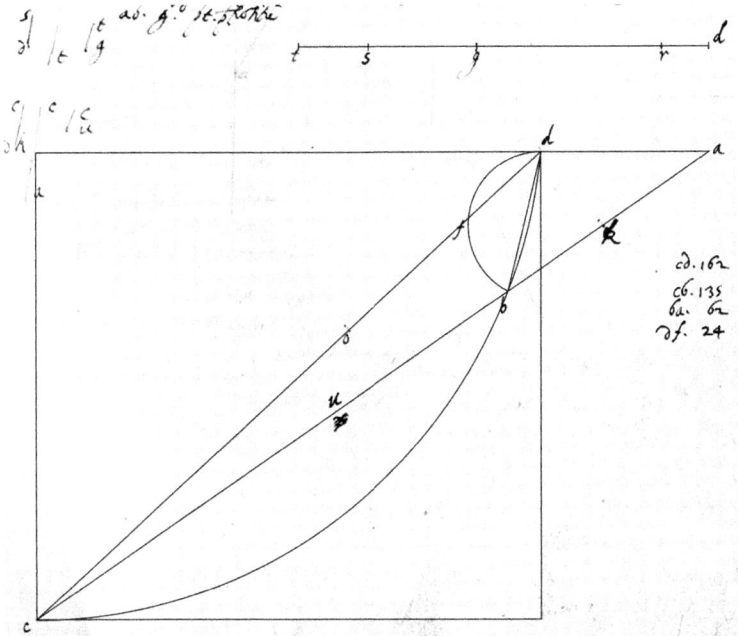

Fig. 13.7 Reconstruction against the background of an image of the folio of the central motion diagram from the marked points. (Courtesy of Ministero dei Beni e le Attività Culturali – Biblioteca Nazionale Centrale di Firenze. Reproduction or duplication by any means prohibited)

of motion along the long inclined plane ac from the given time of motion along the lower broken chord bc from rest in b.[85]

- 150 recto D02A (upper part), conformable timeline constructed (see Fig. 13.7). The timeline dt is constructed conformably, i.e., it preserves the ratios that hold between the times of motion for the motion situation depicted in the central diagram. The scale of the timeline is determined by the choice to have ds representing the time of motion along the chord dc measure one half of the distance dc, i.e., $ds = 1/2 \, dc = t(dc)$ (Galileo could not represent the time of motion along dc by a line of length dc because, had he done so, the completed timeline would not have fitted on the paper).
- 150 recto T2A, what has to be proven is stated:

  ostende co maiorem esse cv.[86]

  Galileo's considerations on 186 verso indeed had left open to show that $co > cv$.
- 150 recto T3, shorthand notation of the construction principle of the timeline:

---

[85] On 186 verso Galileo had first relocated the lower chord cb to start at a and only then constructed a mean proportional ax between ac and the relocated bc. Galileo apparently recognized that this was not necessary and that the law of fall can be applied directly. In his argument on 150 recto ck thus assumes the role that line ax had on 186 verso.

[86] 150 recto T2A.

> sit ut od, dc, ca, av
> ita rd, ds, td, dg.[87]

In this entry Galileo represented in abbreviated textual form what he would, in other cases, have represented with the help of a distance-time ratio diagram. Every ratio that can be formed from two magnitudes in the first row is in proportion to the corresponding ratio formed from the corresponding magnitudes in the second row. It thus holds, for instance, that $dc/av \sim ds/dg$.

- 150 recto T1A-T1C comprehensive description of the construction principle of the timeline:

  > Sit do media inter cd, df; av media inter ca, ab; ck media inter ac, cb; et accipiatur utcumque ds.[88]
  > fiat ut od ad df, ita sd ad dr; ut cd ad do, ita sd ad dr, seu ut bc ad cd, seu ut dc ad ca.[89]
  > Fiat ut bc ad ck, ita sd ad dt; ut autem va ad ab, ita td ad dg.[90]

  This can be paraphrased as follows: Galileo defines $do = mp(cd|df)$, $av = mp(ca|ab)$ and $ck = mp(ac|cb)$. Next, the distances in the timeline are determined in such a manner that their ratios correspond in the manner demanded by the laws of motion to the appropriate ratios of lines of the spatial geometry of the problem: $od/df \sim sd/dr$, $cd/do \sim sd/dr \sim bc/cd/ \sim dc/ca$, $bc/ck \sim sd/dt$, and $va/ab \sim td/dg$. With these definitions it holds that:
  $ds = t(dc)$
  $dr = t(df)$
  $rs = t(fc|d)$
  $dt = t(ac)$
  $dg = t(ab)$
  $gt = t(bc|a) = t(bc|d)$

- 150 recto T1D, reduced statement expressed in terms of distances on the timeline:

  > Probetur gt minorem esse quam sr.[91]

- 150 recto T4 (first part), prove that $rs/gt \sim oc/cv$. Hence $sr > gt \Leftrightarrow co > cv$:

  > Quia enim ut sd ad dr, ita cd ad do, per conversionem rationis et convertendo, ut rs ad sd, ita oc ad cd; ut autem sd ad dt, ita cd hoc est kc, ad ca: et quia est ut td ad dg, ita ca ad av, per conversionem rationis erit quoque ut dt ad tg, ita ac ad cv: ergo, ex aequali, ut rs ad gt, ita oc ad cv.[92]

---

[87] 150 recto T3.
[88] 150 recto T1A.
[89] 150 recto T1B.
[90] 150 recto T1C.
[91] 150 recto T1D.
[92] 150 recto T4.

13.7 Double Folio 149, 150

- 150 recto T4 (second part), attribute geometrical part of the argument to lemmata:

  Ostenditur autem, per lemmata, *co* maior quam *cv*[93];

- 150 recto T4 (third part), argue that *co* > *cv* implies the law of the broken chord:

  ergo tempus *rs* maius est tempore *gt*: est autem *rs* tempus quo peragitur *fc* post *df*, *gt* vero tempus quo peragitur *bc* post *ab*: ergo patet propositum.[94]

  co > *cv* ⇒ *sr* > *gt* and thus the time elapsed during motion over the broken chord is shorter than the time elapsed during motion over the single chord.

- 150 recto C04 incomplete numerical example added:

  cd 162
  cb 135
  ba   62
  df   24.[95]

  The underlying calculations cannot be identified in the *Notes on Motion*.

Folios 149 recto and 150 verso The double page contains one large diagram which is accompanied by neither a textual entry nor by any calculations. Indubitably the diagram belongs to the context of Galileo's broken chord approach as it bears a chord *dc* and a broken chord *dbc*, both spanning the same upright arc. Two aspects of Galileo's construction are particularly noteworthy. First, besides the usual circle in which the chord *dc* and the broken chord *dbc*, respectively, are inscribed, Galileo drew a second larger circle in which the inclined plane *ac* was inscribed, i.e., the extension of the lower of the broken chords to the horizontal through point *d*. With this construction it immediately follows that the ratio of the times of motion over the chords *dc* and *ac* is the same as that of the times of motion over the corresponding vertical diameters. Thus by applying the law of chords also the time of motion along the inclined planes *dc* and *ac* can be related, and Galileo had exploited exactly this type of inference in the context of his pendulum plane experiment (cf. Chap. 4). As the times of motion along *dc* and *ac* could directly be related by means of the length time proportionality, the presence of the particular construction on the double page, moreover, lends additional support to the hypothesis that the length time proportionality had not been established yet by Galileo when he first made his attempt to construct a proof of the law of the broken chord (cf. Chap. 7). The second noteworthy aspect of the construction is the inclined plane parallel to *ac* and running through the point *d* which Galileo represented by a dotted line. This particular geometrical constellation reoccurs in Galileo's elaboration of the tarditas

---

[93] Ibid.
[94] Ibid.
[95] 150 recto C04.

Fig. 13.8 Large diagram extending across the double page 149 recto and 150 verso

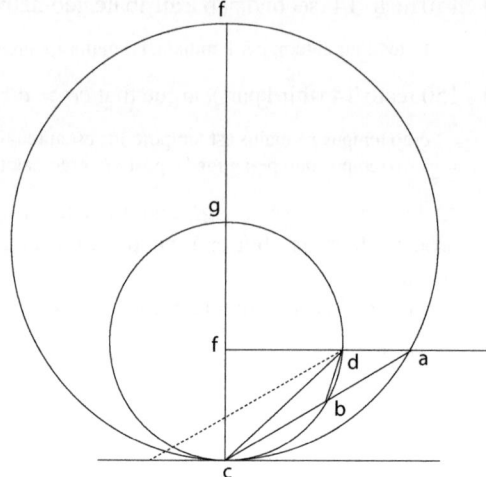

length proportionality on folio 172 verso where it was to assume a crucial role for the argument. Apparently the diagram on the double page served as a blueprint for the diagram based on which Galileo unfolded his consequential considerations documented by the tarditas length proportionality (cf. Chap. 10) (Fig. 13.8).

13.8 Folio 151

Folio 151, according to Drake's assessment, bears the watermark which is referred to here as large star, whereas a working group from the Max Planck Institute for the History of Science who inspected the manuscript in 1996 discerned the small star watermark on the folio. On its recto side, the folio bears the sketch of the first proof idea for the ex mechanicis proof of the law of chords (cf. Chap. 10). Two small tables contained on the page record the results of an application of the double distance rule to a geometrical configuration depicted in the central diagram on folio 131 recto. The obverse side of the folio contains two diagrams which overlap, one related to Galileo's broken chord approach and the other to his consideration regarding isotemporal surfaces, as well as a small doodle of what appears to be two cogged wheels. The purpose of the two scattered calculations remains unclear.[96]

---

[96] For Drake the considerations on the obverse side played a central role. His interpretation is imaginative and so opaque that it cannot possibly be wrapped up in a footnote. See Drake (1987, 43 et sqq.) and Drake (1990, 14–16).

13.8 Folio 151

Folio 151 recto

- C01, sesquialtrum variant of the double distance rule (cf. Chap. 10) applied to geometrical configuration depicted in central diagram on folio 131 recto. Result noted in table:

		temp[us]		temp[us]		temp[us]
p[er]		*dc*		*dg*		*dgc*
ut lin[ea]	*dc*	ad	*dg*	ad	sesqualit[er] *dg*
		temp[us]		temp[us]		[tempus]
p[er]		*bc*		*bl*		p[er] *blc*
–		*bc*	[ad]	*bl*	ad	sesq[ualiter]*bl*[97]

From the distribution of the content over the page, it appears that these two small tables, written with the page upright with respect to its modern orientation in the manuscript, were written on the page first. The content of the tables is unrelated to the central diagram on 151 recto. However, the sparse information conveyed suffices to reconstruct the configuration Galileo is considering which can then be identified as embraced by the central motion diagram on on folio 131 recto.

The scheme of the proportions which find their expression in the two small tables is:

$$t(xz)/t(xy) \sim xz/xy$$
$$t(xy)/t(xyz) \sim xy/1\tfrac{1}{2} * xy$$

The first proportion is compliant with an application of the length time proportionality and the second with an application of the sesquialterum variant of the double distance rule.[98] It is thus implied that xz and xy are inclined planes of equal height and that yz is a horizontal line equal in length to the inclined plane xy, as rendered in Fig. 13.9.

As it turns out, the geometrical configuration thus reconstructed from the content of the tables is part of diagram D01A on folio 131 recto, and Galileo thus was apparently looking at and referring to the diagram on folio 131 recto when he wrote his tables on 151 recto.

- D01B, central motion diagram drawn (see Fig. 8.5). Galileo turns the page 90 degrees clockwise and adopts the diagram from folio 131 recto which he reduces to the elements necessary for the argument he is going to sketch.
- T1A and T1B, central idea of the ex mechanicis proof of the law of chords sketched:

---

[97] 151 recto, C01.

[98] The sesqualterum variant of the double distance rule states that if the accelerated motion of fall over an inclined plane, here plane xy, is diverted into a uniform horizontal motion over a distance yz equal in length to the inclined plane, the overall time for the motion will be one and a half (sesqualiter) times that of the motion of fall on the inclined plane alone. See the discussion in Chap. 10. See also Damerow et al. (2004, 175–179).

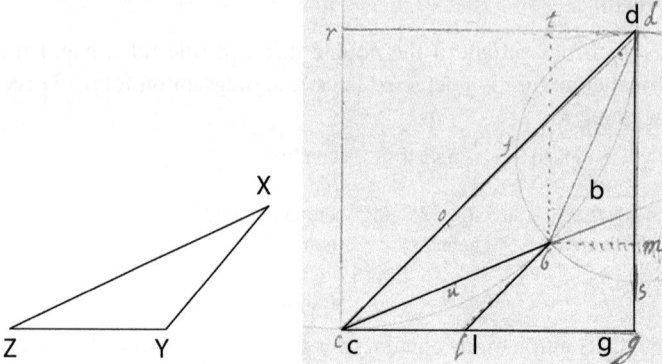

Fig. 13.9 Reconstruction of the geometrical configuration underlying the considerations documented by the two tables on folio 151 recto. Left: universal schema. Right: universal scheme instantiated with the information conveyed by the tables. As a background layer, part of the central diagram on 131 recto is shown which comprises exactly this configuration. (Courtesy of Ministero dei Beni e le Attività Culturali – Biblioteca Nazionale Centrale di Firenze. Reproduction or duplication by any means prohibited)

Sit gd erecta ad orizontem, df vero inclinata: dico, eodem tempore fieri motum ex g in d et ex f in d.[99] Momentum enim super fd est idem ac super contingente in e, quae ipsi fd esset parallela; ergo momentum super fd ad totale momentum erit ut ca ad ab, idest ae. Verum ut ca ad ae, ita id ad da, et dupla fd ad duplam dg; ergo momentum super fd ad totale momentum, scilicet per gd, est ut fd ad gd: ergo eodem tempore fiet motus per fd et gd.[100]

For a detailed discussion of this entry, the reader is referred to Chap. 8.

- D01A, diagram drawn (see Fig. 13.10).

With the accuracy to be expected of a rough sketch, Galileo drew two inclined planes starting at the vertex of and inscribed in the same circle. Thus, according to the law of chords, these two planes are traversed in equal time. Galileo then translated the shorter chord vertically downward such that the endpoints of the two inclined planes came to be positioned at the same height and deleted the chord in its original position.

Folio 151 verso

- D01A, two diagrams drawn, which overlap.

---

[99] 151 recto, T1A.
[100] 151 recto, T1B.

13.8 Folio 151

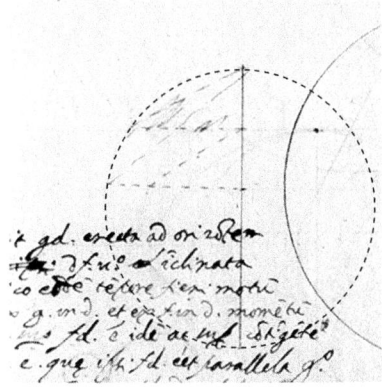

Fig. 13.10 Diagram D01A of folio 151 recto. The circle has been added to indicate that the inclined planes in the diagram are within reasonable accuracy chords in one and the same circle. (Courtesy of Ministero dei Beni e le Attività Culturali – Biblioteca Nazionale Centrale di Firenze. Reproduction or duplication by any means prohibited)

The two unlettered diagrams which overlap in Galileo's original construction have been rendered separate in Fig. 13.11. The diagram rendered in the left half of the figure quite clearly pertains to Galileo's considerations regarding the shape of iso-temporal surfaces for different types of motion (cf. Chap. 8). Galileo included a comparable diagram in the *Discorsi* as part of the scholium succeeding proposition VI, theorem VI of the Second Book of the Third Day in which iso-temporal surfaces of uniform and naturally accelerated motion are being discussed (cf. EN VIII, 224).

The diagram rendered to the right in Fig. 13.11 unambiguously pertains to the general context of Galileo's broken chord approach. Similarity to the diagram Galileo published with the third geometrical lemma associated with Proposition XXXVI of the Second Book of the Third Day of the *Discorsi* (cf. EN VIII, 261) as well as comparison with an almost identical diagram contained on 185 verso strongly suggest that the concrete context in which Galileo drew this diagram was indeed his elaboration of this very geometrical lemma.

What Galileo proved by means of the lemma was that, with regard to the labels of the points provided in Fig. 13.11, it holds for the particular construction that the distance BI is always greater than the distance BO. When providing the argument in the *Discorsi*, Galileo distinguished two cases, the case in which the arc subtends an angle of 90 degrees and the case in which the subtended angle was smaller. Accordingly, in the *Discorsi*, Galileo provided two diagrams to illustrate the two different cases. As it turns out, the diagram on 151 verso corresponds exactly to the first and the one on 185 verso to the second case, and it is thus strongly suggested that both diagrams indeed belong together and served the same purpose, namely, the elaboration of the geometrical lemma in question.[101]

---

[101] In the diagrams on 151 verso and on 185 verso, Galileo drew three as opposed to just one circle that he had drawn in the corresponding diagram in the *Discorsi*. This indeed can be seen

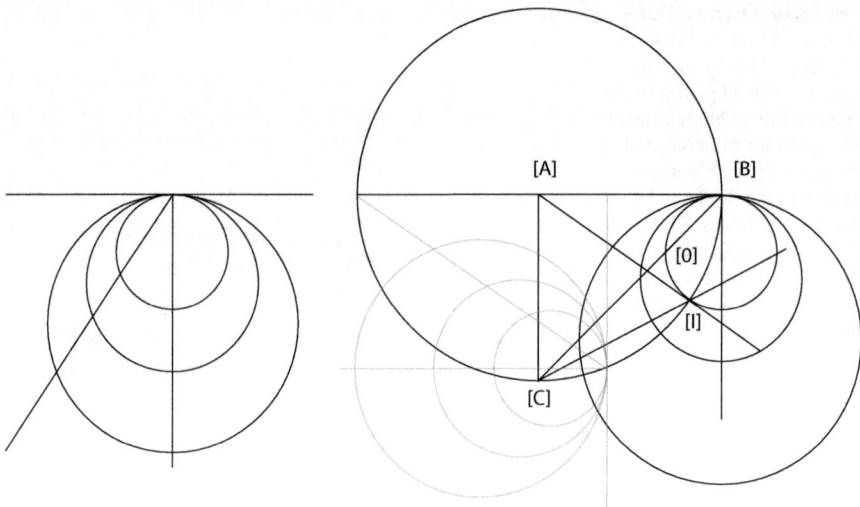

Fig. 13.11 Schematic representation of the two overlapping diagrams on folio 151 verso. The two diagrams have been separated, turned upright, and placed next to each other. The position of the diagram placed on the left in Galileo's original diagram is indicated in light gray in the diagram placed on the right. The letters are not part of Galileo's diagram but have been added for ease of reference and comparison

13.9 Folio 153

Folio 153 bears no watermark; the paper is thin and white. Just as on the preceding folio, folio 152, a corner of the paper is torn off. The shape of the fissure lines are very similar in both cases, and it can be suspected that these two folios once formed a folded double folio that was folded together or else that the two folios were at least placed on top of each other when the fissure occurred.[102]

153 recto The verso side of the folio contains two scratch drawings pertaining to a broken chord motion situation.[103] In the diagrams Galileo varied the angle of the arc spanned as well as the position of the junction point of the two conjugate

---

as corroborating the assumption that Galileo is elaborating lemma three, as the different circles exemplify the universal validity of the lemma for any position of point I on the arc AC. It may in fact have been the construction of internally touching circles thus engendered that have triggered Galileo's consideration regarding the iso-temporal surface of naturally accelerated motion and thus his sketch of the second, overlapping diagram.

[102] Folio 152 documents Galileo's insight that the Sarpi letter principle was erroneous. See Damerow et al. (2004, 184–188).

[103] 153 recto, D01A.

chords. The third diagram on the page resembles the one which Galileo ultimately published together with Proposition XXXII of the Second Book of the Third Day of the *Discorsi* and thus in all likelihood indeed pertains to the elaboration of this least time proposition.[104] In a single calculation contained on the page, Galileo divided the number 158 by 4 resulting in 39.[105]

Folio page 153 verso

- C02–C04, auxiliary calculations for 148 recto. These calculations are discussed together with the content of folio 148 recto.
- C05, rescaling, check of internal consistency:

$$120 * 113\ 2/3 : 141422 = 1339[58].^{106}$$

Results of the broken chord calculations on 148 recto compared to the comparable results listed on 183 recto pertaining to the base 100,000 approach of 166 recto. Galileo calculates the expected time for motion along the broken chord in units of the base unit 100,000 approach exploiting that for the same geometrical configuration, the ratio of the times of motion along chord and along broken chord does not depend on the choice of unit. Thus: $t(broken\ chord)_{148r}/t(chord)_{148r} * t(chord)_{183r} = t(broken\ chord)$. This result has compared to the time of motion along the broken chord 132,593, calculated based on the base 100,000 approach and listed on 183 recto.
- C06, rescaling, check of internal consistency:

$$113\ 2/3 * 254\ 3/5 : 120 = 241.^{107}$$

Results of the broken chord calculations on 148 recto compared to the results pertaining to the base unit 180 approach of 166 recto. Galileo calculates the expected time for motion along the broken chord in the base unit 180 approach by: $t(broken\ chord)_{148r}/t(chord)_{148r} * t(chord)_{166r} = t(chord)$ This result has compared to the time of motion along the broken chord 239 calculated based on the base unit 180 approach. Galileo indeed wrote the corresponding value from 166 recto, 239 next to his calculated result of 241.
- C08, rescaling, check of internal consistency.

$$113\ 2/3 * 85 : 120 = 80\ 1/3.^{108}$$

---

[104] 153 recto, D02A.

[105] 158 is the length of line *scx* on folio 174 recto, yet no particular reason is obvious from the considerations on 174 recto why Galileo should have divided this length by 4.

[106] 153 recto, C05.

[107] 153 recto, C06.

[108] 153 recto, C08.

Results of the broken chord calculations on 148 recto pertaining to the timeline (motion along chord measures 120 units) compared to the results of the calculations pertaining to the spacial geometry on 148 recto (motion along chord measures 85 units): $t(broken\ chord)_{unit\ 120}/t(chord)_{unit\ 120} * t(chord)_{unit\ 85} = t(broken\ chord)$. This result had to be compared to the time of motion along the broken chord of 80 1/2 calculated on 148 recto which Galileo indeed noted next to this calculated result of 80 1/3.

- C07, rescaling, check of internal consistency, erroneous:

$$113\ 2/3 * 30 : 120 = 28.^{109}$$

Galileo tests how the ratio of the times of motion along chord and broken chord relates to the ratio co/cu (see discussion of 149 verso). Calculate cu hypothetically by: $(broken\ chord)_{timeline}/t(chord)_{timeline} * co = cu$. This was to be compared with the actual value of cu of 25 1/2 which Galileo indeed wrote next to the calculation and which should have resulted if the ratios were in fact equal.

13.10 Folio 154

Folio 154 is part of a small group of folios bearing the large star watermark. Besides 154, the group contains the two folios 160 and 149. 160 recto bears the first draft of the ex mechanicis proof. A diagram on 154 recto most likely served to explore the central idea of this proof. The verso side contains merely two calculations that cannot be related unambiguously to any other entry in the *Notes on Motion*.[110] The recto side of the folio bears the record of an observation and theoretical considerations regarding what I refer to as the problem of the cosmological pendulum.

Folio 154 recto

- D01C, diagram drawn (see Fig. 13.12).
  Judging from the distribution of the remaining content, this diagram was drawn on the page first. Characteristically, Galileo drew a chord in the circle as well as a tangent to the circle, parallel to the chord. This geometrical construction is rather unique and was used on folio 131 to determine the mechanical moment

---

[109] 153 recto, C07.

[110] The summing up of numbers in the second calculation C02 is reminiscent of Galileo's summation of times in the case of the pendulum plane experiment.

13.10 Folio 154

Fig. 13.12 Schematic representation of a diagram (D01C) on 154 recto

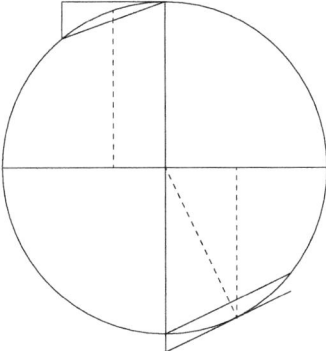

on the tangent to the circle and thus also on the parallel chord. It is strongly suggested that the diagram was sketched in the context of Galileo's elaboration of the ex mechanicis proof. If this assessment is correct, folio 154 recto bears witness to Galileo's familiarity with the law of the pendulum in 1602, the year in which, according to the interpretation given here, the ex mechanicis proof was established.

The cosmological pendulum Galileo assumes that a pendulum whose pendulum length equals the diameter of the earth swings in 6 hours and from this by applying the law of the pendulum calculates the time it would take a pendulum of 16 braccia length to swing.

- Diagrams of the motions of pendulums of different length drawn (D01A and D01B), likely a mnemonic device.
- C01, circumference of the earth calculated in "miglia" from the length of the arc subtending one degree:

$$360 \text{ [degrees]} * 60 \text{ [miglia]} = 21600 \text{ [miglia]}.[111]$$

- C01, the diameter of the earth calculated in "miglia" from circumference:

$$21600 \text{ [miglia]}/22/7 = 6873 \text{ [miglia]}.[112]$$

- C02, the radius of the earth calculated in "braccia" from circumference in "miglia":

$$3436 \text{[miglia radius]} * 3000 \text{[braci per miglia]} = 10308000 \text{ braci semidi[ametro]}$$

$$\text{filo br[aci] } 16.[113]$$

---

[111] 154 recto, C01.
[112] Ibid.

- C04, calculate the root of the circumference:

$$\sqrt{10308000} = 3200.^{114}$$

- C05, the frequency of 16 braccia pendulum calculated by law of the pendulum:

  r[adi]x r[adi]x h[ora]
   3200   4   6

$$3200/4 = 800$$

$$800/6 = 133.^{115}$$

Galileo assumes that a pendulum of the length of the earth's radius has a frequency of one swing in 6 hours. From this he calculated by the law of the pendulum the frequency of a 16 braccia pendulum. Scheme of calculation: $\sqrt{10,308,000}/\sqrt{16}$ = freq. small pendulum/freq. earth pendulum. While the earth pendulum swings once, the small pendulum will swing 800 times. From this Galileo calculates the expected swings per hour for the small pendulum. Scheme of calculation: 800 swings/6 hours = 133 swings per hour.

The earth pendulum is supposed to swing in 6 hours, and hence a frequency of 133 swings per hour results for the 16 braccia pendulum. This can be compared to observed values.

- C03, consistency check (?) of above:

$$800 * 800 = 10240000.^{116}$$

Scheme of calculation:
$freq.^2/freq.^2 \sim length/length.$
Thus:
800 [*swings per 6 h.*]/1 [*swings per 6 h.*] * 16 [*braccia*] = 10,240,000 [*braccia*].

Probably as a consistency check, Galileo calculates the length of the earth pendulum from calculated frequency and the length of the 16 braccia pendulum.
- C08, consistency check (?) erroneous:

$$133 * 133 * 16 = 282624.^{117}$$

---

[113] 154 recto, C02.
[114] 154 recto, C04.
[115] 154 recto, C05.
[116] 154 recto, C03.
[117] 154 recto, C08.

This was probably calculated before C03 and was supposed to follow the scheme:

$freq.^2/freq.^2 = length/length$

Yet, whereas 133 is indeed the frequency of the 16 braccia pendulum in swings per hour, the frequency of the earth pendulum in this unit is not one as implicitly assumed by Galileo in the calculation but 1/6. Thus the square is 1/36, and to yield the length of the earth pendulum, Galileo would thus have had to multiply his result by 36. He did not do so but instead in C03 recalculated based on the frequencies specified in swings per 6 hours.

- C06 and C07, calculation and numbers of unclear purpose:

$$45 * 8 = 360, \ 360 : 6 = 60^{118}$$

$$50, \ 32, \ 4, \ 129.^{119}$$

In the *Dialogue*, in the context of the presentation of his tidal theory, Galileo describes the floating of the water in a sea basin in analogy to a pendulum. Galileo left open whether he really considered the pendulum as the underlying physical explanation of the tides or merely as a nice analogon. The notes on this folio show how seriously he actually took the pendulum model. He assumed that a pendulum whose length corresponds to the radius of the earth would take 6 hours to swing from one end to the other and calculated from this assumption the time a pendulum of 16 braccia length would take according to the law of the pendulum. He arrived at the conclusion that such a pendulum would perform 133 such half swings in the time of 1 hour, that is, roughly one full oscillation in a minute for a pendulum of ca. 10 meters in length. Such a pendulum, although rather long, certainly had been realized in Galileo's day, for instance, by lamps hanging from the ceiling of a cathedral, and the period of such a pendulum could hence be observed by Galileo. He therefore might have realized that the period of pendulum of this length is smaller by about an order of magnitude.

13.11 Folio 155

Drake described the watermark of folio 155 as "mountains 17 × 20 mm." No watermark was identified on the page by a working group from the Max Planck Institute for the History of Science that inspected the manuscript. Folio page 155 recto bears but two diagrams and some scattered numbers. The verso side of the folio remained blank. The purpose of the diagrams was to construct the iso-temporal surface for motions on inclined planes that do not start from rest but have been proceeded by another accelerated motion.

---

[118] 154 recto, C06.
[119] 154 recto, C07.

Fig. 13.13 Schematic representation of the freehand sketch on the left-hand side of folio 155 recto

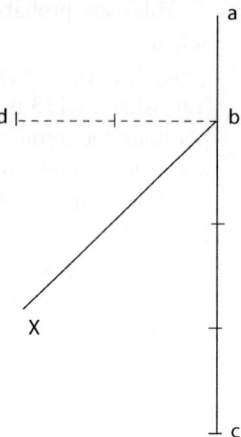

Folio 155 recto

- D01A (left half), freehand sketch of the basic geometrical configuration of the problem (see Fig. 13.13).
  Problem: Given a motion of fall from a to b continued by motion along inclined plane bX. How to determine, depending on the inclination, the distance on bX traversed in the same time as initial fall through ab? The freehand sketch schematically represents the motion situation, including the known solutions to the problem for the bounding cases of continuation of motion along a horizontal db and along a vertical dc (cf. Fig. 13.13). If distance ab is traversed in accelerated motion, then, according to the double distance rule, the distance covered in the same time after the motion has been diverted into the horizontal is twice as long, and so is bd in Galileo's construction. Furthermore, according to the law of fall, the distance bc covered after motion along ab in the same time as motion along ab must be three times the distance ab, and so is bc in Galileo's diagram. What is sought is the position of the point X on any plane intermediate between the horizontal and the vertical that is reached from point b after motion through ab in the time t of fall along ab. The general solution to this problem is provided by a proposition which Galileo issued as Proposition XVII of the Third Book of the Third Day of the *Discorsi*. A first draft of this proposition is contained on folio page 143 recto dating most certainly to Galileo's Paduan period. When working on 155 recto, Galileo had already come up with the solution as can be inferred from the construction on the right.
- D01A (right half), construction of points reached in the same motion along inclined planes of different inclination. (see Fig. 13.14).
  Galileo reproduced the configuration of his freehand sketch, this time using compass and ruler. In-between the horizontal and the vertical, four inclined

13.11 Folio 155

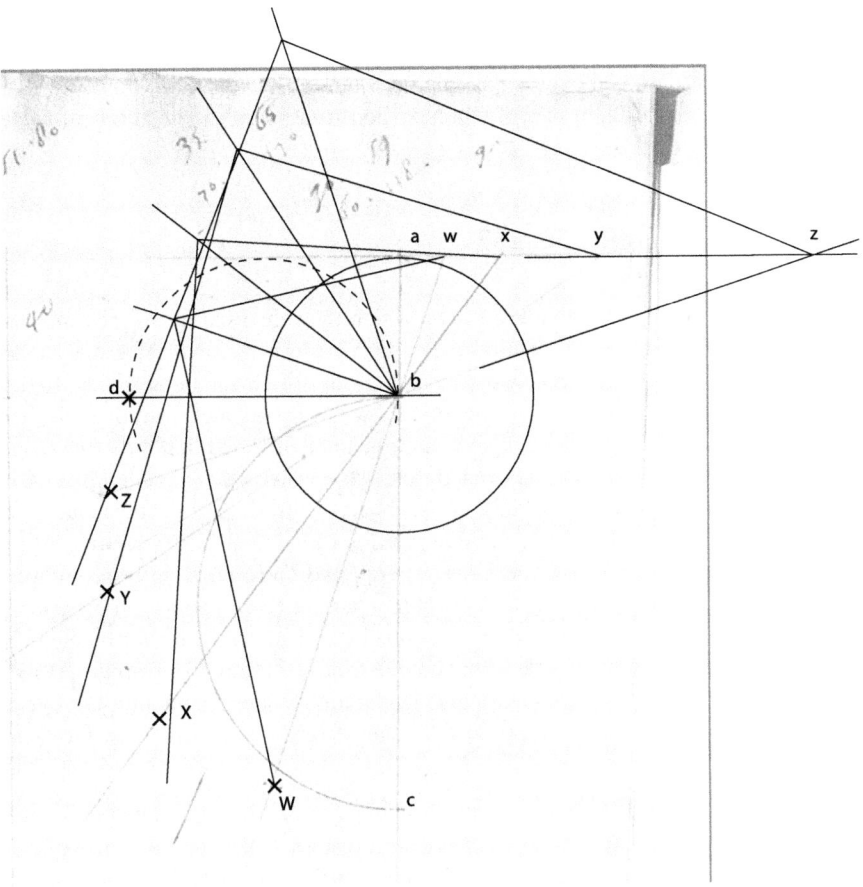

Fig. 13.14 The second construction on folio 155 recto. After fall through ab, the points d, Z, Y, X, W and c are all reached in the same time in motion from b. This time is equal to the time elapsed in fall through ab. The points are not lettered in Galileo's original diagram. The construction for the points Z, Y, X, and W are mine. No construction marks have been identified on the page but are expected to be present. (Courtesy of Ministero dei Beni e le Attività Culturali – Biblioteca Nazionale Centrale di Firenze. Reproduction or duplication by any means prohibited)

planes are drawn on each of which a point is marked. These points were not labeled by Galileo, and in Fig. 13.14, I have accentuated their positions using crosses. In the figure, I have, moreover, provided, the constructions for the positions of points W, X, Y, and Z reached on the different inclined planes after a time of motion that corresponds to the time of the initial fall through the vertical.[120] The positions of W, X, Y, and Z are seen to correspond with

---

[120] To solve the problem, a point a' is constructed on the inclined plane such that $ba' = ab$. Then one needs to find a point X such that xa' is the mean proportional between xb and xX. The way I

reasonable accuracy to the points marked by Galileo. When inspecting the manuscript, I missed the opportunity to check the page for unkinked construction marks. Based on the above reconstruction, I am rather confident, however, that the page must bear the traces of Galileo's construction of the positions of these iso-temporal points.

13.12 Double Folio 156, 157

Folios 156 and 157 form a double folio on which two watermarks are present. The first is a slender crossbow of 39 mm height; the second is a crown. This makes the double folio part of the crown/crossbow paper group which can be attributed to the Paduan period. The slender form of the crossbow is very distinct from that of other crossbow marks in the manuscript, such that it can be affirmed that only one other folio in the *Notes on Motion* bears exactly the same watermark, namely, folio 107. On this latter folio, Galileo attacked the problem of the shape of the catenary.[121] On one side of the double folio, comprising folios 156 recto and 157 verso, Galileo investigated the relation of motion along a chord and corresponding broken chord spanning the same upright arc and elaborated a variant of a hypothesis concerning the ratio of the times of motion along these paths he had originally formulated on folio 166 recto. The opposite side of the double folio, i.e., folios 156 verso and 157 recto, contains a number of geometrical considerations whose exact purpose could not be reconstructed. Galileo reproduced some of these diagrams in an identical form on folio 127 recto.

Folios 156 recto and 157 verso On the double page, Galileo elaborated upon two distinct problems. This first concerns the determination of the ratio of times of fall along a chord and a broken chord and is hence clearly attributable to the context of Galileo's broken chord approach. The elaboration of the second problem may have served as an auxiliary consideration providing a kind of mnemonic device for the purpose of the investigation of motion along the broken chord. The central diagram depicting the spatial geometry of the motion situation under investigation stretches over both folios (see Fig. 13.15). In a short entry, Galileo outlined the problem he was intending to solve. All calculations on the double page as well as the distance time table listing the results pertain to his attempt at a solution. The second, smaller motion diagram and the distance-time ratio diagram underneath

---

have realized this in my reconstruction is by constructing a right-angled triangle in which b is the foot-point of the height with the side above xb being equal in length to xa'. Then the hypotenuse is xX. There are certainly many other ways to realize the same.

[121] See Damerow et al. (2001), in particular 65–82.

it served the elaboration of another problem (see Fig. 13.16).[122] The content of a single line written above the synopsis of the problem is apparently unrelated to the remaining content of the page.

The broken chord problem

- D02A, central motion diagram drawn (cf. Fig. 13.15).

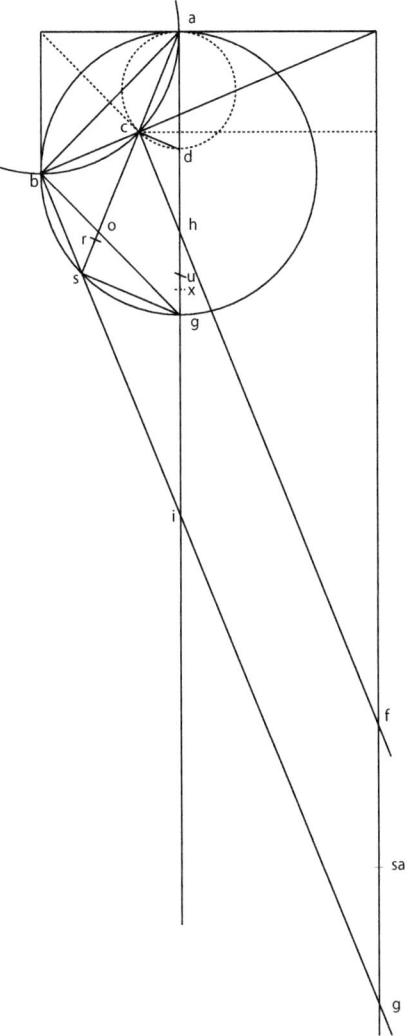

Fig. 13.15 Central motion diagram on double page 156 recto and 157 verso. Galileo tested whether the ratio of the times of motion along broken chord *acb* (same as along *bsg*) and along the single chord *ab* (same as along *bg*) spanning the same upright arc is equal to the root of the root of the ratio of the lines *ao as*

---

[122] 157 verso, D01A, D03A and D04A.

Fig. 13.16 Motion diagram and protect distance-time ratio diagram on double page 156 recto and 157 verso

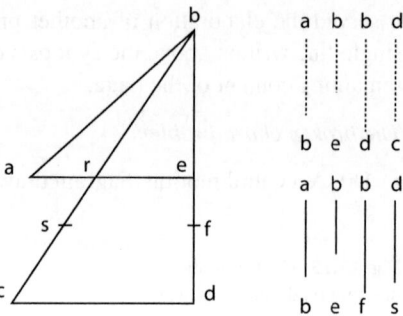

- T2, problem formulated and rudimentary description of geometrical construction given:

  > Inveniendum sit tempus quo conficiuntur [due] *acb* in rationem ad tempus quo conficitur sola *ab* fiat *abg* [angulus] rectus et semicirculus *abg* describatur et protrahatur *ac* ad *s* et connectantur *gs bs*.[123]

  Galileo seeks to determine the ratio of the times to traverse chord *ab* and broken chord *acb*, respectively. As a novelty compared to other constructions in the context of Galileo's investigation of motion along a broken chord, Galileo introduces the circle *abg* in which the broken chord *bsg* is inscribed. This broken chord is similar to the original broken chord *acb* but has a different position and orientation. Kinematically the motion along *acb* and along *bsg* are of course equivalent.
- C01(156 recto), calculate *ar*:

$$[ar =]\sqrt{184776 * 153072} = 168178.^{124}$$

Scheme of calculation: $ar = mp(as|ao)$.
- C04 (156 recto), calculate *gu* such that *ru* is parallel to *sg*:

$$[rs = as - ar = 184776 - 168178 = 16598]$$

$$[gu =]16598 * 2[00000] : 184776 = 17424.^{125}$$

Scheme of calculation: $rs/as \sim gu/ag$ and hence $gu = rs/as * ag$.

---

[123] 157 verso, T2.
[124] 156 recto, C01. 184776, the length of *as* can directly be read from a table of sines.
[125] 156 recto, C01.

- C02 (156 recto), calculate ax:

    media inter ga au[126]

$$[ax =]\sqrt{182576 * 2[00000]} = 191087.$$ [127]

Scheme of calculation: $ax = mp(ag|au)$.
- C01 (157 verso), compile table of results:

    linea as 184776
    ao 153072
    ar 168178
    rs 16598
    gu 17424
    ga 200000   tempus 141422
    ax 191087.[128]

C03 (156 recto), calculate time of motion along broken chord, $t(acb) = t(bsg)$:

$$[t(bsg) =]191087 * 141422 : 200000 = 135118.$$ [129]

Scheme of calculation:
hypothetically $t(ab)/t(bsg) = ag/ax$ and hence $t(bsg) = ag/ax * t(ab)$. Galileo's hypothesis regarding the ratio of times is equivalent (cf. Chap. 5) to the assumption that the time of motion along the broken chord is the same as the time to fall along ar. According to the tested hypothesis, the time of motion along the broken chord should be 135,118 time units. Galileo's earlier calculations, the results of which are noted on 183 recto, had yielded, however, that the time elapsed during motion along the broken chord should be 132,593. Hence the hypothesis could not be upheld. Transcribed in modern algebraic notation, Galileo calculated:
$\sqrt{\sqrt{ao}}/\sqrt{\sqrt{as}} * t(bg) = 135,118 = t(bsg)$
Thus Galileo effectively probed whether the ratio of the times of motion over the chord bg and the broken chord bsg equals the ratio of the root of the root of the line falling from a to o, i.e., to the chord and the root of the root of the line falling from a to s, i.e., to the junction point of the broken chord. This is recognized as being a variant of the root-root hypothesis Galileo had originally formulated on folio 166 recto (cf. Chap. 5). Galileo's hypothesis is equivalent to assuming that motion over the distance ar is completed in the same time as

---

[126] In the electronic representation of Galileo's *Notes on Motion*, "media inter ga au" is transcribed with the content of C01.
[127] 156 recto, C02.
[128] 157 verso, C01.
[129] 156 recto, C03.

motion over the broken chords acb and bsg, respectively. In abbreviated form this can be explicated as follows:

Galileo calculated: $t(bsg) = \sqrt{\sqrt{ao}}/\sqrt{\sqrt{as}} * t(bg)$.
It holds that:
$t(bsg) = \sqrt{\sqrt{ao}}/\sqrt{\sqrt{as}} * t(bg)$
$= \sqrt{\sqrt{ao}}/\sqrt{\sqrt{as}} * t(as)$
$= \sqrt{\sqrt{ao * as}} * as/as * t(as) = mp(mp(ao|as)|as)/as * t(as)$
$= mp(ar|as)/as * t(as) = mp(au|ag)/ag * t(as)$
$= t(ar)$.

Why it was plausible for Galileo to assume that $t(ar) = t(bsg)$ is discussed in detail in Chap. 5 Sect. 5.3.

Mnemonic diagram concerning fall over inclined planes of different height and inclination To the right of the central diagram, Galileo constructed a diagram of the motion situation in which the motions proceed over two inclined planes of different inclination, height, and length, along with an associated distance-time ratio diagram (see Fig. 13.16).[130] Galileo had investigated the same motion situation already on 189 verso where he had likewise availed himself of a distance-time ratio diagram and had come up with a correct solution allowing him to determine the ratio of the times of motion for this particular situation from knowledge of the spacial geometry.

In the distance-time ratio diagrams, to code the proportions that hold for a given motion situation according to the appropriate laws of motion, Galileo represented the times of motion by dotted lines drawn next to each other, keyed according to the respective distance covered. Thus, for instance, the dotted line be represents the time of motion along the distance be. Underneath the dotted lines, Galileo drew the corresponding distances from the spatial geometry of the motion situation. Each adjacent pair of lines codes a proportion between times elapsed and distances covered. Thus, for instance, the four leftmost lines of the distance-time ratio diagram code the proportion $t(ab)/t(be) \sim ab/be$, which holds according to the length time proportionality. Obviously the problem of how to determine the ratio of the times of motion along two inclined planes of different length, height, and inclination had already been solved when Galileo drew the diagrams. It thus appears that the diagrams, in particular the distance-time ratio diagram from which valid proportion can readily be read off, were drawn on the page as a kind of mnemonic device to facilitate work on the much more complex problem of determining the time of motion along the broken chord.

---

[130] 157 verso D03A and D04A.

13.13 Folio 166

The watermark of folio 166 has been described by Drake as three "balls connected with a triangle."[131] Only one other folio, folio 184, shares this watermark and does in fact contain auxiliary calculations pertaining to Galileo's considerations documented on folio 166. The verso side of the folio is empty. The recto side is dominated by a large geometrical construction. The construction is based on a square in which an upright quarter arc is inscribed. A chord and various polygonal lines are inscribed into this arc. Each of the polygonal lines is made up of a series of conjugate chords of equal length. On the page, Galileo investigated the motions made along these polygonal paths to eventually be able to exhaust motion along the arc by motion along a polygonal path with a number of sides growing above all limit. As Galileo assumed that falling along an arc was kinematically equivalent to the swinging of a pendulum bob along the same arc, his considerations were ultimately aimed at establishing a relation between swinging and rolling.

Folio 166 recto The content of 166 recto is pivotal for Galileo's attempt to approximate the motion of a pendulum by naturally accelerated motion along polygonal paths made up of a series of inclined planes (Fig. 13.17). The page bears a heavily annotated motion diagram. Underneath the diagram Galileo listed the results of calculations of the times of motion along the different polygonal paths. Below the list Galileo noted a number of observations and further considerations that had been prompted by his calculations.

A great number of calculations comprised on pages of the manuscript can unambiguously be related to Galileo's work on 166 recto. Most importantly folio 183 recto bears an extended list of results which refers directly to the diagram on folio 166 recto. The entries on the page can be divided into two groups depending on the basic unit of distance and time measure which Galileo employed. In a first set of considerations, Galileo had assumed the radius of the arc to measure 180 distance units and that the time elapsed during fall over vertical distance measuring 180 units would be measured by 180 time units. Galileo's calculations and considerations that pertain to this particular choice of distance and time unit are discussed here under the heading *base 180 approach*. At some point, obviously to increase accuracy, Galileo decided to change his base unit to 100,000, i.e., he now by convention assumed that the radius of the arc measured 100,000 units and that a vertical distance of this length would be traversed in 100,000 time units. The corresponding calculations and considerations are discussed under the heading *base 100,000 approach*.

---

[131] See Drake (1979).

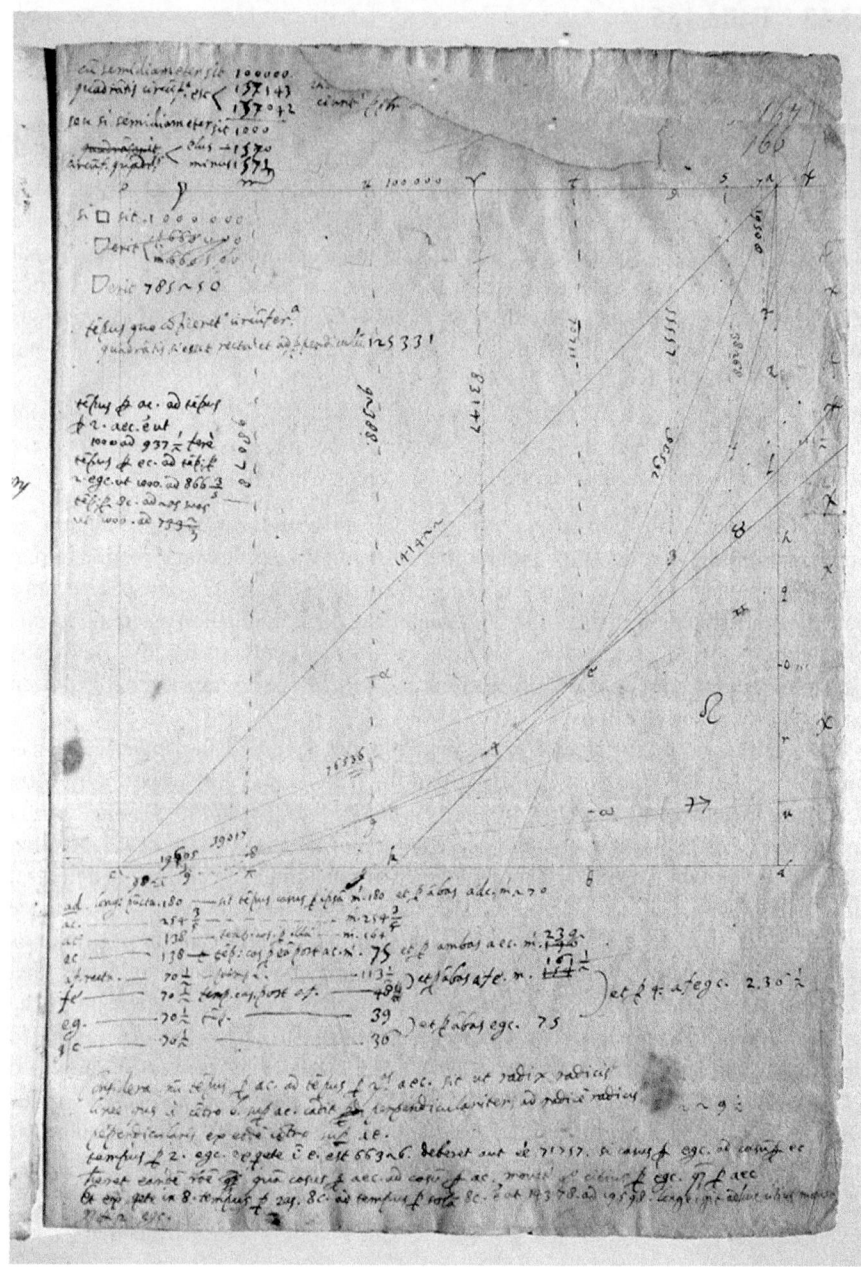

Fig. 13.17 Folio page 166 recto. (Courtesy of Ministero dei Beni e le Attività Culturali – Biblioteca Nazionale Centrale di Firenze. Reproduction or duplication by any means prohibited)

13.13 Folio 166

Base 180 approach

- D01A, central diagram begun.
  Initially the central diagram will have comprised the chord ac, the broken chords aec, and the polygonal path $afegc$ (cf. Fig. 5.7). Interpreted as paths of motion, these paths will be referred to as the single-chord, the two-chord, and the four-chord approximation, respectively, of the arc. All four chords af, ae, eg, and gc subtend the same angel. In other words, e halfs the arc ec, and f and g half the arcs ae and ec, respectively. The side length of the square in which the construction is inscribed measures 174 mm. Dividing this by 180, the side length in Galileo's unit ("longa puncta 180") yields 0.966 mm per one of Galileo's units. This value is in very good agreement with the length unit that has been extrapolated from Galileo's proportional compass and that has become referred to as Galileo's punto.[132] Drawing to scale had the obvious advantage that the results of the calculations of lengths of lines can directly be controlled by means of the diagram.
- T2, table of results noted:

   ad longa puncta 180; sit tempus casus per ipsam m[inutae] 180, et per ambas adc m[inutae] 270.
   ac ——— 254 3/5 ——— m[inutae] 254 3/5
   ae ——— 138 ——— tempus casus per illam m[inutae] 164
   ec ——— 138 ——— tempus casus per eam post ac m[inutae] 75 et per ambas aec m[inutae] 239
   af recta ——— 70 1/2 tempus ——— 113 1/2
   fe ——— 70 1/2 tempus casus post af ——— 48 et per ambas afe m[inutae] 161 1/2
   eg ——— 70 1/2 tempus ——— 39
   gc ——— 70 1/2 ——— 36 et per ambas egc 75
   et per 4 afegc 236 1/2.[133]

For a formatted and thus slightly more accessible version of the table, see Chap. 5, Table 5.1. Galileo assumes the vertical distance ad to be traversed in 180 time units. Arbitrarily he refers to the time units as "minutes" to avoid confusion between numerical values representing distances and values representing times. By the sesquialterum variant of the double distance rule (cf. Chap. 10), it follows that the time to fall through ad and to successively move with the uniform

---

[132] That the construction on 166 recto is drawn to scale has already been assumed by Drake (1978, 88). He gives a value of 29/30 mm, corresponding precisely to my own measurements. From measurements on a proportional compass by Galileo, Naylor (1976) inferred a value of 0.95 mm for the smallest unit. More recently Vergara Caffarelli (2009, 169) has claimed that Galileo's puntus is synonymous to the piccolo which is 1/240 of a Florentine braccio and should thus measure about 2.4 mm. I find his arguments not convincing. In particular it would result in very improbable dimensions for different experimental setups used by Galileo. Such, for instance, the 828 punti Galileo gives on 116 verso as the height of a table would amount to a table of almost 2 meters height instead of the even today still standard height of somewhat less than 80 centimeters which results from the conventional assumption that the punto measures about 0.96 millimeters.

[133] 166 recto, T2.

velocity thus reached through *dc* is one and a half times that of fall through *ad* alone. Hence $t(adc)$ amounts to 270 of Galileo's minutes, and noting this completes the first line of the list. Most of the calculations by which the results listed in the following lines have been achieved are no longer extant. From the calculations of the times along *eg* after motion from *a* and along *gc* after motion from *a* that have survived on folio 184 verso (cf. the discussion of 184 verso in this appendix), the general calculational scheme according to which all other results listed could be achieved is known (cf. Chap. 5).

Once the times of motion over the individual chords have been calculated, the times are added up to yield the overall time of motion along the polygonal paths made up of the individual conjugate chords. The table is completed by taking note of the overall times thus calculated.

- T3A, hypothesis concerning the ratio of the times of motion over the single and over the respective broken formulated:

    Considera num tempus per *ac* ad tempus per [du]as *aec* sit ut radix radicis lineae quae a centro *b* super *ac* cadit perpendiculariter, ad radicem radicis perpendicularis ex eodem centro super *ae*.[134]

Galileo formulates a hypothesis regarding the ratio of the time of motion along a chord and a broken chord spanning the same arc. With regard to Fig. 5.9, Galileo speculates that:

$t(aec) = \sqrt{\sqrt{bz}}/\sqrt{\sqrt{by}} * t(ac)$. Calculating this using Galileo's values results to:

$\sqrt{\sqrt{155\ 9/10}}/\sqrt{\sqrt{\frac{254\ 3/5}{2}}} * 254\ 3/5 = 242$

This is close but does not correspond to the time of 239 units Galileo had calculated for motion along the broken chord *aec* by applying the laws of motion.

- C02, number noted:

$$229\ 1/2.^{135}$$

Galileo had very likely tested whether $t(aec) = \sqrt{bz}/\sqrt{by} * t(ac)$. If this is calculated with the same numbers as above 230 results. The small difference to the number 229 1/2 noted by Galileo would be due to the rounding error introduced when drawing the roots.

Base 100,000 approach To increase the accuracy of the calculations, Galileo switched over to a new, smaller unit. With respect to the new distance unit, the radius of the arc measures 100,000. Galileo writes the lengths of several of the

---

[134] 166 recto, T3A.
[135] 166 recto, C02.

13.13 Folio 166

relevant distances next to the corresponding lines in the diagram. At the same time, as evidenced by the results listed on folio 183 verso, Galileo, by distance time coordination, assumed that the time elapsed during fall along the vertical radius would measure 100,000 time units thus fixing the new time unit.

In the diagram, Galileo adds the junction points 2, 3, 4, and 8 and inscribes the polygonal line $a2f3e4g8c$ composed of eight conjugate chords of equal length into the arc ac. He then calculates the times of fall through each individual chord of each of the polygonal approximations up to the eight-chord approximation from which he can infer the total times of motion along the polygonal paths of each approximation in the new time units. None of these calculations has been preserved in the *Notes on Motion*. However, their results have been listed on folio 183 recto (see the discussion of Sect. 13.16 in this appendix).

- Point 9 introduced that it bisects the arc $8c$. This defines the chord 89 and $9c$ each spanning a 16th of the original quarter arc.
- T3B (first part), relative reduction in time if motion along the chord is compared to motion along the broken chord is considered for different starting points:

> Tempus per 2 *egc* ex quiete in *e* est 66326; deberet autem esse 71757, si casus per *egc* ad casum per *e* haberet eandem rationem quam casus per *aec* ad casum per *ac*: movetur ergo citius per *egc* quam per *aec*.[136]

First of all, the entry shows that Galileo had calculated the time of motion along *egc* from rest in *e*. These calculations are not preserved.[137]

If the ratio of the time of motion along a chord to that along a broken chord composed of two conjugate chords of equal length were independent of the angle subtended by the upright arc in which chord and broken chord are inscribed, it would hold that: $t(ac)/t(aec) = t(ec)/t(egc)$. Galileo calculated the expected time for motion along *egc* if the above relation held: $t(egc)_{expected} = 71{,}757$. Comparing this to the actual time calculated for motion along *egc* $t(egc) = 66{,}326$, it turns out that the relative time reduction if motion over the broken chord is compared to motion over the single chord spanning the same arc is bigger for the case where motion starts from point *e* than for the case where motion starts from *a*, i.e.: $t(ac)/t(aec) < t(ec)/t(egc)$.

- T3B (second part), relative reduction in time if motion along the chord is compared to motion along the broken chord is considered for different starting points:

---

[136] 166 recto, T3B.

[137] On 192 recto Galileo did calculate the time *egc* from rest at point *e* (see the discussion of folio 192 recto in this appendix) yet for the base 100,000 approach. From the time of 66,326 given for motion along *egc* in the entry on 166 recto, it can be inferred that Galileo's calculations on which this entry was based had proceeded from the assumption that $t(ec) = ec = 76{,}536$, thus implicitly defining a time unit not compliant with that of the base 100,000 approach.

Et ex quiete in 8 tempus per [du]as 8*c* ad tempus per solam 8*c* est ut 14378 ad 19598: longe igitur adhuc citius movetur quam per 2 *egc*.[138]

Galileo had calculated the time along 8*c*, and along the broken chord 89*c* ("du]as 8*c*"), the calculations are not preserved.[139] His statement that the body is again therefore moved much faster (" longe igitur adhuc citius movetur") can only refer to the relative saving of time i.e., the comparison of ratios: $t(ec)/t(egc) > t(8c)/t(89c)$.

Galileo's results indicate that if motion along a broken chord composed of two conjugate chords of equal length is compared to motion on a chord spanning the same upright arc, the smaller the angle of the arc considered, the greater the reduction in time. As by the law of chords the motion along all chords is the same, independent of the angle of the arc, this immediately implies at least for the motions along the three broken chords considered *aec*, *egc*, and 89*c* that the smaller the arc, the shorter the time of motion, i.e., $t(aec) > t(egc) > t(89c)$.

- T1D, times of motion along different broken chords related to time of motion along corresponding chords, all times expressed with regard to the same unit:

    Tempus per ac ad tempus per 2 *aec* est ut 1000 ad 937 1/2 fere; tempus per ec ad tempus per 2 egc ut 1000 ad 866 3/5; tempus per 8*c* ad [du]as suas ut 1000 ad 733 2/3.[140]

Because Galileo had used different time units for the calculations pertaining to the different starting points *a*, *e*, and 8 (see above), direct comparison of the results was cumbersome. To allow for better comparability, Galileo recalculated (i.e., rescaled) all times of motions to a time unit defined by the conventional choice to have motion along the single chords be completed in 1000 time units. The calculations have not been preserved.

- T1A, length of 90-degree arc noted for a radius of 1000 and 100,000

    Cum semidiameter sit 100000, quadrantis circumferentia est [plus] 157143 [minus] 157042;
    seu si semidiameter sit 1000, circumferentia quadrantis plus 1570 minus 1571.[141]

In a noteworthy quest for accuracy Galileo gives an upper and a lower limit, respectively. The upper limit results from using a value of 3 1/7 and the lower limit from using a value of 3 10/71 for π.

---

[138] 166 recto, T3B.

[139] From the time of motion along 8*c* assumed to be 19,598 it is clear that these calculations must have been based on assuming by distance time coordination that the time along 8*c* is measured by 19,598, i.e., the length of 8*c* in the length units of the base 100,000 approach. With this choice Galileo implicitly introduced yet another unit of time measurement.

[140] 166 recto, T1D.

[141] 166 recto, T1A.

- T1B area of square compared to area of sector:

  si □ sit 1000000, [sector] erit 785250.[142]

- T1C, time to fall through vertical distance equal in length to length of quarter arc calculated and noted:

  Tempus quo conficieretur circumferentia quadrantis, si esset recta et ad perpendiculum, 125331.[143]

- T4, entry corrupted by damage to paper:

  ra?......?ciuntur per...?......[144]

Galileo's calculations from a modern perspective In the following, for a given approximation, the conjugate chords are indexed from the highest to the lowest chord by the index n where N is the total number of chords for the given approximation. Galileo proceeded from a given approximation the next higher approximation by replacing each individual chord by a broken chord, and hence N takes on the values 1, 2, 4, 8, etc. All conjugate chords of a given approximation have the same length:

$$s_{chord} = 2 * s_{radius} * \sin(1/4\,\pi/N).$$

The inclination of an individual chord measured against the horizontal, which determines the effective force and thus the effective acceleration of a body falling along that chord is given by:

$$\alpha_{chord}(n, N) = 1/4\frac{\pi}{N} + 1/2\frac{\pi\,(N-n)}{N}.$$

With $\alpha_N = 1/2 * \pi/N$ this becomes:

$$\alpha_{chord}(n, N) = 1/2\alpha_N + \alpha_N * (N-n).$$

The vertical height of the starting point of motion on a conjugate chord below the overall starting point, which determines the initial velocity of motion along this particular conjugate chord, is:

$$s_{vertical}(n, N) = \sin(\alpha_N * (n-1)) * s_{radius}.$$

---

[142] 166 recto, T1B.

[143] 166 recto, T1C. The calculation is not preserved. According to the law of fall, the time is simply given by the mean proportional between the radius and the length of the circumference of the quadrant.

[144] 166 recto, T4.

Wit this, the time of motion to fall along an individual conjugate chord becomes:

$$t_{chord}(n, N) = \frac{\left(-\sqrt{2\frac{s_{vertical}(n)}{g}} + \sqrt{2\frac{s_{vertical}(n)}{g} + 2\frac{\sin(\alpha_{chord}(n))s_{chord}(n)}{g}}\right)}{\sin(\alpha_{chord}(n))}.$$

In Galileo's arbitrary units of distance and time fall along the vertical radius of the circle of length $s_{radius} = 1{,}000{,}000\ du$ was completed in $t_{radius} = 100{,}000\ tu$ (where du and tu represent the distance and time units, respectively). In Galileo's units the constant of acceleration can be expressed as:

$$g = 2\frac{s_{radius}}{t_{radius}^2} = 2*10^{-5} du/tu^2.$$

With this, the time of motion along each individual chord can be calculated and by summation also the overall times of motion along the polygonal paths considered. The values thus are compared to Galileo's own results listed on 183 recto (see the discussion of folio 183 recto). I have, furthermore, used this modern formula to calculate the times of motion along the polygonal paths of the next higher approximations. For the 16-chord approximation, this yields a time of 131,111 time units and for the 32-chord approximation a time of 131,105 units. This shows how remarkably close Galileo had already come to the correct solution.

13.14 Folio 167

Like many other folios in the manuscript, folio 167 bears a watermark in the form of a crossbow. On folio 167 the crossbow is encircled and has an additional flower ornament. The only other folio in the manuscript bearing the exact same watermark is folio 183. The size and quality of the paper, as well as the placement in the manuscript, suggest that folio 167 once formed a double page with the preceding folio, folio 166. On its verso side, folio 167 bears the calculations of the times of motion pertaining to a broken chord motion situation which is depicted in a large diagram on the page. The recto side of the folio is empty.

Folio 167 verso Galileo calculated the time to fall along the path *cra* in central diagram D01A (see Fig. 13.18). Galileo's calculations yielded a time of motion of 137,066 time units to traverse the path *cra*, where the radius of the circle had been defined as measuring 100,000 distance units, and, by distance time coordination, the time to fall through this radius had been assumed to measure 100,000 time units. Some of the calculations required to establish this result were obviously carried out on another folio no longer extant. It is conceivable and indeed rather plausible that on such a folio, Galileo also calculated the time of motion

13.14 Folio 167

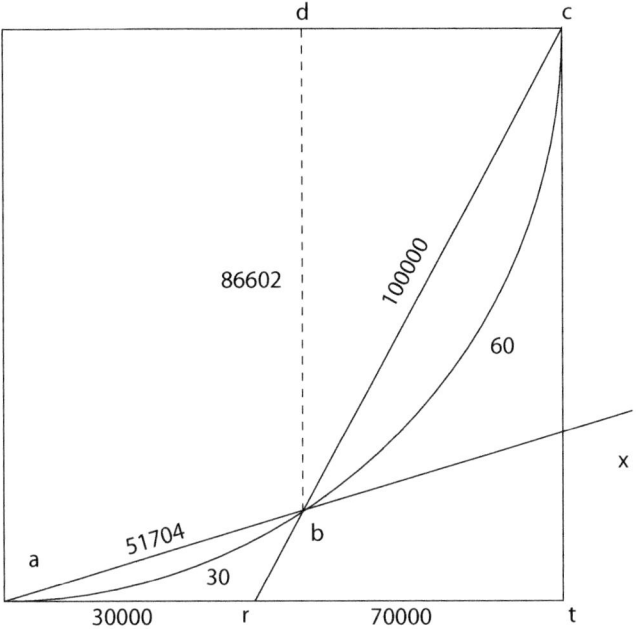

Fig. 13.18 Schematic representation of part of the diagram (D01A) on 167 verso

along the broken chord *cba* likewise represented in the central diagram but not calculated on the page. Indeed the *x* written on the right side of the diagram symbolizes the intersection between the inclined *ab* and the upper horizontal *cd* needed to determine the time of motion over the second part of the broken chord $t(ba|x)$.

- D01A, central diagram drawn. As somewhat of a peculiarity, Galileo noted in the diagram the angles subtended by the two conjugate chords *cb* and *ba*, namely, 60 and 30 degrees.
- C01, results listed:

    tempus per *db* 93060
    tempus per *cb* 107457.[145]

Galileo notes the numerical values for the time of motion along *db* and *cb*. The calculations by means of which these results were achieved are no longer extant, but the way by which Galileo must have determined these numbers is rather obvious. If it is assumed by distance time coordination that $t(radius) = radius$, then by law of fall $t(db)/t(radius) \sim mp(db|radius)/radius$ and hence $t(db) = mp(db|radius)$. Rounded to the next integer, the geometrical mean

---

[145] 167 verso, C01.

between 100,000 and 86,602 is indeed 93,060, as noted by Galileo. From this the time along cb can be determined by the length time proportionality. It must hold that $t(cb)/t(db) \sim cb/db$ and hence that $t(cb) = t(db) * db/cb = 93,060 * 100,000/8660$. If this is calculated and rounded to the next integer, the result is indeed the number 107,457 which Galileo gives for the time along cb. That Galileo calculated the time of motion along cb, the first part of the broken chord cba is a rather clear indication that he intended to and in all likelihood indeed did calculate the time of motion along the broken chord.

- C02, calculation of cr by theorem of Pythagoras:

$$\sqrt{4900000000 + 10000000000} = 122066.^{146}$$

- C01, list of results completed:

  tempus per db 93060
  tempus per cb 107457
  cr 122066.

  By length time proportionality, it holds that $cr = t(cr)$. List refined:

  tempus per db 93060
  tempus per cb 107457
  cr tempus 122066.[147]

- C01, calculation of $t(ra|cr)$:

$$cr \text{ tempus } 122066 + [t(ra|cr)] \, 15000 = [t(cra)] \, 137066.^{148}$$

A body falling from a height $ct = 100,000$ in a time $t(ct) = 100,000$ will, according to the double distance rule, if deflected into the horizontal, traverse in uniform motion with the velocity reached, twice that distance, i.e., 200,000, in the same time. By a kinematic proposition, this uniform motion will then traverse a distance of $ar = 30,000$ in a time $t(ar|c) = 15,000$.

Galileo turned the page upside down and added some additional elements to the construction of the central diagram, among them a chord and an inclined plane parallel to it and tangent to the circle. This particular configuration, which is rather unique in the *Notes on Motion*, played a crucial role in Galileo's construction of the ex mechanicis proof (cf. Chap. 8).

---

[146] Ibid.
[147] Ibid.
[148] Ibid.

13.15 Folio 176

The watermark of folio 176 is a small, thick crossbow (height 34 mm) which due to its comparatively small size can readily be distinguished from other watermarks displaying crossbows.[149] The verso side of folio 176 is blank. The recto side of the folio contains a few calculations by means of which Galileo probed a particular hypothesis concerning the ratio of the times of motion along a single chord and the respective broken chord spanning the same upright arc. It, furthermore, contains a geometrical construction for which a certain geometrical property is claimed to hold. The construction and its claimed property had been exploited by Galileo in the context of his problem of motion along bent planes, which he had begun to investigate on 130 verso. The entry was canceled by Galileo and explicitly marked as false despite being obviously correct.

Folio 176 recto The content of the page falls into two distinct groups formally distinguishable by the fact that they have been written on the page in different writing directions. On the top of the page, written with the page upside down with regard to its modern orientation in the manuscript, Galileo wrote a number of calculations by means of which he tested the modified version of a hypothesis originally noted on folio 166 recto concerning the ratio of the time of motion along a chord and a respective broken chord. The calculations use numbers from that latter page but also the results of related calculations contained on folio 189 recto. Drafted on the page upright with regard to its modern orientation is a diagram and an associated textual entry in which a claim is made regarding a property of the geometrical construction given. The entry was crossed out by Galileo and marked up with the phrase "falsa est," despite being correct from a modern perspective, and even more surprisingly despite the fact that Galileo availed himself of the very geometrical construction, his argument was based on another context. The diagram encompasses a great number of lines scratched into the paper but not drawn out in ink.

Geometrical argument

- D02A, diagram drawn (see Fig. 13.19).[150]
  The circle is merely an auxiliary construction for establishing the right angle at b.
- T1, construction principle outlined, claimed property of construction stated:

    Sit [tri]angulum rectangulum abc, et ab sit aequalis bc, et secetur bifariam ac in d, et connectatur bd, sitque ai ipsi cb parallela; post positaque ae ipsi ab aequali, erunt ca,

---

[149] For a list of the folios sharing this watermark see discussion of folio 130 in this appendix.
[150] 176 recto, D02A.

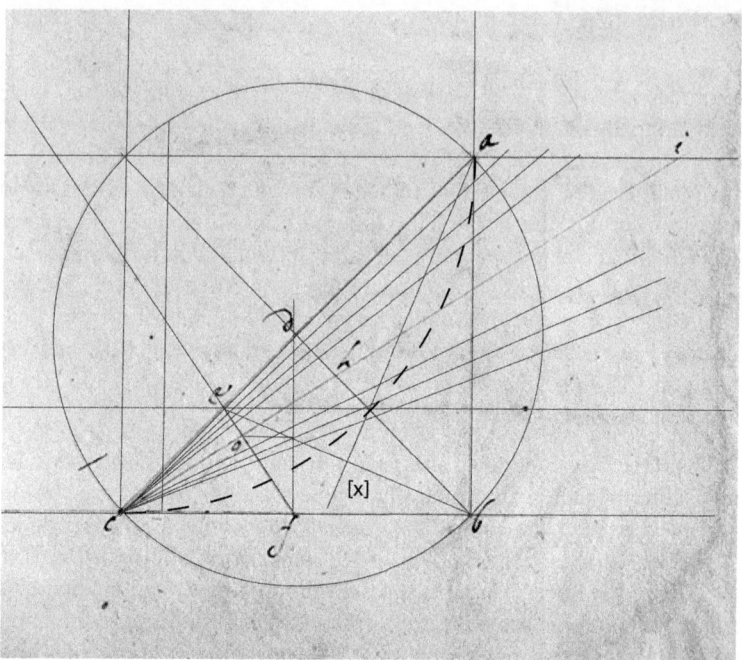

Fig. 13.19 Uninked lines in the central diagram on page 176 recto. The arc represented as a dotted line is not part of Galileo's figure but has been added for analysis of Galileo's construction. (Courtesy of Ministero dei Beni e le Attività Culturali – Biblioteca Nazionale Centrale di Firenze. Reproduction or duplication by any means prohibited)

ae, ad continue proportionales. Secetur *cb* bifariam in *f* et connectatur *ef*: dico, quod si protrahatur quaelibet linea ex puncto *ad* lineam *ai*, ut puta *cghi*, esse proportionales *ci, ig, ih*. Falsa est.[151]

Construction: $\angle abc = 90°$, $ab = bc$, $ad = 1/2\, ac$, $ae = ab, cf = 1/2\, cb$. Galileo firstly claims that $ae = mp(ad|ac)$. This is trivially the case because $ae = ab$ is the leg of the right-angled triangle abc. Hence it is the mean proportional between the hypotenuse ac and the part of the hypotenuse divided by the foot-point d of the height of the triangle directly below the leg. Galileo secondly claims that if a line is drawn from e to f, then for any line ci, its intersection with ef referred to as g will mark the geometric mean between ih and ic, with h being the intersection of ci and db. Hence Galileo claims that whatever the inclination of ci, it holds that $ig = mp(ih|ic)$. Galileo marked this as false and deleted the paragraph by striking it through.

Inspection of Galileo's diagram under raking light reveals that besides the solid lines, it contains a number of uninked lines which are merely scratched into

---

[151] 176 recto, T1.

13.15 Folio 176

the paper, most likely with the tip of the compass. These uninked lines which are represented in Fig. 13.19 as light gray lines reveal a direct relation of the diagram and the related entry on 176 recto, to the geometrical construction and the underlying considerations on 130 verso. The lines which remained uninked are an assembly of inclined planes of different inclination, all emanating from point c representing the different possible orientations of ci and in particular also the line be. For lines running from a to the lower horizontal cb of which Galileo represented just one in the uninked part of the construction, the intersection with this line be (in the construction on 130 verso the corresponding line is $x4$) marks the mean proportional between the lines from a to the intersection with db and from a to the intersection with the horizontal cb.

With reference to the lettering of the diagram on 176 recto, Galileo on 130 verso had posed himself the question for which position h on db of the junction point of a bent plane ahc this bent plane would be traversed in least time. His considerations on 130 verso had yielded that if the time along the first part of the bent plane ah is represented by ax (x not lettered in Galileo's diagram has been added for ease of reference in Fig. 13.19) then the time of motion along the second part hc after motion from rest in a is represented by cg. According to all evidence on 130 verso, Galileo had first calculated the positions of x and g for the various configurations of bent planes he had included in the diagram and only then realized the characteristic geometrical property he explored on 176 recto, namely, that point g was always positioned on ef (in the construction on folio 130 verso the corresponding line is $4z\delta k$). The overall time of motion along a bent plane ahc is given by the sum of ax and cg. On 130 verso Galileo had established that ax, i.e., the time of motion along the first part of the bent plane, assumes a minimal value if the point h is positioned on the upright arc through points a and c. This indeed corresponds exactly to the orientation of the one single first part that Galileo choose to draw as an uninked line. By an argument similar to the one Galileo had employed to demonstrate for which configuration the time of motion along the first part is minimized, it could likewise be demonstrated that the time along the second part of the bent plane becomes minimal for the configuration in which cg is orthogonal on ef. It is telling that of the many different possible configurations for the line cg Galileo drew as uninked lines it was exactly the one orthogonal on ef he inked and lettered.

Why Galileo aborted his considerations and marked them as false is unclear, as his statement is correct. As argued in Chap. 7, finding the configuration for which the sum became minimal exceeded Galileo's means. As likewise argued there he had reasons to surmise (falsely from modern perspective) that it was the broken chord that was traversed in least time and this surmise was challenged. It may have been for this reason that Galileo discarded his correct consideration.

Calculations pertaining to the "root-root hypothesis"

- C01, calculation of roots:

$$\sqrt{100000} = 316$$

$$\sqrt{70711} = 265\ 1/2.^{152}$$

Galileo calculates the roots of the distances bc and te in the central diagram on folio 166 recto.

- C01, calculate ratio of roots:

$$265\ 1/2 * 254\ 3/5 : 316 = 2[14].^{153}$$

Galileo tests whether times of motion along chord $t(ac)$ and along broken chord $t(aec)$ are in the same ratio as the roots of bc and te (which by the law of fall is also the ratio of times to fall over the distances bc and te). The value for $t(ac)$, 254 3/5 is taken from folio 166 recto. Calculation breaks off after the second digit has been calculated likely because Galileo realized that the result would be much smaller than 239, the time noted for motion along aec on 166 recto.

- C02, above calculation repeated for times calculated on 189 recto:

$$265\ 1/2 * 141\ 1/2 : 304 = 12[3]^{154}$$

The number 132 1/2 noted next to calculation. 141 1/2 and 132 1/2 are the times of motion along the chord and broken calculated on folio 189 recto based on assuming that a vertical distance of 100 distance units is traversed in 100 time units.[155] Again, Galileo broke off the calculation before it was finished most likely because it was clear after calculating the first two digits that the result could not be equal to, 132 1/2, the time of motion along the broken chord calculated in the units of 189 recto.

- C02 and C03, calculation repeated for the ratio of the root of roots instead of ratio of roots:

$$141\ 1/2 * \sqrt{265/1/2}/\sqrt{304} = 141\ 1/2 * 16\ 1/4 : 17\ 1/2 = 131^{156}$$

The calculated value of 131 is close to 132 1/2 the time of motion along the broken chord. Yet the roots of the roots calculated are rather small and thus the relative error introduced by rounding comparatively big. It remains unclear whether Galileo saw the result as supporting his hypothesis or not.

---

[152] 176 recto, C01.

[153] 176 recto, C01.

[154] 176 recto, C02.

[155] Galileo explicitly noted a time of 132 for motion on the broken chord on 189 recto, C08, but the result he had calculated in C06, and which he could furthermore read off his list of results on 183 recto, was somewhat greater, namely, slightly more than 132 1/2. Why Galileo changed the value used for the root of 100,000 to 304 remains unclear.

[156] 176 recto, C03.

13.16 Folio 183

The watermark of folio 183 has been described by Drake as a "crossbow in circle 44 mm, spearhead above."[157] The folio shares this watermark with folio 167.[158] The verso side of the folio was left blank. The recto side contains an extensive list of results of the calculations of times of motion over paths composed of conjugate chords inscribed into an upright quarter arc. The times of motion along each individual conjugate chord of the paths made up of one, two, four, and even eight conjugate chords are listed. The lettering of the distances corresponds to the labels used in the diagram on folio 166 recto. Most of the calculations required for establishing the results in the list are no longer extant. The ones that have survived allow, however, to reconstruct the way in which Galileo may in principle have arrived at all results listed.

Folio 183 recto Since the way in which Galileo could have obtained the results listed on the page is clarified sufficiently by the surviving example of such a calculation on 184 verso (cf. Chap. 5), I restrict myself to reproducing the list of results in a slightly more legible fashion. In addition to the values I reproduce here, Galileo also listed the values for times of motion, which in a sense were the by-product of auxiliary calculations required to infer the times of motion along the individual conjugate chords. All values listed have been recalculated (compare the discussion of folio 166 in this appendix). When a value listed by Galileo deviates considerably from the recalculated value, this is indicated in a footnote. Embedded in the very middle of Galileo's list are the results of his calculations of times of motion along path $alekc$ not inscribed into but circumscribing the arc. On this path the motion proceeds first vertically, then on an inclined plane tangent to the arc, and finally finishes with a horizontal motion to the base of the arc. Galileo's results regarding the times of motion along the individual parts of this path are reproduced in a separate table below (Table 13.1).

---

[157] See Drake (1979).

[158] The folios 167 and 166 form a double folio. As the watermark of folio 166 is the same as that on folio 184, it is suggested that folios 183 and 184 were connected as a double folio as well when Galileo wrote on them. This may indeed be the reason why of the pages and pages which Galileo must have filled with the calculations that were required to produce the list on 183 recto and the comparable list on 166 recto, only those calculations written down on 184 have survived. It seems that once the results had been summarized in the lists, Galileo discarded the folios containing the calculations except for 184 as this folio was presumably then still attached to the page comprising the list of results.

Table 13.1 List of results on folio 183 verso rendered as a table

	Length	Time		Time		Time		Time
ad	100,000	100,000						
ac	141,422	141,422						
ae	76,536	91,017[159]	aec	132,593				
ec		41,576[160]						
af		63,045[161]	afe	89,766				
fe	39,017	26,721[162]			afegc	131,319		
eg		21,657[163]	egc	131,319				
gc		19,896[164]						
a2		44,385[165]	a2f	62,873				
2f		18,488			a2f3e	89,605		
f3		14,372	f3e	26,732				
3e	19,604	12,360[167]					a2f3e4g8c	131,078[166]
e4		11,185[168]	e4g	21,652				
4g		10,467[169]			e4g8c	41,473		
g8		10,029[170]	g8c	19,821				
8c		9792[171]						
	Length	Time		Time				
al	41,422	64,360	alekc	135,475				
lek	82,843	50,404						
kc	41,422	20,711						

Remaining content

- T1, time of fall over the vertical distance *de*:

    media inter *ad te* 84090
    tempus per *te* 84090[172]

---

[159] Should be 91,018.
[160] Should be 41,576.
[161] Should be 63,073.
[162] Should be 26,733.
[163] Should be 26,752.
[164] Should be 19,895.
[165] Should be 44,383.
[166] Should be 131,416.
[167] Should be 12,358.
[168] Should be 11,184.
[169] Should be 10,466.
[170] Should be 10,045.
[171] Should be 9,849.
[172] 183 recto, T1.

13.17 Folio 184

Galileo calculates by law of fall the time of fall over the vertical distance te and notes the result.

- C04–C09, as above for vertical distances sf, ug, 72, $2x$, 39, 4Υ and $8\mathrm{M}$:

  media inter $da\ fs$ 61861
  tempus per sf 61861[173]
  tempus per ug 96118[174]
  72 longa 19508 tempus 44168[175]
  $2y$ 20386 tempus[176]
  linea 39 55557 tempus 74536[177]
  4Υ 83147 tempus 91185[178]
  $8\mathrm{M}$ tempus 99030.[179]

- C1, list of numbers noted:

  22, 57, 39, 60, 1377.[180]

  The purpose of these numbers is unclear.

13.17 Folio 184

The watermark of folio 184 shows balls connected by a triangle. It shares this watermark with folio 166. Based on the size and quality of the paper, it can be conjectured that the folio once formed a double folio with 183, just as folios 166 and 167 most likely formed a double folio. On its verso side, folio 184 indeed contains auxiliary calculations for the motion situation Galileo depicted and began to analyze on folio 166 recto, and the results obtained on 184 were indeed noted on 166 recto. The recto side of folio 184 remained empty.

Folio 184 verso Folio 184 verso contains the auxiliary calculations by which some of the times of motion along the various chords of the different broken chord approximations of the arc represented in the diagram on folio 166 recto were

---

[173] 183 recto, C04.
[174] 183 recto, C05.
[175] 183 recto, C06.
[176] Ibid.
[177] 183 recto, C07.
[178] 183 recto, C08.
[179] 183 recto, C09.
[180] 183 recto, C01, incorrectly transcribed in electronic representation of Galileo's *Notes on Motion*.

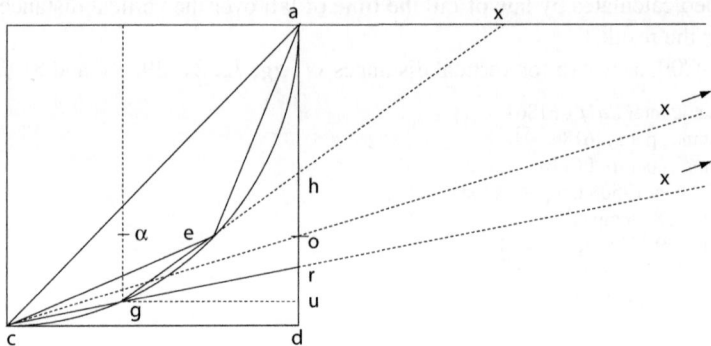

Fig. 13.20 Schematic representation of the diagram on page 166 recto reduced to those elements of the geometry of the construction relevant for Galileo's calculations on page 184 verso

calculated. These calculations are part of the base 180 approach (cf. the discussion of 166 recto in this appendix), delineated by Galileo's choice to have the radius of the arc considered measure 180 units and to have a vertical distance of 180 units be traversed in naturally accelerated motion in 180 time units. Galileo transferred the results of his calculations to the distance time table underneath the central diagram on folio 166 recto. Folio 184 verso is the only page that has been preserved bearing calculations that yielded the results listed in this table, and Galileo must have covered many more pages no longer extant today with calculations similar to the ones on 184 verso to obtain all the results listed. With respect to the diagram on folio 166 recto, the times along the chords *ec* and *gc* after initial fall from *a* are calculated on folio 184 verso. In addition, Galileo calculates the time to fall along the path *arc*. The required calculations of the times of motion along inclined planes all follow the same schema. The inclined plane is extended until it intersects with the horizontal *ab* in a point *x*. The time to fall through the extended plane is calculated by law of fall and length time proportionality. Then by the generalized law of fall, the time of motion along the original chord from rest at *x* is calculated, which by path invariance is the same as if the motion had instead been preceded by motion from *a*, the apex of the quarter arc (cf. Fig. 13.20). Two further calculations on the page served a different purpose and are discussed under a separate heading below.

Calculation of the time along chord eg, t(eg|a)

- T1, restate result known from previous calculation:

    longitudo *ao* 127 tempus per *ao* 151.[181]

---

[181] 184 verso, T1.

13.17 Folio 184

These results have been calculated on 121 recto. The length of the line ao is scribbled next to the point o in the diagram on 166 recto.

- C01, distance xe calculated by rule of three:

$$127 * 70\ 1/2 : 39\ 1/2 = 214.^{182}$$

Scheme of calculation: $eg/\alpha g \sim xe/ao \Rightarrow xe = ao * eg/\alpha g$.
- C02, time along xe calculated by the length time proportionality:

$$214 * 154 : 127 = 254\ 1/2.^{183}$$

Scheme of calculation: $t(ao)/t(xe) \sim ao/xe \Rightarrow t(xe) = xe * t(ao)/ao$.
- C05, calculation of mean proportional between xe and xg:

$$\sqrt{284\ 1/2 * 214} = 246\ 3/4.^{184}$$

- C07, calculation of time along xeg by law of fall:

$$246\ 3/4 * 254\ 1/2 : 214 = 246\ 1/2.^{185}$$

Scheme of calculation: $(xeg)/t(xe) \sim mp(xe|xeg)/xe \Rightarrow t(xeg) = t(xe) * mp(xe|xeg)/xe$.
- C09, calculation of time along eg after fall through xe

$$254\ 1/2 - 246\ 1/2 = 8.^{186}$$

Scheme of calculation: $t(eg|x) = t(xeg) - t(xe)$. The calculated value is obviously much too small which leads to the recognition of a mistake in and the correction of C07.
- C07 and C09, corrections:

$$246\ 3/4 * 254\ 1/2 : 214 = 2931/2^{187}$$

$$293\ 1/2 - 254\ 1/2 = 39.^{188}$$

---

[182] 184 verso, C01.
[183] 184 verso, C02.
[184] 184 verso, C05.
[185] Ibid.
[186] 184 verso, C09.
[187] 184 verso, C07.
[188] 184 verso, C09.

- C20, results achieved noted as list:

  > *ehx* longitudine 214 tempus eius 254 1/2
  > media inter *ehx* et *xhcg* 246 3/4
  > tempus totius *xheg* 293 1/2.[189]

Calculation of the time along chord gc, $t(cg|a)$

- C15, calculation of distance xg by rule of three:

$$66 1/2 * 70 1/2 : 13 1/2 = 869 1/2.^{190}$$

Scheme of calculation: $gc/du = gx/au \Rightarrow xe = au * gc/du$.
- C12, calculation of time along au by the law of fall:

$$\sqrt{180 * 166\ 1/2} = 173.^{191}$$

Scheme of calculation: $(ad)/t(au) = ad/mp(au|ad) \Rightarrow t(au) = t(ad) * mp(au|ad)/ad$.
- C14 calculation of time along xg by length time proportionality:

$$869\ 1/2 * 173 : 166\ 1/2 = 903.^{192}$$

Scheme of calculation: $t(au)/au = t(xg)/xg \Rightarrow t(xg) = xg * t(au)/au$.
- C13, calculation of time along xcg by the law of fall:

$$\sqrt{869\ 1/2 * 940} = 904.^{193}$$

Scheme of calculation: $t(xgc)/t(xg) = mp(xg|xgc)/xg \Rightarrow t(xgc) = t(xg) * mp(xg|xgc)/xg$.
- C16, calculation of time along gc after fall through xg:

$$939\ 1/2 - 903\ 1/2 = 36.^{194}$$

Scheme of calculation: $t(gc|x) = t(xgc) - t(xg)$.

---

[189] 184 verso, C20.
[190] 184 verso, C15.
[191] 184 verso, C12.
[192] 184 verso, C14.
[193] 184 verso, C13.
[194] 184 verso, C16.

13.17 Folio 184

- C19, results achieved noted as list:

  grx longitudine 869 1/2
  tempus motus per au 173
  tempus motus per xrg 903 1/2
  mediam inter $grx\ xrgc$ 904
  tempus per totam $xrgc$ 939 1/3
  tempus per gc 36.[195]

Calculation of the time along inclined plane combination arc, t(arc)

- C21, calculation of xr:

$$869\ 1/2 - 113 = 756\ 1/2\ rx.^{[196]}$$

Scheme of calculation: $xr = xrg - rg$.

- C08 and C21, calculation of rgc:

$$113 + 70\ 1/2 = 183\ 1/2\ rgc.^{[197]}$$

Scheme of calculation: $rgc = rg + gc$.

- C03, calculation of mean proportional

$$\sqrt{756 * 183} = 372$$

Galileo calculated the mean proportional of rc and rx, $mp(rc|xr)$, most likely because he erroneously assumed that the time along rc from rest in r would be measured by rc in which case the calculated value would represent the time of motion along xr from rest in x. Galileo would immediately have recognized that the calculated value of 372 is much too small for $t(xr)$ and corrected his mistake with the next calculation.

- C11, calculation of time along xr by law of fall and length time proportionality:

$$\sqrt{939 * 756} = 842^{[198]}$$

Scheme of calculation: $t(xr) = mp(xr|xrc)$, because $t(xc)=xc$ by length time proportionality and distance time coordination.

---

[195] 184 verso, C19.
[196] 184 verso, C21.
[197] 184 verso, C08 and C21.
[198] 184 verso, C11.

- C17, calculation of time along rc after fall through xr (C17):

$$939 - 842 = 97^{199}$$

Scheme of calculation: $t(rc|x) = t(xrc) - t(xr)$.
- C06, calculation of time along ar by the law of fall (C06, C17):

$$\sqrt{180 * 145} = 161[1/2]^{200}$$

Scheme of calculation: $t(ar) = mp(ar|ad)$, because $t(ad) = ad$ by distance time coordination.
- C17 calculation of time along path arc:

$$97 + 161\ 1/2 = 258\ 1/2^{201}$$

Scheme of calculation: $t(arc) = t(ar) + t(rc|x)$.
- C17, list of results completed:

 tempus $x[r]$ 372[202]
 tempus per rc 97
 tempus per arc m[inutum] 258 1/2[203]

Remaining content

- C05, calculate length of vertical traversed in the same time as motion along the upright arc:

$$235^2/180 = 306.^{204}$$

The number 235 which reoccurs in several calculations represented Galileo's best guess for the time of motion along an upright quarter arc if the time of fall through the radius of this arc was assumed to be measured by 180 time units (cf. the discussion of 115 verso in this appendix).The calculated value of 306 represents the length of a vertical which is traversed in free fall in this same time. Scheme of calculation: $t_{radius}/t_{235} \sim radius/mp(radius|vertical)$; hence, since by

---

[199] 184 verso, C17.
[200] 184 verso, C06.
[201] 184 verso, C17.
[202] Erroneous result noted, see discussion of C03 above. The list is completed by using the recalculated, correct value for $t(xr)$.
[203] 184 verso, C17. Minutum is the name given to the time unit in which the calculations are carried out on 166 recto.
[204] 184 verso, C05.

distance time coordination $t_{radius} = radius$, $t_{235} = mp(radius|vertical)$, and thus $vertical = t_{235}^2/radius$.
- C10, calculate arc length:

$$180 * 11/7 = 283.^{205}$$

3 1/7 is the value Galileo usually used in his calculations for π and hence 11/7 is $\pi/2$. Thus Galileo calculated the length of the quarter arc whose radius measured 180 distance units. Galileo repeated the same calculation for a radius of 100,000 on folio 166 recto.

Thus on the arc a distance of 283 was traversed in 235 time units while falling vertically, as calculated above, a distance of 306 would be traversed in that same time.

13.18 Folio 185

Folio 185 bears a crown watermark and hence can be attributed to Galileo's Paduan period. The content of folio 185 links up with Galileo's considerations on page 186, on which Galileo started to construct a proof for the law of the broken chord and with his considerations on the double folio 150, 149, where the construction of this proof is essentially completed. Just as in the version of the proof of the law of the broken chord which Galileo published in the *Discorsi*, in the draft on double folio 149, 150, the geometrical part of the argument of the proof is attributed to lemmata. In the entry on double folio 149, 150, these lemmata are treated as if they had already been proven or at least as if Galileo was about to prove them without further delay ("[o]stenditur autem, per lemmata"). Folio 185 contains the elaborations of the second and the third lemmata that Galileo used in the published version of the proof of the law of the broken chord in the *Discorsi*. This would indicate that the content of folio 185 was drafted shortly after that of 186 and in close connection to that of double folio 149, 150. This hypothesis is confirmed by the construction and lettering of the motion diagram as well as the timeline.

Folio 185 recto

- Draw central motion diagram (D03A, see Fig. 13.21) in accordance with the motion diagram on 186 verso.

    The upright arc considered subtends an angle of 60 degrees. This was probably a conscious choice, as it would allow to discern how the ratio of the times of

---

[205] 184 verso, C10.

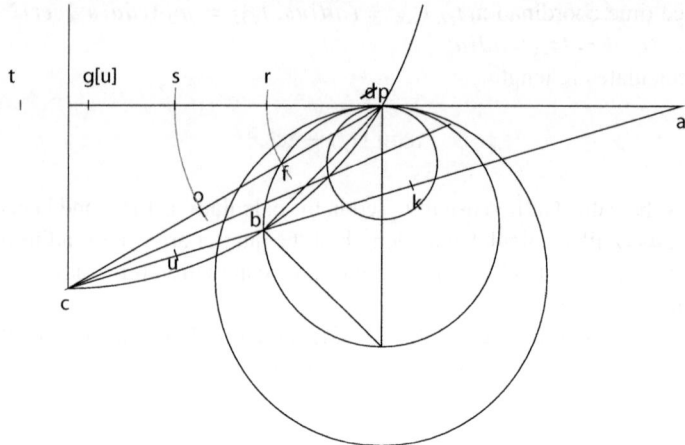

Fig. 13.21 Schematic representation of central diagram (D03A) on folio 185 recto. The light gray arcs are not part of Galileo's diagram but have been added to illustrate how the timeline was constructed. Point g had initially been marked as u

motion along single chord and broken chord spanning the same arc depends on the angle subtended by the arc. That, contrary to Galileo's expectation, the angle subtended by the arc had a crucial influence on the problem at hand had emerged from considerations documented on 166 recto (see the discussion 166 recto in this appendix). The timeline tq which is part of the central diagram is conformably constructed. The construction followed the same scheme by which the timeline had been constructed on folio 186 recto. Indeed the points on the timeline had initially been keyed t, u, s, r, and d exactly as on 186 verso. Only subsequently Galileo renamed points u and d of the timeline to g and q, respectively, most likely in order to avoid confusion between the lines representing times in the timeline and lines that were part of the spatial geometry in motion diagram, which could potentially have arisen from the fact that the letters d and u were also used to mark points of the spatial geometry. As on 186 verso, points s and r were produced by projecting the distances df and do onto the timeline beginning at q. Thus qr represents the time of motion along the chord df and qs the time along dc. For the way which qg and qt representing the times of motion along ab and ac, respectively, can be constructed, see Chap. 7.

- C01, calculate do:

$$\sqrt{101\ 1/2 * 33} = 57\ 100/114.^{206}$$

Scheme of calculation: $cd = 101\ 1/2,\ df = 33,\ do = mp(cd|df)$.

---

[206] 185 recto, C01.

13.18 Folio 185

- C02, calculate ck and au:

$$ab\ 122,\ bc\ 57^{207}$$

$$[ac = ab + bc = 122 + 57 = 179]$$

$$\sqrt{179 * 122} = 147229/294 [= au]$$

$$\sqrt{179 * 57} = 101 [= ck]^{208}$$

Scheme of calculations: $ck = mp(cb|ca)$ and $au = mp(ab|ac)$ The calculations on the page suffice to show that motion along the broken chord dbc is completed in less time than motion along the chord dc, as this, as Galileo had shown on 150 recto, was equivalent to $oc > uc$ which as the results of Galileo's calculations showed, indeed held in the case at hand.

The numerical values for cd, df, ab, and ac are consistent with the geometrical construction. Two things are however noteworthy. As the chord cd subtends an angle of 60 degrees, its length 101 1/2 corresponds to the radius of the arc dbc. The radius was thus apparently not chosen in such a way that a table of sines or chords could be exploited directly. Secondly, with the radius measuring 101 1/2 the length of 57 100/114 for chord cb implies the angle subtend by the chord is 33 degrees and that thus cb is inclined to the horizontal about 16.5 degrees. This indeed corresponds to the angles that can be measured off from Galileo's construction. This, in particular, implies that point b does not exactly half the arc dbc. When inspecting the manuscript, I did not measure any of the distances in the diagram in a modern unit. Yet if in a digital copy the height of the folio is measured in the units of the drawing, the height results to 312 units which can be compared to the height of the folio page of 300 millimeters as it has been measured by a working group of the Max Planck Institute for the History of Science. From these numbers, the unit of the drawing calculates to 0.962 mm, and it can thus be identified with Galileo's punto. It thus emerges as the most likely scenario that Galileo determined the size of his construction more or less arbitrarily and simply retrieved the lengths of the distances used in the calculations by measuring them off the diagram.

- D02A, diagram to be used for the elaboration of the geometrical lemma drawn (Figs. 13.22, 13.23, and 13.24).

Fig. 13.22 Drawing associated to the textual entry on 185 recto

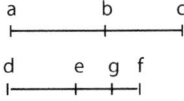

---

[207] This line has not been transcribed in the electronic representation of Galileo's *Notes on Motion*.

[208] 185 recto, C02.

Fig. 13.23 Geometrical construction on 185 verso

- T1, second lemma associated to the law of the broken chord drafted (T1 and D02A):

  > Lemma. Sit linea *ac* maior ipsa *df*, et habeat *ab* ad *bc* maiorem rationem quam *de* ad *ef*: dico, *ab* ipsa *de* maiorem esse. Quia enim *ab* ad *bc* maiorem rationem habet quam *de* ad *ef*; quam rationem habet *ab* ad *bc*, hanc habebit *de* ad minorem quam *ef*. Sit *eg*, et quia *ab* ad *bc* est ut *de* ad *eg*, erit ut *ca* ad *ab*, ita *gd* ad *de*: est autem *ca* maior *dg*: ergo et *ba* ipsa *de* maior erit.[209]

  Galileo proves that if $ac > df$ and $ab/bc > def/df$, then also $ab > df$.

Folio 185 verso Folio 185 verso bears but a single diagram which repeats a construction of 151 verso. The only difference is that the upright arc on which the construction is based does not subtend a right angle as on 151 verso but of about 80 degrees. As has been argued (compare the discussion of 151 verso in this appendix), the construction is that underlying the argument of the third lemma associated with the proof of the law of the broken chord. As the law of the broken chord is claimed to be valid for any upright arc, subtending an angle of 90 degrees or less the lemma likewise has to be valid for arcs smaller than 90 degrees, and the diagram may have served to reassure that this was indeed the case.

---

[209] 185 recto, T1.

13.19 Folio 186

The watermark of folio is a crossbow measuring 35 mm, and the folio can thus, based on the type of paper, be attributed to Galileo's Paduan period.[210] The recto side of the folio contains a diagram of unclear purpose, in which the numbers from 1 to 13 are distributed evenly over the circumference of a circle. The verso side documents Galileo's first attempt to construct a proof for the law of the broken chord.

Folio 186 verso Folio 186 verso contains a central diagram (D01A, cf. Fig. 13.24) with two paragraphs of text above (T1 and T2) and a note added in the margin to the left of the diagram (T3). In addition some sketchy and incomplete notes regarding a numerical example for the motion situation in question are scattered on the page (C01–03). On the folio, Galileo had begun to elaborate a proof of the law of the broken chord. The first two entries represent two different, independent attacks on the problem of the law of the broken chord. The note in the left margin contains an observation Galileo made in the course of these attempts that he considered noteworthy.

- D01A, central motion diagram drawn.
  Notably the upright arc *dbc* is not a quarter arc but subtends an angle of five sixths of a right angle.
- T1A, problem outlined:

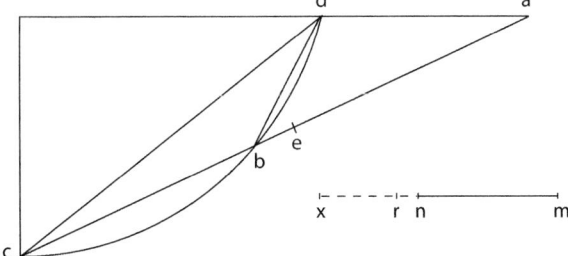

Fig. 13.24 Motion diagram on page 186 verso reduced to the elements necessary for the reconstruction of Galileo's considerations documented by the first paragraph on the same page. For the uninked lines in Galileo's construction, see Fig. 7.1

---

[210] The only other folio of the *Notes on Motion* bearing the exact same watermark is folio 192.

> Dicimus, tempus quo mobile permeat lineas *db*, *bc* brevius esse tempore quo permeat solam *bc*.[211]

It is claimed that body will traverse the *dbc* in a shorter time than the chord *dc*, i.e., the law of the broken chord is stated.

- T1A, first attempt at a solution:

> Sit *ae* aequalis *bc*: si itaque fuerint motus initia puncta *a*, *b*, eodem tempore peragentur lineae *bc* et *ae*. Sit tempus quo conficitur *ae*, vel *bc*, ipsum *mn*, et quam rationem habet *ae* ad mediam inter *ae*, *ac*, hanc habeat *nm* ad *nx*; erit *nx* tempus totius *ac*: quam vero rationem habet *ca* ad mediam inter *ca*, *ab*, hanc habeat tempus *xn* ad *nr*; erit *rn* tempus ipsius *ab*, *rx* vero ipsius *bc* post *ab* (quam *xr* oportet minorem esse ipsa *mn*).[212]

In symbolic notation Galileo's argument can be transcribed as follows:

$t(db) + t(bc|db) < t(dc)$, statement.
$ae = bc$, definition *ae*.
$t(ae) = t(bc)$, by translation invariance.
$t(ae) = nm$, by definition.
$t(bc) = nm$ by above.
$nm/nx = ae/mp(ae|ac)$, definition *nx*.
$t(ae)/t(ac) = ae/mp(ae|ac)$, by law of fall.
$t(ac) = nx$, by above. $nx/nr = ca/mp(ca|ab)$, definition *nr*.
$t(ac)/t(ab) = ac/mp(ac|ab)$, by law of fall.
$t(ab) = nr$, by above.
$t(bc|ab) = t(ac) - t(ab) = rx$, above.
$t(bc|ab) = t(bc|db)$, by path invariance.
$t(dc) = t(bc)$, by law of chords.
$t(bc|db) < t(dc)$, necessary condition by above.
$xr < nm$, necessary condition stated for distances of timeline.

For motion along the broken chord *dbc* to be completed in less time than fall along the chord *dc* as Galileo wants to show, it is of course trivially necessary that the second half of the motion along *dbc*, that is, motion along the lower chord *bc* after motion from *d*, is likewise completed in less time than fall along the chord *dc* as Galileo has argued. In order to be able to restate the law of the broken chord in terms of relation between appropriate distances in the timeline, Galileo would need to construct a distance on the timeline *xm* which represents the time of fall through *db* from rest at *d*. This, however, cannot easily be done because motion along *db* proceeds over a different inclination than the motions over *dc* or *ac* Galileo had regarded thus far and hence the law of fall cannot immediately be applied. At this point Galileo introduced a resourceful argumentative move. By applying the law of chords, he transformed his statement in such a manner that the time through *db*, which had proven difficult to construct in his first approach, could be eliminated from the argument. He introduced a circle through *b* with the sublimity at *d*, cutting

---

[211] 186 verso, T1A.
[212] Ibid.

13.19 Folio 186

the line dc at f (cf. Fig. 5.5 in Chap. 5). Then, by law of chords, the distances df and db were traversed in the same time and what remained to be shown was that that $t(bc|a) < t(fc|d)$.

- T1B, new approach to the problem begun by formulating a reduced statement:

    Ostendatur, citius transiri bc post ab quam fc post df.[213]

    This reduced statement is trivially equivalent to the original statement of the law of the broken chord:
    $t(bc|ab) < t(fc|df)$, statement.
    $t(df) = t(db)$, by law of chords.
    $t(bc|ab) = t(bc|db)$, by path invariance.
    $t(db) < t(dbc)$, statement of law of the broken chord.
- T1B, argument for reduced statement begun:

    Sit ds tempus quo peragitur tota dc, vel bc, et quam rationem habet media inter cd, df ad df, hanc habeat tempus sd ad dr; constat, tempus ipsius fc esse rs: quia vero tempus ipsius bc, seu ae, est idem ds, fiat ut ea ad mediam inter ea et totam ac, ita sd ad dt, eritque dt tempus totius ac. Quod si rursus fiat ut tota ca ad mediam inter ca, ab, ita td ad dv, erit vt tempus ipsius bc post ab: hoc autem ostendendum est, esse minus ipso rs.[214]

In symbolic notion the argument can be paraphrased as follows (for the associated diagram, see Fig. 5.5 in Chap. 5):

$ds = t(dc)$, definition ds.
$ds = t(bc)$, by law of chords.
$ds/dr = mp(cd|df)/df$, definition dr.
$dr = t(df)$, by law of fall and definitions.
$t(dc) = t(df) + t(fc|df)$.
$rs = t(fc|df)$, by above.
$t(bc) = t(ae)$, by translation invariance.
$sd/dt = ea/mp(ea|ac)$, definition dt.
$dt = t(ac)$, by law of fall and definitions.
$dt/dv = ac/mp(ac|ab)$, definition dv.
$dv = t(ab)$, by law of fall and definitions.
$t(ac) = t(ab) + t(bc|ab)$.
$vt = (bc|ab)$, by above.
$vt < rs$, reduced statement by above.

Galileo's approach is rather straightforward, and as a result, he is indeed able to transform the statement in terms of the relationship between times into a statement about the relationship between lines in the external timeline constructed from the spacial geometry of the problem. What remained to be shown after this entry was

---

[213] 186 verso, T1B.
[214] Ibid.

that vt is smaller than rs. Galileo attacked this problem and continued his work on folio 150 recto.

Remaining entries

- T2, noteworthy observation recorded:

  > Nota. Sit in circumferentia utcumque ducta *do*, et iungatur *co*: dico dico [sic], citius moveri ex *d* in *o* quam ex *o* in *c*. Ostensum enim est aequali tempore moveri ex *o* in *c*, atque ex ex [sic] [d] in *c*; verum ex *d* in *o* patet celerius fieri motum quam ex *d* in *c*.[215]

  Galileo states that $t(do) < t(oc)$. This follows trivially since $t(do) < t(doc)$ and by law of the broken chord $t(doc) < t(oc)$. For a discussion of the significance of this note, see Chap. 5 Sect. 5.1.

- C01–C03, numerical example:

  > r[?][adi]x[?] 25 r[4?][adi]x[?] 25 16
  > 25+16=41
  > [$\sqrt{32}$ =]5 7/10
  > [*tempus*]*abc* 16
  > [*tempus*]*ab* 12
  > [er]go. *cb* post *ab* 4 r[?][adi]x4[?] 32 i[dest] 9 7/10

As the main text on the page was written around the entries comprising this numerical example, Galileo seems to have noted it before he started to draft his argument. Galileo determines that if the time along *ab* measures 12 units, the overall time along the broken chord *dbc* is 9 7/10. The information provided is to spare to reconstruct how precisely Galileo came up with the numbers. All that can be said is that they are compliant with the geometrical construction. The ratio of *ac* and *ab* with measures of the lines taken off Galileo's construction is reckoned to 1.785. The square root is 1.336 in accordance with the ratio of the times Galileo gives for *ac* and *ab*. The ratio between *ab* and *db* can be determined from Galileo's diagram to be 2.125. According to the length time proportionality, this should correspond to the ratio of the times for motion along *ab* and *db* which it indeed does.

13.20 Folio 189

Folio 189 bears no watermark. The paper is thin and light brown in color. With regard to content, the notes preserved on folio 189 resolve articulately into three different groups. First of all 189 verso contains the detailed experimental record of the pendulum plane experiment including considerations and calculations related

---

[215] 186 verso, T2.

13.20 Folio 189

to it. Secondly, on the same folio page, albeit in written in a different direction, considerations regarding accelerated motion along inclined planes of different height and inclination are preserved. These considerations eventually resulted in a proposition published as Proposition V of the Second Book of the Third Day of the *Discorsi*. Thirdly, the recto of the folio bears a full numerical example of a broken chord motion situation. In a number of calculations on the same and other folios, Galileo accommodated the theoretical results from 189 recto to his experimentally measured data and thus the content of the page is linked directly to the pendulum plane experiment.

Folio page 189 verso The two groups of notes that are preserved on folio 189 verso pertain to two distinct subjects. Notes belonging to each one of these two groups are easily distinguished as they were written in different writing directions. The two distinct groups of entries are identified in Fig. 13.25 and will be discussed separately in the following under the headings *pendulum plane experiment* and *motion along inclined planes*, respectively. Judging from the distribution of the content, the entries pertaining to the latter group were put on the page second.

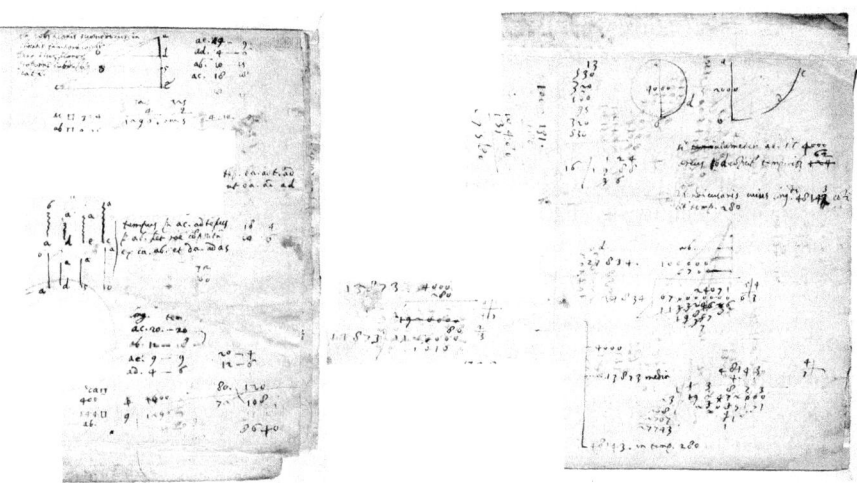

Fig. 13.25 Folio 189 verso. The folio has been artificially cut and reassembled with one of the halfs turned 180 degrees. Galileo's entries pertaining to and discussed under the heading *pendulum plane experiment* are on the right half. Those discussed under the heading *motion along inclined planes* are contained on the left half. (Courtesy of Ministero dei Beni e le Attività Culturali – Biblioteca Nazionale Centrale di Firenze. Reproduction or duplication by any means prohibited)

The pendulum plane experiment Galileo times the swinging of a pendulum as well as the rolling of a ball down along a long gently inclined plane. The quarter period of the pendulum is measured and then compared to the time of free fall along a vertical distance equal to twice the pendulum's length derived by a theoretical inference from the measured data.

In summary Galileo's timing of the period of a pendulum of a length of 2000 units yielded that the time it takes the pendulum to complete a quarter oscillation was 62 time units. The time it took a ball to roll down an inclined plane of a length of 6700 units and of 8 degree inclination was measured to 280 time units. By geometry and an application of the law of chords, Galileo concludes that a vertical measuring 48,142 units would be traversed in naturally accelerated motion in the same time as the inclined plane had been in the experiment. From this Galileo concluded, by applying the law of fall, that a vertical distance of a length of 4000 units, that is, twice as long as the pendulum's length, would be traversed in 80 2/3 time units. By virtue of the law of chords, this would likewise be the time to fall freely along a chord spanning the quarter swing of the pendulum timed in the experiment (cf. Chap. 4, in particular Fig. 4.3).

- D01B, motion diagram. Pendulum of length ab swinging along the arc cdb.[216]
- C01 (first part), time measurement, numbers added up:

$$13 + 530 + 320 + 180 + 95 + 320 + 530 = 1988.[217]$$

The number 1988 represents the time measured for 8 full oscillations or swings of a pendulum, i.e., for 16 half swings. That this number is obtained by adding up smaller quantities is characteristically due to the method of time measurement employed.[218]

- C01 (second part) measured time divided by 16:

$$1988 : 16 = 124$$

Dividing the overall time measured for pendulum motion yielded the time per half swing, i.e., half the pendulum's period.
- T3A, pendulum data recorded: First version:

If the semi-diameter ab is 2000, the arc bd will be traversed in 124.[219]

---

[216] 189 verso, D01B.

[217] 189 verso, C01.

[218] As argued in Chap. 4 Galileo most likely used a water clock for measuring intervals of time in the experiment. If the flow rate is constant over the time of measurement, the weight of the water that has flowed out during measurement can be used as a measure of the time elapsed. The smaller quantities which are being added up could thus be the weights placed on a balance to weigh the water or else the volume of containers filled with the water that had run out.

[219] 189 verso, T3A.

13.20 Folio 189

First correction:

> If the diameter *ab* is 4000, the arc *bd* will be traversed in 124.

Instead of being assigned to the radius, Galileo assigns the label *ab* to be the diameter of the circle representing the pendulum. This was fitting as it is the diameter rather than the radius which would play the crucial role in the successive considerations. Galileo changed his diagram correspondingly (see below).[220]
Second correction:

> If the diameter *ab* is 4000, the arc *bd* will be traversed in 62.

Correction from time for half swing to time for quarter swing.[221]
- D01A, diagram revised. Galileo canceled his original diagram by striking it through and, to the left of it, drew a new diagram in which, according to the correction in the text, *ab* marks diameter instead of the radius of the circle.
- C02, radius calculated:

$$100000 * 6700 : 27834 = 24071.$$[222]

Galileo calculates the radius of a circle in which the inclined plane of his experiment (length 6700 units, inclination 8 degrees) could be inscribed as a chord. He reads off from a table of chords the length of a chord subtended by an arc of 16 degrees in a circle whose radius measures 100,000 to be 27,834. Otherwise Galileo may have used a table of sines in which case he read off the sine of 8 degrees in a circle whose radius measures 100,000 as 13,917 and doubled this value to get the length of the chord.[223]

---

[220] After this correction Galileo used *ab* consistently to refer to the diameter instead of the radius.

[221] Galileo had apparently counted full periods, i.e., the number of times the pendulum bob had returned to his initial position during the measurement. His corrections seem to be due to the realization that dividing the time measure by two times the number of periods counted had given half the pendulum period of the pendulum and not the quarter period he was interested in for his successive theoretical investigations, i.e., the time it took the pendulum to move from one of the turning points of motion to the lowest, the equilibrium position of the bob.

[222] 189 verso,C02A.

[223] The length of a chord subtending an arc of a given angle is twice the sine of half that angle in the same circle. I have not been able to identify which sine table Galileo used concretely. The value Regiomontanus gives for the sines of 8 degrees for a circle of radius 10,000,000 is 139,173. See Peuerbach and Regiomontanus (1541, E3). Hill (1994) assumes that the inclined plane was not 6700 but twice as long, namely, 13400 punti, and was inclined at an angle of about 16 degrees and 10 minutes. This is based on the not further justified surmise that the number 27,834 represents the length of half the chord. Looking up in Copernicus' table of sines (Kopernikus 1542), the value closest to 27,834 Hill read off an angle of 16 degrees and 10 minutes. Whereas a chord length 27,834 can directly be read off from the sine tables, Hill's alleged chord length of twice that value does not correspond to any of the angles listed in steps of 1 minute. Moreover, Hill's interpretation leads together with the assumption that the unit in which Galileo gave the lengths corresponds to about 0.96 mm to the rather implausible conjecture that Galileo used an almost 13-meter-long

Fig. 13.26 Auxiliary diagram (D02A) for Galileo's considerations regarding the pendulum plane experiment on folio 189 verso

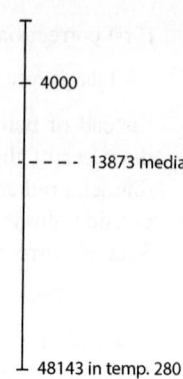

With this information the radius of a circle in which a chord of length 6700 is subtended by an arc of 16 degrees can be determined by the rule of three: $radius_1/chord_1 = radius_2/chord_2$ and hence $radius_1/chord_1 * chord2 = radius_2$

This calculation gave Galileo a value for the radius of 24,071, and multiplication by 2 yielded a diameter of 48,142 which Galileo noted in the next entry.

- T3B, theoretical inference from experimental results recorded:

    perpendicularis cuius longitudo 48143, conficitur tempore 280.[224]

Galileo had measured the time to fall along an inclined plane and a length of 6700 to 280 time units. He had then by geometry determined the length of the diameter of a circle in which the inclined plane of the experiment could be inscribed as a chord sharing the lowest point of the circle. According to the law of chords, this vertical diameter would be traversed in the same time as the inclined plane had been in the experiment, and this is what Galileo noted.

- C03, calculation of the mean proportional between vertical diameter and twice pendulum length:

$$\sqrt{48143 * 4000} = 13873^{225}$$

- D02A, geometrical configuration schematically rendered (see Fig. 13.26).

---

inclined plane, which at an inclination of 16 degrees must have been elevated more than 3 1/2 meters at one end.

[224] 189 verso, T3B. Galileo had originally written 48142, which he corrected to 48143 most likely to account for fractions.

[225] 189 verso, C03. As Hill has already remarked, the actual result ought to be 13877, but Galileo had failed to carry a one when calculating the product.

13.20 Folio 189

Fig. 13.27 Schematic representation of a diagram (D03A) on folio 189 verso

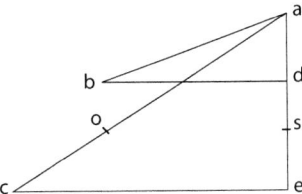

- C06, time of fall through the vertical diameter of the pendulum calculated:

$$4000/13873 * 280 = 80\ 2/3.^{226}$$

Scheme of calculation: $diam._{pend.}/mp(diam._{long.}|diam._{pend.})*t(diam._{long.}) = t(diam._{pend.})$ (by law of fall).

The derived time of 80 2/3 of fall through a vertical distance, equal to twice the length of the pendulum that Galileo had used in the pendulum plane experiment, is the end result of the first stage of the evaluation of the experimental data (cf. Chap. 4). The experimental data and the result of this first stage were further elaborated on by Galileo in a second stage (cf. Chap. 6).

Motion on inclined planes Galileo states an (erroneous) assumption concerning times in naturally accelerated motion along inclined planes of different height and inclination. He then elaborates a numerical example which shows that his assumption is wrong. This eventually leads to the formulation of revised statement, which is in turn tested using the numerical example already established and also using a second numerical example.

- D03A, motion diagram drawn (Fig. 13.27).[227]
- T1, assumption stated:

    In [duo]bus planis quomodocumque inclinatis tempora casuum habent ipsorum planorum proportionem subduplicatam.[228]

Statement of an (erroneous) assumption concerning naturally accelerated motion along inclined planes of different height and inclination:
With reference to the motion diagram, the initial assumption can be expressed as: $\sqrt{t(ab)}/\sqrt{t(ac)} \sim ab/ac$ and hence $t(ab)/t(ac) = ab^2/ac^2$

---

[226] 189 verso, C06.
[227] 189 verso, D03A.
[228] 189 verso, T1.

- C04, (first) numerical example recorded in distance-time table[229]:

	[longitudine]	[tempus]
ae	9	9
ad	4	6
ab	10	15
ac	18	18

- C12, (second) numerical example recorded in distance-time table[230]:

	longitudine	tempus
ac	20	20
ab	12	18
ae	9	9
ad	4	6

- C11, initial assumption tested with values from first numerical example[231]:

  ac □ 324   ab □ 225

  Squares ab^2 and ac^2 calculated with values from first numerical example(C11). Based on these numbers, the initial assumption proves wrong (conjectured) for the first numerical example because: $18/15 = t(ac)/t(ab) \neq ac^2/ab^2 = 324/225$

- C13, initial assumption tested with values from first numerical example[232]:

  ca □ 400   ab □ 144

  $20/18 = t(ac)/t(ab) \neq ac^2/ab^2 = 400/144$

- C05, calculation:

$$324 * 4 = 1296$$
$$225 * 9 = 2025$$

Galileo calculated the products of the squares of the lengths of the inclined planes and the heights of the respective other planes. Based on the solution he will finally formulate it can be conjectured that he wanted to test here whether: $ab^2/ac^2 * ae/ad \sim t(ab)^2/t(ac)^2$.

---

[229] 189 verso, C04. None of the calculations to establish the values in the table can be identified. In view of the simplicity of the example, Galileo could easily have done the required calculations in his head. The values testify that the law of fall and length time proportionality were resorted to in constructing the example. The first entry in the table, the length of the distance *ae*, has been corrected from a now illegible value to 9. As numbers underneath the distance-time table suggest that Galileo had originally chosen *ae* to measure 10 units, but realized that no smaller natural number could be found such that the mean proportional between 10 and this number, representing the time of motion along *ad*, would likewise be a natural number.

[230] 189 verso, C12.

[231] 189 verso, C11.

[232] 189 verso, C13.

13.20 Folio 189

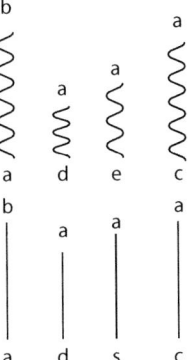

Fig. 13.28 Schematic redrawing of the motion time ratio diagram on 189 verso (D04A)

- C10, same as above for second numerical example:

  1600= 400*4 1296 = 144*9

  Indeed $1600/1296 = t(ac)^2/t(ab)^2 = 20^2/18^2$.
- T4, state proportion ensuing from length time proportionality:

  tempus *ba* ad t[empus] *ad* ut *ba* ad *ad*.[233]

  The length time proportionality is explicated for distance time pairs of the given motion situation.
- D04A, distance-time ratio diagram drawn (see Fig. 13.28).
- T2, revised assumption stated:

  tempus per *ac* ad tempus per *ab* habet rationem compositam ex *ca, ab* et *da* ad *as*[234]

  Correct proportionality inferred from distance-time ratio diagram and stated as a new hypothesis: $t(ac)/t(ab) = ca/ab * da/as$
- C08, test of revised assumption for first numerical example:

  $$18 * 4 = 72$$
  $$10 * 6 = 60^{235}$$

Galileo calculates $ca * da = 18 * 4 = 72$ and $ab * as = 10 * 6 = 60$. According to the revised assumption, the ratio of these two numbers has to be the same as the ratio of the times t(ac) and t(ab) that is of 18 to 15. It is immediately obvious that this holds.

---

[233] 189 verso, T4.
[234] 189 verso, T2.
[235] 189 verso, C08.

- C09 (first part), test of revised assumption for second numerical example:

$$[20 * 4 =]80$$

$$[ab * as = 12 * 6] = 72^{236}$$

As above Galileo, this time for the second example calculates $ca * da = 20 * 4 = 80$ and $ab * as = 12 * 6 = 72$. It is immediately obvious that the ratio of these two numbers is the same as the ratio of the times t(ac) and t(ab) that is the ratio of 20 to 18.
- C09 (second part), test of equivalence:

$$80 * 108 = 8640$$

$$72 * 120 = 8640^{237}$$

Remaining entry Besides the entries pertaining to the two distinct groups of notes discussed above, Galileo squeezed one additional calculation on the page, written in yet another writing direction, namely, with the page upright in its modern orientation.
- C07, rescaling:

$$(131 * 180) : 100 = 235\ 80/100.^{238}$$

For motion along a path composed of eight conjugate chords inscribed into an upright quarter arc of radius 100,000, Galileo had calculated a time of motion of 131,078 given that the radius was traversed in 100,000 (see 183 verso and the discussion in Chap. 5). Thus the ratio of 131 and 100 corresponds to the ratio of the time of motion along the path of the eight chord approximation to the time to fall through the vertical radius, which in the calculation is scaled to the unit of the base 180 approach. It is likely this is the calculation, which is the source of Galileo's choice to accept a value of 235 as the best approximation for the time of motion along the arc. Galileo used this value in his calculations on 115 verso and 184 verso.

---

[236] 189 verso, C09.
[237] 189 verso, C09.
[238] 189 verso, C07.

13.20 Folio 189

Fig. 13.29 Schematic representation of central motion diagram on 189 recto

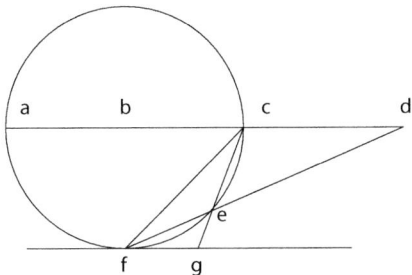

Folio 189 recto 189 recto contains a complete numerical example in which Galileo calculated the time of motion along a broken chord inscribed in the lower upright quadrant of a circle. The radius of the circle measures 100 units, and conventionally by distance time coordination, it is assumed that the time of fall through the radius measures 100 time units. The motion situation is diagrammatically depicted in a motion diagram (D02B); the calculations to determine the times of motion are contained on the page (C03–C06), and the results are neatly written down in the distance-time table (C02). Finally, the time of fall along the broken chord calculated is accommodated to the experimental data to allow for a comparison of the time of motion along a broken chord inscribed in the arc of a pendulum swing with the experimentally measured quarter period of that pendulum.

- D02B, motion diagram (see Fig. 13.29) drawn.
  Galileo aims at inferring the time of motion along the broken chord cef inscribed in the lower upright quadrant of a circle with each of the individual conjugate chords ce and ef spanning arcs of 45 degrees.
- C04, distances fd and ed calculated:

$$141542^2 : 76536 = 261760$$
$$261761 - 76536 = 185225.^{239}$$

Scheme of calculations: $fc * cd/fe = fd$ (by similarity of triangles fcd and fce and $fd - fe = ed$.
- C02, distance-time table (begun)[240]:

---

[239] 189 recto, C04.
[240] 189 recto, C02.

		temp[us]
bf,bc	100000	
cf	141542	
fe	76536	
cd	141542	
bd	241542	
fd	261701	
ed	185225	
ce	76536	

Galileo assigns the radius bf a length of 100,000 units. With this choice most of the relevant distances could trivially be determined with the help of a sine table computed for a circle of radius 100,000. The remaining distances are derived by straightforward calculations. Galileo had already created a column to list the times of motion to be calculated as indicated by the column header "temp[us]." Yet he decided not to calculate the times directly.

- C02, distance-time table (continued):

			temp[us]
100	bf,bc	100000	
141 1/2	cf	141542	
76 1/2	fe	76536	
141 1/2	cd	141542	
241 1/2	bd	241542	
262	fd	261761[241]	
185[242]	ed	185225	
76 1/2	ce	76536	

Galileo scaled all distances by the factor 1/1000 to be able to work with smaller, more convenient numbers when calculating the times of motion. The scaled values were added in a new column of the distance-time table.

- C05 time of fall along de calculated:

$$\sqrt{262 * 185} = 220^{243}$$

By distance time coordination, Galileo assumes the time of fall along the radius bf to be measured by 100 time units. Then, by length time proportionality, the time of motion along df is likewise measured by the length of df, i.e., 262 units. Galileo entered this value in the distance-time table as the time of motion for df. If the time of fall along inclined plane fd is measured by fd, then the time of fall along the first part de of that plane is according to the law of fall measured by the mean proportional of fd and de as Galileo calculated. The resulting value of 220 was added to distance-time table as the time of motion for ed.

---

[241] Wrong in electronic representation of *Notes on Motion*.

[242] Wrong in electronic representation of *Notes on Motion*.

[243] 189 recto, C05.

13.20 Folio 189

- C06, time of fall along *ce* calculated:

$$220 * 76 1/2 : 185 = 90^{244}$$

Calculation of time of fall along first part of broken chord *ce*, $t(ce)$, by application of the length time proportionality according to which it holds that $t(de)/t(ce) = de/ce$. This allows Galileo to calculate the time of motion along *ce*. The resulting value 90 entered in the distance-time table as the time of motion for *ce*.

- Time of fall along *ef* after initial fall through *de* calculated:

$$[262 - 220 =] 42$$

The the trivial substraction needed not to be carried out on paper. The resulting value of 42 was added to distance-time table as the time of motion along *ef* after initial fall through *de*.

- C02, distance-time table (completed):

temp[us]				temp[us]		
[100]	100		bf,bc	100000		
[141 1/2]	141 1/2		cf	141542		
[42]	76 1/2		fe	76536		
	141 1/2		cd	141542		
	241 1/2		bd	241542		
262	262		fd	261761[245]		
220	185[246]		ed	185225	p[er]	ed200
90	76 1/2		ce	76536		

- C08, time of fall along *cef* calculated and noted:

$$[90 + 42 =] temp[us]\ cef\ 132^{247}$$

Assuming path invariance, the times $t(ce)$ and $t(ef)$ are added to yield the time of motion along the broken chord sought for.

- C07, rescaling of time of fall along *cef* to units of the pendulum plane experiment:

$$132/100 * 57 = 75\ 24/100.^{248}$$

---

[244] 189 recto, C06.
[245] Incorrectly transcribed in ERGNM.
[246] Incorrectly transcribed in ERGNM.
[247] 189 recto, C08.
[248] 189 recto, C07.

Scheme of calculation:

$t(rad.)_{189v\ u.}/t(br.\ chord)_{189v\ u.} = t(rad.)_{exp.\ u.}/t(br.\ chord)_{exp.\ u.}$

For similar geometrical configurations, the ratio between the time of motion along the radius and along the broken chord should not depend on unit choices and the absolute size of the system considered. Thus the time along the broken chord in time units of the pendulum plane experiment (in which the time along the radius is measured as determined on 90 verso by 57 units) amounts to 75 1/4. This time has to be compared to the quarter period of a pendulum measured to be 62 time units (cf. Chap. 4).

Remaining entries on 189 verso Besides the elaboration of a numerical example of accelerated motion along a broken chord, folio 189 recto contains some diagrams as well as scratch numbers with no apparent relation to the former considerations.

- D01A-3A and D02C, four diagrams of unknown purpose.[249]

- C01, two numbers:

$$70711,\ 70776[?].^{250}$$

70,711 is the sine of 45 degrees to the base of 100,000. Thus the number measures the distance of point e from the horizontal ad in the distance units Galileo had initially employed. A reading and purpose of the second number are uncertain.

- Two numbers:

$$80,\ 60$$

These two numbers which are added close to the top margin of the page are apparently unrelated to the remaining content on the page and may pertain to the considerations preserved on the obverse side of the folio.[251]

13.21 Folio 192

The watermark of folio 192 is a crossbow of approximately 35 mm in height. It shares this watermark with folio 186, on which Galileo began his attempt at

---

[249] (D01A, D02A, D02C, D03A).

[250] 189 recto, C01. These numbers are incorrectly rendered as 7071 and 7077 in the electronic representation of Galileo's *Notes on Motion*.

[251] These two numbers are not transcribed in the electronic representation of Galileo's *Notes on Motion*.

13.21 Folio 192

constructing a proof of the law of the broken chord. The recto side of the folio contains calculations of times of motion over various paths depicted in the central motion diagram on 166 recto. The verso side of the folio is blank.

Folio 192 recto The page contains calculations related to the motion situation depicted in the central diagram on 166 recto. On folio 192 recto, Galileo calculated the time to from rest at e to traverse the broken chord egc spanning an arc of 45 degrees (distances are referred to according to the point labels of the diagram on 166 recto, see 5.7). The calculations pertain to Galileo's base 100,000 approach characterized by the assumption that the vertical radius of the circle measures 100,000 distance units and the time elapsed during fall over this vertical radius 100,000 time units. The time which Galileo calculated for motion along the broken chord egc from rest at e amounts to 122,557. The value of 866 3/5 written down on 166 recto is obtained from this result by law of three.

- C03, calculation of time along αg (cf. Fig. 13.20) by law of fall:

$$[\sqrt{22677 * 100000}] = \sqrt{2267700000} = 46558.^{252}$$

Scheme of calculation: $t(\alpha g)/t(radius) \sim mp(\alpha g|radius)/radius \Rightarrow t(\alpha g) = mp(\alpha g|radius)$, because by distance time coordination $t(radius) = radius$.

- C04, calculation of time through eg by length time proportionality:

$$\text{tempus } 46558 * 39017 : ga\ 21677 = 83801.^{253}$$

Scheme of calculation: $t(ge)/t(\alpha g) = ge/\alpha g \Rightarrow t(ge) = t(\alpha g) * ge/\alpha g$.
- T4, result noted:

tempus per eg 83801.[254]

- T3, results for cgx, $t(cgx)$, and gx taken from folio 183 recto noted:

$cgx\ ldg$ 511259 tempus 511259
gx 472242[255]

Deleted as not usable.

~~$cgx\ ldg$ 511259 tempus 511259~~
~~gx 472242~~

---

[252] 192 recto, C03.
[253] 192 recto, C04. Galileo carried out the calculation twice because of a slight error in the first attempt.
[254] 192 recto, C03.
[255] 192 recto, T3.

Galileo had sought to exploit results already achieved and noted in the list on 183 recto. Yet these results had been calculated under the assumption that the point x was positioned on a horizontal through the starting point of motion a. What Galileo was seeking to infer on 192 recto, however, was the time of motion along cgx with the point x positioned on a horizontal through e. Thus his prior results could not be reused, and, likely upon noticing this, Galileo canceled his entry.

- C06, recalculation of cgx with respect to the new position of x by law of three:

$$29289 * 39017 : 17612 = 150127.^{256}$$

Scheme of calculation: $cgx/e\theta = cg/gq \Rightarrow cgx = e\theta * cg/gq$.

- C07, calculation of time through $e\theta$ by law of fall:

$$\sqrt{[29289 * 100000]} = \sqrt{2928900000} = 54119.^{257}$$

Scheme of calculation: $t(e\theta)/t(radius) = mp(e\theta|radius)/radius \Rightarrow t(e\theta) = mp(e\theta|radius)$ because by distance time coordination $t(radius) = radius$.

- C05, calculation of time through xgc by length time proportionality:

$$150127 * 54119 : 29289 = 277398.^{258}$$

Scheme of calculation: $t(xgc)/t(e\theta) = xgc/e\theta \Rightarrow t(xgc) = t(e\theta) * xgc/e\theta$.

- C10, calculation of $xg = xgc - cg$ and of time through gx based on values from 183 recto:

$$150127 - 39017 = 111110.$$

$$472242 * 46558 : 21677[= 1014284].^{259}$$

Galileo again erroneously used the value 472,242 for gx from the list of results on folio 183 recto. Midway through the calculation, he realized that the time through gx thus calculated would be much too big and deleted the entry to redo the calculation with the correct values.

---

[256] 192 recto, C06.
[257] 192 recto, C07.
[258] 192 recto, C05.
[259] 192 recto, C10.

$$150127 - \cancel{39017} = \overline{111110}$$

$$472242 * \cancel{46558 : 21677}[= 1014284].$$

- C09, recalculation of time through gx for the new, correct position of point x:

$$46558 * 111110 : 21677 = 238642.^{260}$$

Scheme of calculation: $t(xg)/t(\alpha g) = xg/\alpha g \Rightarrow t(xg) = t(\alpha g) * xg/\alpha g.$
- C08, calculation of time through gc from rest at e:

$$277398 - 238642 = 38756 \text{ per gc.}^{261}$$

Scheme of calculation: $t(gc|xg) = t(xgc) - t(xg)$ and by path invariance $t(gc|xg) =$ t(gc|eg).
- C01, calculation of time through egc:

$$[83801 + 38756] = 122557.^{262}$$

Scheme of calculation: $t(egc) = t(eg) + t(gc|eg).$
- C01, results summarized as list:

   tempus per gc 38756
   tempus per eg 83801
   tempus ambas egc 122557.^{263}

References

Damerow, P., Renn, J., & Rieger, S. (2001). Hunting the white elephant: When and how did Galileo discover the law of fall? In J. Renn (Ed.), *Galileo in context* (pp. 29–150). Cambridge/New York: Cambridge University Press.
Damerow, P., Freudenthal, G., McLaughlin, P., & Renn, J. (2004). *Exploring the limits of preclassical mechanics*. New York: Springer.
Drake, S. (1978). *Galileo at work: His scientific biography*. Chicago: University of Chicago Press.
Drake, S. (1979). *Galileo's notes on motion arranged in probable order of composition and presented in reduced facsimile* (Annali dell'Istituto e Museo di Storia della Scienza Suppl., Fasc. 2, Monografia n. 3). Florence: Istituto e Museo di Storia della Scienza.
Drake, S. (1987). Galileo's constant. *Nuncius Ann Storia Sci, 2*(2), 41–54.
Drake, S. (1990). *Galileo: Pioneer scientist*. Toronto: University of Toronto Press.

---

[260] 192 recto, C09.
[261] 192 recto, C09.
[262] 192 recto, C01.
[263] 192 recto, C01.

Hill, D.K. (1994). Pendulums and planes: What Galileo didn't publish. *Nuncius Ann Storia Sci,* 2(9), 499–515.

Kopernikus, N. (1542). *De lateribvs et angvlis triangulorum, tum planorum rectilineorum, tum sphaericorum, libellus eruditissimus & utilissimus: Cum ad plerasque Ptolemaei demonstrationes intelligendas, tum uero ad alia multa. additus est canon semissium subtensarum rectarum linearum in circulo.* Vittembergæ: Excusum per Iohannem Lufft.

Naylor, R.H. (1976). Galileo: The search for the parabolic trajectory. *Annales of Science, 33,* 153–172.

Peuerbach, J., & Regiomontanus, G. (1541). *Tractatvs Georgii Pevrbachii svper propositiones Ptolemaei de sinubus & chordis. item compositio tabularum sinuum/per ioannem de regiomonte,. adiectae sunt & tabulae sinuum duplices/per eundem Regiomontanum.* Petreius, Norimbergae

Vergara Caffarelli, R. (2009). *Galileo Galilei and motion: A reconstruction of 50 years of experiments and discoveries.* Berlin/New York/Bologna: Springer/Società Italiana di Fisica.

Chapter 14
Appendix: Documents

14.1 Letter from Paolo Sarpi to Galileo, 2 September 1602

PAOLO SARPI a GALILEO in Padova Venezia 2 settembre 1602.

Ecc.mo Sig.re P.rone mio Colen.o

Poichè li 25 miglia, per quanto siamo distanti, m'impedisce il discorrere con V. S., cosa che desidero sopra tutte le altre, voglio tentare di farlo con intermedio delle lettere, et al presente nel proposito ch'incominciai trattare con esso lei quando l'altro giorno fummo insieme, della inclinatione della calamita con l'orizonte.

Il nostro auttore molto raggionevolmente dice, quella non essere una attrattione, ma conversione più tosto, nascendo dalla virtù d'una et dell'altra, che vogliono essere situate in un certo muodo insieme, perilchè il più desiderato muodo di situarsi è quello quando per li poli: imperochè fa l'asse uno, et se ci è moto, ancora tutte le parti participano del moto non solo circa l'asse della grande, ma anco circa il suo; anzi forse si fa talmente uno, che perde il suo equinotiale, et fa accostare quello della grande, perdendo ambi dua li poli in che si congiongano, et facendo come d'un corpo li dua poli estremi. Ma se sono situate per li equinotiali, si vede anco la unione, havendo li assi paralleli et l'equinotiali in un piano, et participando il moto sopra quelli.

PAOLO SARPI to GALILEO in Padua Venice, September 2 1602.

Most Excellent Lord and My Master

Since the 25 miles, by which we are separated, hinder me to speak with Your Lordship, the thing I desire over all others, I will try to make it through letters, and now in reference to the argument [proposito] I began to treat with you when we were together the other day, about the inclination of a magnet with respect to the horizon.

Our author very reasonably says it is not an attraction but rather a conversion because it originates by the force [virtú] of the one and of the other, which want to be placed in a certain way together, moreover, the most desirable way of placing them is along the poles: in this way they have only one axis, and if there is motion, then all parts share the motion and not only towards the axis of the large [magnet], but also towards its own [axis]; it might even become so much one, that it [the small magnet] loses its equinoctial and it makes the one of the large [magnet] come closer, losing the two poles because they touch and so you create one body with two most distant poles. But if they are placed along the equinoctials, one also sees the union because the axes are parallel and

Hora, nelle altre situationi io non so vedere che cosa voglino fare. Andava pensando che accomodassero in qualche maniera insieme il cerchio d'ambe due paralello all'equinottiale et per il vertice della reggione: ma non è così. E ben forza che voglino accomodarsi in qualche maniera pertenente alle sue parti, et che da quelle venga regolata, et denominate le parti: non sono se non poli, asse et cerchi paralelli; come adonque? Forse come il nostro auttore dice? che però non veggo come et a che fine, nè qual parti a quale vogli situare. Ma egli come ha truovato il suo muodo? per esperienze o per raggione? Non per esperienze: perchè, o con la terra, et questo ricercherrebbe viaggio regolato per una quarta; non con la terrella, perchè si ricerca che il versorio non habbia sensibile proportione con la terrella, acciò nell'istesso luoco sii il centro et la cuspide, altrimenti non è fatto niente.

Non mi par manco che per raggione: imperochè bisogna render causse della descrittione de que' cerchi che lui chiama conversionis, che nella piciola dimostratione ne descrive 3: BCL sotto l'equinotiale; ODL di 45; GL di 90. Essendo tutti li tali, come si vede nella figura grande, descritti sopra il ponto della reggione come centro, intervallo una retta da esso centro al polo opposito cerco prima la raggione di questo intervallo; poi, perchè questi cerchi conversionis non sono simili, ma quello del 45 è un quarto, li precedenti più, li seguenti meno. Al che si dà per regola che siino tra il polo opposito L et il cerchio BOG, quale è descritto sopra il centro della balla, intervallo quella che può quanto il semidiametro et il lato del quadrato. Quale è la raggione di pore questo centro et tanto intervallo? poi, perchè debbono essere divisi in tante parti come un quadrante così li grandi come li picioli?

the equinoctials are in the same plane and they share the motion over them.

Now in the other situations I cannot see what they want to do. I went on to think that the circle of both, parallel to the equinoctial and through the vertex of the region would put them together somehow: but it is not so. It is sure they want to take a certain [position] in a certain way according to their parts, and this way is ruled by these parts, and calling the parts: They can only be the poles, the axis' and the parallel circles; hence how? [is the arrangement]? Is it like our author says? But I do not see how and to which end, and which parts he wants to place to which. But how did he find his way of doing it? By experiment or by reason? Not by experiment: because neither with respect to the Earth as this requires a journey through a quarter; also not with the "terrella," because one finds that a versarium does not have such sensitive proportion with the "terrella" so that the same place is the center as well as the cusp; and there is no other possibility.

It seems to me also not by reason: because one needs to explain the description [construction?] of those circles he calls "conversionis," of which in the small demonstration he describes 3: BCL under the equinoctial, ODL of 45, GL of 90 degrees. And they all being such, as one sees it in the big figure, described over the point of the region as the center, the interval being a straight line from this center to the opposite pole. I am first seeking a reason for this interval; because these circles "conversionis" are not similar, that of 45 is a quarter [arc], those preceding more, and those next less [than a quarter arc]. To which it is given as rule that they [circles] have to be placed between the opposite pole L and the circle BOG, which is described around the center of the ball with an interval that can be as [long as] the semidiameter and the side of the square. What is the reason put forth for this center and such an interval? Then, why must they be divided into as many parts as a quadrant, both the large ones and the small ones?

Queste sono le difficoltà. Della spirale non ho difficoltà alcuna, ma è un bel genere di elica, generandosi di dua moti circolari. Prego V. S. che habbia un poco di consideratione sopra le mie difficoltà, et sopplisca al mancamento del nostro auttore, il quale ha taciuto le causse delle più oscure cose che siano: almeno havesse detto come ne è venuto in cognitione. Appresso, perchè desidero far esperienza di questa inclinatione, per levarmi la fatica prego V. S. scrivermi il muodo tenuto in far il versorio, con che li applica li perni, se con fuoco o con cola o come, et di che materia li fa, et sopra che li appoggia, et in soma ogni particolare, perchè non vorrei consumar tempo in esperimentar molte cose, poichè ella ha fatto la fatica. Qui farò fine, pregando V. S. scusare in la mia importunità et non curare di rispondermi se non con suo commodo, sichè non venga impedita nè da' suoi negotii nè dalli studii.

Però li bascio la mano.
Di Vinetia, il dì II Settembre 1602.
Di V. S. Ecc.ma Aff.mo Serv.re F. Paolo di Venezia

Fuori: All'Ecc.mo Sig.re mio Pad.rone Osservan.o Il sig.r Galileo Galilei, Math.co Publico, in Padova, appresso il Santo.[1]

These are the difficulties. I do not have any difficulty with the spiral. It is a nice sort of helix that is generated from two circular motions. I beg Your Lordship to have a bit of consideration concerning my difficulties, and to make up for the failings of our author, who was silent on the causes of the most obscure things: he could at least have said how he acquired knowledge of them. Then, since I wish to make an experiment on this inclination, and in order to avoid a certain effort, I beg Your Lordship to write me about the way of producing a "versarium," how you apply the pivots, whether by fire or by glue or how else, and of which material you make them, over what you place them, and so in conclusion I would like to know every detail, because I do not want to waste time in experimenting with a lot of things since you already made the effort. I finish here begging Your Lordship to excuse my importunity and to write me only when it suits you, since your businesses and studies must not be interrupted.

I kiss your hands.
From Venice, the second day of September, 1602. Most Devoted Servant of Your Most Excellent Lordship Brother Paolo from Venice

Addressed to: Very Excellent Lord and my Master Lord Galileo Galilei, public Mathematician, in Padua, by the Sanct.[1]

---

[1] EN X, 91–93, letter 83.

[1] My translation. Extended thanks to Matteo Valleriani for helping me to produce and check this and other translations.

14.2 Letter from Galileo to Guidobaldo del Monte, 29 November 1602

GALILEO a GUIDOBALDO DEL MONTE [in Montebaroccio] Padova, 29 novembre 1602.
Ill.mo Sig.e e P.ron Col.mo

GALILEO to GUIDOBALDO DEL MONTE [in Montebaroccio]. Padua, 29 November 1602.

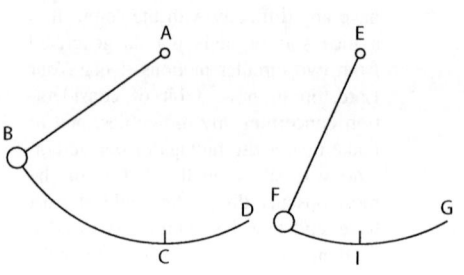 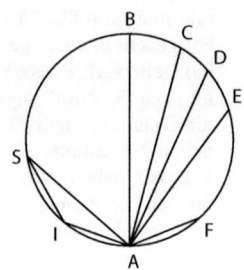

Fig. 14.1 First (left) and second (right) diagram in Galileo's letter to Guidobaldo del Monte sent on 29 November 1602

V. S. Ill.ma scusi la mia importunità, se persisto in voler persuaderle vera la proposizione de i moti fatti in tempi uguali nella medesima quarta del cerchio; perchè, essendomi parsa sempre mirabile, hora viepiù mi pare, che da V. S. Ill.ma vien reputata come impossibile: onde io stimerei grand'errore e mancamento il mio, s'io permettessi che essa venisse repudiata dalla di lei speculazione, come quella che fusse falsa, non meritando lei questa nota, nè tampoco d'esser bandita dall'intelletto di V. S. Ill.ma, che più d'ogn'altro la potrà più presto ritrarre dall'esilio delle nostre menti. E perchè l'esperienza, con che mi sono principalmente chiarito di tal verità, è tanto certa, quanto da me confusamente stata esplicata nell'altra mia, la replicherò più apertamente, onde ancora lei, facendola, possa accertarsi di questa verità.

Piglio dunque due fili sottili, lunghi ugualmente due o tre braccia l'uno, e siano AB, EF, e gli appicco a due chiodetti A, E, e nell'altre estremità B, F lego due palle di piombo uguali (se ben niente importa se fussero disuguali), rimuovendo poi ciascuno de' detti fili dal suo perpendicolo, ma uno assai, come saria per l'arco CB, e l'altro pochissimo,

Most Illustrious Lord and Cultivated Master

Your Lordship, please excuse my importunity if I persist in wanting to persuade you of the truth of the proposition that motions within the same quarter-circle are made in equal times, because having always seemed to me to be admirable, it seems to me [to be] all the more so, now that your Most Illustrious Lordship considers it to be impossible. Hence I would consider it a great error and a lack on my part if I should allow it to be rejected by your speculation as being false, for it does not deserve this mark, and neither [does it deserve] being banished from your Lordship's understanding who, better than anybody else, will quickly be able to retract it [the proposition] from the exile of our minds. And because the experiment, through which this truth principally became clear to me, is so much more certain, as it was explicated by me in a confused way in my other [letter], I will repeat it here more clearly, so that you, by performing it, would also be able to ascertain this truth.

So now I take two thin threads of equal length, each being two or three braccia long, and let them be AB, EF (cf. Fig. 14.1). [I] hang them from two small nails, A and E, and at the other ends, B and F, I tie two equal lead balls (although it would not matter if they were unequal). Then, by removing each of the above-mentioned threads from its

14.2 Letter from Galileo to Guidobaldo del Monte, 29 November 1602

come saria secondo l'arco IF; gli lascio poi nell'istesso momento di tempo andar liberamente, e l'uno comincia a descrivere archi grandi, simili al BCD, e l'altro ne descrive de' piccoli, simili all' FIG; ma non però consuma più tempo il mobile é a passare tutto l'arco BCD, che si faccia l'altro mobile F a passare l'arco FIG. Di che mi rendo sicurissimo così:

Il mobile B passa per il grand'arco BCD, e ritorna per lo medesimo DCB, e poi ritorna verso D, e va per 500 e 1000 volte reiterando le sue reciprocazioni; l'altro parimente va da F in G, e di qui torna in F, e parimente farà molte reciprocazioni; e nel tempo ch'io numero, verbi grazia, le prime cento grandi reciprocazioni BCD, DCB etc., un altro osservatore numera cento altre reciprocazioni per FIG piccolissime, e non ne numera pure una sola di più: segno evidentissimo che ciascheduna particolare di esse grandissime BCD consuma tanto tempo, quanto ogni una delle minime particolari FIG. Or se tutta la BCD vien passata in tanto tempo in quanto la FIG, ancora le loro metà, che sono le cadute per gli archi disuguali della medesima quarta, saranno fatte in tempi uguali.

Or se tutta la BCD vien passata in tanto tempo in quanto la FIG, ancora le loro metà, che sono le cadute per gli archi disuguali della medesima quarta, saranno fatte in tempi uguali. Ma anco senza stare a numerar altro, V. S. Ill.ma vedrà che il mobile F non farà le sue piccolissime reciprocazioni più frequenti che il mobile é le sue grandissime, ma sempre anderanno insieme.

L'esperienza, ch'ella mi dice aver fatta nello scatolone, può essere assai incerta, sì per non esser forse la sua superficie ben pulita, sì forse per non esser perfettamente circolare, sì ancora per non si potere in un solo passaggio così bene osservare il momento stesso sul principio del moto: ma se V. S. Ill.ma pur vuol pigliare questa superficie incavata, lasci andar da gran distanza, come

perpendicular, but one very much [so], as through the arc CB and the other very little, as through the arc IF; I let them go freely at the same moment of time. The one begins to describe large arcs, like BCD, and the other describes small ones, like FIG; but yet the mobile B does not consume more time moving along the whole arc BCD than the other mobile F in moving along the arc FIG. I make absolutely sure of this in the following way:

The mobile B moves along the large arc BCD, returns along the same DCB, and then comes back towards D, and it does this 500 and 1000 times, reiterating its oscillations. Likewise, the other one goes from F to G, and from here returns to F, and will likewise make many oscillations; and in the time that I count, let us say, the first hundred large oscillations BCD, DCB etc., another observer counts another hundred very small oscillations through FIG, and he does not count even a single one more: a most evident sign that each particular of these very large [oscillations] BCD consumes as much time as each particular of those minimal ones [through] FIG.

Now, if all BCD is passed [through] in as much time as FIG, then, in the same way, half of them, these being descents through the unequal arcs of the same quadrant, will be done in equal times. But even without staying on to enumerate other [oscillations], your Most Illustrious Lordship will see that the mobile F will not make its very small oscillations more frequently than the mobile B [will make] its larger ones, but rather, they will always go together.

The experiment which you tell me you have done in the box can be very uncertain, either because its surface has perhaps not been well cleaned or maybe because it is not perfectly circular, and because one cannot observe so well in a single passage the precise moment in which the motion begins. But if your Most Illustrious Lordship still wants to take this concave surface, let the ball

saria dal punto B, liberamente la palla B, la quale passerà in D, e farà nel principio le sue reciprocazioni grandi d'intervallo, e nel fine piccole, ma non però queste più frequenti di tempo di quelle.

Quanto poi al parere irragionevole che, pigliandosi una quarta lunga 100 miglia, due mobili uguali possino passarla, uno tutta, e l'altro un palmo solo, in tempi uguali, dico esser vero che ha dell'ammirando; ma se consideriamo che può esser un piano tanto poco declive, qual saria quello della superficie di un fiume che lentissimamente si muovesse, che in esso non haverà camminato un mobile naturalmente più d'un palmo nel tempo che un altro sopra un piano molto inclinato (ovvero congiunto con grandissimo impeto ricevuto, anco sopra una piccola inclinazione) haverà passato cento miglia: nè questa proposizione ha seco per avventura più inverisimilitudine di quello che si habbia che i triangoli tra le medesime parallele et in basi uguali siano sempre uguali, potendone fare uno brevissimo e l'altro lungo mille miglia. Ma restando nella medesima materia, io credo haver dimostrato questa conclusione, non meno dell'altra inopinabile.

Sia del cerchio BDA il diametro BA eretto all'orizzonte, e dal punto A sino alla circonferenza tirate linee utcumque AF, AE, AD, AC: dimostro, mobili uguali cadere in tempi uguali e per la perpendicolare BA e per piani inclinati secondo le linee CA, DA, EA, FA; sicchè, partendosi nell'istesso momento dalli punti B, C, D, E, F, arriveranno in uno stesso momento al termine A, e sia la linea FA piccola quant'esser si voglia.

E forse anco più inopinabile parerà questo, pur da me dimostrato, che essendo la linea SA non maggiore della corda d'una quarta, e le linee SI, IA utcumque, più presto fa il medesimo mobile il viaggio SIA, partendosi da S, che il viaggio solo IA, partendosi da I.

B go freely from a great distance, such as from point B. It will pass to D, at the beginning producing its oscillations with large intervals, and at the end with small ones; but the latter, however, [will not be] more frequent in time than the former.

With regard now to the unreasonable opinion that, given a quadrant 100 miles in length, two equal mobiles might pass along it, one the whole length, and the other only a span, in equal times, I say it is true that there is something wondrous about it; but [less so] if we consider that a plane can be at a very slight incline, like that of the surface of a slow-moving river, so that a mobile will not have traversed naturally on it more than a span in the time that another [mobile] will have moved one hundred miles over a steeply inclined plane (or otherwise being equipped with a very great received impetus, also over a small inclination). And this proposition does not involve by any adventure more unlikeliness than that in which triangles within the same parallels, and with the same bases, are always equal [in area], while one can make one of them very short and the other a thousand miles long. But staying with the subject, I believe I have demonstrated this conclusion to be no less unthinkable than the other.

In the circle BDA, let the diameter BA be erected on the horizontal, and let us draw from the point A to the circumference any lines AF, AE, AD, AC (cf. Fig. 14.1): I demonstrate that equal mobiles fall in equal times along the perpendicular BA, and along the inclined planes as specified by lines CA, DA, EA, FA; so that, by starting at the same moment from the points B, C, D, E, F, they will reach the end point A at the same time, and let the line FA be as small as we want it to be.

And perhaps even more unthinkable will appear the following, also demonstrated by me; that wherever the line SA being not greater than the chord of a quadrant, and [given] the lines SI and IA, the same mobile, starting from S, makes the journey SIA more quickly than just the journey IA, starting from I.

Sin qui ho dimostrato senza trasgredire i termini mecanici; ma non posso spuntare a dimostrare come gli archi SIA et IA siano passati in tempi uguali: che è quello che cerco.

Al Sig.r Francesco mi farà grazia rendere il baciamano, dicendogli che con un poco d'ozio gli scriverò una esperienza, che già mi venne in fantasia, per misurare il momento della percossa: perquanto al suo quesito, stimo benissimo detto quanto ne dice V. S. Ill.ma, e che quando cominciamo a concernere la materia, per la sua contingenza si cominciano ad alterare le proposizioni in astratto dal geometra considerate; delle quali così perturbate siccome non si può assegnare certa scienza, così dalla loro speculazione è assoluto il matematico.

Sono stato troppo lungo e tedioso con V. S. Ill.ma: mi perdoni in grazia, e mi ami come suo devotissimo servitore.

Di Padova, li 29 Novembre 1602. Di V. S. Ill.ma Serv.re Obblig.mo

Galileo Galilei.[2]

Until now I have demonstrated without transgressing the terms of mechanics; but I cannot manage to demonstrate how the arcs SIA and IA have been passed through in equal times and it is this that I am looking for.

Please do me the favor of kissing the hand of Signor Francesco in return, telling him that when I have a little leisure, I will write to him about an experiment which has already entered my imagination, for measuring the moment of the percussion. Regarding your question, I consider that what your Most Illustrious Lordship said about it was very well put, and that when we begin to deal with matter, because of its contingency the propositions abstractly considered by the geometrician begin to change: since one cannot assign certain science to the [propositions] thus perturbed, the mathematician is hence freed from speculating about them.

I have been too long and tedious for your Most Illustrious Lordship: please excuse me, with grace, and love me as your most devoted servant. And I most reverently kiss your hands.

In Padua, 29 November 1602. From Your Illustrious Lordship's Most Obliged Servant

Galileo Galilei.[2]

---

[2]EN X, pp. 97–100, letter 88. The original letter is not extant, only a copy in a nineteenth-century hand which in turn was made not from the original but from a prior copy by Vivian of the original letter.

[2]Translation adopted from Damerow et al. (2001, 403–405), checked and very sightly adapted.

14.3 Letter to Paolo Sarpi, 16 October 1604

Galileo a Paolo Sarpi in Venezia.
Padova, 16 ottobre 1604.
Molto Rev.do Sig.re et Pad.ne Col.mo.
Ripensando circa le cose del moto, nelle quali, per dimostrare li accidenti da me osservati, mi mancava principio totalmente indubitabile da poter porlo per

Galileo to Paolo Sarpi in Venice.
Padua, 16 October 1604.
Most Reverend Lord and Most Cultivated Master.
Thinking again about the matters of motion, in which, to demonstrate the phenomena observed by me, I lacked a completely indubitable principle to put

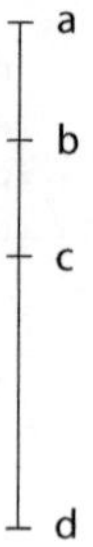

Fig. 14.2 Diagram in the letter to Paolo Sarpi of 16 October 1604

assioma, mi son ridotto ad una proposizione la quale ha molto del naturale et dell' evidente; et questa supposta, dimostro poi il resto, cio' gli spazzii passati dal moto naturale essere in proporzione doppia dei tempi, et per conseguenza gli spazii passati in tempi eguali esser come i numeri impari ab unitate, et le altre cose. Et il principio è questo: che il mobile naturale vadia crescendo di velocità con quella proportione che si discosta dal principio del suo moto; come, v. g., cadendo il grave dal termine a per la linea abcd, suppongo che il grado di velocità che ha in c al grado di velocità che hebbe in b esser come la distanza ca alla distanza ba, et così conseguentemente in d haver grado di velocità maggiore che in c secondo che la distanza da ' maggiore della ca.

Haverò caro che V. S, molto R.da lo consideri un poco, et me ne dica il suo parere. Et se accettiamo questo principio, non pur dimostriamo, come ho detto, le altre conclusioni, ma credo che haviamo anco assai in mano per mostrare che il cadente naturale et il proietto violento passino per le medesime proporzioni di velocità. Imperò che se il proietto vien gettato dal termine d al termine a, è manifesto che nel punto d ha grado di impeto potente a spingerlo sino al termine a, et non più; et quando il medesimo proietto è in c, è chiaro che è congiunto con grado di impeto potente a spingerlo sino al medesimo termine a; et parimente il grado d'impeto in b basta per spingerlo in a: onde è manifesto, l'impeto nei punti d, c, b andar decrescendo secondo le proporzioni

as an axiom, I am reduced to a proposition which has much of the natural and the evident: and with this assumed, I then demonstrate the rest; that is, that the spaces passed by natural motion are in double proportion to the times, and consequently the spaces passed in equal times are as the odd numbers from one, and the other things. And the principle is this: that the natural moveable goes increasing in velocity with that proportion with which it departs from the beginning of its motion; as, for example, the heavy body falling from the terminus a along the line *abcd*, I assume that the degree of velocity that it has at *c*, to the degree it had at *b*, is as the distance *ca* to the distance *ba*, and thus consequently, at *d* it has a degree of velocity greater than at *c* according as the distance *da* is greater than *ca* (cf. Fig. 14.2).

I should like your Honorable Lordship to consider this a bit, and tell me your opinion. And if we accept this principle, we not only demonstrate, as I said, the other conclusions, but I believe we also have enough in hand to show that the naturally falling body and the violent projectile pass through the same proportions of velocity. For if the projectile is thrown from the point *d* to the point *a*, it is manifest that at the point *d* it has a degree of impetus sufficiently powerful to drive it to the point *a*, and not beyond; and if the same projectile is in *c*, it is clear that it is linked with a degree of impetus sufficiently powerful to drive it to the same point *a*, and, in the same way, the degree of impetus in *b* is sufficient to drive it to *a*, whence it is manifest that the impetus at points *d*, *c*,

delle linee da, ca, ba; onde, se secondo le medesime va nella caduta naturale aqquistando gradi di velocità; è vero quanto ho detto et creduto sin qui.

Quanto all'esperienza della freccia, credo che nel cadere aqquisterà pari forza a quella con che fu spinta, come con altri esempi parleremo a bocca, bisognandomi esser costà avanti Ognisanti. Intanto la prego a pensare un poco sopra il predetto principio.

Quanto all'altro problema proposto da lei, credo che i medesimi mobili riceveranno ambedue la medesima virtù, la quale però non opererà in ambedue il medesimo effetto: come, v. g., il medesimo huomo, vogando, communica la sua virtù ad una gondola et ad una peotta, sendo l'una et l'altra capace anco di maggiore; ma non segue nell'una et nell'altra il medesimo effetto circa la velocità o distanza d'intervallo per lo quale si muovono.

Scrivo al scuro: questo poco basti più per satisfare al debito della risposta che al debito della soluzione, rimettendomi a parlarne a bocca in breve.

Et con ogni reverenza li bacio le mani.
Di Padova, li 16 di Ottobre 1604.
Di V. S. molto R.da Ser.re Oblig.mo

Galileo Galilei.[3]

b goes decreasing in the proportions of the lines da, ca, ba; whence, if it goes acquiring degrees of velocity in the same (proportions) in natural fall, what I have said and believed up to now is true.

Concerning the experiment with the arrow, I believe that it does acquire during its fall a force that is equal to that with which it was thrown upwards, as we will discuss together with other examples orally, since I have to be there in any case before All Saints. Meanwhile, I ask you to think a little bit about the above mentioned principle.

Concerning the other problem you proposed, I believe that the same mobiles both receive the same force, which, however, does not create the same effect in both; as for example the same person, when rowing, communicates his force to a gondola and to a larger boat, both being capable of receiving even more of it, but it does not result in the same effect in the one and in the other with regard to the velocity or the distance-interval through which they move.

I am writing in the dark, this little may rather suffice to satisfy the obligation of answering than that of finding a solution of which to speak orally I reserve to a meeting in the near future.

And with all respect I kiss your hands.
In Padua, 16 October 1604
Very Obligated Servant of Your Most Reverend Lordship
Galileo Galilei.[3]

---

[3]EN X, 115–116, letter 105.

[3]Translation adopted from Damerow et al. (2004, 354–355) checked and slightly adapted.

14.4 Draft of a Reply to Baliani Composed ca. 1640

Sopra i principii del Signor Baliani. disteso da me ad mentim G[alileo].

É la nostra intenzione investigare e dimostrare geometricamente accidenti

About the principles of Signor Baliani. Drafted by me according to the intention of G[alileo].

It is our intention to investigate and to demonstrate geometrically the

Fig. 14.3 Diagram on folio page 36 recto of MS. Gal. 74

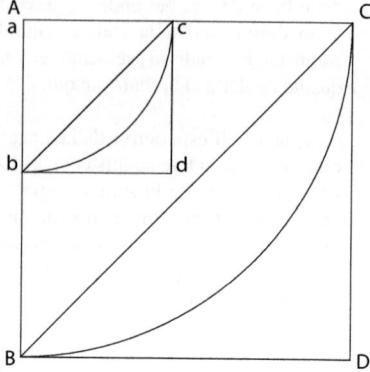

e passioni, che accaggiono ai mobili gravi naturalmente e liberamente descendenti sopra spazii retti differenti o di lunghezza o d'inclinazione o d'ambedue insieme. Nel venir poi alla elezione dei principii sopra i quali deve esser fondata la scienza, prendete come chiare notizie accidenti, i quali niuna connessione hanno con moti fatti sopra linee non rette non d'assegnabile inclinazione nè che in esse le diverse lunghezze operino quello che operano nelle linee rette, ma in tutto é per tutto cose differentissime: lo che mi par grave errore e tanto maggiore quanto che ci sene tira dietro un altro non minore. Mi dichiaro.

Voi pigliate come principio noto e indubitato le vibrazioni del medesimo pendulo farsi tutte sotto tempi equali siano alle di qualsivoglia grandezza. Item supponete i tempi delle vibrazioni di pendoli disequali esser tra di lori in subdupla proporzione delle lunghezze de i lor fili (assunto veramente molto ardito). E da questo che supponete accadere nei mobili descendenti per circonferenze di cerchi vollete raccorre quello che occorre ne i moti retti. Ma se io non erro assai meno obliquamente si poteva ottener l'intento discorrendo cosi.

La line AB intendasi rappresentare il filo pendente e stando fermo il termine supremo A intendasi il mobile posto in B descrive [disegnare?] l'arco del quadrante BC e similmente preso ab come pendolo minore sia l'arco del quadrante

accidents and passions that happen to heavy mobiles falling down naturally and freely over straight spaces of different length or inclination or both at the same time. Then in arriving at the principles upon which the science has to be founded you take as clear knowledge the accidents which have absolutely no connection with the motions made over lines that are non straight, of no assignable inclination, all the more so because in these different lengths work than in those which work in straight lines, but by and large due to very different things: the error is all the bigger in that it entails another not smaller one I say.

Then choosing as a known and undoubted principle that the vibrations of the same pendulum all are made under the same times be they of whatever size. Likewise you assume the times of vibrations of unequal pendulums to be in subduplicate proportion of the lengths of their threads (a very daring assumption). And from what you suppose to happen when mobiles fall through the circumferences of circles you want to infer that what happens in straight motions. But if I don't err much one can arrive at the intended argument less convolutedly in the following way:

The line AB is supposed to represent the thread hanging from above and suppose the highest point A to be fixed, then the mobile carried to B describes the arc of the quadrant BC and similarly I take ab as a smaller pendulum,

14.4 Draft of a Reply to Baliani Composed ca. 1640

bc quello, che descriverebbe il mobile posto in *b* e d'essi quadranti siano le corde suttese *BC. bc.* et intendasi le tangenti orizzontali *BD. bd.* e le perpendicolari *CB. cd.* ora essendo le due declinazioni in tutto e per tutto simile molto ragionevolmente si può prendere e come principio noto supporre che le proporzioni de i moti che le proporzioni de i moti che che [sic!] accadessero farsi sopra le rette *AB. BC.* e per l'arco *CB.* fossero le medesime che nella minor figura per le linee omologhe *ab. bc.* e per l'arco *cb.* onde permutando il moto per l'arco *cb.* al moto per l'arco *CB.* habbia la medesima proporzione che il moto per la perpendicolare *ab.* al moto per la perpendicolare *AB.* e onde pigliando per supposto che i tempi pergli archi siano in subdupla proporzione delle lunghezze de i fili, giá è manifesto che con altra e tanta verità si può suporre che i tempi per le perpendicolari *ab. AB.* siano in subdupla proporzione delle medesimi lunghezze *ab. AB.* E cosi si viene a schivare la supposizione assai dura (come appresso diremo) che i moti per le parte minime delli archi siano come se fosser fatti per linee rette, assunto, come dico assai duro; imperocchè con gran ragione può i lettore domandare che gli sia assegnata la quantità dell'arco che V.S. ichiama minima si chè, per esempio ella intenda l'arco esser minimo fino all metà di un grado. inoltre sarebbe stato necessario dichiarasi quale delle istesse linee rette si deva prendere per arco minimo, cioè se quella che, partendosi dal primo punto dell'arco tocca la circonferenza o pure la sega come corda di esso arco minimo, o pure una delle altre molte che dal medesimo punto primo possono tirarsi. Da queste molte linee pare che venga esclusa la tangente necessariamente, imperocchè considerando nella figura passata la tangente dell'arco *BC.* nel punto *B* che viene ad esser la orizontale *BD.* manifesta cosa è che il mobile posto sopra di essa in nessuna parte si moverà ma bene posto in qualsivoglia punto dell'arco *BC.* remoto dal *B.* descenderà egli in *B.* essendo dunque la discrepanza tra la tangente è l'arco tanto grande per quanto appartie-

the arc of the quadrant *bc* would be the one which the mobile placed in *b* would describe, and of these quadrants let the supporting arcs be *BC. bc.* and let the horizontal tangents be *BD.* and *bd.* and the perpendiculars *CD. cd* (cf. Fig. 14.3). Now since the two declines are all in all similar, one can very reasonably take and assume as a known principle that the proportions of the motions that happen to be made over the straights *AB. BC.* and through the arc *CB.* would be the same as in the smaller figure through the homologous lines *ab. bc.* and through the arc *cb.* therefore by permutation the motion through the arc *cb.* has to the motion through the arc *CB.* the same proportion which the motion through the perpendicular *ab.* has to the motion through the perpendicular *AB.* And therefore if you assume that the times through the arcs are in subduplicate proportion to the lengths of the strings, then it is already manifest that for the same reason one can assume that the times through the perpendiculars *ab. AB.* are in subduplicate proportion to the same lengths *ab. AB.* And in such a way one succeeds in avoiding the very strong supposition (as we will say) that the motions through the minimal parts of the arcs are as if they were made through lines assumed to be straight; as I say very strong, for with good reason the reader can ask the quantity of the arc which Your Lordship calls minimal to be assigned, if for example he intends the arc to be minimal even to the half of one degree. It would further be necessary to determine which of these straight lines one has to take as the minimal arc, be it the one which parts from the first point of the arc touching the circumference or maybe the one cutting as a chord of this minimal arc, or maybe one of the many others that we can draw from the same first point. Of these many lines it seems that the tangent is necessarily excluded, because in the last figure the tangent to the arc *BC* is considered. In *B* it becomes the horizontal *BD.* Then it is a manifest fact that the mobile placed on it does not move at all, but well-placed in any point of the arc *BC.* distant

ne al moto quanto è differente la quiete del moto; con poca ò niuna probabilità si potrà supporre che il moversi dal punto C. per la tangente o per l'arco siano l'istessa cosa. Ma vegghiamo un altra disparità massima; si uno negherà i moti del medesimo mobile fatti sopra piani di diverse inclinazione esser tra di loro differenti, e che in conseguenza un moto, il quale cominciato sopra una tale inclinazione debba di parte in parte tra passare sopra altrettante altre diverse inclinazione sarà sommamente differente da quello che sopra una istessa inclinazione deve andarsi continuando. Ora nella circonferenza del quadrante CB. tante sono le diverse inclinazione quante le tangenti e queste sono quante i punti cioè infinite; per lo che anco in quasivoglia piccola parte della circumferenza si come vi sono infiniti parti vi sono anco infinite inclinazione. per la mutazione delle quali no si può dire che il moto per l'arco possa esser simile non che l'istesso che per una medesima inclinazione sola.[4]

from B, will descend to B. Hence the discrepancy between the tangent and the arc being that big. As far as motion is concerned, just as rest is different from motion one can with little or no probability assume that movement from C. through the tangent or through the arc are the same thing. But let us proceed to another bigger error; No one doubts that the motions of the same mobile made over planes of different inclinations differ from each other and that in consequence a motion which starts over a big inclination must part for part pass over other and other different inclinations which will be very different from that which must continuously go over one and the same inclination. Now in the circumference of the quadrant CB. there are as many different inclinations as there are tangents and these are as much as the points, that is infinitely [many]; Therefore since in every part of the circumference, no matter how small, there are infinitely many points, then there are infinitely [many] inclinations. For the reverse of this one cannot say that motion through the arc can be similar or the same only through one single and the same inclination.[4] disteso da me ad mentim G[alileo].

---

[4]Ms. Gal. 74, f35-f36. The page order is 35 verso, 36 recto, 36 verso, 35 recto. The document can be accessed through the website of the Biblioteca Nazionale Centrale di Firenze at http://www.bncf.firenze.sbn.it/. Accessed 16 Feb 2017. The document is transcribed in Caverni (1895, 313–314). The transcription can be accessed at https://archive.org/stream/storiadelmetodo04cavegoog. Accessed 16 Feb 2017. In my transcription spelling has not been modernized. Some of my readings differ from Caverni's, non of the differences, however is such, that it would affect the interpretation.

[4]I thank Matteo Valleriani for his help in transcribing and translating this document.

References

Caverni, R. (1895). *Storia del metodo sperimentale in Italia* (Vol. IV). New York: G. Civelli.

Damerow, P., Renn, J., & Rieger, S. (2001). Hunting the white elephant: When and how did Galileo discover the law of fall? In J. Renn (Ed.), *Galileo in context* (pp. 29–150). Cambridge: Cambridge University Press.

Damerow, P., Freudenthal, G., McLaughlin, P., & Renn, J. (2004). *Exploring the limits of preclassical mechanics*. New York: Springer.

Index Locorum

33 recto, 128
34 recto, 278
38 recto, 299
47 recto, 202, 217
49 recto, 201, 202
50 recto, 336
57 recto, 336
57 verso, 333, 336
58 recto, 329
59 recto, 115
66 recto, 321
68 recto, 108
77 recto, 182
81 recto, 15, 86
87 verso, 182
88 recto, 352
90 verso, 77, 147, 148, 370–373, 454
91 verso, 322
96 recto, 182
100 verso, 309
107 verso, 81, 85
114 verso, 15, 86
115 verso, 77, 80, 147, 148, 151, 372–374, 434, 450
116 verso, 15, 86, 158, 288, 415
121 recto, 122, 136, 137, 139, 431
121 verso, 135, 140, 375, 376
126 recto, 201, 202, 298
127 recto, 135, 408
127 verso, 203, 382
129 recto, 118, 376–378
130 recto, 385, 386
130 verso, 188–196, 198, 199, 203, 204, 250, 379–386, 423, 425
131 recto, 113, 118, 119, 217–225, 227, 229, 233, 396–398
138 verso, 351
139 verso, 194
140 recto, 203
147 recto, 12, 179, 180, 182, 186, 187, 200, 320–322, 325, 329, 336, 339, 347
147 verso, 234, 235, 278, 330, 331, 333–336, 338, 346, 347, 374
148 recto, 117, 119, 134, 387–391, 401, 402
149 recto, 282, 392, 396
149 verso, 114, 116, 392, 402
150 recto, 112–114, 219, 392–395, 437, 442
150 verso, 282, 392, 396
151 recto, 217, 221–227, 287, 397, 398
151 verso, 399, 438
153 recto, 387, 390, 391, 400–402
154 recto, 92, 221, 402–405
155 recto, 234, 235, 330, 331, 338, 405–407
156 recto, 180, 408–411
156 verso, 408
157 recto, 408
157 verso, 180, 408–412
160 recto, 217, 224–227, 229, 276, 402
163 recto, 35, 115, 117, 118, 128, 141, 217, 274, 275, 286, 287, 292, 387
163 verso, 188, 274, 287, 289, 291, 292, 294, 343
164 recto, 181, 182, 190, 201, 294, 295, 297–300

164 verso, 113, 274, 278, 300–303, 306, 307, 311
166 recto, 80, 117, 120, 122, 124–129, 131, 132, 135, 147, 148, 150, 151, 163, 248, 372–376, 387, 401, 408, 411, 413, 415–419, 423, 426, 427, 429–431, 434–436, 455
167 verso, 120, 421
168 recto, 203, 299, 382
172 recto, 217, 224–227, 229, 273, 275–277, 285, 286, 338, 343
172 verso, 115, 141, 186, 228, 274, 277, 278, 282, 285–287, 329, 339, 392, 396
173 recto, 344
174 recto, 189, 190, 198–201, 295, 378, 401
175 verso, 291
176 recto, 132, 189, 193, 198, 423–426
177 recto, 183
177 verso, 336
179 recto, 351, 352
179 verso, 274
180 recto, 334, 336
183 recto, 126, 134, 135, 248, 249, 401, 411, 413, 417, 420, 426–429, 455, 456
183 verso, 417, 450
184 recto, 124, 376
184 verso, 373, 416, 427, 430–435, 450
185 recto, 114, 391, 436–438
185 verso, 399
186 recto, 114, 436
186 verso, 110–113, 173–178, 301, 391–393, 435, 436, 439–442
187 recto, 193, 195, 198–200, 384
187 verso, 189
188 recto, 118, 119
189 recto, 77, 132, 147–150, 158, 370–372, 423, 426, 443, 451–454
189 verso, 77, 79–81, 88, 93, 180, 320, 327, 339, 370–373, 412, 442–450
190 recto, 250
191 verso, 250
192 recto, 127, 375, 417, 455–457
194 recto, 336

Printed by Printforce, the Netherlands